THE ROUTLEDGE COMPANION TO THE ENVIRONMENTAL HUMANITIES

The Routledge Companion to the Environmental Humanities provides a comprehensive, transnational, and interdisciplinary map to the field, offering a broad overview of its founding principles while providing insight into exciting new directions for future scholarship. Articulating the significance of humanistic perspectives for our collective social engagement with ecological crises, the volume explores the potential of the environmental humanities for organizing humanistic research, opening up new forms of interdisciplinarity, and shaping public debate and policies on environmental issues.

Parts cover:

- The Anthropocene and the domestication of Earth
- Posthumanism and multispecies communities
- Inequality and environmental justice
- Decline and resilience: environmental narratives, history, and memory
- Environmental arts, media, and technologies
- The state of the environmental humanities

The first of its kind, this Companion covers essential issues and themes, necessarily crossing disciplines within the humanities and with the social and natural sciences. Exploring how the environmental humanities contribute to policy and action concerning some of the key intellectual, social, and environmental challenges of our times, the chapters offer an ideal guide to this rapidly developing field.

Ursula K. Heise is the Marcia H. Howard Chair in Literary Studies at the Department of English and the Institute of the Environment and Sustainability at the University of California, Los Angeles, USA.

Jon Christensen is Adjunct Assistant Professor in the Institute of the Environment and Sustainability, the Department of History, and the Center for Digital Humanities at the University of California, Los Angeles, USA.

Michelle Niemann is an independent scholar, writing consultant, and editor, and was a Postdoctoral Fellow in Environmental Humanities and English at the University of California, Los Angeles, USA from 2014 to 2016.

ALSO AVAILABLE IN THIS SERIES

The Routledge Companion to Anglophone Caribbean Literature
Also available in paperback

The Routledge Companion to Asian American and Pacific Islander Literature
Also available in paperback

The Routledge Companion to Experimental Literature
Also available in paperback

The Routledge Companion to Latino/a Literature
Also available in paperback

The Routledge Companion to Literature and Human Rights

The Routledge Companion to Literature and Science
Also available in paperback

The Routledge Companion to Native American Literature

The Routledge Companion to Science Fiction
Also available in paperback

The Routledge Companion to Travel Writing

The Routledge Companion to World Literature
Also available in paperback

THE ROUTLEDGE COMPANION TO THE ENVIRONMENTAL HUMANITIES

Edited by
Ursula K. Heise, Jon Christensen,
and Michelle Niemann

Routledge
Taylor & Francis Group

LONDON AND NEW YORK

First published 2017 by Routledge

2 Park Square, Milton Park, Abingdon, Oxon OX14 4RN
605 Third Avenue, New York, NY 10017

Routledge is an imprint of the Taylor & Francis Group, an informa business

First issued in paperback 2021

Publisher's Note

The publisher has gone to great lengths to ensure the quality of this reprint
but points out that some imperfections in the original copies may be apparent.

British Library Cataloguing-in-Publication Data
A catalogue record for this book is available from the British Library

Library of Congress Cataloging-in-Publication Data
Names: Heise, Ursula K., editor. | Christensen, Jon (Environmental histsorian),
    editor. | Niemann, Michelle, editor.
Title: The Routledge companion to the environmental humanities / edited by
    Ursula K. Heise, Jon Christensen, and Michelle Niemann.
Other titles: Companion to the environmental humanities
Description: Abingdon, Oxon; New York, NY: Routledge, 2016. | Includes
    bibliographical references.
Identifiers: LCCN 2016021347 | ISBN 9781138786745 (hardback: alk. paper) |
    ISBN 9781317660194 (web pdf) | ISBN 9781317660187 (epub) |
    ISBN 9781317660170 (mobipocket/kindle)
Subjects: LCSH: Environmental sciences—Social aspects. | Science and the
    humanities. | Human ecology. | Nature—Affect of human beings on.
Classification: LCC GE105 .R68 2016 | DDC 304.2—dc23
LC record available at https://lccn.loc.gov/2016021347

ISBN: 978-1-138-78674-5 (hbk)
ISBN: 978-1-03-217929-2 (pbk)
DOI: 10.4324/9781315766355

Typeset in Goudy Std
by Swales & Willis Ltd, Exeter, Devon, UK

CONTENTS

CONTENTS

CONTENTS

CONTENTS

ILLUSTRATIONS

Figures

Tables

CONTRIBUTORS

Joni Adamson
English, Julie Ann Wrigley Global Institute of Sustainability, Arizona State University

Stacy Alaimo
English, University of Texas at Arlington

Daniel A. Barber
Architecture, University of Pennsylvania

Hannes Bergthaller
Foreign Languages and Literatures, National Chung-Hsing University, Taiwan

Christian Bök
English, University of Calgary

Brett Buchanan
Philosophy, School of the Environment, Laurentian University

Judith A. Carney
Geography and Institute of the Environment and Sustainability, UCLA

Allison Carruth
English, Institute for Society and Genetics, and Institute of the Environment and Sustainability, UCLA

Jessica R. Cattelino
Anthropology, UCLA

Una Chaudhuri
English, Drama, and Environmental Studies, New York University

Jon Christensen
History, Institute of the Environment and Sustainability, and Center for Digital Humanities, UCLA

Emma Colven
Geography, UCLA

Greg Garrard
Creative and Critical Studies, University of British Columbia, Okanagan Campus

Akhil Gupta
Anthropology, UCLA

Susanna B. Hecht
Luskin School of Public Affairs and Institute of the Environment and Sustainability, UCLA, and International History, Graduate Institute of International and Development Studies, Geneva, Switzerland

Ursula K. Heise
English and Institute of the Environment and Sustainability, UCLA

Heather Houser
English, University of Texas at Austin

Dale Jamieson
Environmental Studies and Philosophy, New York University

Barbara Rose Johnston
Center for Political Ecology, Santa Cruz, California

Dolly Jørgensen
History of Environment and Technology, Luleå University of Technology, Sweden

Rosanne Kennedy
Gender, Sexuality and Culture, Australian National University

Helga Leitner
Geography, UCLA

Stephanie LeMenager
English and Environmental Studies, University of Oregon

Jorge Marcone
Spanish and Portuguese and Comparative Literature, Rutgers University

Emma Marris
Independent scholar and writer

Kathleen D. Morrison
Anthropology and the University of Pennsylvania Museum of Archaeology and Anthropology, University of Pennsylvania

Linda Nash
History, University of Washington

Anahid Nersessian
English, UCLA

Michelle Niemann
Independent scholar, English and Environmental Humanities

James Nisbet
Art History, University of California, Irvine

Lizabeth Paravisini-Gebert
Hispanic Studies and Environmental Studies, Vassar College

Stephanie Posthumus
Languages, Literatures, and Cultures, McGill University

Libby Robin
Fenner School of Environment and Society, Australian National University

Deborah Bird Rose
Environmental Humanities, University of New South Wales, Australia

Catriona Sandilands
Environmental Studies, York University, Toronto

Ronald Sandler
Philosophy, Northeastern University

Eric Sheppard
Geography, UCLA

Stéfan Sinclair
Languages, Literatures, and Cultures, McGill University

Genese Marie Sodikoff
Sociology and Anthropology, Rutgers University, Newark

Sverker Sörlin
History of Science, Technology and Environment, KTH Environmental Humanities Laboratory, KTH Royal Institute of Technology, Stockholm

Heather I. Sullivan
Modern Languages and Literatures, Trinity University

Bronislaw Szerszynski
Sociology, Lancaster University

Thom van Dooren
Environmental Humanities, University of New South Wales, Australia

Robert N. Watson
English, UCLA

Alexa Weik von Mossner
English and American Studies, University of Klagenfurt, Austria

Jennifer Wenzel
English and Comparative Literature and Middle Eastern, South Asian, and African Studies, Columbia University

Richard White
History, Center for Spatial and Textual Analysis, Stanford University

Kyle Powys Whyte
Philosophy and Community Sustainability, Michigan State University

Yuki Masami
Human and Socio-Environmental Studies, Kanazawa University, Japan

Maite Zubiaurre
Spanish and Portuguese and Germanic Languages, UCLA

ACKNOWLEDGMENTS

This volume is based on a series of seminars on "The Environmental Humanities: Emergence and Impacts" held at the University of California, Los Angeles, during the 2014–15 academic year. We would like to thank the Andrew W. Mellon Foundation for making these seminars possible through a grant from its John E. Sawyer Seminar program. Many of the chapters in this volume were first drafted for the Sawyer seminars, and we would like to thank the faculty, students, and members of the public from Los Angeles and around the world who contributed to the seminars through their writing, their presence, and their lively discussion.

We also owe a debt of gratitude to the leaders, faculty, staff, and friends of the Department of English at UCLA for their generous help in matters ranging from the practical and administrative to the intellectual and programmatic, and for their ongoing support for the environmental humanities. The Chair of the English Department, Ali Behdad, and the Dean of Humanities, David Schaberg, in particular, encouraged and supported us in organizing the seminar.

INTRODUCTION

Planet, species, justice—and the stories we tell about them

Ursula K. Heise

The emergence of the environmental humanities

The environmental humanities have emerged as a new interdisciplinary matrix over the last decade, accompanied by programmatic statements, new journals, conferences, research initiatives, and the first few academic programs in Australia, Western Europe, and North America. The label of this research area follows a formula of innovation across a whole range of emergent fields that combines the term "studies" or "humanities" with a concept that has in the past been the purview of disciplines outside the humanities and qualitative social sciences: digital humanities, disability studies, food studies, human–animal studies, and medical humanities, for example. Unlike most of these fields, the environmental humanities do not so much propose a new object of study, a new humanistic perspective on a nonhumanistic field, or a particular set of new methods, as they combine humanistic perspectives and methods that have already developed in half a dozen or so disciplines over the last four decades. Environmental philosophy, for example, emerged in the 1970s, environmental history in the 1980s, and ecocriticism in the early 1990s. Although each of them struggled for a decade or more to be fully accredited by its own discipline, their academic recognition has in recent years opened up the possibility of closer collaborations with neighboring disciplines such as environmental anthropology, cultural geography, and areas in political science and urban studies that converge around the theoretical paradigm of "political ecology."

Such collaborations can build on a set of theoretical works that, even though they were originally written in and for particular disciplines such as anthropology, history, or philosophy, have become shared points of reference across the environmental humanities and social sciences. From anthropology, these "classics" include Sherry Ortner's essay "Is Female to Male as Nature is to Culture?" (1972) and Bruno Latour's publications, particularly *We Have Never Been Modern* (1991, trans. 1993) with its coinage of the term "natureculture." Historical research that has shaped environmental work across disciplines includes Alfred Crosby's *The Columbian Exchange: Biological and Cultural Consequences of 1492* (1972), the first sustained study of how European colonialism in the New World reshaped ecosystems around the world, and William Cronon's "The Trouble with Wilderness, Or, Getting Back to the Wrong Nature" (1995), which shows how sometimes erroneous memories of nature past have shaped the North American environmentalist movement. Criticism of the wilderness ideal has also come from the Indian sociologist Ramachandra Guha, who

1

approached it from the perspective of developing countries in his essay "Radical American Environmentalism and Wilderness Preservation: A Third World Critique" (1989) and elaborated a fuller portrait of the "environmentalism of the poor" in *Varieties of Environmentalism*, the book he co-authored with Joan Martínez-Alier in 1997. The geographer David Harvey's exploration of the connections between advocacy for social justice and for the environment in *Justice, Nature, and the Geography of Difference* (1996) approached the same nexus from a Marxist perspective. Lawrence Buell's *The Environmental Imagination* (1995) and *Writing for An Endangered World* (2001) laid the groundwork, from literary studies, for the analysis of major environmentalist tropes such as place and toxicity. The philosopher Val Plumwood has offered an influential analysis of the cultural roots of ecological crisis in *Environmental Culture: The Ecological Crisis of Reason* (2001), a book that also connects to influential strands of feminist environmental scholarship across disciplines that run from Carolyn Merchant's *The Death of Nature* (1980) to Donna Haraway's work in the history and philosophy of science. In spite of distinct disciplinary trajectories, then, the environmental humanities already share a significant body of theory that has made its influence felt across disciplines.

Unlike most of the other emergent fields of humanistic and qualitative social-science inquiry, the environmental humanities have a clear disciplinary predecessor, the field of environmental studies that took off in many countries through programs, departments, and publications in the 1960s and 1970s. How does the idea of "environmental humanities" modify or improve on the idea of "environmental studies"? As their research has been increasingly recognized as crucial in their own disciplines, quite a few environmentally oriented humanists and social scientists have felt understandably disgruntled with environmental studies programs that, for all their pathbreaking interdisciplinary work, have often limited their reach to the natural sciences, civil engineering, and a few experts on law and policy. Not a few humanists and social scientists who looked forward eagerly to collaborations with natural scientists that they considered essential for their own work have experienced the frustration of interdisciplinary meetings that ended up relegating them to the tasks of communicating and publicizing research agendas and findings mostly shaped by science, engineering, and policy experts (see Neimanis et al. 75).

Environmental humanities as a long-overdue institutional come-uppance, then? Not quite. Most environmental humanists continue to rate the collaboration with scientists as indispensable, even as their own training and research keeps them focused on the differences that divergent histories, cultures, and values make in understanding and solving environmental problems. These differences are more than a matter of acknowledging the "cultural, ethical, and institutional dimensions of environmental crises," in the typical lingo of interdisciplinary programs, international governance offices, and NGOs. They constitute a fundamental challenge to the understanding of environmental crises as basically techno-scientific, with history and culture added on as secondary complications. The environmental humanities, by contrast, envision ecological crises fundamentally as questions of socioeconomic inequality, cultural difference, and divergent histories, values, and ethical frameworks. Scientific understanding and technological problem-solving, essential though they are, themselves are shaped by such frameworks and stand to gain by situating themselves in this historical and sociocultural landscape.

The intractability of some of the most serious global environmental crises has helped to foreground their divergent framing in different communities. Climate scientists, for example, have tended to construe the slow pace of reductions in greenhouse gas emissions as a result of information deficits and organized corporate resistance that improved means of publicity might be able to overcome. But as the sociologist Kari Norgaard, the philosopher Dale Jamieson, and the climatologist Mike Hulme, among others, have shown, simple insistence on the scientific facts remains politically ineffective when it is disconnected from the political, social, cultural, affective, and rhetorical forms that the climate problem takes in different communities. Similarly, attempts to protect endangered species on the grounds that biodiversity in general and so-called "keystone species" in particular are essential for ecosystem functioning have tended to run into trouble when they have ignored the histories, values, and uses of animals and plants as understood and practiced by local communities. Particularly when conservation efforts were organized by activists and institutions in the global North, the assumption that their scientific understanding of relationships between species would be shared by default or could be imposed through education on communities in the global South often turned potential local allies into opponents of conservation and led to disempowerment, expropriation, and displacement (Agrawal and Redford; Dowie; Heise, *Imagining Extinction* ch. 5). Many of the chapters in this collection highlight other scenarios in which the cultural, historical, and social frameworks that shape the understanding of and engagement with environmental crises can make a crucial difference in the way in which scientists, activists, and organizations understand and engage with them. One of our contributors, the Swedish environmental historian Sverker Sörlin, has commented on this shift elsewhere by highlighting that:

> [o]ur belief that science alone could deliver us from the planetary quagmire is long dead. For some time, hopes were high for economics and incentive-driven new public management solutions. ... It seems this time that our hopes are tied to the humanities ... in a world where cultural values, political and religious ideas, and deep-seated human behaviors still rule the way people lead their lives, produce, and consume, the idea of *environmentally relevant knowledge* must change. We cannot dream of sustainability unless we start to pay more attention to the human agents of the planetary pressure that environmental experts are masters at measuring but that they seem unable to prevent.
>
> (Sörlin 788)

The environmental humanities between the Anthropocene and posthumanisms

A great deal of the discussion about the human agents of environmental change over the last decade has revolved around atmospheric chemist Paul Crutzen and ecologist Eugene Stoermer's concept of the Anthropocene. Stoermer coined the term in the 1980s, but it gained public attention in the early 2000s when Crutzen and Stoermer published articles arguing that human impacts on the planet had become so pervasive and enduring that they would leave permanent traces in the Earth's geological strata, which justified postulating the onset of a new geological epoch distinct from the Holocene (Crutzen; Crutzen and Stoermer). Whether geologists will ultimately accept or reject the term is not likely to have much impact on the lively discussions it has generated about the nature of humans' collective agency and their impact. As the philosopher Dale Jamieson argues in this volume, the Anthropocene gives rise to both the impression of human power due to humans' large-scale

3

transformations of planetary ecosystems, and overwhelming feelings of powerlessness because many of these transformations and their consequences were unintended and are difficult to reverse, especially by individuals or small communities.

The biologist Peter Kareiva has compared the Anthropocene to a process of global domestication. The mix of intended and unintended ecological changes, of desirable and undesirable consequences and side effects that this term encompasses lies at the core of the chapters in Part I, "The Anthropocene and the domestication of Earth." Their focus ranges from debates about the pre-Columbian human "footprint" in the Amazon and how to define domestication (Hecht), the cultivation of plants across migrations and diasporas (Carney), and the voluntary consumption of contaminated foods after environmental disasters (Yuki), to the shift of certain nonhuman individuals and species from domestic to feral, wild, and back, and from native, endangered, or protected to introduced, invasive, and targeted for eradication (Marris; Robin; Sandler), and all the way to manipulations of the Earth's climate (Szerszynski). The categories we use to classify these changes ultimately depend on the ideal visions of nature that cultural communities seek to realize, including visions that voluntarily limit human impact by embracing minimalism (Nersessian). By putting the new concept of the Anthropocene into conversation with older ideas about domestication and utopian social visions, the chapters in this section seek to go beyond the public divide over whether the Anthropocene is merely another name for ecological apocalypse or, on the contrary, for new ecological possibilities (Schwägerl), or even for the triumph of human mastery over nature (Ackerman).

The question of how we can or should envision human agency in the Anthropocene informs the essays in Part II, "Posthumanism and multispecies communities." There is no question that the emphasis on humans' transformative ecological power implicit in the Anthropocene concept runs counter to the thrust of environmentally oriented work in anthropology, geography, history, literary studies, and philosophy over the last few decades, which has sought to analyze human cultures and societies in their constitutive relations with nonhuman species, natural processes, ecological systems, and inanimate landscapes and forces. The idea of a geological era marked above all by humans not only underemphasizes dimensions of nature that continue to be outside of humans' influence, from earthquakes to sunlight (cf. Clark), but also the continuous shaping and reshaping of human bodies, minds, and collectives by ecological processes and interspecies relations.

Even more starkly, the central focus on humans' agency collides with the emergence of posthumanist strains of thought in the humanities and social sciences over the last few decades, from Niklas Luhmann's brand of systems theory, Bruno Latour's Actor-Network-Theory, and certain types of media theory to the recent wave of new materialisms, new vitalism, object-oriented ontology, human–animal studies, human–plant studies, and multispecies ethnography. While these paradigms diverge considerably from each other in their founding assumptions, they articulate varying kinds of philosophical and political skepticism toward the integrity and centrality of the human subject in both its individual and collective dimensions. Instead, they highlight how human identity and agency emerge in the context of systems, communities, and "actants" (a term Bruno Latour originally borrowed from Algirdas Julien Greimas's narrative theory) that include nonhuman species all the way from microbes and plant cultivars to animals, and in some theories also objects, physical processes, and social structures. Against the background of these new theoretical frameworks, the environmental humanities have emerged at a moment when the humanities and qualitative social sciences are reinventing what being human means—and by extension, what it means to study human cultures and societies.

"Posthumanism and multispecies communities" seeks to capture some of this productive tension with the Anthropocene and, more generally, with the anthropocentrism that still characterizes most work in the humanities and social sciences. Anthropocentrism itself is a historical phenomenon, as a look at premodern articulations of humans' embeddedness into the natural world through the genre of comedy highlights (Watson). Contemporary thought challenges it through the focus on how humans are constituted through and with other species ranging from viruses, bacteria, and microbes (Sodikoff) to albatrosses and mistletoe (Rose and van Dooren), and through analysis of how human sociality and national belonging are constructed through affective responses to native and invasive species (Cattelino). In the ocean, a space on Earth that remains profoundly alien to human knowledge and experience, alternative conceptions of what it means to be human can be artistically articulated, whether it be through dancing with dolphins (Chaudhuri) or the encounter with species that we have not had a chance to name, let alone to form cultural connections with (Alaimo). Yet, as is clear in all of these analyses, the oceans, too, are being transformed through climate change and ocean acidification in yet another unintended side effect of planetary domestication.

Questioning humans' exceptionality in their relation with other species has dangers of its own, as many chapters in this and other parts of our volume foreground. On the one hand, rethinking the categorical differences between human and nonhuman agents can lead to the well-known "flattening of ontologies" (in a phrase often used in critiques of Actor-Network-Theory) that makes it difficult—though not impossible—to single out humans as uniquely responsible for environmental destruction and restoration. On the other hand, colonialism, racism, and xenophobia have all too often relied on the strategy of declaring as universal certain historical and cultural ways of being human, and relegating all those who cannot or do not conform to these standards to the subhuman or animal sphere outside the human species. The chapters in Part III, "Inequality and environmental justice," confront head-on the difficulties of reconciling an awareness of different kinds of ecological agency, inflected by socioeconomic inequality and political oppression as well as by divergent historical memories, social structures, and cultural practices, with the generalized "species we" implicit in the Anthropocene.

The posthumanist questioning of the human subject similarly needs to situate itself in this context of pervasive inequalities, even as indigenous and anti-colonial perspectives offer new avenues for thinking beyond the human (see DeLoughrey et al.). Several of the chapters in this section—by Joni Adamson, Barbara Rose Johnston, Jorge Marcone, and Kyle Powys Whyte—explore old and new indigenous perspectives in the Americas in order to consider environmentalism and social justice as dimensions of the same project. Akhil Gupta and Jennifer Wenzel explore the colonial and decolonial contexts of environmentalism today, and Wenzel holds out the hope that what we now call "environmentalism of the poor" or the "environmental justice movement" might simply come to be seen as environmentalism through the lens of a new kind of humanism. Helga Leitner, Emma Colven, and Eric Sheppard explore questions of environmental justice in the context of Jakarta, a modern metropolis—a setting that is particularly significant in view of the fact that over fifty percent of humans now live in cities, and that many cities stand to be radically transformed by climate change.

Our contributors' arguments unfold against the background of a lively and genuinely interdisciplinary debate that was sparked by postcolonial historian Dipesh Chakrabarty's seminal essay "The Climate of History: Four Theses." Chakrabarty argues that the global scope and long-term impact of anthropogenic climate change challenge humanists and social scientists, particularly historians, to rethink influential theories based on foundational differences of, for example, class, race, sex, gender, or colonial power relations. Since climate change puts at risk the conditions of human existence regardless of such differences, he argues, critiques of capitalism, colonialism, and patriarchy now need to be reimagined within a context of geological time periods, planetary transformations, and humans' agency as a species. This context calls for a new kind of universalism, in his view, but one that he can only define negatively, as a refusal of the older kind of universalism that set one particular mode of being human as a yardstick. Even this constrained gesture toward universalism, however, has been fiercely criticized by Marxist and postcolonial theorists who argue that, in view of vast and persistent socioeconomic inequality, uneven contributions to planetary change, and uneven exposure to ecological risk on the part of different communities, postulating "species agency" amounts to ideological cover-up (Moore; Žižek 333–334; for a different critique, see Heise, "Comparative Ecocriticism"). The environmental humanities, then, are defined by the productive conceptual tension between humans' agency as a species and the inequalities that shape and constrain the agencies of different kinds of humans, on one hand, and between human and nonhuman forms of agency, on the other.

Narrative, aesthetics, and media

Chakrabarty's essay, with its emphasis on the geological time periods that the Anthropocene calls on us to consider, also foregrounds another defining challenge for the environmental humanities—that of rethinking time, memory, and narrative. Historical memories and retrospective constructions of nature as it once was have decisively shaped ecological science—particularly restoration ecology—as well as environmentalist thought and activism at large. Historians, anthropologists, and literary researchers have critically analyzed the inclusions, exclusions, and creative reinventions of historical ecology ever since William Cronon's seminal essay on "The Trouble with Wilderness" (1995), one of the first studies to argue that the precolonial wilderness North American environmentalists used to envision as the ideal form of nature never existed in the way they imagined. Similar misconstructions of the ecological past have informed environmental thinking about Australia, Latin America (Gammage; Mann), and, as Kathleen Morrison shows in this volume, India. As a consequence, the narrative of the decline of nature under the impact of modern society stands on shaky ground in many contexts, especially when it ignores or overwrites countervailing perspectives that envision humans as improving their natural environments—narratives that, as Richard White argues, have been proposed by some of the environmental movement's foundational thinkers, from George Perkins Marsh to Rachel Carson, but have usually been ignored. As Michelle Niemann shows in her chapter, stories about the decline and improvement of nature are complemented by the dichotomy between hubris and humility as dominant tropes of environmentalist discourse that have been criticized and recuperated in the movement's internal struggles up until the present day.

Narratives of decline and extinction or, conversely, of resilience and improvement in nature always intertwine ecological facts with cultural histories and value judgments, as the chapters by Rosanne Kennedy, Brett Buchanan, and Lizabeth Paravisini-Gebert show through the analysis of writing, philosophy, and art about endangered species, multispecies communities, and seashores at risk from climate change. Such stories often seek to define a particular community's vision of its own place in history and geography, its anxieties over the changes that modernization and colonization impose, and its aspirations for the future. In this context, decline narratives are often a powerful means of expressing political resistance to modernization and colonization, even as they also frequently constrain visions of the socio-ecological future as anything other than a recreation of the past. The expertise of environmental humanists in the critical analysis of fictional as well as nonfictional narrative puts them right at the heart of a vigorous debate between different strains of the environmental movement over what story templates will prove to be most effective in the future. This debate has pitted "eco-modernists" such as Ted Nordhaus and Michael Shellenberger, who claim that "the solution to the unintended consequences of modernity is, and has always been, more modernity—just as the solution to the unintended consequences of our technologies has always been more technology," against the contributors to the volume *Keeping the Wild: Against the Domestication of Earth*, who argue for the accuracy and usefulness of the wilderness idea and the declensionist story template (Wuerthner and Crist). As the chapters in Part IV, "Decline and resilience: environmental narratives, history, and memory," demonstrate, the forms and political functions of environmental narrative vary a great deal more than these two positions allow for.

Not only are they far more varied when we look across different languages and regions of the world, but also when we look across different art forms and media, as the chapters in Part V, "Environmental arts, media, and technologies," do. Environmental aesthetics and communication have long confronted the challenge of mediating between stories and statistics, local experiences and global scenarios, images with instant impact and ideas about long-term transformation, a task that poses problems of scale as well as representation. Environmental art, architecture, and film have sought to convey an understanding of large-scale ecological processes and statistical shifts through techniques that combine realism with abstraction, the animate with the inanimate, and familiar visions with remote or futuristic ones, whether they focus on landscapes, habitats, food, or waste (Nisbet; Weik von Mossner; Barber; Carruth; Zubiaurre). The experimental techniques of the twentieth-century avant-garde in Europe and the Americas—collage, montage, self-reference, combinations of text and image—take on new, environmental functions in this context (Bök).

Over the last four decades, increasingly sophisticated digital tools, from geographic information systems to data visualizations, have generated new maps, images, and narratives about the short- and long-term risks and opportunities of global ecological change (Houser; Sinclair and Posthumus). The challenge for the environmental humanities in this context is not just the study of digital images and artifacts, but the integration of digital tools and methods with older humanistic procedures: the combination of close reading with computational criticism, for example, of thick description with newly accessible statistics about ecological processes and cultural practices, of storytelling with database creation, or of photography with zoomable maps. The role and relevance of these new methods has been fiercely debated over the last decade in the humanities and qualitative social sciences; the environmental humanities, with their dual stakes in histories, cultures, and values, on one hand, and in ecological processes and global risks, on the other, are in a privileged position to showcase the achievements and shortfalls of these innovative approaches and procedures.

In engaging with new archives, tools, and communications venues, environmental humanists seek to make the differences between their own home disciplines productive rather than divisive (cf. Bergthaller et al.), as well as to continue conversations with scientists, activists, policy makers, and urban and regional planners. The interest of environmental anthropologists, geographers, historians, literary scholars, and philosophers in how ecological systems and nonhuman species interact with particular societies and cultures has over the last decade been reinforced by the "material turn" in the humanities and social sciences that emphasizes the otherness and agency as well as the entanglement of nonhuman actants with human ones (Nash; Sullivan; Bergthaller). But just how such Latourian "imbroglios" are theoretically envisioned matters not just for the foundations of humanistic and social-scientific research, but also for the ways in which they can be made to create interdisciplinary networks and public outreach (Nash; Sörlin; Garrard; LeMenager). Such networks in and beyond the university take on particular importance with the new emphasis on urban ecology, prompted by the realization that, sometime around 2010, the proportion of humans living in cities crossed the fifty percent threshold. Cities—and in many cases, very large cities—will be humans' dominant habitat in the future, a partly natural and partly anthropogenic habitat whose study calls for the integration of insights from anthropology, architecture, biology, cultural studies, ecology, geography, political science, sociology, and urban planning (Sandilands; Christensen and Heise).

Critique, a central way of defining the mission of the humanities and qualitative social sciences, certainly remains an important element in these interfaces. But "[a]n important tension is emerging between, on the one hand, the common focus of the humanities on critique and an 'unsettling' of dominant narratives, and on the other, the dire need for all peoples to be constructively involved in helping to shape better possibilities in these dark times. The environmental humanities is necessarily, therefore, an effort to inhabit a difficult space of simultaneous critique and action," the authors of one of the first manifesto-style introductions to the environmental humanities argue (Rose et al. 3). And as one of our contributors, Hannes Bergthaller, has argued elsewhere with a team of collaborators, "A genuinely inclusive and adventurous approach to the Environmental Humanities might also facilitate collaboration with partners outside the academy, where much of the work of adaptation to environmental change, mitigation of ecological damage, and transition to new social structures must take place" (Bergthaller et al.; see also Neimanis et al. 88–90). Encouraging synthetic as well as analytical perspectives and constructive as well as critical thinking is therefore a central task ahead for environmental humanists (Garrard; LeMenager; cf. Bergthaller et al.).

Combining theoretical and analytical work with creative experiments and public engagement was the task that our local community of environmental humanists found itself facing at the end of a year-long Sawyer Seminar, generously funded by the Andrew Mellon Foundation, that took place at the University of California, Los Angeles, in 2014–15. The seminar brought together humanists and social scientists not only from UCLA and its sister universities in the UC system, but also colleagues from around the United States and from Australia, Canada, England, Germany, Japan, Sweden, and Taiwan. The *Routledge Companion to the Environmental Humanities* combines their contributions with those of other researchers who generously agreed to join us in print even when our funds and time schedule did not allow us to invite them in person. The lively discussions and manifest student interest in the monthly seminars led us to create a new undergraduate minor in "Literature and the Environment" that we envision growing into an "Environmental Humanities" major as more faculty and students in the humanities and social sciences discover the connections between their areas of research and unfolding scenarios of global ecological risk and opportunity. It also prompted us to create LENS, the Lab for

Environmental Narrative Strategies, which launched as a new cell of UCLA's Institute of the Environment and Sustainability in the fall of 2016. LENS focuses on research, teaching, and public engagement around the narratives and visions that different cultural communities create to understand and communicate environmental crises and possibilities across a variety of media. In collaboration with activists, artists, researchers, and writers around the world, LENS undertakes and supports experimental strategies for generating new narratives and images, grounded in an understanding of ecological crises as fundamentally cultural processes, that help to create a more sustainable world for humans and the species that coinhabit the planet with us.

References

Ackerman, Diane. *The Human Age: The World Shaped By Us*. New York: Norton, 2014. Print.

Agrawal, Arun and Kent Redford. "Conservation and Displacement: An Overview." *Conservation and Society* 7 (2009): 1–10. Print.

Bergthaller, Hannes, Rob Emmett, Adeline Johns-Putra, Agnes Kneitz, Susanna Lidström, Shane McCorristine, Isabel Pérez Ramos, Dana Phillips, Kate Rigby, and Libby Robin. "Mapping Common Ground: Ecocriticism, Environmental History, and the Environmental Humanities." *Environmental Humanities* 5 (2014): 261–276. Web. 11 Nov. 2014.

Buell, Lawrence. *The Environmental Imagination: Thoreau, Nature Writing, and the Formation of American Culture*. Cambridge, MA: Harvard University Press, 1995. Print.

——. *Writing for an Endangered World: Literature, Culture and Environment in the U.S. and Beyond*. Cambridge, MA: Harvard University Press, 2001. Print.

Chakrabarty, Dipesh. "The Climate of History: Four Theses." *Critical Inquiry* 35 (2009): 197–222. Print.

Clark, Nigel. *Inhuman Nature: Sociable Life on a Dynamic Planet*. London: Sage, 2011. Print.

Cronon, William. "The Trouble with Wilderness; or, Getting Back to the Wrong Nature." *Uncommon Ground: Rethinking the Human Place in Nature*. Ed. William Cronon. New York: Norton, 1995. 69–90. Print.

Crosby, Alfred. 1972. *The Columbian Exchange: Biological and Cultural Consequences of 1492*. 30th anniversary ed. Westport, CT: Praeger, 2003. Print.

Crutzen, Paul J. "Geology of Mankind." *Nature* 415 (2002): 23. Web. 5 Feb. 2016.

Crutzen, Paul J. and Eugene F. Stoermer. "The 'Anthropocene.'" *Global Change Newsletter* 41 (2000): 17–18. Web. 10 Oct. 2013.

DeLoughrey, Elizabeth, Jill Didur, and Anthony Carrigan. "Introduction: A Postcolonial Environmental Humanities." *Global Ecologies and the Environmental Humanities: Postcolonial Approaches*. Ed. Elizabeth DeLoughrey, Jill Didur, and Anthony Carrigan. New York: Routledge, 2015. 1–32. Print.

Dowie, Mark. *Conservation Refugees: The Hundred-Year Conflict between Global Conservation and Native Peoples*. Cambridge, MA: MIT Press, 2009. Print.

Gammage, Bill. *The Biggest Estate on Earth: How Aborigines Made Australia*. Sydney: Allen & Unwin, 2011. Print.

Guha, Ramachandra. "Radical American Environmentalism and Wilderness Preservation: A Third-World Critique." *Environmental Ethics* 11.1 (1989): 71–84. Print.

Guha, Ramachandra and Joan Martínez-Alier. *Varieties of Environmentalism: Essays North and South*. London: Earthscan, 1997. Print.

Harvey, David. *Justice, Nature, and the Geography of Difference*. Oxford: Blackwell, 1996. Print.

Heise, Ursula K. "Comparative Ecocriticism in the Anthropocene." *Komparatistik*. Heidelberg: Synchron, 2014. 19–30. Print.

——. *Imagining Extinction: The Cultural Meanings of Endangered Species*. Chicago, IL: University of Chicago Press, 2016. Print.

Hulme, Mike. *Why We Disagree about Climate Change: Understanding Controversy, Inaction and Opportunity*. Cambridge: Cambridge University Press, 2009. Print.

Jamieson, Dale. *Reason in a Dark Time: Why the Struggle against Climate Change Failed–And What it Means for Our Future*. Oxford: Oxford University Press, 2014. Print.

Kareiva, Peter, Sean Watts, Robert McDonald, and Tim Boucher. "Domesticated Nature: Shaping Landscapes and Ecosystems for Human Welfare." *Science* 316 (29 June 2007): 1866–1869. Print.

Latour, Bruno. 1991. *We Have Never Been Modern*. Trans. Catherine Porter. Cambridge, MA: Harvard University Press, 1993. Print.

Mann, Charles C. *1491: New Revelations of the Americas before Columbus*. New York: Vintage, 2005. Print.

Merchant, Carolyn. 1980. *The Death of Nature: Women, Ecology, and the Scientific Revolution*. New York: Harper & Row, 1983. Print.

Moore, Jason. "The Capitalocene, Part 1: On the Nature & Origins of Our Ecological Crisis." 2014. Web. 6 Feb. 2016.

Neimanis, Astrida, Cecilia Åsberg, and Johan Hedrén. "Four Problems, Four Directions for Environmental Humanities: Toward Critical Posthumanities for the Anthropocene." *Ethics and the Environment* 20.1 (2015): 67–97. Web. 15 Jan. 2016.

Nordhaus, Ted and Michael Shellenberger. "Evolve: The Case for Modernization as the Road to Salvation." *Orion Magazine* (September/October 2011). Web. 6 Feb. 2016.

Norgaard, Kari Marie. *Living in Denial: Climate Change, Emotions, and Everyday Life*. Cambridge, MA: MIT Press, 2011. Print.

Ortner, Sherry B. "Is Female to Male as Nature Is to Culture?" *Feminist Studies* 1.2 (1972): 5–31. Print.

Plumwood, Val. *Environmental Culture: The Ecological Crisis of Reason*. London: Routledge, 2001. Print.

Rose, Deborah Bird, Thom van Dooren, Matthew Chrulew, Stuart Cooke, Matthew Kearnes, and Emily O'Gorman. "Thinking Through the Environment, Unsettling the Humanities." *Environmental Humanities* 1 (2012): 1–5. Web. 5 Feb. 2016.

Schwägerl, Christian. *Menschenzeit: Zerstören oder gestalten? Die entscheidende Epoche unseres Planeten*. Munich: Riemann, 2010. Print.

——. *The Anthropocene: The Human Era and How It Shapes Our Planet*. Trans. Lucy Renner Jones. Synergistic Press, 2014. Print.

Sörlin, Sverker. "Environmental Humanities: Why Should Biologists Interested in the Environment Take the Humanities Seriously?" *BioScience* 62 (2012): 788–789. Print.

Wuerthner, George, Eileen Crist, and Tom Butler, eds. *Keeping the Wild: Against the Domestication of Earth*. San Francisco: Foundation for Deep Ecology, 2014. Print.

Žižek, Slavoj. *Living in the End Times*. Revised and updated edn. London: Verso, 2011. Print.

Part I

THE ANTHROPOCENE AND THE DOMESTICATION OF EARTH

1

THE ANTHROPOCENE
Love it or leave it[1]

Dale Jamieson

One of the central themes of classical philosophy is the persistence and puzzling nature of change. Throughout the history of philosophy this concern appears, disappears, reappears, and never completely goes away. There is a similar oscillation in our sciences between an interest in equilibrium states and a fascination with the sources and persistence of disorder. Which is primary, order or disorder, change or permanence? Perhaps it is a coincidence that the period and place in which I grew up (1950s America) was characterized by an obsession with "progress," but a dislike for change. Both nature and the economy sought equilibrium, and the nation itself was on the way to a more perfect union. Or so it seemed to me that it seemed to my elders.

These human frames and constructions can sometimes obscure the fact that we live in a violent neighborhood in a universe given to cataclysms. Even our own little planet is constantly changing. The forces driving these changes include variation in solar radiation, movements in tectonic plates, volcanic activity, meteor strikes, shifts in orbit, and changes in the tilt of the Earth on its axis. Life itself is among the forces that have changed the Earth, from cyanobacteria that produced the first oxygen on Earth to human beings who are now increasing the concentration of carbon dioxide in the atmosphere.

In 1997, a distinguished group of scientists published an influential article in which they assessed the human impact on the Earth (Vitousek et al.). They calculated that between one-third and a half of Earth's land surface had been transformed by human action; that carbon dioxide in the atmosphere had increased by more than 30 percent since the beginning of the industrial revolution; that more nitrogen had been fixed by humanity than all other terrestrial organisms combined; that more than half of all accessible surface freshwater was being appropriated by humanity; and that about one-quarter of Earth's bird species had been driven to extinction. They concluded that "it is clear that we live on a human-dominated planet" (494).

In recognition of the increasing human domination of the planet, some scientists have proposed that we have entered a new geological era—the Anthropocene. A proposal to declare the Anthropocene a new epoch in Earth's history is under formal review by the International Commission on Stratigraphy (ICS), the authoritative scientific body that makes decisions about the geologic time scale. Even if the ICS declines to declare the Anthropocene a geological epoch, it can still be an important concept for understanding our present condition. There is no guarantee that a record of who we are, how we lived, and how we found meaning in our lives will be encoded in the Earth's crust.

Some humanists (and others) have been suspicious of the concept of the Anthropocene. They see in the very name the glorification of the human domination of the planet. They are not wrong to think that how we categorize can be relevant to what we think is permissible. Categorizing a plant as a weed, an animal as a pest, or a person as a thug has implications about how they may permissibly be treated. But sometimes a category is just a category. Think of triangles. You may treat a triangle as if it were a square, but that would only be wrong in the stupid sense of "wrong," not in any richer normative sense. Geological categories are somewhere in between. The Holocene is not a logical notion in the way that triangle is. The term was not invented until 1867 (and then in French), not proposed as an epoch until the Third International Geological Congress in 1885, and not formally adopted by the US Geological Survey until 1967. Despite the fact that this term was carried over into English from a French term that was constructed from Greek materials less than one hundred fifty years ago, it is not viewed with suspicion in the way that "Anthropocene" is. What accounts for the difference?

Part of the explanation is surely that the historical classifications that we are born into typically form part of an unquestioned baseline: they are "naturalized." When it comes to the Anthropocene, the sausage is being made before our very eyes. And as with actual sausage, it is not hard to think that there is something unseemly about the process. Another concern is the implicit reference to humanity that is contained in the term "Anthropocene." Some seem to think that to name an epoch with a term that refers to humans is to glorify humanity and its domination of nature. But surely this does not follow. We can speak of the Warring States Period in China, as scholars often do, without glorifying warring states.

Another view, often more implied or suggested than asserted, is that the Anthropocene has been around as long as there have been people. There is nothing new about the Anthropocene. What is new is that people are waking up to it.

What matters for the demarcation of the Anthropocene as a morally and culturally interesting category is that some region in time can usefully be spoken of as qualitatively different from some previous region. There can be vagueness between these periods, and the qualitative distinction that marks the Anthropocene can supervene on quantitative differences rather than resting on irreducible differences of kind. It is against this background that Vitousek et al.'s study is a good marker. A study like this, done a thousand years earlier, would have produced radically different results. Yes, life has always affected the Earth, but there is something new and different about the way humanity affects the planet and thereby itself. Whether or not this is of geological interest, it is of great cultural and moral moment.

The most obvious feature of the Anthropocene is the growing human population and its demand for energy, food, goods, services, and information, along with the need to dispose of its waste products. At the beginning of the Holocene there were probably about six million people living as hunter-gatherers. Today there are more than seven billion people, most living in urban societies, many with a command over resources that only the nobility would have had a few centuries ago.

Technology is an important part of the story of the Anthropocene. The humanity that has transformed nature is now organized in highly complex systems bound together by air travel, oil and gas pipelines, electrical wires, highways, train tracks, fiber optic cables, and satellite connections. In 1907 the French philosopher Henri Bergson wrote: "In thousands of years … our wars and revolutions will count for little … but the steam engine, and the procession of inventions of every kind that accompanied it, will perhaps be spoken of as we speak of the bronze or of the chipped stone of pre-historic times: it will serve to define an age" (138–139). Technology enables "action at a distance" that would once have seemed inconceivable, whether as a sexual encounter in virtual reality with someone on the other side of the world, or as the instantaneous transfer of wealth, resources, and power. Some would even say that technology is no longer something that we use; it is an integral part of who we are. On this view our minds extend to devices and configurations of objects far beyond our skulls and even our skin: We are "natural born cyborgs" (Clark).

The conjunction and effects of large population, high consumption, and technology have affected the nature of our relationships and our conception of agency. In some respects we feel empowered. We can save a child in a far-away land by making a phone call and pledging a contribution. The swipe of a credit card can deliver goods and services from remote parts of the world (thank you, Amazon). A few clicks at a computer allow us to register our opinions about everything from Iran's nuclear program to a woman who left her dog in a car with the windows rolled up in a parking lot in Kansas. A man in Las Vegas controlling an unmanned drone can stalk and kill a single individual on the other side of the world. Together we can change climate, drive species to extinction, and acidify the ocean. Never have humans been so powerful. Hence the name "Anthropocene."

Yet, at the heart of the Anthropocene is a widespread sense of the loss of agency. We say money controls politics as if elections are not decided on the basis of who gets the most votes. Many people in both developed and developing countries feel trapped in their classes and social roles. The main obstacle to taking action on climate change is the deep sense of its inevitability and our powerlessness to affect its course.

The Anthropocene has also brought confusion. In a world in which everything affects everything else and no one feels decisive over much at all, the distinction between causation and complicity has become fraught. When I drive my SUV to the 7–11 for a slurpy, am I causing climate change, contributing to it, complicit in it, or does it matter at all what I do? Do these distinctions even matter? Is San Diego Gas and Electric giving me what I want, or manipulating me because I have no real choice but to pay the bill? The list of questions could go on. And on and on. The sense of powerlessness and confusion they reflect and engender is part of why so many of us retreat to private life.

But what is private life in the Anthropocene? Technology has brought with it increased transparency (good) and a loss of privacy (bad). I can read police reports from the jurisdiction of my choice and find the salaries of my colleagues who work in public universities. But if I were to investigate myself, I would be shocked by how much is readily available. And in the technology-driven world of the Internet and social media, the distinction between intersubjective validity and just plain repetition threatens to break down. What does this do to an outlook, such as that of traditional liberalism, in which the distinction between public and private is central?

In a world of greater resources and fewer well-insulated spheres of life, everything becomes increasingly fungible, from wombs to kidneys to artworks to lives themselves. Tradeoffs between the near and dear and the remote and strange become possible in a way they never were before. Sitting in his rooms in Cambridge, Henry Sidgwick could read about famine in Bengal, but there was little he could do about it in real time except decry it. I, on the other hand, can immediately empty my bank account and make something happen on the ground in Nepal (for good or ill). Or instead I could go in for some online gambling on my children's future; or I could just leave everything to my canary. Everything seems possible but nothing seems to matter.

This leads to a crisis in meaning. Human life has traditionally been lived against the background of a nature that is seen as largely independent of human action. The biblical book of Matthew tells us that the sun shines on the just and unjust alike. Once geoengineering is perfected, we may be able to fix this oversight of nature. But what becomes of the message of humility and compassion that this teaching evokes?

What emerges from all this is not clear. Together we are remaking the planet, though no one seems to have ever decided that this was a good idea. Many of us feel as though we have choices over matters that were once fixed, yet feel powerless over things that we once had power over. The old order is shaking. Some new thinking in needed. Welcome to the Anthropocene.

The most fundamental choice the Anthropocene presents us with is whether to love it or leave it. Some writers seem to think that there is no choice here at all; that the Anthropocene is a natural end state—an inevitability—perhaps because they believe that we have always been in the Anthropocene. This is a fundamentally unscientific view. There are no natural teleologies. The Anthropocene will pass away like previous historical epochs, and so will humanity. It is up to us whether to willfully intensify the human domination of nature, accept it, or try to reduce it. To present this as anything other than a choice is bad faith.

However, it is also true that we are living in the Anthropocene and will do so for the foreseeable future even if we desire to exit. Whether we embrace the Anthropocene or want to exit from it, we need to develop ways of life that will allow humanity to flourish in this period. We need an ethics of the Anthropocene.

An ethical system regulates behavior within a social group, helps to coordinate behavior to address collective problems, informs relations with those outside the group, and provides meaning for those who are members of the group. Ethical systems are often dynamic, incomplete, and sometimes incoherent and inconsistent. Most such systems embody ideals of character, criteria for right action, and conceptions of what is morally valuable. Ethical systems are collective constructions, not legislated by single individuals. Agency, its meaning, and its importance are defined by and reflected in an ethical system, and for creatures like us, agency as it is experienced phenomenologically is tightly bound to the proximate: what presents itself to our senses and causally interacts with us in identifiable ways.

From the beginning, ethics has primarily been concerned with the proximate. However, what is proximate is flexible. Stories, music, relics, sacred space, and even the establishment of a common language are all ways of bringing into view what would otherwise be invisible. The expanding circle of ethics, which to a great extent coincides with globalization, has made the distal proximate through new living arrangements, forms of travel, and kinds of imagery enabled by technological innovation. However, there is a limit to what can be made proximate to creatures like us. Carbon dioxide emissions are among those phenomena that are difficult to bring close and make visible. One consequence of this is that if we want to reduce our emissions, we need proxies that will support us in doing that.

A conception of the green virtues—character traits, dispositions, and emotions—can fill that role to some extent. The green virtues are mechanisms that provide motivation to act in our various roles from consumers to citizens in order to reduce our impact on nature, regardless of the behavior of others. They also give us the resiliency to live meaningful lives even when our actions are not reciprocated. They can provide guidance for conducting experiments for how to live in the Anthropocene. A conception of the virtues is at the very foundation of what it is to experience a sense of agency. For to see ourselves as agents is to see ourselves as beings that have both character traits that condition what we do and aspirations to go beyond what has been given to us by the circumstances in which we find ourselves.

A full account of the virtues required to navigate the Anthropocene would involve at least these: love, respect, humility, and responsibility.[2] In what remains in this chapter, I will discuss only respect—in particular, respect for nature. Understanding one of the virtues in some detail will help illuminate just how difficult and subtle the challenge of living well in the Anthropocene really is.

Respect for nature has been celebrated at various places and times to different degrees.[3] It is a persistent if not universal value. There are at least precursors of this idea in Kant and strong assertions of it in the Romantic tradition.[4] It is frequently attributed to indigenous peoples and found in various Asian traditions. While it is difficult to say exactly what this virtue consists in, it is relatively easy to give examples of the failure to express it. If we dominate our planet, as Vitousek et al. claim, then surely we can be said to dominate nature in one important sense of the term. Dominating something can be one way of failing to respect it, so it is plausible to say that in virtue of our domination of nature we fail to respect it.[5]

Domination can be expressed attitudinally in the ways in which we think and feel about nature as well as substantively. We often treat nature as "mere means," as if it does not have any value or existence independent of its role as a resource for us. As a society we seem to treat the Earth and its fundamental systems carelessly, as if they were toys, or as if their functions could easily be replaced by a minor exercise of human ingenuity. It is as if we have scaled up slash-and-burn agriculture to the planetary level.[6]

One of the insights of the social movements of the 1960s was that a vicious circle can take hold with subordinated groups.[7] Mistreatment diminishes respect, which leads to further mistreatment, which further diminishes respect, and so on. The same vicious circle can take hold with nature. Dominating nature both expresses and contributes to a lack of respect, which in turn leads to further domination.

Respecting nature, like respecting people, can involve many different things. It can involve seeing nature as amoral, as a fierce adversary, as an aesthetic object of a particular kind, as a partner in a valued relationship, and perhaps in other ways. These attitudes can exist simultaneously within a single person.

When nature is seen as amoral, it does not constitute a moral resource in any way. One memorable statement of nature as amoral occurs in chapter five of the *Tao Te Ching*, attributed to the Taoist sage Lao-Tse: "Heaven and Earth are impartial; they treat all of creation as straw dogs." In ancient Chinese rituals, straw dogs were burned as sacrifices in place of living dogs. What is asserted here is that the forces that govern the world are as indifferent to human welfare as humans are to the fate of the straw dogs they use in ritual sacrifice. On this view we should respect nature because of its blind, unpurposing force and power.

Seeing nature as amoral can easily slip into seeing nature as an immensely powerful, even malevolent adversary, and humanity as weak, vulnerable, and in need of protection.[8] If humanity and its projects are to survive and thrive, nature must be subdued and kept at bay. Nature, on this view, is the enemy of humanity.

Amoral nature can be respected for its radical "otherness" that cannot be assimilated to human practices. Nature as an adversary can be respected for its power and abilities in pursuing its ends, which are fundamentally at odds with those of humanity. Seeing nature as amoral or as an adversary can provide grounds for respecting nature but can also provide a rationale for dominating nature.[9]

A third way of respecting nature sees profound aesthetic significance in its overwhelming power. This thought is powerfully developed in Edmund Burke's 1757 work, *A Philosophical Enquiry into the Origin of our Ideas of the Sublime and Beautiful*. The human experience of the sublime is, according to Burke, a "delight," and one of the most powerful human emotions. Yet, perhaps paradoxically, the experience of the sublime involves such "negative" emotions as fear, dread, pain, and terror, and can occur when we experience deprivation, darkness, solitude, silence, or vacuity. The experience of the sublime arises when we feel we are in danger, but it is actually not so. Immensity, infinity, magnitude, and grandeur can cause this experience of unimagined eloquence, greatness, significance, and power. The sublime is often associated with experiences of mountains or oceans. Such experiences may occasion wonder, awe, astonishment, admiration, or reverence. In its fullest extent, the experience of the sublime may cause total astonishment.

The idea of the sublime was profoundly influential on nineteenth-century American culture, notably through painters such as Thomas Cole and Frederic Church. It went on to be an important influence on American environmentalism through the writings of John Muir and, more recently, Jack Turner, Dave Foreman, and other advocates for "the big outside." Indeed, the case for wilderness preservation is often made in the language of the sublime.

Finally, there is the idea of nature as a partner in a valuable relationship. People often speak of particular features of nature as if they were friends, lovers, or even parents. People who see elements of nature as friends often feel that they learn from nature as they do from other companions. Some speak of nature in language that is usually reserved for lovers.[10] Indeed, we often speak of those who want to protect nature as "nature lovers." In some people, nature elicits feelings of filial devotion. John Muir wrote that "[t]here is a love of wild nature in everybody, an ancient mother-love" (98). Many of us also associate nature with a feeling of being at home. I grew up in San Diego, California, and the sights, smells, breezes, and quality of light that I experience when I am there are transformative, especially when I step onto the beach at Torrey Pines, just north of the city.

These different ways of respecting nature support somewhat different attitudes toward nature and reasons for respecting it. Rather than discussing the details, I will mention three reasons for respecting nature that seem quite robust across times and cultures. Respect for nature can be grounded in prudence, can be seen as a fitting response to the roles that nature plays in giving our lives meaning, and can also spring from a concern for psychological wholeness.

One reason for respecting nature is that it is in our interests to do so. The geoscientist Wallace Broecker compares our climate-changing behavior to poking a dragon with a sharp stick. Angering the dragon of climate is not likely to be a good business plan for maintaining human life on Earth. Versions of this argument are ubiquitous in the environmental literature, and something like this view is implicit in slogans such as Barry Commoner's "third law of ecology" which states that "nature knows best." It can also be seen as providing the foundation for the precautionary principle.

A second reason for respecting nature is that, for many people and cultures, nature provides important background conditions for lives having meaning. It is easy to think of examples from history, literature, or contemporary culture. William Blake's idea of England as a "green and pleasant land" is important in English literature, history, and identity. The cherry orchard in Anton Chekhov's play of the same name defines the life of everyone in the community. Think of the role that landscape plays in the lives of indigenous peoples. For that matter, think of how the "flatirons" define Boulder, Colorado.

An analogy may help to bring the point out more clearly. Representational painting is not the only kind of valuable painting but it is one very important kind. Indeed, it may be the mother from which other forms of valuable painting emerged. Representational painting exploits the contrast between foreground and background. What is in the foreground gains its meaning from its contrast with the background. What I want to suggest is that nature provides the background against which we live our lives, offering us an important source of meaning. It is thus not surprising that we delight in nature and take joy in its operations, and feel grief and nostalgia when familiar patterns are disrupted and natural features destroyed.[11] In these respects, meaning and mourning are closely related concepts.[12]

A third reason for respecting nature flows from a concern for psychological integrity and wholeness. As Kant (and later Freud) observed, respecting the other is central to knowing who we are and to respecting ourselves. Indeed, the failure to respect the other can be seen as a form of narcissism. Some work in environmental psychology gestures toward a story in which the recognition of nature as an "other" beyond our control is at the root of our self-identity and communal life (for example, see Clayton).

Respect for nature is an important virtue that we should cultivate as part of an ethic for the Anthropocene. Respect can be manifest in many different ways within a single person, sometimes simultaneously. Nature itself is not a single thing and we can respect some elements or dimensions of nature while expressing contempt for others. Respecting nature is respecting ourselves.

The fundamental question the Anthropocene poses for us is whether we love it or want to leave it. However we answer that question, for the foreseeable future we will have to live with it. The immediate challenge we face is how to navigate this period in a way that preserves as much of what we value as possible, and provides us with sources of meaning.

The green virtues are not an algorithm for solving the problems of the Anthropocene. Politics is required, not just virtue. Yet the green virtues can provide guidance for how to live gracefully in a changing world, while helping to restore in us a sense of agency.

Notes

1 This chapter has been influenced by collaborative work with Bonnie Nadzam (see Jamieson and Nadzam). I also thank Michelle Niemann for comments on an earlier version of this chapter, and the participants in a seminar at UCLA for an invigorating discussion.

2 For more on love, see Jamieson and Nadzam; for responsibility, see Jamieson, "Responsibility for Climate Change," and for a fuller (and somewhat different) account generally of the green virtues see chapter 6 of Jamieson, *Reason in a Dark Time*. A fuller account would also have to find a place for the importance of anti-moralism and a rejection of an ideal of moral purity.

3 Respect for nature can be thought of as a duty as well as a virtue, which is how Paul Taylor understands it, and also how I regarded it in "Climate Change, Responsibility, and Justice." And of course, a fuller account of this virtue would require much more to be said about the nature of nature than I can say here.

4 On Kant, see Allen Wood; for an expression of respect for nature in Romantic poetry, see Samuel Taylor Coleridge's poem "The Rime of the Ancient Mariner."

5 There is a sense of "domination" in which it does not imply a lack of respect (e.g., one team can be said to dominate another in a game), but for reasons that are given below (e.g., that our lack of respect for nature expresses itself attitudinally as well as substantively) and for others that are obvious, it is not this sense that is in play here.

6 I owe this image to Jeremy Waldron.

7 This theme was especially prominent in the work of Frantz Fanon and Malcolm X.

8 Werner Herzog's *Grizzly Man* is a wonderful film on this and related themes.

9 John Stuart Mill in his scattered writings on nature is an interesting case of someone who saw nature as amoral but maintained a fundamental respect for nature, in part for its otherness, but also because of its aesthetic qualities and the ways it contributes to human life.

10 There is even a blog "52 Ways to Fall in Love With the Earth" (Wendt), which can be viewed at http://52ways.wordpress.com/.
11 For an articulate example of these feelings regarding the devastation of Utah's red rock canyon country by the creation of Lake Powell, see Edward Abbey's *The Monkey Wrench Gang*. A similar sense of loss and nostalgia can be engaged by urban projects such as Robert Moses's plan to build a highway though Manhattan's Washington Square Park (Caro).
12 I owe this thought to Sebastiano Maffetone.

References

Abbey, Edward. *The Monkey Wrench Gang*. Philadelphia, PA: Lippincott, 1975. Print.

Bergson, Henri. *Creative Evolution*. Trans. Arthur Mitchell. New York: Henry Holt, 1911. Print.

Broecker, Wally. "The Carbon Cycle and Climate Change: Memoirs of My 60 Years in Science." *Geochemical Perspectives* 1.2 (April 2012): 221–340. Web. 1 September 2015.

Burke, Edmund. *A Philosophical Enquiry into the Origin of our Ideas of the Sublime and Beautiful*. 1757. New York: Oxford University Press, 2008. Print.

Caro, Robert A. *The Power Broker: Robert Moses and the Fall of New York*. New York: Knopf, 1974. Print.

Clark, Andy. *Natural-born Cyborgs: Minds, Technologies, and the Future of Human Intelligence*. Oxford: Oxford University Press, 2003. Print.

Clayton, Susan D. *Identity and the Natural Environment: The Psychological Significance of Nature*. Cambridge, MA: MIT Press, 2003. Print.

Commoner, Barry. *The Closing Circle: Nature, Man, and Technology*. New York: Knopf, 1971. Print.

Foreman, Dave. *Confessions of an Eco-warrior*. New York: Harmony, 1991. Print.

Herzog, Werner, dir. *Grizzly Man*. Lions Gate Films, Discovery Docs, and Real Big Production, 2005. Film.

Jamieson, Dale. "Climate Change, Responsibility, and Justice." *Science and Engineering Ethics* 16 (22 October 2009): 431–445. Springer Link. Web. 1 September 2015.

——. *Reason in a Dark Time: Why the Struggle against Climate Change Failed—and What It Means for Our Future*. New York: Oxford UP, 2014. Print.

——. "Responsibility for Climate Change." *Global Justice: Theory Practice Rhetoric*, in press.

Jamieson, Dale, and Bonnie Nadzam. *Love in the Anthropocene*. New York: OR, 2015. Print.

Muir, John. *Our National Parks*. Boston: Houghton Mifflin, 1901. Print.

Taylor, Paul. *Respect for Nature: A Theory of Environmental Ethics*. 1986. Princeton, NJ: Princeton University Press, 2011. Print.

Turner, Jack. *The Abstract Wild*. Tucson: Univeristy of Arizona Press, 1996. Print.

Wendt, Tamara. "52 Ways to Fall in Love With the Earth." *Hooked on Nature*. 2009. Web. 18 December 2015.

Wood, Allen W. "Kant on Duties Regarding Nonrational Nature." *Aristotelian Society Supplementary Volume* 72.1 (June 1998): 189–210. Wiley Online Library. Web. 1 September 2015.

Vitousek, Peter M., Harold A. Mooney, Jane Lubchenco, and Jerry M. Melillo. "Human Domination of Earth's Ecosystems." *Science* 277 (25 July 1997): 494–499. JSTOR. Web. 1 September 2015.

2

DOMESTICATION, DOMESTICATED LANDSCAPES, AND TROPICAL NATURES

Susanna B. Hecht

Domestication politics and the human footprint

This chapter explores how competing models of domestication and human impact are currently playing out in debates over indigenous "footprints" in the Amazon basin in pre-Colombian times. These questions might be viewed as arcana in a scientific debate, except that these controversies illuminate divergent ideologies about nature, domestication, and social impacts on landscapes. From the perspective of science and technology studies, this debate involves differing epistemic communities, their sociologies, and their explanatory framings. These analytics also pertain to broader questions such as the implications of historic land use for climate change, dating the Anthropocene, civilizational discourses about native peoples, current development, and indigenous land politics in Amazonia.

The Brazilian constitution of 1988—which has been widely copied throughout the Amazon countries—recognizes land claims based on historical markers and explanations of use, which include forest activities. Thus, traditional and indigenous peoples need forms of evidence including ethnohistorical and archival documents, landmarks like historical villages, and historical ecological information as part of the legal dossier to claim lands. While showing land use legacies is important for such territorial claims, showing the absence or minimal presence of humans has become increasingly necessary for land assertions associated with "strong" conservation for biodiversity, for carbon offsets and watershed control, and ultimately for Reduced Emissions from Deforestation and Degradation (REDD) payments (De Barros et al.) as these are inscribed in the Paris Climate accords. Since conservation is a regional land use "development" option, it is also often contested.

Further, the sustainability arguments that infuse regional development debates privilege local knowledge systems and their derived practices, arguing that these hold the keys to long-term ecological viability of Amazonian land uses under a range of climate and economic regimes, especially compared with many of the monocultural land uses on offer (Brondizio, "Agricultural Intensification"). These sustainability arguments had their roots in the search for alternatives to the highly destructive Amazon development models of the authoritarian period (1964–1985) and focused on traditional technologies and social formations (Hecht and Cockburn; Posey and Balée; Redford and Padoch). This has produced a cohort of researchers concerned with theoretical dimensions of knowledge systems, settlement, land use, and history as well as socially situated discussions about Amazonian research.

Underpinning these debates are assumptions about the civilizational capacities of Amazonians that have been in play for centuries. Were Amazonians even able to influence their environments at large scale or on the terra firma? Many nineteenth-century development thinkers, such as Henry Buckle, argued that the exuberant Brazilian nature swamped the civilizational capacity of locals and condemned them to underdevelopment, a position echoed by many nineteenth-century naturalists, ethnographers, casual observers, and politicians who used this idea, along with concepts of racial debility and the "miscegenation degeneration problem," to promote European immigration (Hecht, *Scramble*; Skidmore; Weinstein). Indigenous and traditional populations have been subject to structural and overt racism for hundreds of years, with natives as wards of the state during the much of the nineteenth and twentieth centuries and with highly contested claims to territory (Hemming; Garfield).

Mid-century functionalist cultural ecological anthropology, as for example Napoleon Chagnon's description of the Yanomami, fed into negative views of natives. Because of Chagnon's sociobiology of the "fierce people," Brazil's authoritarian regimes interpreted the Yanomami as violent brutes lacking the skills to even act as citizens, let alone partake in their own autonomy, as Davi Kopenawa, a long time Yanomami activist, describes in his moving autobiography. The functionalism of Betty Meggers's "soil limitation" model, which I discuss further on, also rendered many Amazonian populations "invisible." She insisted that there were environmental limitations on native culture, and the resultant trope of "demographic void" became one of the pretexts for Amazonian occupation as a central political, geopolitical, and cultural modernization project under Brazil's authoritarian regimes (1964–1985) (Hecht and Cockburn; Silva; Becker). Ironically, Meggers thought ecological limits would preclude Amazonian interventions, and did not expect that her arguments—sparse population and cultural insufficiency—would stimulate rather than restrain the developmentalist Generals. Later in this essay, I will return to Meggers's civilizational arguments and the "modernization" of her pedological arguments, as she serves as the muse for a recent set of debates that rehearse her premises and blind spots, but first, the questions of domestication, why now, and why environmental humanities?

Domestication and the environmental humanities

Domestication has been a durable research arena and analytic trigger for most of the last two millennia. The practice of agriculture was taken as a central proof of the Earth as an "Abode for Man" in classical and medieval thought (Glacken), and specifically, agricultural surplus was considered a necessity for more complex divisions of labor, political complexity, and urbanization. Domestication fundamentally conflates with sedentary lifeways and civilization and has often, incorrectly, been portrayed as part of a linear progression from nomadism to sedentarism, associated either with domesticated organisms or exceptionally rich "natural" concentrations of wild resources. The globalization of food and industrial crops has been a regular feature of imperial movements, an icon of improvement, and, in modern political ecology analytics, the axis for analyzing an array of environmental problems. In this sense, agricultural differences became markers of the exotic and fulcrums of resistance that emblematize cultural aptitudes, differences, virtues, vices, and land transformations and continuities.

Environmental historians and cultural geographers have been especially interested in landscapes and social changes associated with agricultural and agrarian systems, and the institutions, such as botanical gardens, royal charters, ecclesiastic missions, markets, and slavery, that mediated these transformations. These institutions made some agricultural systems prominent while others sank from view, thus contributing to the invisibility of indigenous cultivars and techniques even if they did not fade from practice (Mann; Schiebinger and Swan; Endersby; Driver and Martins; Brockway; Carney; Carney and Rosomoff; Carney and Voeks). One of the central foci of much of current ethnoagronomy and ethnoecology has been to recover and analyze these practices.

Many landscapes "read" as wild have much stronger human signatures than generally assumed. Many land uses, cultigens, and activities have fallen outside what observers have understood or could even *see* as agriculture, as the recent attention to nontimber forest products, successional management, forest tending, and an array of arboreal management has shown (Brookfield; Brookfield and Padoch; Ellen et al.; Chomitz and Kumari; Aguilar-Støen et al.; Freitas et al.; De Jong; Lewis; Mathews; Kennedy). Problems of classifying subsistence strategies revolve around such invisibilities and hamper understanding of different paradigms of agriculture, especially when fundamentally dichotomous classifications hold sway, including those of wild versus domesticated, forest versus garden, and tended versus planted. These dichotomies obscure the complex management regimes spread over time and space that constitute the relations of people with landscapes at many scales and that construct agrodiversity in tropical livelihoods (Kennedy; Brookfield; Padoch and De Jong). These epistemic blinders problematically shape the ways in which these landscapes are represented.

The resurgence of interest in paleo-agriculture and early domestication, with several recent themed issues in journals like *Holocene* (2015), *Diversity* (2010), and *Proceedings of the National Academy* (2014), suggests that there are other issues in play. Partly this may reflect anxiety about the vast scale of simplifications of agriculture landscapes in the US, Latin America, China, and Southeast Asia, the simultaneous concerns over heritage food production, loss of agrodiversity, biodiversity, and *in situ* conservation, the privatization of germplasm and the rise of GMOs, the rapid and drastic changes in agrarian communities, and perhaps most centrally, the larger issue of the viability of production systems in relation to climate change, with agriculture as both driver and victim of climatic events. But why should there be an explosion of domestication debates, and why should this be of interest for the environmental humanities? While climate change is in the foreground, I think there are several other reasons as follows.

First, domestication and crop lands, as a highly humanized landscape or "second nature" and, in highly manipulated GMO landscapes, even a third nature or "neo-nature," have implications for thinking about cultures, representations and constructions of nature, science ideologies, indigenous knowledge, technology, and political ecologies.

Second, domestication and landscape models are sites for new methods of environmental history that deploy quantitative biophysical data from the "natural archive"—agronomic, botanical, pedological, and palynological evidence—in historical debates. While the "natural archive" is widely used in archeology, its application in environmental history is relatively recent. Scientific data, ideologies of science, and critique thus become part of the arsenal of analysis for constructing environmental histories. These debates are becoming more salient because, as in the climate sciences, "rerunning the models" through environmental change is a means of analyzing current framings and exploring biophysically mediated cultural transformations (Wood; Davis).

Third, discourses about forms of domestication and the extent of agricultural or "intervened" landscapes engage sharp political histories and narratives of rights over germplasm and territory. Lands and futures are contested on the basis of historical biotic landscapes, and whether they are "natural." The debates have roiled over intellectual property of organisms and genomes, but also more generally over questions of common property, cultural recognition, and sociocultural development alternatives, especially in the developing world. Domestication also operates at scales from that of the seed to complex ecologies and thus engage definitional questions (Kennedy; Clement; Fraser et al.; Clement et al.; Pautasso et al.; Rival and McKey; McKey et al., "Ecological Approaches"; Pujol et al.).

Fourth, scientists have always had an important role in colonial and modernization policy (Schiebinger and Swan; Bowd and Clayton; Markham), but especially in tropical environmental and land politics since the mid-twentieth century (Foresta; Garfield). The tropical development literature is replete with interventions from scientists with stakes in development controversies, because they have an interest in the actions and scale of what constitutes agrarian landscapes. In short, scientists have been entangled in mythologies of development (the idea of progress) and conservation (lost Edens).

At another, deeper level, these debates reflect landscapes as texts that inform philosophies of nature. This is relevant in terms of the paradigms of normal science, whether positivist or inductive, the "co-production models," the emergent analytics of Actor Network and object theory, and other models that recognize nonhuman elements as actors in human systems (Latour). Amazonian ethnographic research on what might be called "nature philosophy," rooted in indigenous epistemes, is enjoying significant prestige and figures prominently in emergent "posthuman" or nonhuman paradigms in environmental thought as exemplified by the theoretical work of Philippe Descola, Eduardo Viveiros de Castro, and Eduardo Kohn. Researchers more rooted in archeological and ecological framings also engage alternative, integrative human/nature epistemes involving material approaches rather than the semiotic approaches mentioned above (Balée, "Culture"; Heckenberger and Neves; Heckenberger et al.; Balée and Erickson; Erickson, "Domesticated Landscapes"; Posey and Plenderleith). Thus, at issue in the "footprint debate" are practices usually categorized as "normal" and bench sciences where the sample and the transect become authoritative representatives of much larger systems, versus those approaches that contextualize and use an array of materials, including the natural sciences as well as local knowledge systems, archives, ethnography, ethnohistory, and ethnobiology, and include native Amazonians as authorities as well.

The great Amazon wilderness debate: Meggers, "Neo-Meggersians," and the "Denevan School"

How do ideologies of nature, scientific practices, and the framing of domestication play out in current Amazonian pre-Columbian impact debates? Was Amazonia, as early observers like Walter Raleigh, gazing over savannas which actually had a significant human signature, saw it, a pristine "nature never sackt"? Or was it instead, as Carvajal, the chronicler of the first careening voyage by Europeans down the Amazon, put it, a populous "place of teeming shores" (Carvajal et al.)? In the great "footprint" debate, I divide the controversy into two "camps." The first camp I call the "Neo-Meggersians" because it takes on Meggers's premises; the second is the "Denevan School," which I discuss further on.

Betty Meggers's functionalist ecological approach to Amazonian cultural development was initially seen as a vanguard analysis using equilibrium systems ecology informed by soil characteristics. She argued that ecological limitations, though Amazonian natives adapted to them in different ways, precluded civilizational development in terra firme environments because the agriculture could not generate sufficient surpluses (Meggers, "Environmental Limitations;" Meggers and Evans). Meggers's work concretized cultural limitations within the easily intelligible dynamics of tropical soil fertility, which she generalized from large-scale assessments and whose documentation was just developing in the 1950s and 1960s. Tropical forest nutrient cycling was a highly funded and central research concern in this period (e.g., Jordan) and an early intellectual fetish of tropical systems ecology.

Meggers's powerful theory of nature and cultural limitation in the tropics held sway for decades because its intellectual coherence coincided with academic fashion and the emergent attention to equilibrium modeling and ecological explanations. Her work inspired an explosion in ecologically inflected field studies on tropical adaptation, especially in Amazonia (Hames and Vickers; Gross). While not exactly detaching her work from the more egregious colonial ideologies, Meggers scientized "stone age continuities" in Amazonia in a modern register (Amazonia; "Continuing Quest"). Even dramatic counter-evidence of anthropogenic soils could not shake her implacable rejection of autochthonous cultures of Amazonia (Meggers, "Mystery"), in spite of a plethora of soil studies on high-fertility Amazonian dark earths (ADEs) presented by a set of scholars from diverse intellectual lineages, institutions, and disciplinary backgrounds (Schmidt et al.; Lima et al.; Lehmann et al.; Woods et al.; Peterson et al.; Rebellato et al.).

The other camp, in contrast to the "constraint" theorists, is that of the "Amazon possibilists," which I would place in a "Denevan School" of Latin American tropical analysis. The possibilists hold that Amazonian environments with their high primary productivities are largely able to develop complex human systems. William Denevan is significant because he pioneered diverse elements of this counter-paradigm, including, for example, his pathbreaking contributions in recasting pre-Columbian demography (Native Population); the documentation of enormous archeological sites of indigenous production systems in Bolivia's Llanos de Mojos and beyond (Aboriginal Cultural Geography); his research on tropical forest swidden cultivation, particularly successional management and forest ecosystem management (Denevan et al.; Denevan, "Pre-European Forest"; Clement et al.); and his magisterial compendium on indigenous cultivation systems in the Andes and the Amazon and forest upland management (Cultivated Landscapes). Because of this attentiveness to the range of indigenous conditions of possibility, one did not have to be his student or colleague to feel his influence. It is almost impossible to find an article on traditional Amazonian historical and pre-Columbian settlement and land use that does not reference him. His work on pre-Columbian—not "pre-historic," as the Neo-Meggersians would have it (Bush et al.; McMichael et al.)—production systems as well as their intellectual and landscape legacies are key lodestars in Amazonian analytics.

In the possibilist school, though it involves rigorous science, authority and insight are not uniquely available through the canons of Western science, but also through experiential relations with native Amazonian experts and extensive field lives. Situating knowledge means understanding how social solidarities and values affect the generation of knowledge and using this understanding to create more diverse forms of explanation and insight (Jasanoff). Denevan "partisans" include scholars from all the major Amazon and Brazilian academic institutions, British, Dutch, French, and several US institutions, and embrace a much wider array of disciplines. The Neo-Meggersians concentrate in a handful of labs and share similar training and major professors.

The controversies between these two camps pivot on what Denevan has called the pristine myth ("Pristine Myth") and also pertain to how domestications are conceptualized. Is domestication understood to include only specific forms of fire-based agriculture and specific annual crops, or is human impact assessed through a range of co-produced landscapes? Rather than the forest primeval of Amazonian conservation lore, the region's landscapes are the outcome not only of shifting cultivation, but also of other fire-based systems, human water management, household planting experiments, gathering histories, and extensive movement through and within these landscapes. These human production systems incorporate, harness, and reflect wild and natural systems as well, but not within those categorical frameworks (Erickson, "Artificial" and "Domesticated Landscapes"; Raffles and WinklerPrins). Fire is not the only tool in the management repertoire that affected vegetation. Periodic treks for military, ritual, spiritual, surveillance, medical, trade, and collecting purposes have been shown to affect forests (Politis; Balée, "Indigenous Adaptation"). Further, light management fires often used in forest understories, or to clean forest campsites, might not leave much of a charcoal signature, which the Neo-Meggersians consider the only legitimate sign of human impact.

Indigenous peoples' travels depended on resource islands that recent research suggests were largely anthropogenic and were certainly tended landscapes (Erickson, "Domesticated Landscapes"; Posey and Plenderleith; Rival). Management activities included camp clearing, planting (whether intentionally or otherwise), ritual activities, transfer of germplasm, extraction, and the casual pruning and weeding of resources islands. And even extensive, intensively managed agricultural systems may not always use fire (Iriarte et al.). Considering movement on secondary and minor tributaries, not just the massive rivers that form the central imaginary of human settlement among the Neo-Meggersians, can also help scholars understand settlement better. The rubber period reveals many things, among them how extensively permeable riverine travel was throughout the Amazon basin, and how significant the impact on hinterlands can be from urban systems and transregional trade (Hecht).

The Neo-Meggersians argue, based on a very small sample set over half of the Amazon basin, that human impact on Amazonia was sparse (Bush et al.; McMichael et al.). Based on fifty-five clustered sites and 245 soil cores in the western Amazon—a quantity of samples that would not be adequate for understanding local shifting cultivation sequences—the Neo-Meggersians argue that upland forests in western Amazonia were occupied by small, shifting human populations. They claim that these small populations had minor effects on forests and cleared very little land, and thus one cannot assume that Western Amazon forests were resilient after pre-Columbian disturbance. They also argue that oligotrophic forests—forests dominated by one species, usually one that is very useful to humans—reflect mainly natural dispersal. This argument is surprising in light of the well-documented fact that the palm Buriti (*Maritius*), Brazil nuts (*Bertholetia*), açai (*Euterpe*), Piqui (*Caryocar*), and Babassu (*Orbignia*), for example, are often intensively used by local people and have figured in regional and international markets. Some, like Brazil nuts and cacao, are often reported in pure stands on the banks of the Purus and Madeira and have been in international export markets for centuries (Roller). There is extensive ethnobotanical evidence of these trees' use, management, and planting (Rival; González-Pérez et al.; Shepard and Ramirez; Souza et al.; Paiva et al.; Clement).

In the Amazon, there is evidence of widespread raised-field agriculture, mounds, development of forests islands, and extensive distribution of terra preta associated with historical urbanism; intensive agroforestry; more than four hundred sites of ring-ditched formations, which are a striking signature throughout the uplands of extensive geo-engineering; a "tropical Stonehenge"; and artistic masterpieces of many kinds. Given all this, how is it that the Neo-Meggersians' model rejects human impacts on forest systems? While referencing the entire Amazon Basin in their titles, the Neo-Meggersians' model has stayed intact by hiving off sites that do not fit their analysis, and only accepting as evidence of human intervention shifting cultivation as exemplified by charcoal signatures in soils, phytoliths, and palynological data. Many of Amazonia's most significant domesticates, such as manioc, Brazil nuts, and cacao, do not have these signatures to a significant degree. Neo-Meggersians conclude that if their (sparse) samples do not show charcoal on the uplands, human impact was negligible.

The Denevan School, as I have explained, suggests that human intervention was much more widespread (Clement et al.). While pristine versus anthropogenic understandings of Amazonian nature as well as declensionist versus possibilist interpretations of human intervention are in play here, another bifurcation pertains to models of domestication. Is domestication wholly defined by the "domestication syndrome," or does it reflect the hybridity and expansiveness of the "landscape model?"

Models of domestication

Annual crops in the temperate zone epitomize "the domestication syndrome," a term initially used by Harlan to describe a suite of characteristics outlined in Table 2.1. Ideas about domestication have largely unfolded in the context of some "model crop" systems: most of these model crops are annuals, including many genera of the Poaceae (grasses) and a few annual legumes, and very few of them are trees (Meyer et al.). This model of domestication, which has a venerable history, reflects the durability of these kinds of plant materials, especially in more arid environments, and their dominance in temperate zone diets and now in world commerce. Of the fifty-four "key plants" in an important review of domestication studies, only five were tree crops, and such shrubs as coffee, cacao, and rubber, among other crops with long histories and immense current markets, were not included (Larson et al.). Analysts of the ecological dynamics of domestication of root and tropical crops, especially manioc, make rather different arguments (McKey et al., "Ecological Approaches"; McKey et al., "Pre-Columbian"; Pujol et al.). Many useful trees and their array of products do not necessarily reflect domestication criteria that are based on short-cycle annual grass cropping systems (McKey et al., "Ecological Approaches"; Rival and McKey; Clement). The lack of easily traceable ethnoarcheological features has obscured how extensive manioc cultivation was and how its cultivation and phenotypic plasticity in mixed systems was widespread. This limited the understanding of manioc as a foundational carbohydrate for complex societies (Isendahl; Heckenberger). But manioc was far from the only kind of production system. The ensemble of varying production regimes mediated by a range of non- or semi-domesticated organisms is predicated on a different kind of interaction with nature and positioning of one's place in it. Table 2.1 suggests some of these differences.

Table 2.1 Contrasting models of domestication and the characteristics of species and ecosystems associated with them.

Domestication Syndrome	Landscape Domestication
"Wild" vs. "domesticated"	Extensive range of use between "wild" and "domesticated"
Annual seed plants as central model; temperate zone grains	Tropical tubers and trees
Tendency to monocrop	Some monocropping, some multispecies cropping
Plants have reduced ability to disperse seeds without human intervention	Most plants not dependent on human dispersal
Reduction in plants' toxic chemical compounds for palatability	Toxins removed in processing; toxins act to reduce predation
Reduction in seed dormancy	Seed dormancy variable
Predictable and synchronous germination	Much less synchronous germination
Larger inflorescences	Depends on degree
Reduction in size of plant	Variable, but plant size mostly not reduced
Single use (food for humans)	Multiple use (food, sap, artisanal, and industrial uses)
Narrow ecological function (food for humans)	Diverse ecological functions including wildlife support
Ecosystem simplifications	Multiple ecosystem manipulations at different scales; beta diversity

The "landscape domestication" model addresses historical ecology and human impacts in the context of complex ecosystems. It turns our attention away from specific plants and frames a set of domestication processes more around "group selection" and human effects at large scales. Analysts generally assert that domestication of landscapes occurs before, during, and after the emergence of full-scale agriculture (Balée and Erickson). This approach reassesses paradigms of human intervention in tropical landscapes as well as more general processes of civilizational development in six major ways:

1. The landscape domestication model diverts attention away from the Neolithic revolution and grain-based agriculture as the most transformative events in shaping environments, especially in the tropics where a long history of landscape interventions may be more critical.
2. It emphasizes cultural activities that influence the presence, availability, and productivity of a wide range of species, rather than focusing on a clear cultivation agriculture.
3. It shifts attention from individual species to landscapes and their contingencies as historical outcomes of a "co-produced" landscape with human and nonhuman signatures.
4. The landscape domestication model rejects the idea of a linear evolution from foraging to agriculture, and looks at organisms with longer temporal scales, as well as short cycle agronomies and complex civilizations, as constitutive of places.
5. In landscape domestication, the production of the system is not uniquely a function of human agency. Nonhumans also perform "work" of various kinds. While this is true of many kinds of systems, in tropical areas these effects are so profound as to make these landscapes seem wild. But nonhumans can be beneficiaries as well, so the idea of "utility" has a more relational resonance.

6. Such a paradigm moves us away from pristine tropics with noble savages into a more complex framing of interactivity and conditions of possibility at the level of landscapes. It not only helps us understand the past and current plant distributions as part of landscape legacies (Hecht et al.), but also can inform study of forest transitions now.

The Neo-Meggersian position is basically that culture can be understood without reference to culture, but only to soil chemistry, pollen, starch grains, and phytoliths as the central explanatory elements whose goal in the service of tropical ecology.

By contrast, long-time field researchers armed with soils samples and phytoliths as well as with historical accounts, ethnographic data, historical ecology, and a wide range of ethnobotanical sources, tell a story about landscapes as artifacts and habitats, one that includes Amazonians as important sources of knowledge who go beyond what the laboratory can say.

This brief overview of a fairly acerbic debate between the optics of constraint, declension, and wildness, and those of possibilist innovation in the tropics—innovation that rivaled that of any place in the world in pre-Columbian (not "pre-historical") times—is an exercise in examining scientific practices, explanations of land patterns, loci of intellectual authority, and the kinds of proof that are invoked in these arguments. This ongoing debate leaves us with important questions: What constitutes domestication? Do different epistemes of nature, as many Amazonian ethnographers argue, conceive of ways of being in the natural world that diverge from the reductionisms of normal science? How did the nonhuman world figure in the shaping of Amazonia in the past, and how will the nonhuman world shape its future? By discounting the human in Amazonian landscapes, we run the risk of blinding ourselves to history, knowledge systems, management possibilities, and ultimately to the natures of Amazonia itself.

References

Aguilar-Støen, Mariel, Arlid Angelsen, Kristi-Anne Stølen, and Stein R. Moe. "The Emergence, Persistence, and Current Challenges of Coffee Forest Gardens: A Case Study from Candelaria Loxicha, Oaxaca, Mexico." *Society & Natural Resources* 24 (2011): 1235–1251. Taylor & Francis Online. Web. 14 April 2016.

Balée, William. "Indigenous Adaptation to Amazonian Palm Forests." *Principes* 32.2 (1988): 47–54. Print.

——. "The Culture of Amazonian Forests." *Advances in Economic Botany* 7 (1989): 1–21. Print.

Balée, William and Clark L. Erickson, ed. *Time and Complexity in Historical Ecology: Studies in the Neotropical Lowlands.* New York: Columbia University Press, 2006. Print.

Becker, Bertha K. *Geopolítica da Amazônia: a nova frontiera de recursos.* Rio de Janerio: Zahar, 1982. Print.

Bowd, Gavin and David Clayton. "Tropicality, Orientalism, and French Colonialism in Indochina: The Work of Pierre Gourou, 1927–1982." *French Historical Studies* 28 (2005): 297–327. Duke Journals. Web. 14 April 2016.

Brockway, Lucile. *Science and Colonial Expansion: The Role of the British Royal Botanic Gardens.* New York: Academic Press, 1979. Print.

Brondizio, Eduardo S. "Agricultural Intensification, Economic Identity, and Shared Invisibility in Amazonian Peasantry." *Culture and Agriculture* 26 (2004): 1–24. Print.

Brookfield, Harold, ed. *Cultivating Biodiversity: Understanding, Analysing and Using Agricultural Diversity.* London: ITDG Publishing, 2002. Print.

Brookfield, Harold and Christine Padoch. "Managing Biodiversity in Spatially and Temporally Complex Agricultural Landscapes." *Managing Biodiversity in Agricultural Ecosystems.* Ed. D. Jarvis, C. Padoch, and H. Cooper. New York: Columbia University Press, 2007. 338–361. Print.

Buckle, Henry T. *History of Civilization in England.* New York: D. Appleton, 1867. Print.

Bush, Mark, Crystal McMichael, Dolores Piperno, Miles Silman, Jos Barlow, Carlos Peres, Mitchell Power, and Michael Palace. "Anthropogenic Influence on Amazonian Forests in Pre-history: An Ecological Perspective." *Journal of Biogeography* 42 (2015): 2277–2288. Wiley Online Library. Web. 8 April 2016.

Carney, Judith. "African Rice in the Columbian Exchange." *The Journal of African History* 42 (2001): 377–396. JSTOR. Web. 16 April 2016.

Carney, Judith and Richard Rosomoff. *In the Shadow of Slavery: Africa's Botanical Legacy in the New World.* Berkeley: University of California Press, 2009. Print.

Carney, Judith and Robert A. Voeks. "Landscape Legacies of the African Diaspora in Brazil." *Progress in Human Geography* 27 (2003): 68–81. SAGE Journals. Web. 16 April 2016.

Carvajal, Gaspar de, José Toribio Medina, Gonzalo Fernández de Oviedo y Valdés, Bertram T. Lee, and H. C. Heaton. *The Discovery of the Amazon According to the Account of Gaspar de Carvajal and Other Documents.* New York: American Geographical Society, 1934. Print.

Castro, Eduardo Viveiros de. *The Relative Native: Essays on Indigenous Conceptual Worlds.* Chicago: University of California Press, 2016. Print.

Chagnon, Napoleon. *Yanomamö: The Fierce People.* 1968. New York: Holt, Reinhart and Winston, 1977. Print.

Chomitz, Kenneth and Kanta Kumari. "The Domestic Benefits of Tropical Forests: A Critical Review." *World Bank Research Observer* 13 (February 1998): 13–35. Web. 16 April 2016.

Clement, Charles R. "1492 and the Loss of Amazonian Crop Genetic Resources." *Economic Botany* 53 (1999): 188–216. JSTOR. Web. 16 April 2016.

Clement, Charles R., William M. Denevan, Michael J. Heckenberger, André Braga Junqueira, Eduardo G. Neves, Wenscelau G. Teixeira, and William I. Woods. "The Domestication of Amazonia before European Conquest." *Proceedings of the Royal Society B-Biological Sciences* 282 (2015): 32–40. Web. 8 April 2016.

Davis, Mike. *Late Victorian Holocausts: El Niño Famines and the Making of the Third World.* London: Verso, 2001. Print.

De Barros, Alan E., Ewan A. Macdonald, Marcelo H. Matsumoto, Rogério C. Paula, Sahil Nijhawan, Y. Malhi, and David W. Macdonald. "Identification of Areas in Brazil that Optimize Conservation of Forest Carbon, Jaguars, and Biodiversity." *Conservation Biology* 28 (2014): 580–593. Wiley Online Library. Web. 8 April 2016.

De Jong, Will. "Tree and Forest Management in the Floodplains of the Peruvian Amazon." *Forest Ecology and Management* 150 (2001): 125–134. Print.

Denevan, William. *The Aboriginal Cultural Geography of the Llanos de Mojos of Bolivia.* Berkeley: University of California Press, 1966. Print.

——. *Cultivated Landscapes of Native Amazonia and the Andes.* New York: Oxford University Press, 2001. Print.

——. *The Native Population of the Americas in 1492.* Madison: University of Wisconsin Press, 1976. Print.

——. "Pre-European Forest Cultivation in Amazonia." *Time and Complexity in Historical Ecology: Studies in the Neotropical Lowlands.* Ed. William Balée and Clark L. Erickson. New York: Columbia University Press, 2006. 154–164. Print.

——. "The Pristine Myth: The Landscape of the Americas in 1492." *Annals of the Association of American Geographers* 82 (1992): 369–385. Print.

Denevan, William, John M. Treacy, Janis B. Alcorn, Christine Padoch, Julie Denslow, and S. Flores Paitan. "Indigenous Agroforestry in the Peruvian Amazon: Bora Indian Management of Swidden Fallows." *Interciencia* 9 (1984): 346–357. Print.

Descola, Philippe. *Beyond Nature and Culture*. Trans. Janet Lloyd. Chicago: University of California Press, 2013. Print.

Driver, Felix and Luciana Martins, ed. *Tropical Visions in an Age of Empire*. Chicago: University of California Press, 2005. Print.

Ellen, Roy, Peter Parkes, and Alan Bicker, ed. *Indigenous Environmental Knowledge and Its Transformations: Critical Anthropological Perspectives*. London: Routledge, 2000. Print.

Endersby, Jim. *Imperial Nature: Joseph Hooker and the Practices of Victorian Science*. Chicago: University of California Press, 2008. Print.

Erickson, Clark L. "An Artificial Landscape-scale Fishery in the Bolivian Amazon." *Nature* 408 (2000): 190–193. Web. 8 April 2016.

——. "Domesticated Landscapes of the Bolivian Amazon." *Time and Complexity in Historical Ecology: Studies in the Neotropical Lowlands*. Ed. William Balée and Clark L. Erickson. New York: Columbia University Press, 2006. 187–233. Print.

Foresta, Ronald. *Amazon Conservation in the Age of Development: The Limits of Providence*. Gainesville: University of Florida Press, 1991. Print.

Fraser, James Angus, Alessandro Alves-Pereira, André Braga Junqueira, Nivaldo Peroni, and Charles R. Clement. "Convergent Adaptations: Bitter Manioc Cultivation Systems in Fertile Anthropogenic Dark Earths and Floodplain Soils in Central Amazonia." *PLOS One* 7. 29 August 2012. Web. 18 April 2016.

Freitas, Carolina de, Glenn Shepard Jr., and Maria T. F. Piedade. "The Floating Forest: Traditional Knowledge and Use of Matupa Vegetation Islands by Riverine Peoples of the Central Amazon." *PLOS One* 10. 2 April 2015. Web. 18 April 2016.

Garfield, Seth. *In Search of the Amazon: Brazil, the United States, and the Nature of a Region*. Durham: Duke University Press, 2013. Print.

Glacken, Clarence. *Traces on the Rhodian Shore: Nature and Culture in Western Thought from Ancient Times to the End of the Eighteenth Century*. Berkeley: University of California Press, 1976. Print.

González-Pérez, Elizabeth, Márlia Coelho-Ferreira, Pascale de Robert, and Claudia Leonor López Garcés. "Knowledge and Use of Babassu (*Attalea speciosa* Mart. and *Attalea eichleri* (Drude) A.J. Hend.) among Mebengokre-kayapó from Las Casas Indigenous Land, Pará state, Brazil." *Acta Botanica Brasilica* 26 (2012): 295–308. Web. 18 April 2016.

Gross, Daniel R. "Protein Capture and Cultural Development in the Amazon Basin." *American Anthropologist* 77 (1975): 526–549. AnthroSource. Web. 8 April 2016.

Hames, Raymond B. and William T. Vickers, ed. *Adaptive Responses of Native Amazonians*. New York: Academic Press, 1983. Print.

Harlan, Jack R. *Crops and Man*. Madison, WI: Crop Science Society of America, 1992. Print.

Hecht, Susanna B. *The Scramble for the Amazon and the "Lost Paradise" of Euclides da Cunha*. Chicago: University of California Press, 2013. Print.

Hecht, Susanna B. and Alexander Cockburn. *The Fate of the Forest: Developers, Destroyers, and Defenders of the Amazon*. London: Verso, 1989. Print.

Hecht, Susanna B., Kathleen Morrison, and Christine Padoch, ed. *The Social Lives of Forests: Past, Present, and Future of Woodland Resurgence*. Chicago: University of California Press, 2014. Print.

Heckenberger, Michael. "Manioc Agriculture and Sedentism in Amazonia: The Upper Xingu Example." *Antiquity* 72 (1998): 633–648. Cambridge Journals. Web. 8 April 2016.

Heckenberger, Michael and Eduardo Goés Neves. "Amazonian Archaeology." *Annual Review of Anthropology* 38 (2009): 251–266. Web. 18 April 2016.

Heckenberger, Michael, J. Christian Russell, Carlos Fauto, Joshua Toney, Morgan Schmidt, Edithe Pereira, Bruna Franchetto, and Afukaka Kuikuro. "Pre-Columbian Urbanism, Anthropogenic Landscapes, and the Future of the Amazon." *Science* 321 (29 August 2008): 1214–1217. Web. 18 April 2016.

Hemming, John. *Amazon Frontier: The Defeat of the Brazilian Indians*. Cambridge, MA: Harvard University Press, 1987. Print.

Iriarte, José, Mitchell J. Power, Stéphen Rostain, Francis E. Mayle, Huw Jones, Jennifer Watling, Bronwen S. Whitney, and Doyle McKey. "Fire-free Land Use in Pre-1492 Amazonian Savannas." *Proceedings of the National Academy of Sciences of the United States of America* 109 (2012): 6473–6478. PNAS. Web. 8 April 2016.

Isendahl, Christian. "The Domestication and Early Spread of Manioc (Manihot Esculenta Crantz): A Brief Synthesis." *Latin American Antiquity* 22 (2011): 452–468. JSTOR. Web. 8 April 2016.

Jasanoff, Sheila, ed. *States of Knowledge: The Co-Production of Science and the Social Order.* London: Routledge, 2004. Print.

Jordan, Carl F. *Nutrient Cycling in Tropical Forest Ecosystems.* New York: Wiley, 1985. Print.

Kennedy, Jean. "Agricultural Systems in the Tropical Forest: A Critique Framed by Tree Crops of Papua New Guinea." *Quaternary International* 249 (2012): 140–150. Web. 18 April 2016.

Kohn, Eduardo. *How Forests Think: Toward an Anthropology Beyond the Human.* Berkeley: University of California Press, 2013. Print.

Kopenawa, Davi and Bruce Albert. *The Falling Sky: Words of a Yanomami Shaman.* Trans. Nicholas Elliott and Alison Dundy. Cambridge, MA: Harvard University Press, 2013. Print.

Larson, Greger, Dolores R. Piperno, Robin G. Allaby, Michael D. Puruggaanan, Leif Andersson, Manuel Arroyo-Kalin, Loukas Barton, Cynthia Climer Vigueira, et al. "Current Perspectives and the Future of Domestication Studies." *Proceedings of the National Academy of Sciences of the United States of America* 111 (2014): 6139–6146. PNAS. Web. 8 April 2016.

Latour, Bruno. *Science in Action: How to Follow Scientists and Engineers through Society.* Cambridge, MA: Harvard University Press, 1987. Print.

Lehmann, Johannes, Dirse C. Kern, Bruno Glaser, and William I. Woods, ed. *Amazonian Dark Earths: Origin, Properties, Management.* New York: Springer, 2003. Print.

Lewis, J. A. "The Power of Knowledge: Information Transfer and Acai Intensification in the Peri-urban interface of Belem, Brazil." *Agroforestry Systems* 74 (2008): 293–302. Web. 18 April 2016.

Lima, H., C. Schaefer, J. Mello, R. Gilkes, and J. Ker. "Pedogenesis and pre-Columbian Land Use of 'Terra Preta Anthrosols' ('Indian Black Earth') of Western Amazonia." *Geoderma* 110 (2002): 1–17. Print.

Mann, Charles. *1493: Uncovering the New World Columbus Created.* New York: Knopf, 2011. Print.

Markham, Clements. *Travels in Peru and India: While Superintending the Collection of Chinchona Plants and Seeds in South America, and Their Introduction into India.* London: Murray, 1862. Print.

Mathews, Andrew. *Instituting Nature: Authority, Expertise, and Power in Mexican Forests.* Cambridge, MA: MIT Press, 2011.

McKey, Doyle, Marianne Elias, Benoit Pujol, and Anne Duputié. "Ecological Approaches to Crop Domestication." *Biodiversity in Agriculture: Domestication, Evolution, and Sustainability.* Ed. Paul Gepts et al. New York: Cambridge University Press, 2012. 377–406. Print.

McKey, Doyle, Stéphen Rostain, José Iriarte, Bruno Glaser, Jago Jonathan Birk, Irene Holst, and Delphine Renard. "Pre-Columbian Agricultural Landscapes, Ecosystem Engineers, and Self-organized Patchiness in Amazonia." *Proceedings of the National Academy of Sciences of the United States of America* 107 (2010): 7823–7828. PNAS. Web. 8 April 2016.

McMichael, Crystal H., Dolores Piperno, Mark Bush, Miles Silman, Andrew R. Zimmerman, Marco Razca, and L. Lobato. "Sparse Pre-Columbian Human Habitation in Western Amazonia." *Science* 336 (2012): 1429–1431. PubMed. Web. 8 April 2016.

Meggers, Betty. *Amazonia: Man and Culture in a Counterfeit Paradise.* Chicago: Aldine Altherton, 1971. Print.

——. "The Continuing Quest for El-Dorado: Round Two." *Latin American Antiquity* 12 (2001): 304–325. Print.

——. "Environmental Limitations on the Development of Culture." *American Anthropologist* 56 (1954): 801–823. Print.

——. "The Mystery of the Marajoara: An Ecological Solution." *Amazoniana-Limnologia Et Oecologia Regionalis Systemae Fluminis Amazonas* 16 (2001): 421–440. Print.

Meggers, Betty and Clifford Evans. *Archaeological Investigations at the Mouth of the Amazon*. Smithsonian Institution; Bureau of American Ethnology. Washington, DC: Government Printing Office, 1957. Print.

Meyer, Rachel, Ashley DuVal, and Helen Jensen. "Patterns and Processes in Crop Domestication: An Historical Review and Quantitative Analysis of 203 Global Food Crops." *New Phytologist* 196 (2012): 29–48. Wiley Online Library. Web. 8 April 2016.

Padoch, Christine and Will De Jong. "Diversity, Variation, and Change in Ribereño Agriculture." *Conservation of Neotropical Forests: Working from Traditional Resource Use*. Ed. Kent Redford and Christine Padoch. New York: Columbia University Press, 1992. 158–174. Print.

Paiva, Paulo Marcelo, Marcelino Carneiro Guedes, and Claudia Funi. "Brazil Nut Conservation through Shifting Cultivation." *Forest Ecology and Management* 261 (2011): 508–514. ScienceDirect. Web. 18 April 2016.

Pautasso, Marco, Guntra Aistara, Adeline Bernaud, Sophie Callon, Pascal Clouvel, Oliver T. Coomes, et al. "Seed Exchange Networks for Agrobiodiversity Conservation: A Review." *Agronomy for Sustainable Development* 33 (2013): 151–175. Web. 18 April 2016.

Peterson, James, Eduardo Neves, and Michael Heckenberger. "Gift from the Past: Terra Preta and the Prehistoric Occupation of the Amazon." *Unknown Amazon*. Ed. Colin McEwan, Christina Barreto, and Eduardo Neves. London: British Museum, 2001. 86–108. Print.

Politis, Gustavo. *Nukak: Ethnoarchaeology of an Amazonian People*. Walnut Creek: Left Coast Press, 2007. Print.

Posey, Darrell and William Balée. *Resource Management in Amazonia: Indigenous and Folk Strategies*. Bronx, NY: New York Botanical Garden Press, 1989. Print.

Posey, Darrell and Kristiana Plenderleith. *Indigenous Knowledge and Ethics: A Darrell Posey Reader*. New York: Routledge, 2004. Print.

Pujol, Benoît, François Renoux, Marianne Elias, Laura Rival, and Doyle McKey. "The Unappreciated Ecology of Landrace Populations: Conservation Consequences of Soil Seed Banks in Cassava." *Biological Conservation* 136 (2007): 541–551. ScienceDirect. Web. 8 April 2016.

Raffles, Hugh and Antoinette WinklerPrins. "Further Reflections on Amazonian Environmental History: Transformations of Rivers and Streams." *Latin American Research Review* 38 (2003): 165–187. Project Muse. Web. 8 April 2016.

Raleigh, Walter. *The Discoverie of the Large, Rich and Bewtiful Empyre of Guiana*. 1596. Ed. Neil L. Whitehead. Manchester: Manchester University Press, 1998. Print.

Rebellato, Lilian, William Woods, and Eduardo Neves. "Pre-Columbian Settlement Dynamics in the Central Amazon." *Amazonian Dark Earths: Wim Sombroek's Vision*. Ed. William Woods, Wenceslau Teixeira, Johannes Lehmann, Christoph Steiner, Antoinette WinklerPrins, and Lilian Rebellato. New York: Springer, 2009. 15–31. Print.

Redford, Kent and Christine Padoch, ed. *Conservation of Neotropical Forests: Working from Traditional Resource Use*. New York: Columbia University Press, 1992. Print.

Rival, Laura. "Amazonian Historical Ecologies." *Journal of the Royal Anthropological Institute* 12 (March 2006): S79–S94. Wiley Online Library. Web. 18 April 2016.

Rival, Laura and Doyle McKey. "Domestication and Diversity in Manioc (Manihot esculenta Crantz ssp esculenta, Euphorbiaceae)." *Current Anthropology* 49 (2008): 1116–1125. Web. 8 April 2016.

Roller, Heather. *Amazon Routes: Indigenous Mobility and Colonial Communities in Northern Brazil*. Stanford: Stanford University Press, 2014. Print.

Schiebinger, Londa and Claudia Swan, ed. *Colonial Botany: Science, Commerce, and Politics in the Early Modern World*. Philadelphia: University of Pennsylvania Press, 2005. Print.

Schmidt, Morgan, Anne Py-Daniel, Claide de Paula Moraes, Raoni Valle, Caroline Caromano, Wenceslau Texeira, Carlos Barbosa, et al. "Dark Earths and the Human Built Landscape in Amazonia: A Widespread Pattern of Anthrosol Formation." *Journal of Archaeological Science* 42 (February 2014): 152–165. Web. 18 April 2016.

Shepard, Glenn and Henri Ramirez. "'Made in Brazil': Human Dispersal of the Brazil Nut (*Bertholletia excelsa*, Lecythidaceae) in Ancient Amazonia." *Economic Botany* 65 (2011): 44–65. Web. 18 April 2016.

Silva, Golbery do Couto e. *Geopolítica e poder*. Rio de Janeiro: UniverCidade, 2003. Print.

Skidmore, Thomas E. *Black into White: Race and Nationality in Brazilian Thought*. Durham, NC: Duke University Press, 1992. Print.

Souza, Mércia, Cristina Monteiro, Patricia Figueredo, Flavia Nascimento, and Rosane Guerra. "Ethnopharmacological use of Babassu (*Orbignya phalerata* Mart) in Communities of Babassu Nut Breakers in Maranhão, Brazil." *Journal of Ethnopharmacology* 133 (2011): 1–5. Web. 18 April 2016.

Weinstein, Barbara. *The Color of Modernity: Sao Paulo and the Making of Race and Nation in Brazil*. Durham: Duke University Press, 2015. Print.

Wood, Gillen D'Arcy. *Tambora: The Eruption That Changed the World*. Princeton: Princeton University Press, 2014. Print.

Woods, William, Wenceslau Teixeira, Johannes Lehmann, Christoph Steiner, Antoinette WinklerPrins, and Lilian Rebellato, ed. *Amazonian Dark Earths: Wim Sombroek's Vision*. New York: Springer, 2009. Print.

3

"THEY CARRY LIFE IN THEIR HAIR"

Domestication and the African diaspora

Judith A. Carney

Suriname, 1711: they carry life in their hair

For all the blacks that get crucified or hung from iron hooks stuck through their ribs, escapes from Surinam's four hundred coastal plantations never stop. Deep in the jungle a black lion adorns the yellow flag of the runaways. For lack of bullets, their guns fire little stones or bone buttons; but the impenetrable thickets are their best ally against the Dutch colonists. Before escaping, the female slaves steal grains of rice, corn, and wheat, seeds of bean and squash. Their enormous hairdos serve as granaries. When they reach the refuges in the jungle, the women shake their heads and thus fertilize the free land.

—Eduardo Galeano (8)

The first generations of enslaved Africans found themselves in environments not yet tamed by European colonization. New World plantations and mines fronted vast expanses of unexplored territory, which promised sanctuary to those willing to take the risk. Some runaways joined free communities of other escapees in wildernesses that largely lay beyond the reach of colonial control. They escaped to terrains whose inaccessibility thwarted pursuit and capture. By the beginning of the eighteenth century, more enslaved Africans had achieved freedom as runaways than by legal manumission (Higman 386). In the montane regions of tropical America, in sheltering jungles above Amazonian river rapids, and in swamps and other hinterlands of mainland North America, communities of fugitive slaves tenaciously defied white authority. Survival depended on the ability to evade detection and critically, in Eduardo Galeano's evocative words, on the plant seeds with which slaves made their escape. Food, and the ability to wrest it from their harsh surroundings, was indispensable to maroon survival and freedom.

Galeano's prose embellishes an oral history widely recounted in maroon-descended communities across much of northeastern South America from the Guianas to Amapá, Pará, and Maranhão, Brazil. The broader narrative concerns rice—and no other seeds—that escaping female ancestors sequestered in their hair. Each variation attributes rice beginnings to the deliberate act of an enslaved woman. In the version from Suriname, she tucks the grains into her hair while fleeing a plantation rice field. Maroon accounts from French Guiana and Amazonic Brazil place her aboard an arriving slave ship. Realizing that she is about to be taken off the ship, the enslaved woman steals some unhusked rice grains from the ship's larder and hides them in her hair. The risk of discovery is enormous, but she knows something important that makes the danger worthwhile: that unmilled grains are also seeds and that from these rice can be grown. In this telling, rice is explicitly tied to a slave ship and, by inference, to Africa (Vaillant 520–529; Price, *First-Time* 129; Carney, "'With Grains in Her Hair;'" Carney, "Rice and Memory"). The seeds link an African past to contemporary maroon community and culture, for rice is the gift that fed (and continues to feed) the woman's descendants. These narratives also function as a foundation story for maroons, situating life and survival as runaways in an African woman's daring act of defiance and vision.

Rice is a crop that was introduced to the Americas; moreover, it is not just a crop of Asian origin. Scholars now recognize a separate species, *Oryza glaberrima*, which was domesticated some three thousand years ago in West Africa. Rice has long been a "women's crop" in the region, whether planted solely by females or in a gendered system where women do the sowing, weeding, and processing. *Glaberrima* may well have been the first rice to reach New World shores (Carney, *Black Rice*). Recently, African rice was found growing in a maroon food garden in Suriname (Van Andel; Carney and Rosomoff). This discovery complements the maroon narrative that rice arrived on slave ships and fugitive slaves pioneered its cultivation in the New World wilderness.

The oral history of rice serves as a point of departure for examining other food staples of African origin that were brought to the Americas during the transatlantic slave trade. By combining several methodologies from the environmental humanities, it is now possible to unambiguously trace the early presence of these African foods in plantation societies and to recognize evidence of slave agency behind their introduction. The multidisciplinary approach aims to overcome the limitations of written documentation, which records the words and deeds of those who participated in, or benefited from, slavery but largely leaves silent African voices and contributions. Indeed, a major challenge to the modern scholar is the limited ability of colonial and national archives to capture the contributions of peoples whose marginal social position render them largely invisible. New evidence from oral histories, food plant dispersals, historical linguistics, archaeobotany, genetics, literature, and the visual record can be synergistically combined to clarify what once was obscure.

The forced migration of some 12.5 million Africans to the Americas registers few slave testimonies, and the written record on African food crops is sparse. An institution that lasted some 350 years has left a persistent legacy that acknowledges little more than slaves' rote performance of their masters' directives. Traditional sources fail to appreciate that enslaved Africans were stolen from societies with sophisticated farming and herding practices adapted to the tropics. Their forced transatlantic migration settled them in tropical and subtropical environments not much different from those they had left. History emphasizes the labor slaves performed for their owners and the commodities they produced for export. Less consideration is given to the crops they chose to plant and the foods they ate. New attention to the role of African food crops in the transatlantic slave trade, and the sites where they were established in tropical America, expands our understanding of the intercontinental crop transfers known as the Columbian Exchange.

The Columbian Exchange: migration and crop transfers

Alfred W. Crosby's *The Columbian Exchange* (1972) brought scholarly attention to the intercontinental exchange of crops, animals, and microbes that occurred in the period of

European maritime expansion, from the sixteenth to the nineteenth centuries. Columbian Exchange species wrought enormous environmental transformations around the globe, as new foods were adopted and others became commodities in new lands. Colonial archives substantiate the significance of these biotic transfers but place the diffusion narrative in the words, experiences, and perceptions of the Europeans who participated in shaping the early modern world. The "New World" was forged by the subjugation of Amerindian peoples and the appropriation of their lands; it was also indelibly shaped by the forced migration of Africans to colonial plantations and mines. As the substantial body of Columbian Exchange scholarship makes clear, European hegemony was built not only on firearms and military dominance but also on the plants, livestock, and pathogens European ships and commercial interests transplanted to new lands. But these accounts ignore species domesticated in Africa, the narrative focusing instead on the Amerindian and Asian crops Europeans introduced to Africa and elsewhere. Scholarship emphasizes, for example, the revolutionary role of Amerindian crops such as peanuts and maize on African food systems (Crosby, *Columbian*; Alpern). To the extent Africa plays a part in the Columbian Exchange, it is as a passive recipient.

Crosby later drew attention to the contributions of migrants to the environmental histories of selected regions around the globe. His book *Ecological Imperialism* (1986) discusses how emigrants to New Zealand, Australia, and South Africa carried familiar Old World crops and animals to the new lands they settled. With these temperate-zone species, migrants deliberately transformed distant and exotic outposts into landscapes resembling the ones they had left behind. In this way, ordinary people—not merely plutocrat planters and agents of empire—actively remade unfamiliar environments into approximations of their birth countries, transforming them into *Neo-Europes*. Crosby's innovative research calls attention to the ways that migrants alter the environmental histories of the lands they settle.

The massive influx of Africans to the Americas, which until the nineteenth century exceeded the number of European arrivals, is not portrayed in the same way, in part because the historical association of Africans with slavery treats them as passive actors, incapable of independent action or self-conscious agency. African food crops were transferred to the Americas during the era of plantation slavery, yet Columbian Exchange scholarship is silent on their impact. This chapter identifies the crop transfers that came *out* of Africa through the conduit of the transatlantic slave trade and were enabled by *enslaved* migrants.

Slave agency in crop introductions

The history of the early plantation period in the Americas features assertions by a number of eminent Europeans who credit slaves with the introduction of specific foods, all previously grown in Africa. There are at least two dozen Old World tropical plants that naturalists and visitors to plantation societies claimed slaves introduced. These include yams, sorghum, millet, sesame, black-eyed and pigeon peas, the kola nut, oil palm, and okra (Grimé; Carney and Rosomoff 136–137).

It could be sensibly argued that commentaries by seventeenth- and eighteenth-century historical personages, such as Hans Sloane (the founder of the British Museum), Willem Piso and Georg Marcgraf (members of the Dutch scientific expedition to Brazil in the 1630s), Thomas Jefferson, and others who made these claims, are simply not very convincing. After all, Africans arrived in plantation societies as chattel, without personal belongings.

Crediting enslaved Africans with plant introductions attributes to them an agency that is entirely at odds with the long-standing view that slaves contributed little but muscle to the agricultural history of the Americas. The question is why such prominent historical witnesses would claim otherwise.

Many of the plants Europeans attributed to slave introduction are, in fact, of African origin. These African food staples arrived in the New World tropics between the sixteenth and eighteenth centuries. Plantation records and historical accounts from the period also mention the castor bean, melegueta pepper, hibiscus, *Bambara* groundnut, watermelon, and ackee—all domesticated in Africa. Other Old World staples, such as the banana, plantain, taro, and the pigeon pea came to the Americas from Africa but originated in Asia. They reached tropical Africa millennia before the Columbian Exchange, arriving through overland and maritime trade networks in prehistory, when Africans adopted them into existing agricultural systems.

Africans participated fully in the process of plant and animal domestication that occurred in different parts of the world beginning some 10,000 years ago. African contributions to global food supplies include nine cereals, half a dozen root crops, five oil-producing plants, several forage crops and as many vegetables, three fruit and nut crops, coffee, and the bottleneck gourd. Most of the African domesticates are tropical species and not widely known to Western consumers, and they are often overlooked because several them are incorrectly considered to be of Asian origin. Nevertheless, Africa harbors Old World species of sesame (*benne*), eggplant, the pigeon pea and, significantly, the indigenous *glaberrima* rice species (Carney and Rosomoff 20–21). Consideration of Africa's agricultural history belies the common perception of a continent that is perpetually hungry.

African food crops in the transatlantic slave trade

The transatlantic slave trade vitally depended on food grown in Africa to provision the captives destined for shipment to the Americas. Despite the removal of able-bodied youth from the population, Africa routinely produced surplus food during the Atlantic slave trade. We know this from slave-ship manifests as well as the logs and drawings of ship captains. While slave ships carried some food stores from Europe, captains relied in no small part on African victuals to provision their human cargoes during the Middle Passage. Slavers acquired food for the transatlantic crossing from African merchant middlemen, supplies stocked by European forts along the Guinea Coast, and local markets (Carney and Rosomoff 49–55). The food slavers purchased in Guinea's ports included not only the Amerindian introductions—maize, peanuts, and the sweet potato emphasized by Columbian Exchange scholarship—but also indigenous African crops, such as millet, sorghum, rice, yams, and black-eyed peas (Harms 279–281; Klein 94; Newson and Minchin 81–82, 320). Ancient Asian transplants additionally figured among slaver food purchases, notably the Asian yam, taro, and plantains. Slave-ship captains often showed a distinct preference for traditional African dietary staples because they commonly believed that mortality rates across the Middle Passage diminished when captives were given food to which they were accustomed (Carney and Rosomoff 68–69).

No African crop attracted European attention as much as rice. Portuguese mariners first encountered African rice among *Mande*-speaking peoples in the mid-fifteenth century and even exported it to the metropole. (The Portuguese would not come into contact with Asian rice societies until Vasco da Gama's epochal 1497 voyage to India.) In Guinea, Captain Samuel Gamble described a mangrove rice farming system that furnished provisions for his transatlantic slave voyage of 1793–1794 (Carney and Rosomoff 43; Mouser). By the time of his account, rice had become a commercial crop in the Atlantic world, transforming landscapes and the economies of colonial Carolina, Brazil, and Suriname. Plantations dedicated to rice earned fortunes for generations of planters and slaveholders.

How did European interest in rice develop into the vast and highly profitable plantation system that first emerged in the Carolina colony? The commonly accepted answer is given by Carolinian planters and their descendants, who credited their own ingenuity with discovering a crop so exceptionally suited to cultivation in the coastal wetlands (Doar; Heyward). Such claims must be assessed against the founding planters' utter lack of experience with this tropical crop and its sophisticated water-management regimes. Moreover, other comments from the historical record call into question the prevailing narrative. In 1648, decades prior to Carolina's founding, a letter from the initial period of Virginia's settlement reports: "The governor *Sir William* [Berkeley], caused half a bushel of Rice … to be sowen, and it prospered gallantly … for we perceive the ground and Climate is very proper for it … as our *Negroes* affirm, which in their Country is most of their food" (Littlefield 100). The comment notes the presence of slaves from the West African rice-growing region, acknowledges their agricultural expertise, and presents them more as collaborators than unskilled laborers. A countervailing narrative begins to emerge: perhaps slaves contributed more to the agricultural history of the New World than they have been credited with by historians to date.

While rice did not become a prominent plantation crop in Virginia, it did find footing in the Carolina colony within a decade of its founding in 1670. The introduction of rice to Carolina is not well documented. However, one archival reference from the 1690s finds rice the unexpected dividend of a slave ship arriving from West Africa: "a Portuguese *vessel* arrived with slaves from the east, with a considerable quantity of rice, being the ship's provision" (Collinson). The comment exemplifies the role of rice—rice from Africa—in provisioning a slave ship. It also reveals that the rice was unmilled, which meant it could serve as seed for planting. A Swiss correspondent, writing in 1726, elucidates the colony's early rice history, claiming that "it was by a woman that Rice was transplanted into Carolina" (Wood 35–36). In these instances, the historical record parallels the maroon oral history that opened this chapter: both emphasize the primacy of a slave ship, leftover grains, and the impetus of an enslaved African woman for the beginnings of rice cultivation.

From its beginnings, Carolina rice culture bore unmistakable resemblance to African mangrove, tidal, inland swamp, and rain-fed rice systems. Beyond similarities of topography and micro-environment, the introduced rice culture relied upon African methods of sowing, growing, harvesting, milling, winnowing, and cooking the grain. Rice cultivation symbolized not only the transfer of African seed to the colony, but also the simultaneous migration of a highly sophisticated African agricultural and processing technology. From humble beginnings the Carolina rice economy was built upon what enslaved rice growers knew—that from field to kitchen, rice was distinctly African.

The origins of Carolina rice culture are perhaps best understood by recognizing that slaves were anything but passive participants. Evidence of their agency can be found in the plots and physical spaces where the enslaved cultivated their own food. Slaves grew food not only for those who held them in bondage, but also, out of necessity, for themselves. When sugar failed as an export commodity in Carolina's early years, planters looked to exploit other crops in their midst. Two African foodstaples—rice and black-eyed peas—were already being grown in slave food plots. These commodities, along with salted beef, were shipped to the British West Indies and quickly became profitable exports. After first establishing itself as a larder to the English Caribbean, the Carolina colony came into its own by the mid-eighteenth century as the Atlantic world's chief supplier of rice (McWilliams 133–136).

Subsistence and slave food fields

The presence of rice and other African food staples in slave food fields holds the key for understanding the process by which slaves instigated the broader cultivation of these crops. The cereal grains and root crops occasionally remaining aboard slave-ship voyages provided enslaved Africans opportunities to access familiar dietary staples and to begin planting them in their dooryard gardens and subsistence plots. The plants they grew attracted the attention of planters and naturalists, to whom some were new species. A few of these plants were adapted to export markets and became commodities; others found their way onto planters' tables, significantly amending white cuisines.

In the initial period of plantation development, European colonizers knew little about growing food in tropical America. The principal food that concerned them was sugar, a manifestly profitable commodity that required significant apprenticeship for growers eager

Figure 3.1 Habitação de negros (Dwelling of the Blacks) by Johan Moritz Rugendas, 1820s, Brazil. Note the agroforest planted around the slave dwelling, which includes New World papaya and pineapple along with the African castor plant (to the right) and Old World plantain and banana trees.

Figure 3.2 Militia units setting fire to an eighteenth-century maroon village in Dutch Guiana, with fugitive slaves fleeing recapture. Detail from the Alexandre de Lavaux map of Suriname (*Generale caart van de provintie Suriname*) *c*.1737 (Bubberman). Inset is a drawing of a maroon settlement in northeast Dutch Guiana destroyed by the militia of Captain John Stedman in November 1776 (Stedman vol. II, 128–129). Numbers in sketch refer to: 8) Maroon defensive position; 9) rice and maize field; 11, 16) rice fields; 12) maroon settlement; 14) abandoned settlement; 15, 17) manioc, yams, plantain fields; 18) protective swamp.

to learn its cultivation and processing methods. With their attention focused on commodity production, European planters for the most part left the crucial matter of producing food to their slaves. There was a simple logic behind this. The peoples that the colonizers subjugated—initially Amerindian, then African—were for the most part already expert tropical farmers, unlike their owners. Tropical agriculture relies on entirely different crops and cultivation methods than those used in temperate climates. It requires expertise in adapting agricultural practices and cultivars to high temperature and precipitation regimes, soils of often limited fertility, and the year-round menace of insect pests. For the uninitiated, the production of food in the tropics presented considerable challenges met only by radically new practices and paradigms.

The subsistence food fields allotted to slaves were known variously as provision grounds, yam grounds, or slave gardens. They included the small yards surrounding dwellings that slaves intensively cultivated (Figure 3.1; Parry; Mintz). Slaves also grew food staples in plantation areas where sugarcane had exhausted the soil, transforming marginal land into food forests. Old World tropical tubers, such as yams, plantain-banana, and taro, produced prolifically in poor soils. Demanding minimal labor inputs, they could be continuously harvested as needed. The extraordinary range of Old World and New World tropical crops slaves planted made their food fields resemble, in the words of one eighteenth-century observer of Saint Domingue, "*une petite Guinée*" (Tomich).

Maroon food fields also demonstrated a preference for African diets. Drawings from military campaigns in Dutch Guiana show food fields planted with rice, yams, and plantains (Figure 3.2). Eighteenth-century reports from the colony indicate that the Saramaka maroons grew millet, the *Bambara* groundnut, okra, pigeon pea, watermelon, and sesame (Price, "Subsistence"; Price and Price 128–129). Elsewhere in tropical America where militias destroyed many fugitive communities in the same century, descriptions similarly illustrate food systems and land use of an African provenance (Duvall; Carney and Rosomoff 85–86).

African plants and the transatlantic commodity chain

Documentation now records that at least 35,000 slave voyages carried the eleven million Africans who were disembarked in the Americas (Eltis and Richardson; Emory University). The commerce in slavery operated across vast geographical space as a kind of transnational commodity chain. Food, and the locations where it could be purchased, was an indispensable part of the process. Captains of slavers arriving along the African coast learned where African food surpluses were sold and which subsistence staples were available. By the time the captives reached the auction blocks of the New World, the only persons in the commodity chain experienced in cultivating the foodstaples that accompanied them into bondage were the enslaved Africans themselves.

For slave-ship captains, the utility of African crops ended when their victims were disembarked and sold. For the landed Africans, that utility was never lost; it was recast and transformed in the plantation and mining societies of the New World. The familiar foods and medicines that accompanied them across the Middle Passage could now forestall hunger, treat ailments, and preserve life—not to mention arouse a social memory of lost homelands.

In the early colonial period, plantation owners encountered many new plants growing in their slaves' food plots. Europeans referred to some species by geographical descriptors that indicated their African provenance. Many of these dietary staples are still known in the Portuguese, Spanish, French, and English languages by the place name "guinea," the name slave traders generically applied to the African continent. In English we have guinea corn (sorghum), guinea sorrel (*Hibiscus sabdariffa*, whose flower comprises the tropical cranberry-flavored beverage known as *jamaica* in Mexico), guinea squash (*Solanum aethiopicum*), guinea melon (*Cucumis melo*), guinea pepper (*Xylopia aethiopica*), guinea grass (*Panicum maximum*), and the guinea fowl (*Numida meleagris*).

Other Old World tropical introductions were named after specific African regions or slave ports, where surplus food was usually available and purchased for the Atlantic crossing. Examples include the African pigeon pea, called Congo or Angola pea in English, *pois d'angole* in French, *loango pesi* in Dutch. One type of cooking banana or plantain is still known in Brazil as *banana de São Tomé*. In the former plantation areas of eastern Cuba, along Colombia's Caribbean coast, and in El Salvador, bananas are known as *guineos*, underscoring their arrival from the African continent, where Europeans first encountered this Old World tropical species (Van Andel et al.). As such toponyms indicate, slave vessels served as an unplanned conduit for the introduction of African food staples to the Western

Hemisphere. Even though not every slave ship stocked adequate food supplies, and most victuals were consumed en route, provisions in the form of seeds and rootstock occasionally remained from Atlantic crossings.

Many African plants grown in tropical America are known in European languages by their African vernacular names. For the crops that had no existing words in European languages, plantation owners and European naturalists borrowed the names by which they were known to the slaves who grew and prepared them. The adoption of African loan words into the colonial languages of plantation societies powerfully suggests slave agency. "Yam," for example, is a plant name borrowed from African languages (*ñame, inhame*). The Gullah descendants of slaves who worked the rice plantations of South Carolina and Georgia used the Mande word *malo* for rice well into the twentieth century (Turner 128). Tropical America refers to a number of crops by various African names: *guandu, guandul, wando* (the pigeon pea in Portuguese, Spanish, and Dutch); *quiabo, quingombó, oko* (okra in Portuguese and Spanish); *bissy* and *eddo* (kola nut and taro in the Caribbean, respectively); *pindal* and *goober* (for the Amerindian peanut) and *benne* (for sesame) in South Carolina and the English Caribbean; and *abbay* and *dendê* (the Jamaican English and Brazilian Portuguese words) for the African oil palm (Cassidy and Le Page; Schneider). *Dendê* defines not only Afro-Brazilian cuisine but also a New World landscape shaped by Africans and maintained by their descendants. A 30-kilometer portion of the Atlantic coastal lowlands of Bahia is generally known as *Costa do Dendê* or the Oil Palm Coast. The Brazilian use of an African vernacular word for the oil palm and as geographical descriptor for a cultural landscape transformed by enslaved Africans provides a rare acknowledgment of their agency in New World environmental history (Watkins).

African names are also given to many dish preparations of tropical America. The one-pot stews known in the Caribbean as *callalou* and in Louisiana as *gumbo* are well-known examples. In Bahia *acarajé* is a popular fritter made with black-eyed peas. A regional specialty of Maranhão, Brazil, is *arroz de cuxá*, rice with sorrel. The loan word *cuxá* comes from the West African Mande-language word *kucha*. Rice-growing Mande speakers in Senegambia and Guinea-Bissau still cultivate and make several food preparations with sorrel (*Hibiscus sabdariffa*), another West African domesticate. Maranhão's eighteenth-century rice plantation economy was supported in part by enslaved Mande speakers drawn from West Africa's rice region (Carreira). The social memory of their New World presence is vested in a humble dish of *arroz de cuxá* and in the oral history passed down through the generations: that an African woman came off a slave ship and into a new land with grains of rice in her hair.

The social and racial prejudice that divided slaveholders from their bondsmen failed to keep separate the foods they ate. In the tropics, presumed walls of culinary segregation disintegrated over time as signature ingredients of the African diaspora stealthily made their way into white kitchens and dining rooms. Today we recognize the infiltration of these African staples in the distinctive regional foodways of former slave societies that extend from South Carolina and Louisiana through the Caribbean to Mexico's Gulf Coast and Bahia, Brazil and along Colombia's Pacific lowlands.

Conclusion

The migrations of African plants in the period of plantation slavery are ineluctably tied to the institution and processes of the transatlantic trade in human beings. Slave ships carried Africa's botanical heritage, which gave uprooted Africans opportunities to establish them anew in the spaces allowed them for food cultivation in plantation societies. In their dooryard gardens and food fields, slaveholders discovered many of these novel crops and, at times, exploited their commercial potential. Driven by the basic human need for food, slaves transformed neotropical environments into sites of food production. From such modest foundations, maroons sourced the African staples that facilitated their survival in fugitive communities.

An examination of the subsistence systems of New World Africans illuminates how slaves and maroons fought oppression and persevered against great odds. The environmental humanities bring new methodologies to bear on obscured episodes of our common past. These extend beyond written and oral accounts to include findings from historical linguistics and botany, art history and literature, and human ecology's attention to food crops and cultural landscapes. Together, they allow us to read the hidden transcripts of African contributions to New World food systems while offering a corrective to scholarship on the Columbian Exchange.

References

Alpern, Stanley B. "The European Introduction of Crops into West Africa in Precolonial Times." *History in Africa* 19 (1992): 13–43. Print.

Bubberman, F. C. et al. *Links with the Past: The History of the Cartography of Suriname 1500–1971.* Amsterdam: Theatrum Orbis Terrarum B.V., 1973. Print.

Carreira, António. *As companhias pombalinas de Grão Pará e Maranhão e Pernambuco e Paraíba.* Lisbon: Editorial Presença, 1983. Print.

Carney, Judith. *Black Rice: The African Origins of Rice Cultivation in the Americas.* Cambridge, MA: Harvard University Press, 2001. Print.

———. "Rice and Memory in the Age of Enslavement: Atlantic Passages to Suriname." *Slavery and Abolition* 26.3 (2005): 325–347. Print.

———. "'With Grains in Her Hair': Rice History and Memory in Colonial Brazil." *Slavery and Abolition,* 25.1 (2004): 1–27. Print.

Carney, Judith and Richard Rosomoff. *In the Shadow of Slavery: Africa's Botanical Legacy in the Atlantic World.* Berkeley, CA: University of California Press, 2009. Print.

Cassidy, F. G. and R. B. Le Page. *Dictionary of Jamaican English.* Kingston, Jamaica: University of the West Indies Press, 2002. Print.

Collinson, Peter. "Of the Introduction of Rice and Tar in our Colonies." *Gentleman's Magazine* 36 (June 1766): 278–280. Print.

Crosby, Alfred W. *Ecological Imperialism: The Biological Expansion of Europe, 900–1900.* New York: Cambridge University Press, 1986. Print.

———. *The Columbian Exchange: Biological and Cultural Consequences of 1492.* Westport, CT: Greenwood Press, 1972. Print.

Doar, David. *Rice and Rice Planting in the South Carolina Low Country.* 1936. Charleston, SC: Charleston Museum, 1970. Print.

Duvall, Chris S. "A Maroon Legacy? Sketching African Contributions to Live Fencing Practices in Early Spanish America." *The Singapore Journal of Tropical Geography* 30 (2009): 232–247. Print.

Eltis, David and D. Richardson. *Atlas of the Transatlantic Slave Trade.* New Haven: Yale University Press, 2010. Print.

Emory University. "Voyages: The Trans-Atlantic Slave Trade Database." 2009. Web. 1 August 2015.

Galeano, Eduardo. *Faces and Masks: Memory of Fire*. Vol. 2. Trans. Cedric Belfrage. New York: Norton, 1998. Print.

Grimé, William. *Ethnobotany of the Black Americans*. Algonac, MI: Reference Publications, 1979. Print.

Harms, Robert. *The Diligent: A Voyage through the Worlds of the Slave Trade*. New Haven, CT: Yale University Press, 2002. Print.

Heyward, Duncan. *Seed from Madagascar*. Chapel Hill: University of North Carolina Press, 1937. Print.

Higman, Barry. *Slave Populations of the British Caribbean, 1807–1834*. Kingston, Jamaica: University of the West Indies, 1995. Print.

Klein, H. S. *The Atlantic Slave Trade*. Cambridge: Cambridge University Press, 1999. Print.

Littlefield, Daniel C. *Rice and Slaves*. Baton Rouge, LA: Louisiana State University Press, 1981. Print.

McWilliams, James E. *A Revolution in Eating: How the Quest for Eating Shaped America*. New York: Columbia University Press, 2005. Print.

Mintz, Sidney W. "Houses and Yards among Caribbean Peasantries." *Perspectives on the Caribbean*. Ed. Philip W. Scher. Malden, MA: Wiley-Blackwell, 2010. 9–24. Print.

Mouser, Bruce L., ed. *A Slaving Voyage to Africa and Jamaica: The Log of the Sandown, 1793–1794*. Bloomington: Indiana University Press, 2002. Print.

Newson, L. A. and S. Minchin. *From Capture to Sale: The Portuguese Slave Trade to Spanish South America in the Early Seventeenth Century*. Boston, MA: Brill, 2007. Print.

Parry, John H. "Plantation and Provision Ground." *Revista de historia de America* 39 (1955): 1–20. Print.

Price, Richard. *First-Time: The Historical Vision of an Afro-American People*. Baltimore, MD: Johns Hopkins University Press, 1983. Print.

——. "Subsistence on the Plantation Periphery: Crops, Cooking, and Labour among Eighteenth-Century Suriname Maroons." *Slavery and Abolition*, 12.1 (1991): 107–127. Print.

Price, Richard and Sally Price, ed. *Stedman's Surinam: Life in an Eighteenth-Century Slave Society*. Baltimore, MD: Johns Hopkins University Press, 1992. Print.

Schneider, John T. *Dictionary of African Borrowings in Brazilian Portuguese*. Hamburg: Helmut Buske Verlag, 1991. Print.

Stedman, John Gabriel. *Narrative, of a Five Years' Expedition, against the Revolted Neroes of Surinam, in Guiana*. 2 vols. London: J. Johnson, 1813. Print.

Tomich, Dale. "Une petite Guinée: Provision Ground and Plantation in Martinique, 1830–1848." *Cultivation and Culture: Labor and the Shaping of Slave Life in the Americas*. Ed. Ira Berlin and Philip D. Morgan. Charlottesville: University Press of Virginia, 1993. Print.

Turner, Lorenzo Dow. *Africanisms in the Gullah Dialect*. Columbia: University of South Carolina Press, 2002. Print.

Vaillant, M. "Milieu cultural et classification des variétés de Riz des Guyanes française et hollandaise." *Revue internationale de botanique appliquée et d'agriculture tropicale* 28 (1948): 520–529. Print.

Van Andel, Tinde R. "African Rice (*Oryza glaberrima* Steud.): Lost Crop of the Enslaved Africans Discovered in Suriname." *Economic Botany* 64 (2010): 1–10. Print.

Van Andel, Tinde R., et al. "Local Plant Names Reveal that Enslaved Africans Recognized Substantial Parts of the New World Flora." *Proceedings of the National Academy of Sciences* 1 December 2014. Web. 1 August 2015.

Watkins, Case. "Landscapes in Diaspora: The Development of African Oil Palm Groves in Bahia, Brazil." *Environment and History* 21.1 (2015): 13–42. Print.

Wood, Peter H. *Black Majority: Negroes in Colonial South Carolina from 1670 through the Stono Rebellion*. New York: Norton, 1974. Print.

4

DOMESTICATION IN A POST-INDUSTRIAL WORLD

Libby Robin

The concept of the Anthropocene unsettles ideas about time and place. The local scales of the village and the nation are challenged by planetary narratives across geological timescales. Who belongs where, when people are all "Out of Africa" at this level? What is civilization and whose civilization is "progressive"? At the heart of literatures of belonging, from fiction to conservation biology, is the concept of agricultural settlement as civilization. A world history narrative has identified domesticating animals and sowing crops as the first big step forward (the agricultural revolution) and the tapping of energy sources beyond human and animal bodies as the second (the industrial revolution). Domestication of animals in Western civilization frames understandings of the way nature is conceptualized and controlled. The civilization narrative also affects questions of who has the right to speak for nature and what moral rights are accorded to nonhuman others.

This chapter begins with domestication and agriculture in the Middle East and Europe, an old-fashioned linear narrative of civilization, and considers why it has become pertinent in our present environmental crisis. It then considers the legacies of the historical moment when Europe expanded and the associated "ecological imperialism" (Crosby) that transferred animals and plants to the neo-Europes. In the post-imperial era, terms such as *native* and *alien* have become controversial, bringing together discussions from invasion biology, anthropology, and environmental psychology. As the ecological limits of place have affected the behaviors of animals, the domesticated become *de-domesticated* or feral, and enemies of the agricultural project. Finally I return to Europe where *re-domestication* is emerging as a civilizing art in our post-industrial era.

The domesticated life and its history

The idea that the planet has entered a new geological epoch, the Anthropocene, has refocused on the story of the epoch we are leaving, the Holocene. As global change modelers measure biophysical factors that define planetary limits for "a safe operating space for humanity" (Rockström et al. 472), they observe that the Holocene was the time when Earth's regulatory capacity created particular conditions that suited human development. The Holocene began with the warming after the last glacial maximum that allowed agriculture and domestication to develop. Later came the industrial revolution that began to upset Earth's regulatory capacity, creating conditions for the next new epoch. Steffen et al. argue that "the relatively stable, 11,700-year-long Holocene epoch is the only state of the Earth System that we know for certain can support contemporary human societies" (*Global Change* 736). As global system modelers seek to identify factors that "affect the capacity of the Earth system to persist in a Holocene-like state under changing conditions" (Steffen et al., "Planetary Boundaries"), they also scrutinize the history of civilizations in the Holocene to look at what environmental factors have enabled growth in human societies (Robin and Steffen).

The global focus of earth system science means that the tale of "progress" is often told as if human civilizations unfolded the same way everywhere: the European story, with an Out-of-Africa prehistory, is often assumed to be global. This version of world history goes as follows: humanity distinguished itself from other animals through the use of stone tools, gradually moving from simple biface tools three million years ago to finely sharpened arrowheads and stone axes perhaps forty to fifty thousand years ago (Schwägerl 130). Between ten and six thousand years ago, as the climate warmed after the Last Glacial Maximum, much of humanity turned to settled agriculture and domesticating animals, which in turn set up the framework for the rise of the state (Brooke 121–164). Yet, as Jared Diamond has observed, domestication for agricultural purposes has only occurred in limited places where the climate is warm and stable, and where the animals and plants have an appropriate suite of qualities. Potential species for domestication must be genetically and behaviorally flexible, adaptable to different environmental conditions, and productive enough in the wild state to be attractive to humans (Diamond 114–175). An annual breeding cycle is essential to the logic of agricultural development. Domesticated animals come from geographically limited areas of the world, particularly the Levant and central Asia (Brooke; Diamond). As global climate historian John Brooke has argued, "the pull of particular biotic resources profoundly shaped the move to food production" (126). Agriculture had some disadvantages over hunting: each calorie yielded from the land cost more effort. Early farmers had far less free time than hunters and gatherers. Even so, settled agriculture enabled people to store food against hard times, and to expand their populations. In northern Europe with its bitterly cold winters, agriculture proved a particularly worthwhile investment. As Europe expanded first to the north as the ice of the Last Glacial Maximum retreated, then beyond to the New World, it took its particular brand of agriculture with it (Crosby).

So the history of the Holocene opens with domestication and agriculture. The beginning of its end comes as city states form and bring with them a rise of mercantilism. From the mid-1700s, the industrial revolution gathered pace through technologies like the steam engine and the energy provided by coal and other fossil fuels; scientific and medical innovations enabled longer lives and bigger populations. The last century brought a profligate use of energy, on a scale never seen before: John McNeill called his environmental history of the twentieth century *Something New Under the Sun*. From 1950, there has been a global Great Acceleration of population, wealth, carbon emissions, ocean acidification, and biodiversity loss (Steffen et al., *Global Change*; Costanza et al.).

Domestication is a major highlight in the early history of the Holocene and of civilizations, because it enabled sedentary living and agriculture. While cows, sheep, and goats were the animals that provided protein, horses, oxen, donkeys, and camels provided energy and transport. Most of these animals evolved in the Middle East and moved north into Europe as northern ice sheets melted (Diamond). Recent DNA evidence confirms that the chicken (Red junglefowl) was first domesticated in China about 8000 BCE, and was documented as present in Greece by 5000 BCE, but other archaeological evidence documents fowl kept for eggs in Egypt centuries earlier (Carter). Only llamas and alpacas originate in the southern hemisphere (Diamond).

One of the most prominent domestication narratives comes from the cooler north, where the wolf was tamed to create the dog. Indeed some 343 officially recognized dog breeds have been created through selective breeding (Hemmer). Some argue that the cat, so prominent in Egyptian mythology, was never fully tamed and still reverts readily to its wild state (*Threat Abatement Plan*). Companion animals, particularly dogs and cats, co-located with the humans that fed and sheltered them over the Holocene. These were the working animals of villages and settlements—dogs working to round up sheep and other domestic animals, and cats to keep places free from the mice and rats that brought disease. Their role as family pets came later. Domestication and agriculture enabled humans to spend less time hunting and foraging, but the downside was that closely settled populations were vulnerable to diseases. By the end of the Holocene, when medicine and hygiene practices had defeated many diseases, settling in one place had become the norm for many, and being sedentary has come to entail its own health issues, such as posture-related back problems and obesity in Western societies.

De-domestication: European expansion and settler anxiety

Toward the end of the Holocene, as the city states expanded, Europeans moved out into the worlds beyond Europe, first the New World and then the neo-Europes of the southern hemisphere. In Australia, British invaders brought a simultaneous agricultural and industrial revolution in 1788, as steam-engine technologies accompanied the great agricultural experiments. Since the eighteenth century, domesticated animals and plants have been systematically borrowed from other places to civilize the land and to "improve" its value and the place of the Australian colonies, and later the nation, in international trade. In a place that yielded no recognized agricultural products, the newcomers imported foodstuffs familiar to them and grew Old World crops and livestock. In Australia and places like it, people have become anxious about whether agriculture is progressive and whether, when agriculture fails, the European settlement has become "uncivilized" (Muir, *Broken*). Agriculture and pastoralism remained part of the Australian identity long after the nineteenth century. Even when the land was running on mineral wealth in the 1960s, I was taught at school that the country was the land of the Golden Fleece, and we lived "off the sheep's back" (Robin, *Continent* 162). The Toison D'Or—the Χρυσόμαλλον Δέρας or golden fleece of Greek myth—was an important symbol of ancient civilization in a land where no hard-hooved animals had grazed before the arrival of the British, and where the settlers were slow to recognize the even more ancient civilization of the Aboriginal people they displaced.

"Australian colonists were haunted by the spectre of degeneration," historian Tom Griffiths writes. "The dangers of the strange, ancient environment could only be overcome by subverting the natural order, making it anew, acclimatising imported species, destroying indigenous nature, sponsoring aggressive biological imperialism" (12). Australian nature was besieged with aggressive imperatives for agricultural progress and competition from acclimatized species. In the twenty-first century, conservation biology and concerns about invasive animals are prominent in Australia, as well as in New Zealand and South Africa, all places where animals and plants from other places have often outcompeted the local biota. Science has spent a century or more trying to deal with invasions of exactly the hardy animals and plants whose cosmopolitan qualities historically made them suitable for domestication. Invasion biology attempts to eliminate the feral animals, now *de-domesticated*, undermining the civilizing project of agriculture.

A government authority once rejoiced in the name "The Vermin and Noxious Weeds Destruction Board." It administered the Vermin and Noxious Weeds Destruction Act of 1958 (Public Record Office Victoria). The task of the Destruction Board was to clear the country of animals and plants problematic to the civilizing forces of Western agricultural and pastoral enterprises. In other places, vermin are usually rodents, but in Australia the term was used to refer to rabbits and even marsupials like wombats (because they dug up fencing). "Noxious" was not a synonym for "invasive": it simply described plants that impeded agriculture. Both the Destruction Board in the 1950s and the invasion biologists in the twenty-first century expressed concern about "weeds," but they were not targeting the same plants. The earlier term "vermin" is now more commonly "pests," "threats to biodiversity," or "ferals," with the last becoming increasingly common.

Native, *non-native*, and *alien* are irrelevant to agriculture but crucial categories for invasion biology. Ecologists Matthew Chew and Andrew Hamilton argue cogently that history and cultural expectations, rather than science, determine "biotic nativeness." Their reading of "nativeness" in the botanical literature shows that history rather than science underpins this taxonomic category. Nativeness is, they assert, "the *sine qua non* invoked by many management policies, plans and actions to justify intervening on prevailing ecosystem processes" (38), underpinning practices like ecological restoration. The tacit assumptions about nativeness in the minds of ecologists are that a taxon *belongs* where it occurs, and that such *belonging* signifies a "morally superior claim to existence, making human dispersal tantamount to trespassing" (Chew and Hamilton 41). The human rights of "natives" become rights for native plants, and defend a particular form of ecological purity in management that is constructed on doubtful logic and poor science (Pearce).

Livestock and crops, since these were purposeful introductions, were not considered aliens and fell beyond the purview of invasion biology. Indeed, Lesley Head and colleagues have argued that wheat is no longer even a plant. If it is a crop it is excluded from maps of vegetation. Its status in the economy subsumes its "plantiness" (Head et al.). This hyper-separation of economy and ecology makes agricultural pasture plants effectively invisible, and even where they become invasive weeds in threatened native grasslands, managers concentrate on showy escaped garden plants, rather than agricultural weeds (Kendal et al.). Similarly, feral animals, once-domesticated animals that have gone wild, have become the focus for whole landscape conservation programs. Not just feral cats, foxes, and wild dogs, but also horses, donkeys, pigs, camels, and goats are among the "feral animals" subject to aerial culling conservation programs over vast areas. These programs are so expensive that new artificial intelligence drone technologies are being developed to cull specific animals like feral pigs.

In an eerie echo of the metaphor of "battling the land" to wrest from it agricultural and pastoral produce (Robin, *Continent*), feral control programs abound with military metaphors: biosecurity, threats, and invasions. Ferality is a category constructed in contradistinction to domesticity and domestication. Feral animal control exposes different understandings of "nature" and what belongs or does not in a landscape. The management literature surrounding

49

feral animals abounds with military metaphors (Larson, "War"). Invasion biology "combats" feral animals, controlling the undomesticated for the sake of wild landscapes. There is complex cultural baggage in terms like "natural resource management" (NRM).

Aspirations for managing nature vary between nations, not only because of different ecologies, but also because of the cultural strength of metaphors like "wilderness" in policy and identity making (Robin, "Wilderness"). Since the 1990s, invasion biologists in Australia have focused on the invasive species that compromise biodiversity. There is a growing fear that alien plants and animals—the "New Nature" (Low, *New Nature*)—will alter the course of evolution, compromising local endemic plants and animals. Ecologists like Tim Low urge that Australia should resist the globalisation of ecological systems. In a continent with high biodiversity values, the fear is that feral animals will simplify the genetic diversity of the ecosystems and limit their evolutionary futures (Low, *Feral Future*).

"Biodiversity" is the latest way to speak of nature and to measure it in numbers (Robin, "Idea of Biodiversity"). Those species deemed by the International Union for the Conservation of Nature (IUCN) or national governments to be "threatened" are a particular focus of natural resource managers, but there is also a continuing, and sometimes contradictory, pressure to protect plants, soils, and water points for agricultural enterprises and urban living.

Sedentary agriculture is not a good adaptation for every ecological environment. A migratory lifestyle works better in a low nutrient, arid land like Australia, where "dwelling in place" is neither possible nor desirable (Robin, "Seasons and Nomads"). Aboriginal people—present in Australia for at least fifty thousand years, well before the Last Glacial Maximum—did not domesticate animals, but this was not a failure of their civilization. Animals suitable for domestication were not adapted to Australia's ecological systems, and variability of climate made year-round settlement, particularly in the arid zone, impracticable (Smith 2013). While the global story is important for planetary narratives, civilization stories in different places vary for climatic and biogeographic reasons.

In Australia, when domesticated animals arrived with Europeans, the most famously successful "Cattle King," Sidney Kidman, made his fortune not because he was a settler, but rather because he was a *nomad*. He drove his herds thousands of miles from western New South Wales all the way to the Kimberley in northwestern Australia (Idriess). Cattle remain outback Australia's most successful industry because they can avoid droughts and floods with transhumance. Kidman's success was that he had vast acreages and used a series of stations to follow the available pasture, rather than using the pasture to "settle" his family in European style. Indeed the very word *station*, which Australians use where Americans use *ranch*, is itself telling: a station is a point on a journey, not a destination. A land of very un-European ecological systems like Australia exposes the Eurocentricity of ideas about domestication and settlement. Australia's Western society still has much to learn from the long history of Indigenous peoples living in this low-nutrient land.

The Food and Agriculture Organization (FAO) is a global institution that was born just as the Great Acceleration began in 1945. The FAO was an important sponsor of the "Green Revolution" in agricultural research and development as well as of technology transfer initiatives up to the late 1960s. The new modeling it sponsors considers fast-expanding Asian markets, including one model where all the planet's current cropping land is planted to soy beans (Mauser). This modeling suggests that no further land clearance or GMOs are needed to achieve sufficient food. Surely this report is good news for planetary health, provided the West is willing to eat tofu? Yet the "tofu solution" is an example of a single-crop global quick fix that lacks biogeographical nuance. Even a cursory look at the modelers' global map of "cropping lands" reveals questions. The northern 30 percent of the Australian continent, a very large area, is significant in these calculations. History shows that these are not cropping lands. Although generations of boosters have talked up the agricultural promise of the north in Australia, crops have repeatedly failed (Robin, *Continent* 123–151). There is a lot of land, but not much soil; water supplies are irregular and the rivers do not flow year round. Even with massive subsidies and the best Western science, this land is probably not a place to grow crops such as soybeans. It may in fact be a place for something entirely different. New work is reimagining northern Australia as an energy landscape where solar, wind, and other renewable energy supplies can be generated on a vast scale for Asian markets (Muir, "Powering").

There may be other untapped potential for protein growing in Australia, but developing it will take locally informed and culturally nuanced imagination. It has never been wise to focus only on the biggest markets or to attempt to solve global problems at a global—that is, undifferentiated and unecological—scale. Nor is industrial agriculture the only way to grow food (Muir, *Broken*).

As "Key Threatening Processes," animals lose their right to existence. The metaphors are compelling (Larson, *Metaphors*; J. Griffiths). A rethinking of the culture of culling might lead to what Thom van Dooren terms a more *ethical taxonomy*. He urges a taxonomy, or categorization, that avoids terms setting animals up as mere economic commodities, and rather creates ways for them to be perceived as sentient beings and managed in ways that allow them moral consideration (see also Clark). Since "biodiversity is a nebulous term, and one that lacks clarity in important ways" (van Dooren 289), it is the culture of policies supporting conservation biology, not the science, that determines which species are protected (Pearce). The whole kingdom of fungi seldom gains attention in red lists or threatened species plans (Pouliot).

The uncharismatic cane toad, *Bufo marinus*, is an alien invasive species, deliberately introduced from South America by Queensland sugar farmers to defend their cane crops against beetles. It is not a pretty animal; even its body length is measured negatively: "snout to vent." Cane toads are number one on the government's *Feral Animals* control list ("Feral Animals"). Cane toad "control" is increasingly getting too expensive for governments and falling to volunteer groups of "Toad Busters," as the cane toad population has spread right across the tropical top end, reaching the Kimberley (northwestern Australia). The Kimberley Toad Busters catch toads in car headlights and crush them with heavy vehicles. By "community involvement in controlling cane toad population numbers in a given area you minimise the chance of native predators attacking and consuming a toad, which in turn reduces the number of native animals dying as a the [sic] direct impact of the cane toad" (Kimberley Toad Busters).

But cane toad populations continue to grow and spread, and larger animals are frequently "busted." It is far more a "twisted sports idea," accompanied by drinking sprees, than a conservation strategy. Wry singer-songwriter Dana Lyons, wearing a combat-camouflage singlet and Akubra hat, croons that the Cane Toad Muster is an opportunity for "ornery cowboys" to "unify the nation/to fight a plague they all fear." The Muster is a "violent brand new sport" in which participants kill cane toads any way they can, especially with heavy vehicles: "Have you got the nerve to swerve?" (Lyons). The club activity enables people of very different political views to drink beer together, celebrate dead toads, and exaggerate ideas of borders and aliens.

This is not good environmental management, not just because the cane toads survive and spread anyway, but because it confuses machismo with good outcomes for the natural world.

Its only virtue is that it does not cost the government and private conservation groups the high price aerial shooting programs do. In a deeply divided nation, cane toads are something to hate together. They "threaten native species" (that eat them!), so have no rights. It is an extraordinary argument where *the prey is accused of threatening the predator*. The feral status of toads justifies deeply cultural activities of human choice that reinforce a nature divided into "aliens and natives" and become associated with patriotism and Top End identity. It is of doubtful worth for the humans participating in it and for the native animals it purports to protect.

Conservation biology in Australia is about policy and management (control), particularly political forms of ecology. While ecologists often choose to step away from the more-than-scientific aspects of biodiversity control, it behoves the environmental humanities to reflect on the conceptual leakages between ideas about animals and ideas about people where they affect the moral standing of either.

Australia's tough policy regimes in relation to alien people arriving from elsewhere are distinctive. Refugees who do not go through formal channels are "jumping the queue" by arriving on Australian shores in leaky boats. The questions of who "belongs" in a nation where 97 percent of people are non-indigenous and 25 percent were born overseas are alive and contentious. Making a sporting club out of busting toads is a symptom of the insecurity of this nation of immigrants, not a morally defensible measure to protect threatened or native species. There is in Australia a particular tendency to confuse ideas about nature and nation. If we disentangle "civilization" and "domestication" from ideas of national progress, we enable new ways to manage nature, especially feral (or formerly domesticated) animals.

Re-domestication?

The geography of history matters. It may not be *more* civilized to practice agriculture and import domesticated animals to a low-nutrient land. Nor might it be *less* civilized to return to pre-industrial practices like transhumance and organic farming in Europe. In this final section we consider the value to humans of domesticating animals in the twenty-first century, even where the economic benefits are not obvious.

Domestication has created a particular relationship between people and their animals, a relationship that has made both less dependent on environmental conditions, something that might be important in times of rapid change. Living closely with animals may also be a way of relearning to be fully human: hunters (Western and otherwise) have been leaders in restoring animal habitats and other conservation initiatives. Anthropologists Michel Meuret and Fred Provenza argue that the practical work of domesticating sheep and shepherding skills can enable twenty-first century Europeans to re-engage with nature. Good tasting, cheap lamb imports from New Zealand broke the expensive sheep feed-lot industry in France in the 1990s. French-grown lamb could only compete if farmers did not have to buy purpose-grown feed for sheep confined to industrial feed-lots. New twenty-first century shepherds have emerged, often from urban, non-farming backgrounds, who follow earlier practices of transhumance, moving sheep to the free grasses of the high country in the summers (Despret and Meuret). The new shepherds have had to find historic paths and travel with their sheep. Relearning to be a "flock" (not a commodity) has not been easy for the sheep or the people. In just a few generations of living in feed-lots, sheep had forgotten how to graze. They had to learn again that grass is food. Restoring sheep grazing has also recreated the signature rural mountain landscapes of southern France, opening up areas overgrown with decades of neglect.

Restoring cultural landscapes has also been the motivation for subsidising organic lamb farming in Sweden. Economic outcomes are essential, yet the organic farmers at *Alsike Prästgård* near Uppsala, portrayed in Chris Potter's documentary film, have worked to ensure the lives of the animals there are worth living, however long or short those lives may be. Silage is home-grown, and the animals are pastured outside for as long as the severe Swedish winters allow (Potter). This is far from a FAO program that will "feed the world" on a global scale, but it is not just agricultural production. It also works to restore landscapes, human-animal relations, and meaning to new agricultural economies in post-industrial Sweden. Farmer Amelie freely admits that European subsidies are "essential" to her endeavor, but observes that agriculture is not the only enterprise to receive government subsidies. Taking the industrial out of agriculture and instead valuing the land and the quality of food is also worthwhile. Using agriculture to create landscapes is a cultural choice the Swedes have made before: they reintroduced cows to one of their first national parks, Ängsö, in the 1920s because it was losing its character as an archipelago farm landscape (Saltzman et al.). Re-learning the quiet skills of domestication may even prove *more* economic than industrial agriculture, if health and other factors are included in the "economy" of living.

In the Anthropocene era where human and other-than-human worlds are intertwined and co-constructive, humans have moved beyond a trajectory of "civilizing" the planet. Rather than asking questions about what humans can do *for* nature—the traditional concerns of controlling invasives, improving ecosystem function and restoring Edenic past natures—humans are "inside" nature. When nature is *in* us and in our behaviors, new ethical questions emerge in the relations between humans and other-than-human nature. Although humans need a global imagination and have to consider the planet as a whole, we also have personal ethical responsibilities to other-than-human nature and an aspiration to continue to live with the principles of civil societies. No longer in "control" of an anthropocentric and colonizing global economy where animals and plants are commodities, humanity needs a new civilization narrative that focuses on improving ecological connectivity in the brave, new post-industrial world.

References

Brooke, John. *Climate Change and the Course of Global History: A Rough Journey.* New York: Cambridge University Press, 2014. Print.

Carter, Howard. "An Ostracon Depicting a Red Jungle-Fowl." *The Journal of Egyptian Archaeology* 9 (April 1923): 1–4. Print.

Chew, Matthew K. and Andrew Hamilton. "The Rise and Fall of Biotic Nativeness: A Historical Perspective." *Fifty Years of Invasion Ecology: The Legacy of Charles Elton.* Ed. David M. Richardson. London: Blackwell, 2011. 35–47. Print.

Clark, Jonathan L. "Uncharismatic Invasives." *Environmental Humanities* 6 (2015): 29–52. Web. 4 Nov. 2015.

Costanza, R., L. J. Graumlich, and W. Steffen, eds. *Sustainability or Collapse? An Integrated History and Future of People on Earth.* Cambridge, MA: MIT Press, 2007. Print.

Crosby, Alfred. W. *Ecological Imperialism. The Biological Expansion of Europe.* 1986. Cambridge: Cambridge University Press, 2004. Print.

Despret, Vinciane and Michel Meuret. "Cosmo-ecological Sheep and the Arts of Living on a Damaged Planet." KTH Stockholm. 2–4 December 2014. Conference Presentation.

Diamond, Jared. *Guns, Germs and Steel: The Fates of Human Societies.* New York: Norton, 1997. Print.

"Feral Animals in Australia." Department of the Environment. Australian Government. Web. 7 June 2015.

Griffiths, Jay. "Forests of the Mind." *Aeon.* 12 October 2012. Web. 20 November 2015.

Griffiths, Tom. *Hunters and Collectors: The Antiquarian Imagination in Australia.* Cambridge: Cambridge University Press, 1996. Print.

Head, Lesley, Jennifer Atchison, and Alison Gates. *Ingrained: A Human Bio-Geography of Wheat.* Farnham, Surrey: Ashgate, 2012. Print.

Hemmer, Helmut. *Domestication: The Decline of Environmental Appreciation.* 1983. Trans. Neil Beckhaus. Cambridge: Cambridge University Press, 1990. Print.

Idriess, Ion. *Cattle King.* Sydney: Angus & Robertson, 1936. Print.

Kendal, D., L. M. Pearce, J. W. Morgan, et al. *Melbourne's Native Grasslands: Guiding Landscapes and Communities in Transition.* Melbourne: ARCUE, May 2015. Web. 15 July 2015.

Kimberley Toad Busters. www.canetoads.com.au/. Web. 7 June 2015.

Larson, Brendon M. H. *Metaphors of Environmental Sustainability,* New Haven: Yale University Press, 2011. Print.

——. "The War of the Roses: Demilitarizing Invasion Biology." *Frontiers in Ecology* 3.9 (2005): 495–500. Print.

Low, Tim. *Feral Future: The Untold Story of Australia's Exotic Invaders.* Ringwood: Viking, 1999. Print.

——. *The New Nature: Winners and Losers in Wild Australia.* Ringwood: Viking, 2002. Print.

Lyons, Dana. "Cane Toad Muster." *Cows With Guns: The Music of Dana Lyons.* 26 May 2013. Web. 7 June 2015.

Mauser, Wolfram. "Global Food Futures: Agriculture and What Is Left Over." Rachel Carson Center for Environment and Society, LMU Munich. 10 December 2014. Presentation.

McNeill, John. *Something New Under the Sun: An Environmental History of the Twentieth Century World.* New York: Norton, 2001. Print.

Meuret, Michel and Fred Provenza. *The Art and Science of Shepherding: Tapping the Wisdom of French Herders.* Austin, TX: Acres, 2014. Print.

Muir, Cameron. *The Broken Promise of Agricultural Progress.* London: Routledge, 2014. Print.

——. "Powering Asia: The Battle Between Energy and Food." *Griffith Review* 49 (2015): 287–303. Print.

Pearce, Fred. *The New Wild: Why Invasive Species Will Be Nature's Salvation.* Boston: Beacon Press, 2015. Print.

Potter, Chris. *Alsike Prästgård: An Organic Farm.* Documentary film. 30 January 2015. YouTube. Web. 15 July 2015.

Pouliot, Alison. "A Forgotten Kingdom." *Wildlife Australia* 51.3 (Spring 2014): 14–19. Print.

Public Record Office Victoria. "Vermin and Noxious Weeds Destruction Board." Web. 6 June 2015.

Robin, Libby. *How a Continent Created a Nation.* Sydney: University of New South Wales Press, 2007. Print.

——. "Seasons and Nomads: Reflections on Bioregionalism in Australia." *The Bioregional Imagination: Literature, Ecology, and Place.* Ed. Tom Lynch, Cheryll Glotfelty and Karla Armbruster. Athens: University of Georgia Press, 2012. 278–294. Print.

——. "The Rise of the Idea of Biodiversity: Crises, Responses and Expertise." *Quaderni* 76 (2011): 25–38. Print.

——. "Wilderness in a Global Age, Fifty Years On." *Environmental History* 19 (2014): 721–727. Print.

Robin, Libby and Will Steffen. "History for the Anthropocene." *History Compass* 5.5 (2007): 1694–1719. Wiley Online Library. Web. 15 July 2015.

Rockström, Johan, Will Steffen, Kevin Noone et al. "A Safe Operating Space for Humanity." *Nature* 461 (2009): 472–475. Print

Saltzman, Katarina, Lesley Head, and Marie Stenseke. "Do Cows Belong in Nature? The Cultural Basis of Agriculture in Sweden and Australia." *Journal of Rural Studies* 27.1 (2011): 54–62. ScienceDirect. Web. 15 July 2015.

Schwägerl, Christian. *The Anthropocene: The Human Era and How It Shapes Our Planet.* 2010. Trans. Lucy Renner Jones. Santa Fe: Synergetic Press, 2014. Print.

Smith, M. A. *The Archaeology of Australia's Deserts.* New York: Cambridge University Press, 2013. Print.

Steffen, W., A. Sanderson, and P. D. Tyson et al. *Global Change and the Earth System: A Planet under Pressure.* Berlin: Springer, 2004. Print.

Steffen et al. "Planetary Boundaries: Guiding Human Development on a Changing Planet." *Science* 347.6223 (13 Feburary 2015). Web. 15 July 2015.

Threat Abatement Plan for Predation by Feral Cats. Department of the Environment. Australian Government. PDF. 2015. Web. 15 July 2015.

van Dooren, Thom. "Invasive Species in Penguin Worlds: An Ethical Taxonomy of Killing for Conservation." *Conservation and Society* 9 (2011): 286–298. Print.

5

MEALS IN THE AGE OF TOXIC ENVIRONMENTS

Yuki Masami

Originating in the Latin word "domus," which means "house," domestication is related to the act of making one's home. Meals are one of the indispensable and most distinctive elements of home. You are what you eat—or, as Michael Pollan aptly says, it is more like "you are what what you eat eats" these days due to the longer and more complicated food chain from farm to table (Pollan 84)—and meals shape you not only physically but also mentally. There is an important difference between a meal and food: food refers to what we eat, whereas a meal involves not only the food we eat but also the occasion on which we eat, what we usually call breakfast, lunch, and dinner. Whether it is special or ordinary, having a meal involves a certain degree of privacy. It is rare to have a stranger at a meal unless we are willing to accept the person; as John Steinbeck's short story "Breakfast" illustrates, inviting the other to join in a meal implies an acceptance. Thus meals play no small role in a discourse of domestication (which is related to the alteration and exploitation of the environment in which one's home is made) and a domestic discourse (which is concerned with the ideas, values, and concepts of home and family). Among the number of conceivable topics regarding meals and domestication, I will pay attention to meals made from and consumed in environments which, while perceived as home, have nonetheless been damaged, contaminated, and polluted with toxic substances from anthropogenic sources. I will particularly focus on literary representations of such meals, because literary works demonstrate different degrees of consciousness, awareness, and values regarding human relationships with the environment and thereby provide a multilayered picture in which different values come into contact, conflict, and possibly negotiate with each other.

Thinking of the issue of literary representations of meals that are contaminated with toxic substances, some locales come immediately to mind. One is Minamata, a city in southern Japan where a chemical plant operated by the Chisso corporation flushed mercury-laden waste water directly into the nearby sea for more than three decades starting in 1932. The local marine ecosystem was contaminated, and due to the bioaccumulation of mercury in the local food chain, those who largely lived on locally harvested fish and other marine products—that is, those who lived by the sea and enjoyed a self-sufficient, sustainable life— were poisoned and suffered what is known as Minamata disease. Another example is the radioactively contaminated area downwind of Chernobyl. The explosion of reactor number four at the Chernobyl Nuclear Power Plant in Ukraine on April 26, 1986, contaminated vast areas in Ukraine, Russia, and Belarus, a region home to six million people, with radioactive dust. Some such areas have restricted public access or are distinguished as exclusion zones.

Yet in those areas, there were, and still are, those who return to their place, resettle, and sustain themselves on what they grow and harvest from their now contaminated lands. A more recent example is Fukushima. Since the triple disaster of earthquake, tsunami, and meltdown on March 11, 2011, radioactive contamination from the Tokyo Electric Power Company's Fukushima Daiichi Nuclear Power Plant has been continuing to contaminate the surrounding land and sea. Fukushima and its neighboring prefectures have a much higher percentage of their population working in primary industries, including agriculture and fishery, than the national average. While about one hundred and fifty thousand residents have been evacuated from Fukushima, some have decided to stay and live within the radioactively contaminated zone, either by choice or simply out of necessity. In literature, Japan's nuclear exclusion zone has gained a new reality, which is similar to the case of Chernobyl, but with some crucial differences that I would like to discuss later.

Minamata, Chernobyl, and Fukushima are home to many people while at the same time being sociopolitically marginalized and exploited, thus presenting a conflicted picture of domestication. In her *Refuge: An Unnatural History of Family and Place*, American writer Terry Tempest Williams shows how the perception of the Great Basin as a blank spot on a map legitimized its exploitation as a "home for used razor blades, toxins, and biological warfare" (244). Similarly, Minamata, Chernobyl, and Fukushima—all located away from national, social, political, and economic centers just like the desert towns in Nevada and Utah—may have been perceived as industrial and economic wildernesses, or ideal places for environmentally harmful industrial enterprises. Such political manipulation and industrial alteration of rural environments suggest that, contra received opinion, cities may not "represent the most domesticated landscapes on the planet" (Kareiva et al. 1868). Peter Kareiva and his fellow researchers claim that "urban regions reflect the endpoint of landscape domestication, showing trends that may soon appear in other areas," but the endpoint of landscape domestication could instead be industrially exploited landscapes, which mostly correspond with rural regions. If places are altered according to the desires and values of people in cities, rural environments may be domesticated more deeply and thoroughly than urban areas.

As locales of environmental pollution, social justice, and alternative values, Minamata, Chernobyl, and Fukushima attract literary and artistic attention. Among the number of literary works that address these locales, I will mainly focus on the following three, which are rich with accounts and descriptions of residents' meals: Ishimure Michiko's novel *Kugai jōdo* [*Paradise in the Sea of Sorrow: Our Minamata Disease*] (Japanese original in 1969; English translation in 1990); Svetlana Alexievich's *Voices from Chernobyl: The Oral History of a Nuclear Disaster* (Russian original in 1997; Japanese translation in 1998; English translation in 2005), which collects personal accounts of victims of the disaster at Chernobyl; and Taguchi Randy's *Zōn nite* [*In the Zone*], a series of novellas on Fukushima after the nuclear meltdowns, which first appeared in a literary magazine in 2011 and 2012 and as a book in 2013.

Paradise in the Sea of Sorrow, *Voices from Chernobyl*, and *In the Zone* all detail people's everyday meals, which are based on what they gather and catch from, or grow in, nearby environments that are dangerously contaminated due to anthropogenic causes. These literary descriptions of people eating contaminated food from, and in, contaminated environments suggest a notion of "food as gift," or, to use a phrase from Ishimure's novel, food as "a gift from heaven" (207). Food as gift refers to food that is perceived as given, rather than produced by humans. In other words, it is food that is not entirely appropriated by human management or control.[1] This does not necessarily mean that food as gift comes from a wild environment; a domesticated landscape such as a garden is described as a site in which food is perceived as gift, too. Ishimure's novel depicts people who fish as thankful for the fish and shellfish they catch in the nearby sea, which they affectionately refer to as their "garden," and stories collected in Alexievich's *Voices*, as well as one of Taguchi's novellas, illuminate the trust that those who resettled in an exclusion zone have in the harvest from their gardens. The sense of food as gift in a domesticated, toxic landscape is noticeable in other literary works as well. For instance, in *The Sky Unwashed*, a novel which narrates people's lives in a farm village in Ukraine before and after the Chernobyl nuclear disaster, an elderly woman, who returns to her homeland after it has been turned into a forbidden, radioactively contaminated zone, starts gardening in the no-man's-land in order to feed herself. Her garden is depicted as a source of life for her and the food produced there as a gift. Also, the garden was made possible by seeds the woman had preserved from her harvest prior to the nuclear accident and the resultant evacuation. The fact that the preserved seeds literally save her life helps remind us of seeds' reproductive ability and implicitly questions the validity of hybrid and genetically engineered seeds, especially those designed not to reproduce. In addition, the novel demonstrates a sense of humility provoked by the act of gardening: "Marusia believed that her life was saved because of the garden. Whenever she felt the unbearable loneliness weigh her down and the sinking feeling of betrayal, she worked in her little patch of land, and its capacity for life humbled her" (Zabytko 163). The perception of food as gift evokes a sense of human smallness in an interrelated and interdependent more-than-human world.

As opposed to the more common notion of food as commodity, the concept of food as gift does not involve assessing risk and safety when deciding what to eat and what not to eat. This does not mean that no victims in contaminated places questioned the safety of the food; in reality, as Sarah Drue Phillips discusses, in post-Chernobyl Ukraine people's attitudes to food were based on the opposing categories of "safe" or "contaminated." But the concept of food as gift, and the fact that some people do eat food grown in contaminated areas, provides ground for literary exploration of human relationships with the environment. Perhaps because it involves an assumption of human connectedness with the environment, the idea of food as gift could be dismissed as nostalgic or naively romantic. But the fact that it continues to resurface in contemporary literature suggests that there are larger implications than simply an apparent endorsement of a pre-modern sense of belonging in a particular place.

Literary focus on an intense sense of interconnectedness between the human body and the environment seems to have some similarities with Stacy Alaimo's idea of "trans-corporeality" in her *Bodily Natures: Science, Environment, and the Material Self*. Alaimo argues that "as a particularly vivid example of trans-corporeal space, toxic bodies insist that environmentalism, human health, and social justice cannot be severed" (22). This holds true for the victims in Minamata as well as those downwind of Chernobyl and Fukushima. However, while Alaimo, and other scholars including environmental historian Nancy Langston, highlight the movement of toxic substances from plants and factories to the environment and into human bodies, the act of eating contaminated food as a gift, as described in the works of Ishimure, Alexievich, and Taguchi, emphasizes the trans-corporeal circulation of life.

Rachel Carson demonstrated how easily a web of life turns into a web of death once an environment is contaminated with toxic pollutants. Ishimure, Alexievich, and Taguchi show how the tightly knit relationship between people, food, and place yields both life and death, both health and illness, in an age of toxic environments. They describe people whose relationships with their surroundings are so strong that they continue to eat the food they grow and harvest themselves regardless of possible contamination. However, unlike Rachel Carson's *Silent Spring*, one of the first works to articulate a risk scenario of environmental contamination, the works of Ishimure and Alexievich, and in a more subtle way Taguchi's, do not emphasize risks in their depiction of people eating locally grown or harvested food while knowing that it is likely to be contaminated with toxic elements. This divergent emphasis reflects the difference between focusing on the trans-corporeal movement of toxic substances and focusing on the trans-corporeal circulation of life.

There is no doubt that scholars in environmental humanities, such as Alaimo and Langston, have contributed considerably to the understanding of the interconnectedness between humans and the nonhuman environment by delineating a toxic flow without end. But at the same time, scholars in the environmental humanities should also attend to the life that continues to circulate among human and more-than-human bodies and environments, despite the presence of toxic substances. What I find significant in the works of Ishimure, Alexievich, and Taguchi is their literary attempt to articulate the intense degree of life evident in those who continue to eat food that they know is likely to be contaminated and to perceive it as a gift. The following excerpts are some exemplary passages that demonstrate the writers' attention to an extraordinary sense of life in toxic mealscapes. First, from Ishimure's *Paradise in the Sea of Sorrow*:

> Those who liked seafood in my village had been quick to find out that around Koiji Island, not far from the mouth of the wastewater channel at Hyakken, small sardines and *wakame* kelp were proliferating again, and that the folks from other villages who went there always returned weighed down by their catches. Whether polluted by mercury or not, *wakame* is one of the delicacies of spring. I decided to use it as an ingredient for a *miso* soup.
>
> (280–281)

According to social and medical accounts of Minamata disease, the majority of the victims, who lived close to the sea, knew that the water was contaminated *and* did not stop catching and eating fish, shellfish, and marine plants. The second to last sentence of the passage above in particular demonstrates the life-centered values of those who fish; these values are exemplified in their eating what they catch, which they perceive as "a gift from heaven," even though they sense that something is wrong with the sea. Also, the very last sentence— Ishimure's attempt to experience the foodways of the Minamata disease victims, the majority of whom depend on fishing—shows Ishimure's desire to understand their extraordinary trust and faith in life, indicating that what Ishimure tries to do as a writer is not really to speak *for* those people but to struggle to understand them. In other words, the notion of food as gift has such a big impact that the writer herself is unsettled and questions her own value system in order to articulate the underlying value embedded in the idea of food as gift.

Life-centered values similar to those of the Minamata disease victims permeate the personal accounts collected in Alexievich's *Voices from Chernobyl*. The following is from the testimony of a couple living in a radioactively contaminated village:

> [Y]ou weren't allowed to use anything from your garden in the first year, but people ate it anyway, cooked it and everything. They'd planted everything so well! Try telling people that they can't eat cucumbers and tomatoes. What do you mean,

"can't"? They taste fine. You eat them, and your stomach doesn't hurt. And nothing "shines" in the dark … In the beginning people would bring some products over to the dosimetrist, to check them—they were way over the threshold, and eventually people stopped checking them. "See no evil, hear no evil. Who knows what those scientists will think up!" Everything went on its way: they turned over the soil, planted, harvested. The unthinkable happened, but people lived as they'd lived. And cucumbers from their own garden were more important than Chernobyl.

(118)

The last sentence in the quoted passage clearly echoes the attitude of those in Minamata who are described as having a firm faith in life even when their home environment is contaminated. Statements such as "cucumbers from their own garden were more important than Chernobyl" in Alexevich's book and "whether polluted by mercury or not, *wakame* is one of the delicacies of spring" in Ishimure's novel shed light on a trans-corporeal flow of life materialized in the foodways.

Ishimure's and Alexievich's works demonstrate a way of living developed and matured in a particular place which is called home. Because of their unusual nonchalance toward risk and food safety, these works serve as ground from which to question common perceptions of and attitudes toward food in a risk society. While concepts of food safety are relative, still the idea of "safety" has a considerable impact and influence on public food consumption (Nestle). Ishimure's and Alexievich's literary approaches to a consistent faith in life in the mealscapes of toxic home environments unsettles modern common sense and urges readers to think about what they have faith in.

The life-oriented values represented in Ishimure's and Alexievich's works are lacking in Taguchi's work on Fukushima after the accident at Fukushima Daiichi. This is perhaps because, unlike Ishimure who was brought up in Minamata and has developed a sensitive understanding of its fishing community, Taguchi is not a native to Fukushima or a dedicated settler in that place. Also, unlike Alexievich, whose collection of oral histories represent victims' testimony, Taguchi writes fiction in which the author invents rather than recounts stories. Instead of foregrounding the values shared among those who consider food as gift, *In the Zone* displays the uncertainty that characterizes the narrators' behavior and thinking. Put in a different way, *In the Zone* demonstrates what Ursula Heise has drawn attention to as the ironies and ambiguity involved in nuclear risk scenarios. In her examination of German texts on the explosion in Chernobyl, Heise claims that "the complexity of the nuclear risk scenario arises not only from unfamiliar scientific concepts … but also from the unexpected double meanings and ironies it creates for nonscientific discourse," and continues, "a word such as 'radiation,' for example, acquires an odd ambiguity as it refers both to the radioactivity that might cause cancer and the procedure used to fight it" (184). Likewise, Taguchi's novellas show the ambiguous nature of such concepts as "radiation," "zone," "environmental consciousness," and "life." For instance, in a story entitled "A Paradise for Cows," radiation is depicted as a hopeful material for cancer treatment *and* the cause of suffering for ex-residents in areas designated as an exclusion zone. In the stories "In the Zone I" and "In the Zone II," Taguchi allows the very notion of zone to become porous, suggesting that it could be either a dystopia that is filled with the terrible smells of the corpses of livestock or a utopia characterized by the intensity of life. Unlike Ishimure and Alexievich, Taguchi does not appear to shun nuclear risk scenarios. Instead, her narrative negotiates conflicts among the words and concepts that characterize the filters through which we see the world.

In the process of finding such balance, the act of eating is highlighted. "Guinea Pigs," the last novella collected in *In the Zone*, for instance, tells a story of an elderly woman who remains in a nuclear exclusion zone in Fukushima, in a former commune for Japanese hippies that everybody else has left. Like the downwinders of Chernobyl whose trust in, dependence on, and appreciation of the generative power of land are powerfully described in literary works such as *Voices from Chernobyl* and *The Sky Unwashed*, the old woman in Taguchi's novella cultivates a garden, grows vegetables, hunts animals, and gathers edible grass and mushrooms. The following passage is narrated in the voice of a young woman who is attracted by and starts to live with the elderly woman and continues to stay there for seven years until the old dweller dies:

> In Nekozoko [a fictional village in a nuclear exclusion zone], I ate everything but the mushrooms, an act that perhaps helped me keep a "balance" with which I could feel safe.
>
> I ate a variety of things, except for mushrooms. It was this place that had nurtured me for the past seven years … It occurred to me that all animate beings are part of the earth, for they live by eating and being eaten by each other on this planet. Each being has its own law, and what is important is to keep a good balance; if you keep your balance, you'll be able to hold yourself on a thin rope.
>
> (224; my translation)

Here, again, ambiguity characterizes the narrative: the long-term inhabitant's total acceptance of her life in the zone is contrasted with the young person's hesitation that eventually marks her decision to leave the zone. As is illustrated in the quoted passage, the elderly resident eats everything grown, hunted, and gathered in the zone, but her younger disciple avoids mushrooms, which are commonly thought to absorb radioactive substances at higher rates than other foods such as grains and root vegetables. The faithful inhabitant and her young disciple appear to share common values, and that is why they have a peaceful life together in a contaminated and forgotten place; and yet the difference in their criteria for food to eat suggests that they stand out as individuals.

In fact, each of the four novellas collected in *In the Zone* highlight the commonalities and differences between a pair of characters. In the first two novellas, Taguchi compares and contrasts a victim in Fukushima and a visitor from Tokyo; in the third story, "A Paradise for Cows," an evacuee from a nuclear exclusion zone and a man under radiation treatment for cancer; and, in "Guinea Pigs," an ex-hippie dancer who became a lone resident in the zone and her young disciple. The conceptual negotiations and dialectic interactions involved in Taguchi's representation of these paired characters eventually suggest a foothold from which to accommodate the world of human control and domestication of physical environments. Domestication accelerates a sense of separation between humans and biophysical nature on the one hand, and emphasizes inevitable connections between them due to prevailing human activities on the other hand. The young woman's decision to eat everything but mushrooms is not so much an indication of the victory of risk scenarios as a shifting, contemporary version of food as gift in the age of toxic environments.

It is easy to claim that the idea of food as gift is outdated and obsolete; in fact, literary representations of food as gift, such as that in *Paradise in the Sea of Sorrow*, tend to be interpreted as characteristically pre-modern. But, as is implied in Taguchi's work, the act of eating contaminated food and perceiving it as a gift—that is, the act of eating life—challenges our complacent norms of food safety and thereby opens up a conceptual realm of human interactions with the environment. The idea of food as a gift, even in toxic environments, is not a thing of the past, but one that keeps turning up in our thoughts, values, and beliefs, however subtle it may seem. Literary examinations of meals that reflect a belief in food-as-gift help expose the ambiguities of domestication, suggesting the need for reflection and reconsideration of individual and societal attitudes toward the environment, which includes the places we call "home."

Note

1 My usage of "food as gift" and "food as commodity" is not necessarily the same as sociologist Deborah Lupton's in her 1996 book *Food, the Body, and the Self*. Lupton analyzes food as gift mainly in a social context of human relationships with other humans, especially in a family context (47–49). In my usage, however, food as gift refers to that which is produced beyond human control and management. It hardly concerns social relationships with other humans; rather, it suggests human attitudes to the more-than-human world. Although I share with Lupton a belief that food as gift is characterized by acts of love, caring, and duty, my focus is more on human–nature relationships than on human–human relationships. Likewise, Lupton focuses on food as commodity in terms of the construction of subjectivity, discussing how people purchase and consume food not only for its edibility but also for inedible cultural values and social roles (22–24). My usage of food as commodity does not go that far, simply referring to food as a consumer good, something which people choose, buy, and consume.

References

Alaimo, Stacy. *Bodily Natures: Science, Environment, and the Material Self*. Bloomington: Indiana University Press, 2010. Print.

Alexievich, Svetlana. *Voices from Chernobyl: The Oral History of a Nuclear Disaster*. 1997. Trans. Keith Gessen. New York: Picador, 2005. Print.

Carson, Rachel. *Silent Spring*. 1962. Boston: Houghton Mifflin, 1994. Print.

Heise, Ursula K. *Sense of Place and Sense of Planet: The Environmental Imagination of the Global*. New York: Oxford University Press, 2008. Print.

Ishimure, Michiko. *Paradise in the Sea of Sorrow: Our Minamata Disease*. Trans. Livia Monnet. 1990. Ann Arbor: Center for Japanese Studies, University of Michigan, 2003. Print.

Kareiva, Peter, Sean Watts, Robert McDonald, and Tim Boucher. "Domesticated Nature: Shaping Landscapes and Ecosystems for Human Welfare." *Science* 316.5833 (29 June 2007): 1866–1869. Web. 1 September 2015.

Langston, Nancy. *Toxic Bodies: Hormone Disruptors and the Legacy of DES*. New Haven: Yale University Press, 2010. Print.

Lupton, Deborah. *Food, the Body and the Self*. London: Sage, 1996. Print.

Nestle, Marion. *Safe Food: The Politics of Food Safety*. Berkeley: University of California Press, 2010. Print.

Phillips, Sarah Drue. "Half-lives and Healthy Bodies: Discourses on 'Contaminated' Food and Healing in Post-Chernobyl Ukraine." *The Cultural Politics of Food and Eating: A Reader*. Ed. James I. Watson and Melissa I. Caldwell. Malden: Blackwell, 2005. 286–298. Print.

Pollan, Michael. *The Omnivore's Dilemma: A Natural History of Four Meals*. New York: Penguin, 2006. Print.

Steinbeck, John. "Breakfast." 1938. Rpt. in *The Portable Steinbeck*. Rev. ed. New York: Viking, 1971. 417–420.

Taguchi, Randy. *Zōn nite [In the Zone]*. Tokyo: Bungeishunju, 2013. Print.

Williams, Terry Tempest. *Refuge: An Unnatural History of Family and Place*. 1991. New York: Vintage, 1992. Print.

Zabytko, Irene. *The Sky Unwashed*. Chapel Hill: Algonquin, 2000. Print.

6

HYBRID AVERSION
Wolves, dogs, and the humans who love
to keep them apart

Emma Marris

Wolves are, for many people, the living embodiment of wildness. Dogs, on the other hand, are the nonhuman animal we have most closely adopted as one of our own: man's best friend, the apogee of domestication. And yet, the two creatures can and do mate and have fertile offspring. The wolf's scientific name is *Canis lupus*. The dog's scientific name is *Canis lupus familiaris*. That is to say: they are the same species.

It startles us that two kinds of creatures we perceive to be in polar opposition as symbols are in fact the same species. I am curious about what we can learn from the cultural confusion, anxiety, and strange human behavior that results when wolves and dogs do mate.

Human reaction to the event seems to depend partially on whether the mating was controlled and wished for by humans or the result of the animals' own actions and desires. In the case of human-mediated matings, the resulting hybrid "wolfdog" is generally sold as a pet.[1] In this case opinion divides, but the majority of people seem unnerved and distrustful of the result. For matings outside of human control, the disapproval seems nearly unanimous.

Why such aversion? I believe the answer lies in our fetishizing the wolf as the preeminent icon of wildness. As the symbol of wildness, wolves must be absolutely free of human influence—pure. By mating them with dogs, we sully this wild purity with grubby DNA from the dog, the icon of domestication.

Wolves as *symbols* of wildness are so culturally important that we humans will go to great lengths to protect their purity, even if doing so involves restricting the freedom of actual animals. Thus we create a paradox: in order to protect the wildness of the wolf, it must be controlled, but wildness is often defined as that which is not controlled.

Conservationists, in particular, seem to value what we might call "wild DNA"—a genome untainted with domesticated genes—above behavioral or functional wildness of the actual animals.[2] The very existence of the field of "wildlife management" suggests that we are remarkably comfortable exerting some kind of control over organisms and places we designate as "wild."

I will present some case studies involving free-ranging and captive wolfdogs below, many of which pit different kinds of wildness—behavioral freedom, DNA wildness, aesthetic wildness—against one another. Behavioral freedom never seems to come out on top. When push comes to shove, we prefer animals that look wild to those we can not control.

Finally, I will discuss the ethics of purity policing in animals. It is likely that the value that humans place upon "wildness" has hurt many individual animals.

Thelma and Louise

In the winter of 2013–2014, two wolves in Northeastern Washington State started "frequenting a residence that had a dog," according to Donny Martorello, the Carnivore Section Manager for the Washington Department of Fish and Wildlife (WDFW). The two wolves were both females and probably young wolves dispersing from the nearby Ruby Creek pack. Washington State had at the time just fifty-two wolves in thirteen packs and were watching the population carefully ("Gray Wolf").

Martorello's agency worked with the family who owned the dog to install a 9-foot fence around its yard. Still, the female wolves hung around and were felt to be "coaxing" the dog out. Finally, the 90-pound sheepdog managed to scale the fence and, in Martorello's words, "decided to be a wolf for a month." Local ranchers named the two females Thelma and Louise, after the characters in the 1991 film.

WDFW officials did not want the dog to impregnate either of the female wolves. For one thing, dogs are property, whereas wolves are an endangered species. The rules and regulations governing them are wildly different, and neither set of rules is clear about how to handle hybrids. For another thing, the "genetic integrity" of the wolves was at stake. "Genetic integrity" is a phrase commonly used in the conservation biology literature to evoke the value of a genome that is "pure" and not "polluted" with the genes of related species or subspecies. Typically, the notion that a pure genome is intrinsically valuable is taken for granted, despite the fact that there is little clear ethical support for this idea (Rohwer and Marris).

As Martorello put it in this case, "What the [Endangered Species Act] and the state counterparts try to preserve is that natural selection process without diluting that or tainting that with a set of genes that were selected by humans."

One of the female wolves was wearing a radio collar as part of the state monitoring program. This made it a relatively simple matter to find the two wolves and the dog and, using helicopters, dart the wolves with tranquilizers. The wolves were examined (while the dog stood by, furiously barking and guarding the females) and one ("Thelma") was found to be pregnant.

After consultation with the state wolf advisory group, the decision was made to recapture the pregnant wolf and spay her—that is, to remove not only the hybrid embryos but also all reproductive parts. Martorello explained that spaying was deemed to be safer and easier than an abortion.

The female was thus recaptured by helicopter, knocked out, and her embryos and womb removed. The dog was returned to its owners. A few months later, the now sterile female Thelma was struck by a car and killed.

Louise returned to the house where the sheepdog lives. Efforts to drive her away from the dog's enclosure proved unsuccessful. "She thinks that male dog is a pack member," says Martorello. "Over the summer and early fall we did everything to haze the animal: rubber bullets, chasing it, trying to catch it up. We couldn't do it. She is still a wolf and she is wily."

In February 2015, Louise was knocked out by a tranquilizer dart shot from a helicopter and moved to Wolf Haven, a sanctuary for pet wolves and wolfdogs near Olympia, Washington. Here she will spend the rest of her life in captivity. She has been "partially altered" so she can never breed, and placed, for company, with a similarly altered male.

Wendy Spencer, the director of animal care at Wolf Haven, told me something that suggested that by hanging out around the dog for so long, Louise had lost some of her inherent wildness, at least in the eyes of the humans around her. "Based on all the reports we have from the field about how habituated she was, she is probably a good candidate for life in captivity," she says. "It was that or be shot."

The primary reason for intervening in the hybrid pregnancy and in removing the second female to captivity was policy-based. WDFW has a mandate to protect the gray wolf as an endangered species; hybridization would create animals of uncertain status that would entail uncertain obligations, possibly including the obligation to kill the pups as a threat to the genetic integrity of nearby wolf packs or as vermin.

However, the decision also potentially reveals certain values held by the agency and, given the incident's quiet reception by the conservation community and the public, possibly hints at wider attitudes. It suggests that they value a "wild" genetic lineage uncorrupted by domesticated genes more than they value the instantiation of wildness in unrestrained behavior. It also suggests that maintaining the distinction between wolves and dogs is very important to the agency. A wolfdog in the wild is worse than useless; it is the seed of chaos, the harbinger of the hybrid swarm, in which all order breaks down and wolves lose their value.

Free-ranging wolfdog hybrids

The history of interbreeding between wolves and dogs naturally stretches back to the origin of dogs, since the domestic dog derives from the wolf, as many as 40,000 years ago (Skoglund et al). Whether humans domesticated dogs by intentionally breeding tractable wolves or whether the animals domesticated themselves because of the existence of a fruitful ecological niche in human spaces remains up for debate (Hare and Woods).

Since the domestication of the dog, wolves and dogs have mated and produced offspring on their own. The coat color of North American black wolves derives from a trait that first evolved in the domestic dog. Thus at least some wolfdogs were able to reproduce in the wild, and a substantial percentage of wild American wolves probably have a dog ancestor or two.

More recently, Italian wildlife managers have had to cope with what they call a "hybrid swarm"—a pattern of matings between dogs and wolves so pervasive and long-standing that most Italian "wolves" are wolfdogs of varying percentage. Genetic studies suggest many of the hybridization events took place generations back, as the Italian wolf population expanded into new areas, which in turn suggests that such events may increasingly become an issue in Northern Europe and North America, where wolf populations are actively expanding into new territories filled with humans and dogs (Randi et al.).

Luigi Boitani has studied Italian wolves for decades, and says that, as you might expect, there is no black-and-white prescription for dealing with hybrids. "Of course, you do not want to kill all the black wolves of North America, that would be ridiculous," he says. "But you would probably want to intervene if you saw a wolf mating with a dog in your garden. Between these two extremes, there is a continuum. And it is very hard to draw a line."

Boitani cheerfully admits that wolves and dogs are the same species, that they are very difficult to tell apart, even with genetic tools, and that hybrids do as well or better than wolves on the human fringes where they live in Italy. In 1975, he radio-collared three free-ranging hybrids and followed them for a year. "They behaved 100% like wolves," he says. They defended home ranges; they killed livestock. The only difference was that the male dogs made lousy fathers, not bringing bellies full of meat back to the den to regurgitate for the pups in their early months. Nevertheless, "a she-wolf that mated with a big sheep-guarding dog" that Boitani observed raised six pups all alone. All dispersed and went on to lead wolfy lives.

And yet, he wants to weed them out. "I like to think that I am a pragmatic person," he says. "I would intervene if you have an isolated case. The Finnish did that. They had one case and they went in and killed everything. If you have hybrid swarms, where the level of introgression is widespread, you can't go back unless you kill everything. You can't send the army in the forest and kill everything they find. You have a huge ethical and legal issue."

So what do you do? You just kill the ones that look like dogs, Boitani says. "If they look like wolves, if they behave like wolves, if they have an ecological function like wolves, then leave them. On the other hand, if you have a wolf that looks like a dog, then I would have a problem. I would like to have around a wolf that looks like the wolf and not polluted with our dogs."

Here, the recommended management strategy is to keep the wolves looking like wolves. The aesthetic appearance and ecological role of the wolf trump both the instantiation of wildness in the freedom of individual wolves and "wild DNA."

Free-ranging hybrids have, if possible, a worse reputation than wolves. "There is a wide-spread fear among rural publics that hybrids are more dangerous than real wolves—that they are as big as a real wolf but lack the shyness and the fear of a wolf and that that would make them more dangerous," says carnivore specialist John Linnell, who adds that whether folk wisdom is right or not has not been thoroughly examined by science.

And why not? Because biologists prefer to study real wolves. "[Hybrids are] not sexy. It is not the Romantic world; it is not the sexy world. And lots of biologists basically don't like people and civilized places; they want to go out into the purer wilderness," Linnell says. "And people who like livestock and tame animals don't like hybrids and wild dogs because they are not tame. People reject them from both sides."

Perhaps the only people who embrace them are those who invite them into their homes.

Wolfdog hybrids as pets

Domestication of dogs from wolves could have occurred multiple times, and in the early days, there was likely much backcrossing with wolves (Lescureaux and Linnell 234). Throughout history, dogs and wolves have bred, sometimes at the will of humans—as when the ancient Gauls tied their female dogs to trees to make them available to local male wolves (235). Several breeds of dogs are known for recent additions of wolf genes. In their review of the relations between wolves and dogs, Nicolas Lescureux and John Linnell mention "the Saarloos wolf dog, the Czechoslovakian wolf dog, the Lupo Italiano, [and] the Kunming wolf dog" (234).

Humans continue to breed wolves with dogs intentionally, and wolfdogs are available for sale in the states where they are legal (and in the states where they are illegal). There are more captive wolves and wolfdogs in the United States than wild wolves (Terrill 251). According to an article on wolfdogs as pets in the doglover's magazine *The Bark*, the appeal of the wolfdog is the idea of "a dog's friendly companionship paired with a wolf's good looks and untamed nature. Buy a wolfdog, the thinking goes, and live out your Jack London fantasies, even if you're in Akron rather than Anchorage" (Connors). The piece goes on to quote the author of a guide to owning wolfdogs on the motivations of her readers: "They want to own a piece of the wild, and they often say that the wolf is their spiritual sign or totem animal." The paradox of "owning" part of the wild must either appeal to such people or not occur to them.

Wolfdogs are much frowned upon, however, chiefly because it is felt that their temperaments are poorly suited for life as pets, creating a miserable situation for animal and owner. Here is a list of traits typical of "high content" wolfdogs, given by Howling Woods Farm in New Jersey, a facility that rescues the animals. Wolfdogs, they say,

> react poorly to standard dog training; may dig large holes in their pens or the backyard, especially if bored or tied down; can jump or climb a six-foot high fence; are smart and learn commands easily, but often decide to ignore them; will roll over and over again in the most obnoxious smelling substance known to the human nose; do not respond to discipline the way most dogs do. (Training a wolf dog is about as "easy" as training a house cat. You must earn their respect if you expect them to listen to you.) [They] do not like to be alone (they need a canine or human companion); require a high protein diet; may be fearful of people outside of the family pack; require a very high amount of socialization, often and repeatedly, from a very young age through adulthood, if you expect to take them to public places, a dog park, or even for a walk; are not good off leash; bark very little, but boy do they howl.
>
> (Hodanish)

In addition, wolfdogs are nineteen times more likely than a dog to kill a human (Terrill 251).

Decidedly, wolfdogs seem like poor pets.

Diane Gallegos, the Executive Director of Wolf Haven, where Louise lives now, told me that she speaks out against breeding wolves and dogs, despite friction it causes with donors and supporters of the sanctuary who own and love wolfdogs themselves. The sanctuary has several wolfdogs that owners could not manage. One ripped the bumper off her owner's SUV before the owner cried uncle. Gallegos estimates that 80 percent of wolfdogs are dead by age two—euthanized because owners could not handle them and shelters would not take them. Gallegos adds that many wolfdog owners are young men who are interested in a "macho" pet, or women who see wolves as their "spirit animal."

The Bark itself appended to the piece on wolfdogs as pets an editorial note in which the magazine comes out against wolfdogs, saying that "despite their undeniable beauty and appeal, deliberately breeding or purchasing wolfdogs as companion animals does a disservice to both *Canis lupus* and *Canis lupus familiaris* as well as to the individual animal."

And in the comments section on the article, a reader named Sasha comments, "[W]olf dogs are not just some breed of dog, they are the result of humans reckless [*sic*] mixing a wild and domesticated animal without knowing the outcome. Essentially, people are recklessly undoing thousands of years of domestication." In her comment, it is possible to read something more than a pragmatic objection to the animals as poor pets. There's a sense in which Sasha is saying that it is wrong to mingle the wild and the domestic.

The reaction of many to wolfdogs is to see their very existence as a tragedy. Patricia McConnell, an animal behavior expert and author, writes on her blog that "they can't live in the wild, and many of them can't live with us in our homes, and so they are trapped in a never-never land of never being comfortable in their surroundings."

Zweiweltenkind

Ceiridwen Terrill had just ended an abusive relationship with a man when she began one with Inyo, a wolfdog she raised from a pup. All was not smooth sailing, as she explains in her 2011 memoir, *Part Wild*. The hybrid was almost impossible to live with, a *Zweiweltenkind*, a child of two worlds. Because she raised it, rather than wolves, it would have been helpless if released into the wild. But it was unhappy at home too, relentlessly trying to dig into the wooden floor, eating electrical cords, escaping and keeping neighbors awake with its howls. It constantly tried (and often succeeded) to escape from its kennel and yard, but it had nowhere to go. Eventually, as Inyo began to bite dogs and people, Terrill was forced to euthanize her.

Our present Western culture famously loves categories and especially dualities. Man and Nature. East and West. Wild and tame. Wild and domesticated. In large part, what seems to be a near-universal horror of wolfdogs, both captive and free-roaming, must spring not only from the real danger they can pose and the real suffering they feel in captivity but also from a sense of displeasure at a blurring or breaking of the wild versus tame duality.

But I think another fear in the case of free-roaming wolfdogs is that boring old dogs will swamp the exciting, rare wolf and we will have to live in a poorer world, where the deer are taken down by some kind of mangy, flop-eared wolfdog (probably with a healthy splash of coyote) that is not distinct enough from Fido to excite our admiration.

What would have happened if the owners of the big sheepdog in Washington and WDFW let the dog answer his own *Call of the Wild* and form a pack with Thelma and Louise?

Certainly, their children would have been *Zweiweltenkinder*, and if they ever stepped beyond the bounds of space and behavior we have (unilaterally) set for wild animals, it would have been more confusing for us humans to know what to do. Was it a feral dog that killed this sheep, and should it now be taken to the pound or put down? Was it a wolf that rather needs to be hazed away from all things human and reminded of its wild role?

But what if the hybrid family never crossed those lines? What if they just existed out there in the rolling mountains of the Colville National Forest, breasting snow drifts in pursuit of a yearling from the last US herd of caribou, denning in spring and playing with new pups each summer? Would these animals, socialized by their mother in the ways of wolfkind, be as happy as pure wolves? Certainly, we humans would not be harmed at all by their lack of purity. They would be wild. But would we humans be less excited to see them than a pack of pure wolves? Thelma and Louise were black wolves, the dog white. Would their pups look wolfy enough to make us feel we were seeing wild wolves?

Wildness at what cost?

There are ethical problems with the purity policing we humans engage in with wolves and dogs. In some cases, such as that of Thelma and Louise, it seems clear that individual animals suffered—or at least had their own desires thoroughly thwarted—to keep our human categories neat and tidy. (Though there is also a compelling argument that not intervening in that case risked very poor outcomes for all animals involved due to the risks of becoming habituated to humans: being run over by cars, being shot, etc.)

Current wildlife laws and policies make it difficult for state and federal officials to let the creatures alone, but I think it can plausibly be argued that aggressive efforts to prevent free-ranging animals from hybridizing are ethically dubious and also reduce the very wildness we so admire in these animals. I do not pretend to have a clear answer for how such cases should be handled, but I do think we should open a discussion on the topic, as it seems quite likely there will be more such decisions to be made as wolves re-expand across a crowded continent.

Conversely, *forcing* hybrid breeding for the pet trade seems to be a much clearer case: it is ethically wrong, as a very high percentage of hybrid pups kept as pets die young and have a frustrated, unfulfilling life.

Finally, we must ask ourselves whether the very category of "wild" is a value that serves the animals in which we seek it. If wolves are to be controlled by humans so they can remain purely wild, paradoxically, does it improve or reduce the quality of their lives? If they are culled because they look too doglike or are bred as pets because they are appealingly wild looking, it seems that they suffer precisely *because* of our fondness for wildness.

After all, "wildness" is not an entity that can feel pleasure or pain or have desires. We might value it, but, as an abstract concept, it does not have rights itself, nor can we owe wildness any obligations. But we do owe the wolf and the dog, Thelma and Louise and Inyo and the many others. It is right to first ask how our actions will affect actual creatures prior to worrying about whether wildness will be increased or reduced by our actions. This does not mean we must retire wildness as a valuable property. But it does suggest that care for sentient individuals might well trump it. As pet owners or wildlife managers, we might want to "own" or "protect" the wild, but at what cost to wolves, dogs, and wolfdogs themselves?

Notes

1 Some object to the term "hybrid" being used to apply to wolfdogs, since the two parents are technically of the same species. However, I believe the word "hybrid" is commonly used for a much wider range of mixtures and mash-ups beyond two species, so I will use it here. It is certainly often used by scientists and laypeople alike to refer to the offspring of wolves and dogs.
2 I originally wanted to call this "genetic wildness" but that phrase is already in use to refer to the lack of tameness that is encoded in a wild animal's genes and cannot be overcome by socialization. Wolfdogs make poor pets because even intensive, thoughtful socialization cannot overcome genetically coded traits like low thresholds for stress hormone release and, perhaps, fear and instinctive avoidance of humans.

References

Boitani, Luigi. Personal Interview. 27 January 2015.
Connors, Martha Schindler. "Do Wolfdogs Make Good Pets?" *The Bark*. November/December 2010. Web. 10 February 2015.
Gallegos, Diane. Executive Director, Wolf Haven. Personal Interview. 13 December 2014.

"Gray Wolf Conservation and Management: Wolf Packs in Washington." *Washington Department of Fish and Wildlife*. Washington Department of Fish and Wildlife, 2015. Web. 1 February 2015.

Hare, Brian and Vanessa Woods. "Opinion: We Didn't Domesticate Dogs. They Domesticated Us." *National Geographic*. National Geographic Society, 3 March 2013. Web. 10 February 2015.

Hodanish, Michael. "About Wolf Dogs." *Howling Woods Farm*. Howling Woods Farm, 2007. Web. 10 February 2015.

Lescureux, Nicolas and John D. C. Linnell. "Warring Brothers: The Complex Interactions between Wolves (Canis lupus) and Dogs (Canis familiaris) in a Conservation Context." *Biological Conservation* 171 (2014): 232–245. ScienceDirect. Web. 10 February 2015.

Linnell, John. Personal interview. 7 September 2014.

Martorello, Donny. Carnivore Section Manager, Washington Department of Fish and Wildlife. Personal Interview. 23 January 2015.

McConnell, Patricia. "The Tragedy of Wolfdogs." *The Other End of the Leash*. McConnell Publishing, 6 July 2013. Web. 10 February 2015.

Randi, Ettore et al. "Multilocus Detection of Wolf x Dog Hybridization in Italy, and Guidelines for Marker Selection." *PloS one* 9.1 (22 January 2014): e86409. Web. 10 February 2015.

Rohwer, Yasha and Emma Marris. "Is There a Prima Facie Duty to Preserve Genetic Integrity in Conservation Biology?" *Ethics, Policy & Environment*. Forthcoming.

Skoglund, Pontus, Erik Ersmark, Eleftheria Palkopoulou, and Love Dalén. "Ancient Wolf Genome Reveals an Early Divergence of Domestic Dog Ancestors and Admixture into High-Latitude Breeds." *Current Biology* 25 (2015): 1515–1519. Web. 19 November 2015.

Spencer, Wendy. Director of Animal Care, Wolf Haven. Personal Interview. 24 January 2015.

Terrill, Ceiridwen. *Part Wild: One Woman's Journey with a Creature Caught Between the Worlds of Wolves and Dogs*. New York: Scribner, 2011. Print.

7

TECHNO-CONSERVATION IN THE ANTHROPOCENE

What does it mean to save a species?

Ronald Sandler

An array of novel and interventionist species conservation and ecosystem management strategies have been proposed and, in some cases, pursued in recent years: assisted colonization, ecosystem engineering, Pleistocene rewilding, de-extinction, frozen zoos, gene-drives, and conservation cloning. At the same time, the idea that the Earth has entered a new geological age in which humans are the dominant force on the planet has been rapidly gaining proponents. These two trends are connected. Accelerated rates of ecological change caused by human activities are what endangers so many species and creates the need for more interventionist conservation; and that we have entered a period of natural history defined by humanity's influence is thought to create a responsibility to correct past wrongs as well as provide license for more actively designing the ecological future. We must take a more "hands-on" approach if we want to conserve species in the Anthropocene (Minteer and Collins; Camacho et al.; Donlan et al., "Re-Wilding North America"), and we need to become comfortable with that since we are now living in the Anthropocene (Marris; Ackerman; Ellis).

"The Anthropocene" is a deceptively thick concept. It is often presented as merely descriptive, and whether we have entered a new geological period is taken to be an empirical question. For example, a working group comprised largely of geologists, ecologists, and climate scientists has been convened to make a recommendation to the next International Geological Congress on whether the Earth is in a new epoch, a new age within the Holocene, or nothing new at all—the scientific facts about human impacts will settle whether we are in the Anthropocene. However, the concept also has an evaluative and normative valence (Crist; Caro et al.). To declare the Anthropocene would be confirmation of human power, influence, and dominance within ecological, climatic, and geological systems, and it would claim this period of natural history for us. It would be our time of influence, officially. How could we not take control of ecological and climatic systems in that case, when, after all, we already will have been? Is it not better that we design them rationally rather than destroy them thoughtlessly?

In this chapter, I critically assess the view that there is a responsibility to embrace more hands-on approaches to conserving species in the Anthropocene. In the first section I provide some background on the general case offered in support of interventionist species conservation. In the second section I suggest that rather than providing a creative and optimistic conservation vision as an alternative to mitigating extinction and loss, we ought to understand the "need" for these strategies as an indictment of the Anthropocene. I also argue that, in most cases, the strategies are actually inadequate for conserving species. While they can maintain individuals of endangered (or even extinct) species in existence, they typically cannot conserve what matters most about species, what makes them valuable, wonderful, and meaningful. I then suggest an alternative path forward that involves less human control and learning to live with loss and uncertainty: a path in which we do not embrace the Anthropocene.

Climate change, species extinctions, and conservation strategies

The background or historical rate of extinction is estimated to be less than one species per million per year (Baillie et al.; De Vos). On most estimates, there are ten to twenty million eukaryotic (plant and animal) species (Vié et al; Strain). Thus, a "normal" number of extinctions would be less than twenty extinctions per year. However, it is estimated that species extinction rates are currently thousands per year due to, for example, habitat destruction, extraction, pollution, and introduced species (Baillie et al; IUCN). Anthropogenic climate change is expected to increase species extinction rates still further. A recent study found that 24–50 percent of bird species, 22–44 percent of amphibian species, and 15–32 percent of coral species have traits that make them "highly vulnerable" to climate change (Foden et al.). An earlier study projected that 15–37 percent of species will be committed to extinction by 2050 on mid-level climate change scenarios (Thomas et al.), with significantly increased rates of extinction even on very optimistic scenarios (IPCC). Over all, the Intergovernmental Panel on Climate Change concludes that "a large fraction of species face increased extinction risk due to climate change during and beyond the 21st century, especially as climate change interacts with other stressors (high confidence)" (IPCC 10).

Macroscale anthropogenic change, and global climate change in particular, also undermines the historically dominant approaches to conserving species: creation of reserves and ecological restoration (Sandler, *Ethics of Species*). The difficulty for reserve-oriented conservation is that it depends on ecological systems in the protected location remaining sufficiently intact. However, global climate change represents non-normal spatial and temporal change, in comparison to the recent historical past (i.e., about the last twelve thousand years). As a result, current habitats and species assemblages are coming apart at unusually high rates, and they are doing so in ways and for reasons that cannot be addressed by ecosystem managers, since these changes are the product of global climatic processes and not stressors local to the system that might be reduced or eliminated (e.g., pollution, habitat destruction, and extraction). Thus, to the extent that species are climate-threatened, place-based or reserve-oriented protections will be less effective for preserving species and maintaining species communities.

As with place-based protection, anthropogenic climate change undermines traditional ecological restoration as an effective ecosystem management strategy. The difficulty for ecological restoration, given global climate change, is that the ecological past of a place is a less good approximation of its ecological future than it has been in recent history. Therefore, historical ecosystems (and associated reference conditions) will in general be less good proxies for ecological integrity, and native species less good proxies for what is ecologically beneficial or suitable. Too strong a commitment to historicity could actually be a form of insensitivity to ongoing ecological changes.

That reserve-oriented preservation and ecological restoration are undermined (or diminished) as effective species conservation strategies under conditions of rapid and open-ended anthropogenic ecological change contributes to a *species conservation dilemma*. On the one hand, the number of threatened species is dramatically increasing. On the other hand, the core ecosystem management strategies for conserving them are less effective.

The dilemma is being felt in conservation communities, as is evident in the robust discourse regarding assisted colonization. Many conservation biologists have begun to advocate for translocating individuals of climate-threatened species beyond their historical range in order to avoid extinction. Assisted colonization is controversial because it involves intentional creation of independent non-native species populations and is, therefore, in tension with maintaining historical continuity, non-intervention, and native species prioritization. Nevertheless, proponents of assisted colonization argue that given global climate change, "the future for many species and ecosystems is so bleak that assisted colonization might be their best chance" (Hoegh-Guldberg et al. 346), and unless such novel strategies are adopted, conservation biology will become a field of "managing extinctions" (Donlan et al., "Re-Wilding North America" 913–914).

The case for assisted colonization, rewilding, conservation cloning, and other interventionist conservation strategies is that if we cannot accomplish species conservation with the traditional ecosystem management paradigm, then we ought to give up the paradigm. On this view, anthropogenic climate change is the final "nail in the coffin" of park and reserve approaches to ecosystem management, since they are "mismatched to a world that is increasingly dynamic" (Camacho et al. 21; Donlan et al., "Re-Wilding North America"). Proponents of interventionism believe that unlike park and reserve conservation, which is doomed to mitigating and documenting loss, these strategies provide an optimistic, creative, and hopeful agenda for conservation under conditions of rapid anthropogenic change (Donlan et al., "Pleistocene Rewilding"; Brand; Ackerman).

The tragedy of the Anthropocene: from passenger pigeons to polar bears

It is estimated that when Europeans arrived in what is now the United States there were three to five billion passenger pigeons, constituting 25–40 percent of the bird population. However, habitat destruction and commercial hunting enabled by technological innovation and dissemination in the nineteenth century (e.g., guns, railroads, and the telegraph) resulted in a rapid decline in their numbers. The last passenger pigeon, Martha, died in the Cincinnati Zoological Garden just over one hundred years ago, alone and in a cage.

There is interest in reviving the passenger pigeon, bringing it back from extinction. This is de-extinction, and it may be possible by means of synthetic genomics and interspecific surrogacy. The plan is to reconstruct a passenger pigeon genome so far as possible using DNA fragments from existing specimens; fill in missing segments using genetic information from the closely related band-tailed pigeon; synthesize the genome; insert it into enucleated band-tailed pigeon stem cells that differentiate into germ cells, which would then be injected into developing band-tailed pigeons. If the band-tailed pigeons then mate successfully, the result would be offspring with a high level of genetic similarity to formerly existing passenger pigeons. Repeating this enough times with sufficient genetic variation could produce a founder population, from which a larger population could be rebuilt (Novak).

Proponents of passenger pigeon de-extinction believe that it would help to make up for the wrong done in causing the species to go extinct, as well as provide a boost to conservation generally (Brand). It would be an act of species recovery, rather than loss, which could galvanize public support for conservation "in a way that 40 years of doom and gloom has beaten out of them" (Ben Novak, quoted in Yeoman). Extinction would no longer be forever. As long as a species' genetic information and an appropriate surrogate persist, we could recover it (see Sandler, "Ethics of Reviving," for a comprehensive evaluation of the case for de-extinction).

Suppose that it is possible to bring back passenger pigeons in this way. Many conservation biologists doubt that it would be viable or advisable to introduce them into ecological systems in order to create independent populations (Ehrenfeld; Temple). One reason for this is that the form of life of the passenger pigeon—the way in which it made its way through the world—involved enormous flocks (on the concept of "forms of life," see Rose and van Dooren in this volume). It may not be possible to have a small flock of passenger pigeons that successfully breeds and resists predators. Another reason is that the ecological systems in North America have "moved on," such that many of the pigeon's food sources are no longer prevalent, for example. Moreover, accomplishing adequate genetic diversity and learned behaviors without existing birds as models (and without significant hybridization) are seen as enormously challenging. For these reasons, it is perhaps more likely that scientists recreate Marthas—small numbers of conservation-dependent birds, enclosed and monitored, a techno-scientific spectacle—than flocks of passenger pigeons migrating throughout the United States.

There already has been a "successful" de-extinction. The *bucardo*, an ibex subspecies that went extinct in 2000, was cloned by researchers in 2009 using genetic material from preserved tissue samples and a domestic goat surrogate. Most of the embryos created in the effort failed, but one was born alive. It survived for several minutes before dying from lung abnormalities (Folch et al.). The subspecies has now gone extinct twice. It is difficult to see this as a symbol of conservation hope, just as it would be if researchers managed to create more Marthas. They seem more like stark reminders of the wrongs that have been and continue to be done.

Now consider polar bears, a charismatic but climate-threatened existing species. Sea ice is a crucial component of polar bear habitat. The bears depend upon it as a platform for hunting seals and other marine mammals, which are their primary food source. As the climate has warmed, the sea ice has begun breaking up earlier in the year, so bears have less time to build up the fat reserves that they need during the period of food scarcity until the sea ice reforms. They also must swim longer distances between ice platforms, further depleting their energy reserves. The result has been decreases in the average body weight of the bears in some populations. This has, in turn, led to higher mortality rates, lower percentages of bears having litters, and smaller litter sizes. Consequently, bear numbers are declining in those populations (Molnár et al.; Regehr et al.). Because climactic change and not just local factors, such as hunting or mining, are driving the decreases in population sizes, local management plans alone are inadequate for protecting them in their current locations, since they will not limit increases in surface air temperature.

An amazing and distinctive animal is imperilled, and the greenhouse gas emissions from our industrial activities are the primary cause. It seems as if there is an ethical responsibility to help the polar bear. But it cannot be preserved by creating wildlife refuges alone, since this will not prevent the polar ice losses. If we want to preserve polar bears in the wild, we are going to have to do something more interventionist. Some have suggested delivering food to them to help maintain their body weights. Others have proposed translocating them to Antarctica. Still others have proposed freezing tissue samples so that they can be cloned and reintroduced when suitable habitat re-emerges.

However, each of the proposed measures would be inadequate. The reason for this is that they will fail to preserve what is most important about polar bears, their unique biological form and way of going about the world: that they are up north roaming the Arctic, hunting seals, and swimming frozen seas. The problem for polar bears is the same as the problem for passenger pigeons. There is not space for their form of life in the Anthropocene. This is the deeper problem with the Anthropocene. Not only are so many species at risk, but for many of them there is no way to conserve what matters most about them—that is, their unique form of life, evolved independently of humans. The human-modified planet is inimical to them, and probably permanently so. Large-scale extinctions have always been followed by expansions of biological diversity, but only over long periods and not with re-creation of the same species that went extinct. Climatic conditions, ecological systems, and evolution do not repeat themselves.

Elizabeth Kolbert captures this when relaying her conversations with researchers working on amphibian conservation in Panama. "Everyone I spoke to at EVACC [the El Valle Amphibian Conservation Center] told me that the center's goal was to maintain the animals until they could be released to repopulate the forests, and everyone also acknowledged that they couldn't imagine how this would actually be done. 'We've got to hope that somehow it's all going to come together,'" one herpetologist tells her. "'We've got to hope that something will happen, and we'll be able to piece it all together, and it will all be as it once was, which now that I say it out loud sounds kind of stupid.'" According to another researcher, "The point is to be able to take them back, which every day I see more like a fantasy" (Kolbert 14–15).

The idea that someday it will be viable to put polar bears back into their Arctic ecological system is a fantasy. The system is disappearing, and when it is gone it will not return. At some point, ice will reform at similar rates in that region of the planet, but when it does there will be a newly configured ecological system. If human beings are around, and if there are polar bear specimens in captivity or if it is possible to clone them from frozen tissue or synthesized DNA, introducing them into that system would likely be ecologically irresponsible. It would probably be bad for the polar bears (who would not be adapted to that system) and detrimental to other species that have co-evolved in it. The bears would be an exotic species.

Species are valuable in numerous ways. They have ecological value to the systems of which they are part. They provide ecosystem services and natural resources to us and to other species. They can have cultural and religious significance. They have intrinsic value in virtue of being unique accomplishments of human-independent evolutionary processes, each an irreplaceable component of the natural history of the planet. However, these types of value are not based only on the genomes or phenotypes of individual organisms. They depend as well on their ecological, evolutionary, cultural, and economic properties, which are relational (on valuing species relationally, see Whyte in this volume). The difficulty with interventionist conservation strategies is that, in general and for the most part, the intervention would fail to maintain or re-establish the crucial value grounding relationships.

De-extinction does not bring back the extinct species' habitat. Assisted colonization moves species beyond their historical ranges. Pleistocene rewilding introduces non-native proxy species to novel locations. As a result, they do not conserve or re-establish the value of the species—even if they keep individuals of the species (or their genomes) in existence. What is valuable and awesome are polar bears roaming the Arctic, passenger pigeons in massive flocks, and cheetahs stalking prey on the Serengeti.

There might be good reasons to engage in de-extinction, to clone endangered species, to create frozen zoos, or to translocate species, in some cases. It might be culturally significant, it might be scientifically productive, it might be economically beneficial, or it might be prudent, for example. But it is a mistake to think of these strategies as robust conservation activities (Sandler, "Value of Species" and "Ethics of Reviving"). These strategies do not conserve what is valuable about a species, and still less do they do so by addressing the causes of extinction, such as climate change, pollution, extraction, and habitat destruction—that is, the Anthropocene itself. Today the conservation problem is not that there are no passenger pigeons. It is that there is no longer a place for them within ecological systems. The same is true of the Yangtze River dolphin and Monteverde golden toad, as well as, sadly, probably the polar bear, Bengal tiger, eastern hemlock, and cloud forest orchids. These species are phenomena. Human activities destroyed them, or are in the process of doing so. If scientists create genetically similar organisms, engineer spaces where individuals of the species might persist, or store their genetic information in perpetuity, the phenomena are still lost. They are functionally and valuatively extinct. Moreover, the Anthropocene is detrimental to nonhuman life beyond that of threatened and extinct species. Recent studies estimate that the total global population of vertebrates has declined by 52 percent since 1970 (WWF) and that humans appropriate approximately 25 percent of biospheric or net primary plant production (Krausmann et al.). It is terrible that there are no passenger pigeons in the United States or fresh water dolphins in China. But what is even more terrible is that this is no longer a world for them.

What is the alternative?

Over time, some ecosystems may undergo state changes such that managing for resilience will no longer be feasible. In these cases, adapting to climate change would require more than simply changing management practices—it could require changing management goals. In other words, when climate change has such strong impacts that original management goals are untenable, the prudent course may be to alter the goals. At such a point, it will be necessary to manage for and embrace change.

(US Climate Change Science Program ch. 9, p. 3).

I argued above that highly interventionist species conservation strategies do not address the source of the problem with the Anthropocene, which is that it is a world in which many species and biological phenomena, such as migrations and bioabundance, have no place. The species' habitats no longer exist; the conditions they depend upon no longer obtain; they cannot adapt quickly enough. Interventionist strategies do not conserve what is valuable about them, and often presuppose a fantastical future. They also do not begin to scale to the size of the extinction crises. They are intensive efforts focused on individual or small numbers of species, when there are potentially tens of thousands of species at risk annually. I also suggested that the strategies are not really for the benefit of the target species, the individual organisms involved, or ecological systems. Instead, they are for those of us humans who would engage in and support them. They are about our remorse for what has been done, our techno-scientific ingenuity, our attachments to particular species, and the sort of ecological futures that we desire. They are more of the Anthropocene.

There is an alternative way "to manage and embrace change" besides pursuing techno-conservation. One aspect of this alternative is continuing to try to increase the resilience and adaptive capacity of species populations and communities. Although anthropogenic climate change diminishes the effectiveness of parks and reserves for preserving some species populations, species assemblages, and ecosystems as they currently are, they are likely to maintain comparatively high species preservation value when measured against non-protected areas. Protected areas provide some adaptive space, and so more adaptive possibilities, for populations and systems. Moreover, more biodiverse places, often the target of protection, are likely to have more species with sufficient behavioral and evolutionary adaptive potential to persist under conditions of rapid ecological change. Therefore, identifying and protecting biologically diverse and rich habitats (including diverse physical environments), crucial or productive wildlife corridors and ecological gradients (particularly "climate-connection corridors"), and promoting landscape permeability continue to be crucial under conditions of anthropogenic change (Smith et al.; Barnosky; Wapner; Rands et al.). In addition, familiar stressors of ecosystems and species populations (e.g., pollution, extraction, and habitat fragmentation) decrease their resistance and resilience. Reducing or managing such factors can increase the adaptive potential of species and ecosystems, again by removing anthropogenic impediments, rather than by more interventionist activities (NPS; CCSP). Moreover, lightly managed spaces will continue to have value—sometimes called "natural value"—in virtue of being places where ecological and evolutionary processes play out comparatively independent of human intention, design, and manipulation.

For these reasons, climate change is not the final "nail in the coffin" of the parks and reserve model for ecosystem management. However, appropriate goals for such places must shift away from preservation of particular species and assemblages to promoting adaptive capacity and allowing for ecosystem reconfigurations. Under conditions of rapid ecological change, place-based protection, rather than being valuable for maintaining a space largely as it is, is valuable for the processes of change that occur in it, including human-independent adaptation and reconfiguration. We must change our expectations about what these approaches can (and cannot) accomplish and the time-scales in which they can do so. We also must refrain from designing ecological systems as we think they ought to be, engaging in intensive efforts to prop up dwindling populations when the decline is associated with changing background conditions, and introducing species that we find particularly charismatic. Human-independent ecological and evolutionary processes and their products have been valuable in the past, and they will be so again in the future given enough time and space.

Managing for and embracing change also requires learning how to lose species and communities in cases where taking heroic measures to preserve them is not appropriate. The costs and risks associated with attempting to maintain them might be too high, and the attitude that we can engineer their continued existence as they were may be hubristic and ecologically insensitive. For this reason, the significance of reconciliation is increased under conditions of anthropogenic climate change. Reconciliation, in environmental contexts, is the disposition to accept and respond appropriately to ecological changes that, though unwanted or undesirable, are not preventable or ought not be actively resisted. Reconciliation has always been relevant to ecological practice. Even independent of global anthropogenic change, ecosystems are always dynamic, and individuals, species, and abiotic features are always coming into and going out of existence. Good ecological engagement requires accepting and not resisting too strongly such changes and losses. The increased rate and magnitude of ecological change and loss associated with climate change makes reconciliation still more necessary.

Reconciliation is not indifference. Species are rapidly going extinct, ecological relationships are being disrupted, and human activities are the cause. The result is an enormous loss of value in the world. Recognition of the magnitude of the loss and remorsefulness for it are appropriate. The fact that we are now at the point that we should not actively aim to prevent or undo some losses, and instead need to reconcile ourselves to them, is terrible. As David Foster puts it when reflecting on the fate of the eastern hemlock: "This is most definitely a tragedy … We will lose hemlock. We will lose our flagship for old-growth and primeval forest in the Northeast. We'll lose distinctive variation in our landscape. And we will lose the history and experiences embodied in these woods" (228–229).

In this chapter, I have raised several concerns regarding highly interventionist species conservation strategies. These strategies do not address the causes of extinction. They typically do not conserve what is valuable and matters most about species, that is, their ecological relationships and form of life. They can be relatively costly, harmful to individual organisms, ecologically risky, and difficult to accomplish. They are not scalable to the extinction crisis. Thus, even when they are justified, as they sometimes are, they are something of a conservation sideshow; and they have the potential to be a distraction. We need to decrease our impacts on ecological systems so that we do not "need" to create frozen zoos, engage in cross-continent translocations, and develop interspecific cloning. We need less of us, not more. We should not embrace the Anthropocene. The reality is that there is not anything approaching an adequate adaptation strategy for widespread extinction. From the perspective of the value of species and biodiversity, the only reasonable option is to try to reduce the scale of the problem, and the only way to do that is by consuming a smaller share of planetary resources by changing our consumption patterns, reducing our population, and innovating technologies.

References

Ackerman, D. *The Human Age: The World Shaped By Us.* New York: Norton, 2014. Print.

Baillie, J. E. et al. *A Global Species Assessment.* Cambridge: IUCN, 2004. Print.

Barnosky, A. *Heatstroke: Nature in an Age of Global Warming.* Washington, DC: Island Press, 2009. Print.

Brand, Stewart. "The Dawn of De-extinction: Are You Ready?" February 2013. TED Talk. Web. 1 August 2015.

Camacho, A. E. et al. "Reassessing Conservation Goals in a Changing Climate." *Issues in Science Technology* 26 (2010): 21–26. Print.

Caro, T. et al. "Conservation in the Anthropocene." *Conservation Biology* 26.1 (February 2012): 185–188. Wiley Online Library. Web. 1 August 2015.

Crist, E. "On the Poverty of Our Nomenclature." *Environmental Humanities* 3 (2013): 129–147. Web. 1 August 2015.

De Vos, J. M. et al. "Estimating the Normal Background Rate of Species Extinction." *Conservation Biology* 29.1 (April 2015): 452–462. Wiley Online Library. Web. 1 August 2015.

Donlan, J. et al. "Re-Wilding North America." *Nature* 436 (18 August 2005): 913–914. Web. 1 August 2015.

Donlan, J. et al. "Pleistocene Rewilding: An Optimistic Agenda for Twenty-First Century Conservation." *The American Naturalist* 168.5 (November 2006): 660–681. Print.

Ehrenfeld, D. "Extinction Reversal? Don't Count on It." TEDx DeExtinction Talk. National Geographic, Washington, DC. 15 March 2013. Web. 1 August 2015.

Ellis, Erle. "Stop Trying to Save the Planet." *Wired.* 6 May 2009. Web. 1 August 2015.

Foden, W. et al. "Identifying the World's Most Climate Change Vulnerable Species: A Systematic Trait-Based Assessment of All Birds, Amphibians and Coral." *PLOS One.* 12 June 2013. Web. 1 August 2015.

Folch, J. et al. "First Birth of an Animal from an Extinct Subspecies (Capra Pyrenaica Pyrenaica) by Cloning." *Theriogenology* 71 (2009): 1026–1034. Print.

Foster, D., ed. *Hemlock: A Forest Giant on the Edge.* New Haven, CT: Yale University Press, 2014. Print.

Greely, H. "De-extinction: Hubris or Hope." TEDx DeExtinction Talk. National Geographic, Washington, DC. 15 March 2013. Web. 1 August 2015.

Higgs, E. et al. "The Changing Role of History in Restoration Ecology." *Frontiers in Ecology and the Environment* 12 (2014): 499–506. Print.

Hoegh-Guldberg, O. et al. "Assisted Colonization and Rapid Climate Change." *Science* 321 (2008): 345–346. Web. 1 August 2015.

Intergovernmental Panel on Climate Change (IPCC). *Climate Change 2014: Synthesis Report.* Geneva: IPCC, 2014. Print.

International Union for the Conservation of Nature (IUCN). "Biodiversity." IUCN, 2011. Web. 1 August 2015.

Kolbert, Elizabeth. *The Sixth Extinction: An Unnatural History.* New York: Henry Holt, 2014. Print

Krausmann, F. et al. "Global Human Appropriation of Net Primary Production Doubled in the 20th Century." *PNAS* 110 (2013): 10324–10329. Web. 2 September 2016.

Marris, Emma. *Rambunctious Garden: Saving Nature in a Post-wild World.* New York: Bloomsbury, 2013. Print.

Minteer, B. and J. Collins. "Move It or Lose It? The Ecological Ethics of Relocating Species Under Climate Change." *Ecological Applications* 20 (2010): 1801–1804. Print.

Molnár, P. K. et al. "Predicting Climate Change Impacts on Polar Bear Litter Size." *Nature Communications* 8 (2011): 186. Web. 2 September 2016.

National Park Service (NPS). *National Park Service Climate Change Response Strategy.* Fort Collins, CO: NPS, 2010. Print.

Novak, B. J. "How to Bring Passenger Pigeons All the Way Back." TEDx Deextinction Talk. National Geographic, Washington, DC. 15 March 2013. Web. 1 August 2015.

Rands, M. et al. "Biodiversity Conservation: Challenges Beyond 2010." *Science* 329 (2010): 1298–1303. Web. 1 August 2015.

Regehr, E. V. et al. "Survival and Breeding of Polar Bears in the Southern Beaufort Sea in Relation to Sea Ice." *Journal of Animal Ecology* 79 (2010): 117–127. Pub Med. Web. 2 September 2016.

Sandler, Ronald. "The Ethics of Reviving Long Extinct Species." *Conservation Biology* 28 (2013): 354–360. Wiley Online Library. Web. 1 August 2015.

——. *The Ethics of Species.* Cambridge: Cambridge University Press, 2012. Print.

——. "The Value of Species and the Ethical Foundations of Assisted Colonization." *Conservation Biology* 24 (2010): 424–431. Wiley Online Library. Web. 1 August 2015.

Sherkow, J. and H. T. Greely. "What if Extinction Is Not Forever?" *Science* 340 (2013): 32–33. Web.

Smith, T. B. et al. "Biodiversity Hotspots and Beyond: The Need for Preserving Environmental Transitions." *Trends in Ecology and Evolution* 16 (2001): 431. Print.

Strain, D. "8.7 Million: A New Estimate for All the Complex Species on Earth." *Science* 333 (2011): 1083. Web. 1 August 2015.

Temple, S. "De-extinction: A Game-changer for Conservation Biology." TEDx Deextinction Talk. National Geographic, Washington, DC. 15 March 2013. Web. 1 August 2015.

Thomas, C. D. et al. "Extinction Risk from Climate Change." *Nature* 427 (2004): 145–148. Web. 1 August 2015.

United States Climate Change Science Program (CCSP). *Preliminary Review of Adaptation Options for Climate-Sensitive Ecosystems and Resources*. Washington, DC: Environmental Protection Agency, 2008. Print.

Vié, J-C., C. Hilton-Taylor, and S. N. Stuart (eds.). *Wildlife in a Changing World: Analysis of the 2008 IUCN Red List of Threatened Species*. Gland, Switzerland: IUCN, 2009. Print.

Wapner, P. *Living Through the End of Nature: The Future of American Environmentalism*. Cambridge, MA: MIT Press, 2010. Print.

World Wildlife Fund (WWF). *Living Planet Report 2014: Species and Spaces, People and Places*. Switzerland: WWF, 2014. Print.

Yeoman, B. "Why the Passenger Pigeon Went Extinct." *Audubon Magazine* May–June 2014. Web. 1 August 2015.

8

COLORING CLIMATES
Imagining a geoengineered world

Bronislaw Szerszynski

> Then one winter evening they were sitting on the westernmost bench, in the hour before sunset. … Maya looked up … and clutched Sax by the arm, "Oh my God, look," … Sax swallowed … "Ah," he said, and stared. Everything was blue, sky blue, Terran sky blue, drenching everything for most of an hour, flooding their retinas and the nerve pathways in their brains, long starved no doubt for precisely that color, the home they had left forever.
>
> (Robinson 672–673)

In a scene toward the end of the final book of Kim Stanley Robinson's trilogy on the settling and terraforming of Mars, *Blue Mars*, two surviving members of the first one hundred settlers, now ageing despite their longevity treatments, have got into the habit of sitting looking at the sky. Using color charts, Sax and Maya put names to the new colors that they see slowly emerging in the Martian sky, as the planet is altered in order to make it more hospitable to life. With their senses tuned by months of observation, they share wordlessly the moment when the color of an Earth sky finally appears.

In this chapter I will discuss the contemporary sociotechnical imaginary of climate geo-engineering. If in the Anthropocene our own planet, as much as any other planet we may come to inhabit, becomes an intended world, one in which the majority of once-natural processes and systems are not just accidentally but *deliberately* shaped by human action, what would be the best word to capture its climates? "Engineered," or "managed"? "Synthetic," "made," or "fabricated"? "Assembled," "composed," or "designed"? Each candidate word has a slightly different resonance and set of micromeanings. In his lecture to the 2008 conference of the Design History Society, Bruno Latour made a case for "design" as a fruitful way of describing human *poiesis*, making. He suggested that, unlike alternative words such as "building" and "constructing," the notion of design implies a modest, post-Promethean theory of action suitable for a context of ecological crisis. To support his case he highlighted five features of design: its humility; its attention to details; its focus on signs and meaning; its working with what already exists; and its normative distinction between good and bad design. According to Latour, then, the climate of an Anthropocene Earth would—or at least should—not be *made*, or *constructed*, but *designed* ("A cautious Prometheus?").

Yet Robinson's passage above suggests that the word "design" might not be quite the right one to orient our thinking about what it would mean to make climates. In its linguistic origins in the Italian *disegno*, the notion of design was deeply shaped by the rivalry in the early and middle Renaissance between two regional aesthetic approaches to the visual arts, those of Florence (*disegno*, "design") and Venice (*colorito*, "coloring"). To simplify a complex history, the Florentine style emphasised the artist's conceptual mastery and technical skill, while the Venetian style focused more on observation and the conveying of vivid reality. The art of Florence thus typically started with outlines, to which were added individual hues using fresco and tempera paints, whereas the artists of Venice such as Titian—influenced by its maritime openness to the Byzantine East as well as by its damp climate—more typically used oils to build a composition from diffuse patches of color in relationship to one other.

In *Blue Mars* it is striking that the moment when these two ageing humans are convinced that Mars has somehow reached an Earth-like state is not through temperature readings, atmospheric gas concentrations, or estimates of biomass but through an experience of color—and then, not a technical matching of light frequencies but a visceral shock of recognition, the kind of speechless wonder that the ancient Greeks called *thaumazein*. In the end it is not the designing of Mars—the measuring of future sea levels, the building of colonies, the engineering of ecologies—but the emergent coloring of an evening sky that convinces the body that Mars has become a dwelling place for humans. This is why debates about geoengineering need to be informed by an environmental humanities sensibility. If humans ever come to make Earth climates, then we will need an expanded way of thinking about two things, each of which should inform the other: what it means to make or compose something, and what it would be like to live in a made or composed world.

Geoengineering and the disciplines

Geoengineering, or climate engineering, is a prospective suite of techniques to control the climate in order to counteract the warming effects of raised carbon dioxide levels in the Earth's atmosphere. Global climate control, as opposed to local, short-term weather control, was promoted by scientists such as John von Neumann and Edward Teller after World War II, during a period of heightened technological optimism and global sensibility. But it was an article by Nobel Prize-winning chemist Paul Crutzen in 2006 and a subsequent Royal Society report in 2009 that put geoengineering firmly on the research and policy agenda (Crutzen; Royal Society). Since then, scientific and "grey" publications have grown steadily (Belter and Seidel; Oldham et al.). Proposals under discussion generally fall into two classes: some techniques are about extracting CO_2 from the atmosphere, either biologically or chemically; others are about making the Earth more reflective, whether by spraying sulphate particles into the stratosphere or brightening tropospheric clouds or the Earth's surface, and thereby reducing the amount of solar energy getting through to warm the atmosphere.

Geoengineering can provoke strong reactions, for reasons which have deep cultural roots. It can seem Promethean in the extreme to claim to be able to manage something as vast and complex as the global atmosphere (Hamilton). Proponents of the idea of a possible "good Anthropocene," in which humans would aspire to new levels of the Baconian project of the mastery of nature and an "age of humanity," often include geoengineering as a signature technology (e.g., Lynas). Geoengineering was shaped by a post-WWII, cold war, and ultimately military imaginary (Fleming "The Climate Engineers" and *Fixing the Sky*; Masco), and can feel alienating and anti-democratic in character (Szerszynski et al.). Perhaps prompted by these concerns, policy debates about geoengineering were quicker to try to incorporate nontechnical or "social" issues than those around earlier technological controversies (Schäfer and Low).

Yet the research literature is dominated by science and engineering (Oldham et al.), and although there have been a number of interdisciplinary research projects on geoengineering, their style of interdisciplinarity has generally meant that natural-science framings of the issue still dominate (Szerszynski and Galarraga). The critical social-science research that has been carried out is often motivated by a desire to correct this tendency, by pointing out the danger of adopting a narrow set of problem definitions and expert-analytical assessment methods (e.g., Bellamy et al.), or by seeking to widen debates through techniques of public engagement (for a summary, see Bellamy and Lezaun).

In the humanities, there have been publications about geoengineering in domains such as ethics (e.g., Jamieson; Gardiner), political theory (e.g., Dalby) and theology (e.g., Clingerman; Kearnes). Some leading public commentators on geoengineering have themselves drawn on philosophical and historical forms of reasoning to try to deepen the debate (Hamilton; Hulme, *Can Science Fix Climate Change?*). Generally, contributions from the humanities have been deeply suspicious of geoengineering as a sociotechnical imaginary. But some have argued that it opens up a potentially interesting way of thinking about planetary politics, with Nigel Clark suggesting that it offers an opportunity "to imagine a new kind of geologic politics in which identity, citizenship, and governance are construed ... in the relation to a dynamic and stratified earth" (Clark 2831).

What happens when you look at an issue that has been largely framed by science and engineering, using social science techniques, but informed by a humanities sensibility? In the rest of this chapter I will reflect on a line of work that explores geoengineering as the "making of worlds"—and not just in the sense that it would be the fabrication of climates. Geoengineering would involve us in a deeper complicity in processes of anthropogenic climate change in the guise of promising to halt it, and thereby engender a new kind of relation between humans and the weather (Szerszynski). As Bill McKibben has argued, the meaning of everyday weather events has already changed due to unintended anthropogenic climate change: "Yes, the wind still blows—but no longer from some other sphere, some inhuman place" (McKibben 44). How much more would that be the case with intentional "climate control"? How would geoengineering color our relationship with the air, and what sort of world would it bring about?

Coloring the sky

In 2012 Maialen Galarraga and I published a philosophical paper on geoengineering. Using the term "making" in an inclusive way to describe all forms of *poiesis*, the deliberate en-forming of matter, we tried to approach making as something that has to be grasped as a whole rather than decomposed into a set of technical questions on the one hand and a set of ethical, political, or aesthetic questions on the other. We drew on the philosophy of technology to develop three very different accounts of how making takes place, which we called respectively *production* (imposing existing forms onto matter), *eduction* (drawing forms out of the potentiality of matter), and *creation* (creating radically new forms by rearrangements of matter). We argued that each of these models of making implies a particular version of human agents and their responsibility, which in the case of geoengineering we called the "climate architect," the "climate artisan," and the "climate artist" respectively. These are not concrete individual people but archetypes that shape imaginations and actions and thus the way the future unfolds.

The *climate architect* is our name for the picture of the maker of climate that currently dominates the contemporary discourse of geoengineering. The climate architect is an idealised, imagined figure who "designs" in advance the form that they want the climate to take, who can identify the process whereby they can provoke the climate to take it, and who can carry out that process and bring the matter of climate into the desired form (for more on this model of making, see Simondon, *L'Individu et sa genèse physico-biologique* 48–49; Protevi 8). This way of thinking about making climates is encouraged by the centrality of computer models in climate science, including geoengineering research, which has the effect of rendering climate as pure information *in silico*—as form stripped of matter. This dematerialised, formal climate can then be imagined as something that can be recombined with matter and thus made actual.

But it is a mistake to imagine that a predetermined form can be imposed onto the metastable climate system. In Gilbert Simondon's well-known analysis of brick-making, the process of en-forming the clay with the help of a mold is dependent on the clay having been purified—for example, by removing any clots or stones that would act as "parasitic singularities" and disrupt the process of en-forming (Simondon, *Du mode d'existence des objets techniques* 42). Yet the climate, continually in formation, cannot be purified in this way. The uncertainties in climate models are not mere "noise" to be erased, but the result of potentialities intrinsic to the way that the atmosphere maintains and develops its form over time, in interaction with incoming solar energy and the other dynamic "compartments" of the Earth including the biosphere.

So second, we discussed an alternative way to imagine the making of climates, not "production" but "eduction": coaxing out the latent forms in matter. The imaginary figure who would make climates in this way, the *climate artisan*, would focus less on the final form to be taken by the climate than on the process whereby the en-forming of climate takes place. Adopting a greater humility toward the desires and tendencies of the more-than-human world, both biotic and abiotic, out of which climate occurs as a collective achievement, they would allow the "form" of the made climate to emerge out of their interactions with matter. They would thus emphasize recursive learning and treat computer models not as "truth machines" which reveal the future, but as experimental arenas in which the beginnings of a "feeling for climate" might be cultivated.

But we suggested that even the artisanal approach has its limitations. As ideal types, neither the climate architect nor the climate artisan is oriented toward the radical novelty that making climates might entail. So we developed a third figure, the *climate artist*. This was a deliberately provocative move, in that a climate artist might become a "climate auteur," even more hubristic than the climate architect about their capacity to design every feature of a geoengineered world, while also being unconstrained by the demands of technical effectiveness. But we tried to avoid this by loading onto this third figure an even greater set of responsibilities—not just to the final form of a made climate, or to the matter out of which it would be composed, but also to the way in which major technological innovations like geoengineering can fundamentally alter the human condition. In our normative typology the climate artist would thus approach the making of climates as an act of "creation" in the sense used by Cornelius Castoriadis. For Castoriadis, "producing" artifacts in the sense discussed above is not really making anything new but merely imitating, because it is simply re-producing existing forms (Castoriadis 197). It is only if we create a new form or *eidos* that genuine novelty or "ontological genesis" occurs. Understanding making as creation also requires us to understand time as genuinely historical; just as some works of art inaugurate a new way of looking at the world and even at earlier works of art, according to Castoriadis each new form of society is the emergence of a new *eidos*, or societal imaginary—a radically new way of organizing thought and action. Being a "climate artist" would thus involve not only an artisanal awareness of the need to collaborate with one's material, but also an artistic awareness that making climate would inevitably involve creating *climatically novel states*—climates and therefore biomes with no historical analogues—and also *historically novel states*: a new kind of society, with a new articulation of the atmosphere and how we relate to it.

The implications of this are profound. Both the climate architect and the climate artisan as we have imagined them assume that it is possible to maintain some kind of continuity and consistency between a goal formed in advance and the final achievement of a made climate. But the climate artist would recognize that the creation of a new *eidos* produces a historical rupture, a new context in which ways of thinking and forming intentions can be utterly transformed. Seeing geoengineering as creation means that it could never simply be judged as a means to an end, and thus as capable of being deemed "successful" or "unsuccessful" by criteria set in advance in the way that is envisaged by most scientists and policy actors. Instead, its deployment would have "changed the end in changing the means" (Latour, "Morality and Technology" 252); it would create a new kind of society, in which geoengineering would take on new meanings, be put to new uses, and be judged in new ways, and in which the very nature of climate, the sky, and the weather for us would be altered. This is where the sensibilities of *colorito* become relevant. Just as the great *colorito* paintings of the Renaissance, or the skies later painted by Turner, are concerned less with extensive outline and shape and more with intensive atmosphere, so too would the climate artist as we imagined them be concerned not so much with the achievement of specific climate parameters as with wider features of society, culture, and subjectivity in a geoengineered world. To reflect on these aspects of climate engineering one has to move, as it were, from Florence to Venice, from tempera to oils.

Reflecting (on) the sky

If Sax and Maya were to look at the sky of a geoengineered Earth, what would they see? Perhaps stratospheric aerosol injection would have turned the deep blue skies of rural areas into a Parisian-style white haze—but also have made dramatic fiery sunsets like the Krakatoa-induced one painted by Edvard Munch in *The Scream* entirely routine. Perhaps continuous sprays of seawater droplets would have turned the grey stratocumulus clouds that flank western-facing continental coastlines a dazzling white—or, more disturbingly, the leaden monsoon skies of South Asia may have become less common. But let us also expand our gaze beyond the literally visual: what of the wider "colors" of a geoengineered world? What kind of people and societies would live under and be illuminated by these altered skies? And might they view *all* weather events differently? I explored such questions in a project with the science and technology studies scholar Phil Macnaghten, in which we adapted public engagement methods in order to explore questions about the governability of geoengineering in the context of a collective, phenomenological exploration of what it would be like to live in a world shaped by geoengineering.

This approach was informed by the philosophy of technology and art as discussed above, and also by historical and sociological insights about technological change. Major shifts in technology are not simply the insertion of a new tool into an existing society; they change that society, subtly or drastically altering what people want and feel entitled to, and how social practices and social relations are organized (Nordmann). If you change the *means*, you can change the *ends* (Latour, "Morality and Technology"). We thus felt that asking the public what they thought about adding geoengineering to the existing world would be the wrong question, one which was likely to get us a familiar but misleading answer. By adding geoengineering to the world, you are likely to change that world in a profound way.

We carried out seven focus-group discussions in three UK cities in December 2011, each lasting three hours, and each with six to eight participants sharing a particular set of life-world characteristics. What first emerged was a phenomenon that is familiar from other studies of public attitudes to new technologies—"conditional acceptance." Participants expressed a reluctant acceptance of research into geoengineering as a necessary evil, given the geopolitical realities around emissions-reduction negotiations. But they were only happy for this research to go ahead if they were given reasons to have greater confidence in climate science, in geoengineering research, in geoengineering governance structures, and in wider political institutions.

We then asked our participants to put those conditions of acceptance—and the dominant imaginaries of key scientists, policymakers, and civil society actors involved in geoengineering debates—to a plausibility test by imagining future geoengineered worlds. Drawing on earlier experiments with storytelling techniques in public engagement about the future (Roberts), participants were encouraged to imagine the worlds that might be brought about by these technologies—their political economy, their institutions, their inhabitants, their lifeworld. Thus, rather than asking them to accept or reject the implicit imaginaries of policymakers and scientists, they were encouraged to carry out their *own* imaginings; and rather than simply picturing geoengineering as being added to the world that is, they tried to render the interconnected colors of a geoengineering world.

In such a world, in which the reality of geoengineering had percolated through to many aspects of society and culture, the "climate architect" imaginary of policymakers became implausible. In a geoengineering world, the existing problems identified with the public meanings of climate mitigation—alienation, dependence on science, and truth claims that are seen as at odds with the everyday experience of the limits of human knowledge and control (Wynne)—were intensified. And the possible side effects of geoengineering that were highlighted were not confined to the realm of the physical; our visioning exercises enabled us to paint a picture of a world in which immense challenges were being posed to the institutions of liberal democratic politics.

We summarized and expanded on these in a subsequent paper (Szerszynski et al.). We argued first that with geoengineering, the attribution of cause and effect, and of liability and accountability, would be impossible to carry out in any definitive way, putting strains on the international system and further politicizing scientific knowledge in ways that would affect the broader politics of climate change (see Hulme, "Climate Intervention Schemes" for an imaginative rendering of a possible geoengineered future). Second, we suggested that, because the emerging "social constitution" of geoengineering—its implied social relations—would be global in scale and technocratic in character, it would be experienced as incompatible with democratic control. Third, we argued that, in the transition to a world of "made" climates, the relationship between intention, deployment, and consequences would always be unstable: a new, geoengineered and geoengineering world would be one with new senses of possibility, in which geoengineering technologies would be used for new purposes such as improving agricultural yields or undermining regimes, leading to new kinds of conflict and controversy. Fourth, we suggested that geoengineering would become conditioned by economic forces, with artificial markets, promissory, "vision-based" dynamics and "bubbles" of hope and hype, generating further problematic effects on the practice and authority of science. The novelty of a geoengineered world when fully imagined thus lies not just in the geophysical realm, or in the felt subjective experience of the weather, but also in the geopolitical sphere.

Conclusion

We have seen that geoengineering raises complex issues about the unfolding of sociotechnical futures, ones for which a narrow "climate architect" imaginary is inadequate. Yet in the mainstream debates about geoengineering, it is still that very imaginary which prevails. If we genuinely want creatively to explore the space that geoengineering has opened up for new forms of politics suitable for living on a dynamic planet, we will have to develop a broader palette of imaginaries. For a start, we will need to explore more artisanal techniques of climate alteration, with very different implied social relations—for example, locally implemented and easily reversible "soft geoengineering" techniques such as building soils and increasing the reflectance of cities (Olson), or more "enchanted" options such as rewilding landscapes and creating biophilic cities (Buck). But we also need a sense of climate responsibility with deeper and richer hues, a wider culture of climate artistry.

Where can we find the sort of cultural politics of climate-making that could generate the reflections that we need? Geoengineering has featured in popular news media, but coverage has been limited both in scale and in the range of narratives and metaphors deployed (Luokkanen et al.). In wider popular culture it has received more sustained and complex treatment. It figured prominently in the 1982 graphic novel *Le Transperceneige* (Lob and Rochette), later translated into English and adapted as the 2013 film *Snowpiercer* (Bong), though this focused more on the (in this case disastrous) possible after-effects of geoengineering. The kind of judgments that would be involved in real-time climate modification itself have been foregrounded in a number of interactive games. In the computer game *Fate of the World* (Red Redemption), players can try out various options including geoengineering in order to reduce or adapt to global warming over the period of two centuries (Red Redemption). In the transmedia role-playing game *Bluebird*, sponsored by the Australian Broadcasting Corporation in 2010, Bluebird was the name of a "rogue" geoengineering project initiated by fictional billionaire "Harrison Wyld"; players were able to live life for six weeks as if geoengineering was really happening and could engage in the struggle between the "Go Bluebird" PR campaign and the "Stop Bluebird" group set up by whistlebower "Kyle Vandercamp." Artists such as Bigert and Bergström, Karolina Sobecka, and Weather Permitting have also produced works that raise more open-ended questions about what a geoengineered world would be like.

If we do ever come to make climates, it will never be the simple commanding of matter, since it will entangle us even more deeply in the endless becoming of the more-than-human world. Making climates will thus have to be a form of what Tim Ingold calls "textilic" rather than "architectonic" making, "a weaving of, and through, active materials" (93). But it will also have to be sensitive to the ontological dimension of making climates, which will necessarily be more than the mere rearrangement of existing elements; it will be the creation of a new world, in which the very nature of air as "matter-that-takes-form" will have been altered. We may never be able to "design" a planet—but we may yet come to "color" one—and, if so, we should learn to do it well.

Acknowledgments

I am particularly grateful to Maialen Galarraga and Phil Macnaghten, my primary collaborators in the research discussed in this chapter, for all the conversations we have had over the years, without which my thinking about geoengineering would have been much the poorer. I also thank Monika Bakke, Holly Buck, Forrest Clingerman, Alan Cottey, Matt Edgeworth, Andrea Gammon, Ursula Heise, Anne Kull, Michelle Niemann, Lisa Sideris, and Heather Sullivan for extremely helpful comments on an earlier draft.

References

Bellamy, Rob and Javier Lezaun. "Crafting a Public for Geoengineering." *Public Understanding of Science* (2015): 1–16. Print.

Bellamy, Rob, Jason Chilvers, Naomi E. Vaughan, and Timothy M. Lenton. "A Review of Climate Geoengineering Appraisals." *Wiley Interdisciplinary Reviews: Climate Change* 3.6 (2012): 597–615. Print.

Belter, Christopher W. and Dian J. Seidel. "A Bibliometric analysis of Climate Engineering Research." *Wiley Interdisciplinary Reviews: Climate Change* 4.5 (2013): 417–427. Print.

Bong, Joon-Ho, dir. *Snowpiercer*. 2013. Movie. CJ Entertainment.

Buck, Holly Jean. "On the Possibilities of a Charming Anthropocene." *Annals of the Association of American Geographers* 105.2 (2014): 369–377. Print.

Castoriadis, Cornelius. *The Imaginary Institution of Society*. Trans. Kathleen Blamey. Cambridge: Polity Press, 1987. Print.

Clark, Nigel. "Geoengineering and Geologic Politics." *Environment and Planning A* 45.12 (2013): 2825–2832. Print.

Clingerman, Forrest. "Geoengineering, Theology, and the Meaning of Being Human." *Zygon: Journal of Religion and Science* 49.1 (2014): 6–21. Print.

Crutzen, Paul J. "Albedo Enhancement by Stratospheric Sulfur Injections: A Contribution to Resolve a Policy Dilemma?" *Climatic Change* 77.3–4 (2006): 211–220. Print.

Dalby, Simon. "Geoengineering: The Next Era of Geopolitics?" *Geography Compass* 9.4 (2015): 190–201. Print.

Fleming, James Rodger. "The Climate Engineers: Playing God to Save the Planet." *Wilson Quarterly* 31.2 (2007): 46–60. Print.

——. *Fixing the Sky: The Checkered History of Weather and Climate Control*. New York: Columbia University Press, 2010. Print.

Gardiner, Stephen. "Is "Arming the Future" with Geoengineering Really the Lesser Evil? Some Doubts About the Ethics of Intentionally Manipulating the Climate System." *Climate Ethics: Essential Readings*. Ed. Gardiner, Stephen et al. Oxford: Oxford University Press, 2010. 284–314. Print.

Hamilton, Clive. *Earthmasters: The Dawn of the Age of Climate Engineering*. New Haven, CT: Yale University Press, 2013. Print.

Hulme, Mike. *Can Science Fix Climate Change? A Case Against Climate Engineering*. Oxford: Polity Press, 2014. Print.

——. "Climate Intervention Schemes Could Be Undone by Geopolitics." *Yale Environment 360*. 2010. Web. 27 September 2015.

Ingold, Tim. "The Textility of Making." *Cambridge Journal of Economics* 34.1 (2010): 91–102. Print.

Jamieson, Dale. "Ethics and Intentional Climate Change." *Climatic Change* 33.3 (1996): 323–336. Print.

Kearnes, Matthew. "Miraculous Engineering and the Climate Emergency: Climate Modification as Divine Economy." *Technofutures, Nature and the Sacred: Transdisciplinary Perspectives*. Ed. Celia Deane-Drummond, Sigurd Bergmann, and Bronislaw Szerszynski. Farnham: Ashgate, 2015. 219–237. Print.

Latour, Bruno. "A Cautious Prometheus? A Few Steps Toward a Philosophy of Design (with Special Attention to Peter Sloterdijk)." *Networks of Design*. Ed. Fiona Hackney, Jonathan Glynne, and Viv Minton. Boca Raton, FL: Universal Publishers, 2008. 2–10. Print.

——. "Morality and Technology: The End of the Means." *Theory, Culture and Society* 19.5/6 (2002): 247–260. Print.

Lob, Jacques and Jean-Marc Rochette. *Le Transperceneige*. Tournai: Casterman, 1982. Print.

Luokkanen, Matti, Suvi Huttunen, and Mikael Hildén. "Geoengineering, News Media and Metaphors: Framing the Controversial." *Public Understanding of Science* 23.8 (2014): 966–981. Print.

Lynas, Mark. *The God Species: How the Planet Can Survive the Age of Humans*. London: Fourth Estate, 2011. Print.

Masco, Joseph. "Bad Weather: On Planetary Crisis." *Social Studies of Science* 40.1 (2010): 7–10. Print.

McKibben, Bill. *The End of Nature*. London: Penguin, 1990. Print.

Nordmann, Alfred. "Responsible Innovation, the Art and Craft of Anticipation." *Journal of Responsible Innovation* 1.1 (2014): 87–98. Print.

Oldham, Paul, Bronislaw Szerszynski, Jack Stilgoe, Calum Brown, Bella Eacott, and Andy Yuille. "Mapping the Landscape of Climate Engineering." *Philosophical Transactions of the Royal Society A* 372.2031 (2014): 20140065. Print.

Olson, Robert L. "Soft Geoengineering: A Gentler Approach to Addressing Climate Change." *Environment: Science and Policy for Sustainable Development* 54.5 (2012): 29–39. Print.

Protevi, John. *Political Physics: Deleuze, Derrida and the Body Politic*. London: Athlone Press, 2001. Print.

Red Redemption. *Fate of the World*. Vers. 2011.

Roberts, Thomas Campbell. "Tales of Power: Public and Policy Narratives on the Climate and Energy Crisis." Ph.D. Lancaster University, 2010. Print.

Robinson, Kim Stanley. *Blue Mars*. London: HarperCollins, 1996. Print.

Royal Society. *Geoengineering the Climate: Science, Governance and Uncertainty*. London: The Royal Society, 2009. Print.

Schäfer, Stefan and Sean Low. "Asilomar Moments: Formative Framings in Recombinant DNA and Solar Climate Engineering Research." *Philosophical Transactions of the Royal Society of London A* 372.2031 (2014): 20140064. Print.

Simondon, Gilbert. *Du mode d'existence des objets techniques*. 3rd ed. Paris: Aubier, 1989. Print.

——. *L'Individu et sa genèse physico-biologique*. Paris: Presses Universitaires de France, 1964. Print.

Szerszynski, Bronislaw. "Reading and Writing the Weather: Climate Technics and the Moment of Responsibility." *Theory, Culture & Society* 27.2–3 (2010): 9–30. Print.

Szerszynski, Bronislaw and Maialen Galarraga. "Geoengineering Knowledge: Interdisciplinarity and the Shaping of Climate Engineering Research." *Environment and Planning A* 45.12 (2013): 2817–2824. Print.

Szerszynski, Bronislaw, Matthew Kearnes, Phil Macnaghten, Richard Owen, and Jack Stilgoe. "Why Solar Radiation Management Geoengineering and Democracy Won't Mix." *Environment and Planning A* 45.12 (2013): 2809–2816. Print.

Wynne, Brian. "Strange Weather, Again: Climate Science as Political Art." *Theory, Culture & Society* 27.2–3 (2010): 289–305. Print.

9

UTOPIA'S AFTERLIFE IN THE ANTHROPOCENE

Anahid Nersessian

Utopia has always been a dirty word. Recent developments, both environmental and theoretical, have made it effectively anathema, and for good reason. Here we may rehearse a well-worn roll call of present and impending catastrophes—sea level rise, mass extinction, ocean acidification—to ground a claim that, under the grim-faced aegis of the Anthropocene, to speak of the perfect world is vapid, credulous, or a form of disavowal: deck chairs on the *Titanic* and all that. Whatever the pros and cons of "Anthropocene" as a descriptive or discursive term, those may be hotly debated without the debate itself making a dent in the circumstances of anthropogenic climate change. In other words, the nomenclature of the crisis might change, but the crisis itself would seem to remain incommodious to anything that smacks of utopianism, if by utopian we mean *optimistic*. This chapter leaves the question of the Anthropocene's rhetorical propriety somewhat to the side to ask: how might an awareness of planetary fragility, and of the role of the human species in intensifying it, prompt a redefinition of the long-lived political concept of utopia such that it might join and help steer creative, credible conversations about the future, even in that future's apparent intractability?

In this chapter's first section, I revisit some political arguments against the vocabulary of the Anthropocene. In its second, I discuss the generic dependence of the Anthropocenic picture on utopian allegory, and suggest some modifications of utopian thought derived from an idiosyncratic archive of literature and theory. To be explicit, I argue for the advancement of utopian minimalism as an aesthetic mode that affirms the formal practice of limitation, and that takes "limitation" to allegorize the sorts of ethical and ecological commitments necessary for imagining utopia in the Anthropocene. The last section offers brief remarks on how literary scholarship might inform the larger rubric of the environmental humanities, in particular as it responds to the moral and epistemic pressures of scientific fact.

Anthropocene and utopia

The argument against Anthropocenic discourse from Marxist and postcolonialist perspectives is familiar, but worth rehearsing.[1] In Andreas Malm and Alf Hornborg's succinct phrasing, "if climate change represents a form of apocalypse, it is not universal, but uneven and combined" with a complexity that the term "Anthropocene," which means to telegraph the claim that "the power to shape planetary climate has passed from nature

91

into the realm of humans," handily suppresses (66–67). This might be called the "blunt instrument" objection, and it censures the Anthropocenic framework for the very reasons it would seem, to other minds, to be helpful: because it "mix[es] together the immiscible chronologies of capital and species history" (Chakrabarty 220). At the heart of these debates lies an anxiety about the severance of ecological from social-justice concerns, about the sidelining of class struggle, racism, sexual violence, and hemispheric inequality as though attention to these other large-scale phenomena will distract from the arguably more urgent problem of circumventing ecological doom. Thus the specific grievance that "the Anthropocene makes for an easy story … because it does not challenge the naturalized inequalities, alienation, and violence inscribed in modernity's strategic relations of power and production" (Moore 2) is not, I think, directed toward the *terminology* of the Anthropocene or, really, toward its "story," but rather toward the political *program* that terminology and that story imply.

It should be uncontroversial that environmental impacts are at least as "unequally distributed" on both local and global scales as any harm normally "rendered invisible" to more protected populations (Nixon 160). More remains to be said about the lack of fit between an ecologically minded abstraction like the Anthropocene and this state of affairs; even more remains to be said about how what might be broadly characterized as oppositional politics might be cut to fit new goals designated by an awareness of ecological crises, insofar as those crises are inseparable from extant protean forms of injustice. Present conditions ought to inspire constructive assessments of well-established political idioms, evaluating them as variously appropriate or ill-suited to the advancement of robust ecological objectives. This work is well underway in, for example, Dipesh Chakrabarty's effort to rethink the significance of "history" in classical Marxist thought, which seems to assume a casual but absolute division between natural and human chronologies. It is underway, too, in postcolonial studies, which is evolving a "postcolonial ecology" whose "complex epistemology … recuperates the alterity of history *and* nature" (DeLoughrey and Handley 4). Meanwhile, the interdisciplinary insights offered by animal, food, and science studies have been enlarging the contours of terms like "production" and "consumption" by applying them to the contexts of industrialized agriculture and laboratory as well as factory labor; this sort of work also revives fundamental philosophical questions about the nature of life and the parameters of its biopolitical use and abuse.

This chapter focuses on another foundational philosophical concept, namely utopia, or the perfect world. It asks how both the idea and the value of "perfection" might be calibrated to a planetary situation of amplified instability and attenuated possibilities. My purpose is to offer a model of what I call utopian minimalism, one that is derived—as that name might suggest—from essentially formal paradigms borrowed from literary and aesthetic theory. Thinking about utopia in this way requires disrupting multiple historical and theoretical conventions, among them the identification of the utopian with the perfect, the use of utopian discourse to fetishize the impossible and unachievable, and the narrow understanding of utopia as a genre as opposed to a political supposition. More importantly, it requires distinguishing utopian minimalism from liberalism's protocols of moderation. In a contemporary political culture that endlessly defers social and economic justice, not to mention serious, sustained action on climate and other crises, it may seem reckless to hold up minimalism as a standard or ideal. Nonetheless, my goal is to uncover the radicalism of being minimal in a way that resists the acquisitive dictates of neoliberal modernity, from the colloquial craze for "having it all" to neoliberalism's own involvement in the unchecked assault on the planet and its inhabitants. It is also to name as utopian the prospect of "sacrific[ing] or severely restrict[ing]" those "resources of gratification and self-realization" that, in Kate Soper's frank appraisal, "it seems very difficult simply to dismiss as false" (*What is Nature?* 168). That said, utopian minimalism is best understood as what Soper elsewhere defines as an "alternative hedonism," one that exposes the "spiritual deprivations" as well as the material harms of modern economic and political systems ("Alternative Hedonism" 578). Aesthetic experience offers an exemplary instance of such a hedonism, with pleasures that range from the cognitive to the sensuous, and which expand the category of pleasure per se by treating it as part of rather than inimical to lively, even joyous practices of moderation and restraint.

Utopia at the limit

Although the word "utopia" was coined by Thomas More in his sixteenth-century romance of the same name, the imaginative situation to which it refers has been around, figuratively speaking, forever. Any culture with a version of human history that divides into two parts—the first, an idyllic existence anchored in the mythological past, and the second, a miserable and deteriorating set of circumstances continuous with the present—has a default model of utopia. A compound of the Greek word for "place" (*topos*) and the negatory prefix *u-*, which sounds punningly like another prefix, *eu-*, or "good," utopia is both a good place and a no-place, a nowhere. It is an umbrella term covering both lost estates, like the Garden of Eden, and the better prospects that lie beyond or invisibly adjacent to this one: the Kingdom of Heaven, the Land of Cockaigne, or the Peach Blossom Spring. In each case, utopia is persistently defined against and as a back-formation of things as they are: laboring, dog-tired, we conjure up an Earth that gives away its fruits for free; mortal, we imagine immortality; complicit in the predations of capital, we dream of cleaner air and the renaissance of the passenger pigeon.

The story of the Anthropocene is—in one dominant interpretation—a declensionist narrative, and it is bound up in a long history of other such narratives, offering a contemporary iteration of a very old tale of the fall of man from a state of better grace. Pointing this out does not dismiss the reality or seriousness of anthropogenic climate change and other global ecological crises. Rather, it helps situate Anthropocenic plots on a cultural continuum with utopian ones, which have always had a close relationship to lapsarian themes. Such intimacy is evident in the nearly tautological rapport between utopia and dystopia, utopia's lurid other. Think of Margaret Atwood's *Oryx and Crake*, which imports the memorable image of poultry growing, already cooked, on trees from Bruegel's 1567 painting *The Land of Cockaigne* and turns it into Chickienobs, chicken parts bred in the form of "animal-protein tuber[s]" (246); or of the moment in *Paradise Lost* when, after the Fall, God's angels are sent to "turn askance/The poles of Earth twice ten degrees and more/From the Sun's axle," thus creating the novel phenomenon of bad weather (X.668–670). Anthropocenic fictions like Atwood's may secularize and scientize the theological drama of Milton's poem, but both texts are bound by a literary tradition for which the present is a negative image of the past, and where even the dream of a better world cannot be untangled from the recognition of its impossibility. While critics of the explanatory presumptions of the Anthropocene emphasize their apparent erasure of the history of capitalism from the history of human transformations of the planet, the trouble with Anthropocenic models is generic as much as it is conceptual: by adopting the postlapsarian orientation of both conventional utopian and dystopian narrative, they commit to a mythic perspective according to which things are as they are, with little room for modification or resistance.

It is worth noting, too, that in the twenty-first century utopian writing is often taken to be synonymous with science fiction. The *Cambridge Companion to Utopian Literature* (Claeys) is weighted heavily toward the novels of H. G. Wells and George Orwell, and merely nods at the work of agrarian reformer Gerrard Winstanley and feminist socialist Charles Fourier—an emphasis that participates in an ongoing, large-scale weakening of utopia as a social hypothesis via its reduction to a set of themes and tropes. When Wells's *Men Like Gods* is grouped with Michael Hardt and Antonio Negri's *Empire* trilogy—as it is in Lyman Tower Sargent's *Utopia: A Very Short Introduction*—essential distinctions between allegorical and analytic narratives about the world are too neatly elided, and the nonidentity of their premises overlooked. Meanwhile, Ursula K. Heise observes that nonfiction narratives of the Anthropocene draw heavily on science-fiction conceits (like terraforming), an act of cultural borrowing that underscores the Anthropocene's reliance on the muddled legacies of literary utopianism (*Imagining Extinction*, ch. 6). From Genesis to geoengineering, utopia too has become a conceit, a shorthand for the mistakes of humankind and the implausibility of their amendment.

The circumstances described by the language of the Anthropocene demand a new paradigm for articulating opposition to the state of the present and the probable state of the future. Instead of asking, "why are we still letting our rhetorical tools for describing and conceiving ideal and non-ideal worlds be limited to such antiquated means?" we ought to insist that the Anthropocenic context requires us to adapt the cultural inheritance of utopia to suit new and urgent purposes. Such an inquiry might begin with a reclamation of utopia as a concept in excess of any particular generic form, most obviously the genre of lapsarian fable that continues to shape so much of the environmentalist conversation. Unhitching utopia from genre allows this conversation to become open to experimental redefinitions of utopian vocabularies, which need no longer participate in the rhetorical segregation of an apocalyptic *after* from a prehistoric *before*. As this idea of redefinition might suggest, this is a task for literary criticism as much as it is for progressive political discourse, which—as the critique of the Anthropocene reveals—has not yet entered into a fully collaborative relationship with the exigencies of environmental crisis.

A cursory look at classical utopias reveals their dependence on two sets of tropes—tropes of politics and tropes of pleasure. The political tropes usually exist to propound a roomier idea of equality than the one predominating in the real world: in More's sixteenth-century utopia, property is held in common; in Sarah Scott's eighteenth-century *Millenium Hall* [sic] (1762) and James Lawrence's 1811 *Empire of the Nairs*, women are in charge of their community or their state; in William Morris's *fin de siècle News from Nowhere* (1890), subtitled "An Epoch of Rest," the overgrown brush of the bureaucratic state, with its "army, navy, and police," has been replaced by a regime of "habit" that compels each individual into "acting on the whole for the best" (126). When it comes to pleasure, each of these exemplary utopias has it in spades—food, sex, leisure time—and in each case the hedonic payoff of living in an ideal world "communicates," in Robert Appelbaum's phrase, "the value of general welfare that utopia has been designed to satisfy" (145). In this sense, utopian pleasures would seem to do precisely what Fredric Jameson says all politicized pleasures ought to do, namely appear not only as goods in themselves but as figures for "the systemic revolutionary transformation of society as a whole" ("Pleasure" 13). It is not enough for

More or Morris to show people who are not hungry; the "material need" of their hunger must be sated as a "symbolization of the need and its satisfaction" by utopian governance (Appelbaum 45).

Satisfaction, systematicity, wholeness, "the best." The watchwords of classical utopia hold up plenitude as a primary value and goal, seldom asking if it is the *right* one. Is it possible that the orthodox hedonism of classical utopias has kept them from accommodating rigorously ecological concerns? In a dire planetary situation fueled and exacerbated by overconsumption, would utopias be more utopian if they abandoned their fixation on satiety and contentment? Critics of utopian literature seldom raise these sorts of questions, and indeed tend to disparage utopias that fail to present unconstrained human flourishing as their supreme design. Consider Jameson's complaint against Ursula K. Le Guin in his *Archaeologies of the Future*, which unsympathetically describes Le Guin's representational technique as one of "world reduction." "Based on a principle of systematic exclusion," world reduction is "a kind of surgical excision of empirical reality, something like a process of ontological attenuation in which the sheer teeming multiplicity of what exists, of what we call reality, is deliberately thinned and weeded out through an operation of radical abstraction and simplification" (276). The effect, according to Jameson, is "a fantasy realization of some virtually total disengagement of the body from its environment or eco-system"—a fantasy that in turn exposes Le Guin's fiction as utopian in name alone (269).

It is a curious thing that, "although [Jameson] refer[s] to Le Guin's world reduction as an 'experimental ecology,'" he does not "explore its significance in terms of ecology per se" (Prettyman 61). As Heise puts it, "Jameson's reading must ultimately remain unconvincing" from a literary as well as environmentalist point of view, for in assigning world reduction "a purely metaphorical function," he misses the opportunity to contemplate its significance as ecological allegory ("Reduced Ecologies" 101). The argument against world reduction rests on the assumption that representational practices of exclusion, excision, attenuation, thinning, weeding, abstraction, and simplification constitute an evasion of political analysis rather than a mode of political analysis in their own right. Perhaps these techniques figure another kind of revolutionary social transformation: the necessary but no less ethical rejection of plenitude as the premise of utopian achievement, and the affirmation of "dispossession and sacrifice," along with scarcity, as utopia's first "condition" (Nadir 40).[2] The suggestion is backed up by Le Guin's novel *The Dispossessed*, in which the barren, communist planet Anarres, with its difficult living conditions and limited biodiversity, stands in stark moral contrast to lush and lovely Urras, where social injustice and economic inequality sponsor the conspicuous consumption of the planetary one percent. From Le Guin's opposition of these two societies emerges a grammar of distinction that is also a theory of utopia, one that has nothing to do with her novel's generic status as science fiction. It is a grammar that forms the basis of an "ecological political theory" founded upon conditions of scarcity, under which "resource use and disposal" and the complexity of securing "relative stability and longevity" under duress become intrinsic to the imagination of an ideal world, no longer difficulties that must be surmounted before it can be secured (Stillman 57).

Anarres is a utopia because it is a hard place to live—a place where humanist values of autonomy and self-expression are denigrated, labor is physically demanding, mind-numbing, or both, and where the world has very little in the way of visual beauty. This question of beauty is significant, for the sheer unsightliness of Anarres, as an exemplum of world reduction, has precisely the opposite effect from the one Jameson predicts. Far from functioning as "a symbolic affirmation of the autonomy of the organism" from its environment

(*Archaeologies* 269), these radically reduced aesthetic circumstances announce the maximum dependence of every organism on the human species' capacity and willingness to diminish—to reduce—its impacts. Subtitled "An Ambiguous Utopia," *The Dispossessed* cashes in on the etymological root of ambiguity by arguing that the human concern called utopia must be pursued by taking two separate but mutually dependent paths at once: toward greater perfection and toward greater loss.

I have elsewhere characterized this concept of utopia as "utopia, limited," and described its unique grammar with reference to the nineteenth-century linguist Wilhelm von Humboldt's gloss of language, in *On Language* (1836), as an "infinite use of finite means" (qtd. in Nersessian 29). As Le Guin's experimental portrayal of utopia as a set of diminished circumstances suggests, the semantics of finitude may be elaborated across multiple aesthetic contexts, belying the usual association of utopia with either lapsarian or science-fiction narrative or, for that matter, with narrative, period. In its refusal to legitimize ease and self-sovereignty as indispensible social values, is *The Dispossessed* more like *News from Nowhere*, so often taken as a exemplum of utopian writing, or is it more like George Herbert's 1633 poem "Paradise," in which five rhymed tercets addressed to God proceed through a series of lexical amputations—for example, "Inclose me still for fear I START./Be to me rather sharp and TART,/Then let me want thy hand and ART"—emblematic of the spiritual and compositional discipline Herbert calls "pruning" (7–9)? Does its insistence on associating collective political justice—where "collective" encompasses a trans-species and interplanetary franchise—with the diminishment and mitigation of human interest nudge Le Guin into greater solidarity with Huxley and Wells or with John Clare? Clare's "Lament of Swordy Well" uses the literary device of prosopopoeia to speak from the diffuse point of view of a quarry unmasking the ills of enclosure, including the extirpation of flowers and, with them, "butterflyes" and bees (93). Does the mode of dispossession articulated by the little girl who tells her weeping father, "You can share the handkerchief I use" (instead of "You can use my handkerchief") (Le Guin, *The Dispossessed* 253) seem akin to a rejection of "ownlife," the Newspeak word "meaning individualism and eccentricity" in George Orwell's *Nineteen Eighty-Four* (85)? Or does it share in those "hints of a theory of use present in the Pauline letters, in particular in 1 Corinthians 7:20–31, in which using the world's resources[,] ... not abusing" them defines an ethical response to living in a world whose present "form ... is passing away" (Agamben 139)?

My purpose is not to dismiss speculative fiction, but to suggest that utopia's genealogies be mapped across alternative data points, namely those which mark, by way of formal and rhetorical tendencies, a sensibility in alignment with Michel Serres's sober credo, "I wish for, and practice, the dispossession of the world" (73). The obvious affinity between this remark and the claims made in Corinthians offers just one instance of a salvaged intellectual tradition that overflows the simple descriptor of asceticism, Christian or otherwise. What Serres and Saint Paul are advocating is both a universalization of scarcity and, more importantly, a desiring or "wishful" relationship to it, such that dearth is no longer a trope of dystopia but, on the contrary, the condition and goal of moral and political choice. This seems to be Le Guin's argument, too, at least in *The Dispossessed*, and it is also an appropriate extension of Marx's critique of private property, according to which property wrongly trains "each man to see in other men not the realization but the limitation of his own freedom" (229–230). What is not often observed about this well-known claim is that it implicitly recuperates "limitation" as the basis and not the circumvention of freedom, whose purchase is unthinkable outside of a positive investment in dispossessive habits of use without abuse. The salience of such an investment to an Anthropocenic moment is clear, but tracking earlier instances of it, centuries and even millennia before the notion of "'the environment" as we understand it now might be said to pertain, reveals an ecological utopian tradition organized by the favorable valuation of loss, bereavement, and exacting habits of doing-with-less.[3]

Does placing Marx in this tradition smooth some of the perceived tensions between ecocritical and Marxist thought? I think so, but it is not a point I want especially to belabor. What I would like to point out is that this particular reading of Marx, and for that matter of Le Guin and Saint Paul, is just that—a reading, which is to say the yield of an interpretive practice trained on the broadly figurative elements of a text or even an idea, which it treats as a thing open to reinflection and tweaking. Reading is the labor that prepares Marx for the Anthropocene, and it is the labor that prepares Saint Paul for Marx; it is also what literary criticism has to contribute to the environmental humanities. This contribution is made possible not despite of but on account of literary criticism's affinity for interpretation against the grain of its object. The next and final section expands upon this claim as part of a nonstandard defense of reading in the context of the environmental humanities, whose own promise will depend on its capacity to secure the meaningful coexistence, if not always the cooperation, of scientific and humanist methodologies.

Weak utopia

I have already asked how a heightened awareness of planetary hazards invites a creative engagement with the cultural inheritance of utopia and its representation. Literary criticism is also a cultural inheritance, and one equally in need of imaginative adjustments. The diverse set of practices now grouped under the heading of "ecocriticism" has offered a number of such adjustments on multiple fronts, providing new insights into the mainstays of literary-critical interest: narrative, genre, characterology, tone, and topic, putting together what Eve Kosofsky Sedgwick might call a "nonce taxonomy of texts" (23) both obviously and less obviously engaged with ecological questions. One thing that unites the various sorts of interpretation we call ecocritical is, paradoxically, their "divergent perceptions of how the sciences should inform cultural inquiry." Should the sciences serve as "the foundation of literary study," or should literary study stress that "our perception of the environment is culturally shaped" and ask "how that perception is mediated through language" (Heise, "Hitchhiker's Guide" 509, 511)? The question might be phrased yet more starkly: either we believe our theories should be in line with the relevant scientific data, or we don't.

I am not sure if there are data to which my idea of a minimalist utopianism might be said to correspond. The trouble is not that this idea may be uncovered in texts written thousands of years apart, under wildly divergent circumstances; it is not even that utopia, that idealized no-place, by definition exceeds the scope of empirical explanation. It is, rather, that this is an idea whose ecological propriety seems plain but that has come into being through a frankly biased reading of particular texts, many of which—von Humboldt's essay on language, Paul's letters to the Corinthians—were never intended to be read this way. A familiar construct from literary studies could be invoked here: we could call this mode of reading, in Roland Barthes's terms, "the writerly," an interpretive posture that opens each text to a "plurality of entrances" in recognition of "the infinity of languages" on which every artwork draws (5). I wonder, though, if we might modify Barthes's phrasing, putting emphasis not on "plurality" or "infinity" but on a more constrained view of the text as precarious and undefended. Such a view might include the assertion that what makes an aesthetic object unique—that is, its formal elements and attributes—is also what leaves it vulnerable to being read (again) against the grain of its apparent surface. Not only does interpretation have no pressing responsibility to "the sciences" and what they know, it might also seek to surprise the object itself with what it knows, without having realized it.

Recent polemics against interpretation in this sense—as an "activist" mode of engagement—have allied themselves with a "modesty" that seeks to dispense "with some of the overblown assertions of literary theory" held over from the 1970s and 1980s while also expressing "the shrunken expectations of academe, particularly of the humanities" (Williams). The implicit assumption here is that modesty is antithetical to the making of strong claims, and that what Sedgwick calls "weak theory" wisely washes its hands of the authority to make political comment. It is worth pointing out, however, that Sedgwick comes up with the idea of "weak theory" in response to the AIDS activist Cindy Patton's coolly militant observation that even if we could learn all sorts of terrible things about the world, including "the likelihood of catastrophic environmental and population changes," we would not know much more than we "already know": acting, in other words, is quite another thing from being informed. To be weak *is* a political stance, one that shrugs off the promise of mastery over data, facts, figures, and projections to immerse itself into a relentlessly creative practice of reading for what is in excess of the probable or predictable. This is not far at all from what Le Guin seems to mean when she notes that "we've had enough words of power," and so instead may need "some words of weakness" ("Left-Handed Commencement Address" 115). Utopia is one such word to add to our vocabulary.

Notes

1 These are the perspectives that concern me here, but they are not, of course, the only ones from which one might contest the Anthropocenic proposition.
2 I have a more optimistic reading of Le Guin's depictions of scarcity and sacrifice than Nadir, who seems to believe that they can only represent a means of social control.
3 In some sense, such practices join up with a long and particularly an American tradition of ecological asceticism from Henry David Thoreau to "No-Impact Man" Colin Beavan, whose year-long experiment in living, ecologically speaking, as lightly as possible is documented in his 2009 book by that same name. That said, the moral ostentation and (as many critics have pointed out) barely suppressed machismo of this tradition situates it at a remove from the practices of rigorous effacement I associate with the alternative utopianism of Paul, Marx, Serres, and especially of LeGuin, whose body of work may plausibly be described as a feminist correction of its least useful impulses.

References

Appelbaum, Robert. *Aguecheek's Beef, Belch's Hiccup, and Other Gastronomic Interjections: Literature, Culture, and Food Among the Early Moderns*. Chicago: University of Chicago Press, 2008. Print.

Agamben, Giorgio. *The Highest Poverty: Monastic Rules and Form-of-Life*. Trans. Adam Kotsko. Stanford: Stanford University Press, 2013. Print.

Atwood, Margaret. *Oryx and Crake*. New York: Random House, 2003. Print.

Barthes, Roland. *S/Z: An Essay*. Trans. Richard Miller. New York: Hill and Wang, 1974. Print.

Chakrabarty, Dipesh. "The Climate of History: Four Theses." *Critical Inquiry* 35 (2009): 197–222. JSTOR. Web. 1 August 2015.

Claeys, Gregory, ed. *The Cambridge Companion to Utopian Literature*. Cambridge: Cambridge University Press, 2010. Print.

Clare, John. *John Clare: Major Works*. Ed. Eric Robinson and David Powell. Oxford: Oxford University Press, 2004. Print.

DeLoughrey, Elizabeth and George B. Handley. "Introduction: Toward an Aesthetics of the Earth." *Postcolonial Ecologies: Literatures of the Environment*. Ed. Elizabeth DeLoughrey and George B. Handley. Oxford: Oxford University Press, 2011. Print.

Heise, Ursula. "The Hitchhiker's Guide to Ecocriticism. " *PMLA* 121 (2006): 503–516. Project Muse. Web. 1 August 2015.

——. *Imagining Extinction: The Cultural Meanings of Endangered Species*. Chicago: University of Chicago Press, 2016. Print.

——. "Reduced Ecologies: Science Fiction and the Meanings of Biological Scarcity." *European Journal of English Studies* 16.2 (2012): 99–112. Print.

Herbert, George. *The English Poems of George Herbert*. Ed. Helen Wilcox. Cambridge: Cambridge University Press, 2010. Print.

Jameson, Fredric. *Archaeologies of the Future: The Desire for Utopia and Other Science Fictions*. London: Verso, 2005. Print.

——. "Pleasure, a Political Issue." *Formations of Pleasure*. London and Boston: Routledge and Kegan Paul, 1983. 1–14. Print.

Le Guin, Ursula K. *The Dispossessed: An Ambiguous Utopia*. New York: Avon, 1974. Print.

——. "A Left-Handed Commencement Address." *Dancing at the Edge of the World: Thoughts on Words, Women, Places*. New York: Grove Press, 1989. 115–117. Print.

Lewontin, Richard and Richard Levins. "Organism and Environment." *Biology under the Influence: Dialectical Essays on Ecology, Agriculture, and Health*. New York: Monthly Review Press, 2007. 31–34. Print.

Malm, Andreas and Alf Hornborg. "The Geology of Mankind? A Critique of the Anthropocene Narrative." *The Anthropocene Review* 1.1 (2014): 62–69. SAGE Journals. Web. 1 August 2015.

Marx, Karl. "On the Jewish Question." *Early Writings*. Trans. Rodney Livingstone. New York: Penguin, 1992. 211–242. Print.

Milton, John. *Paradise Lost*. Ed. Gordon Teskey. New York: Norton, 2004. Print.

Moore, Jason W. "The Capitalocene: Part I: On the Nature & Origins of Our Ecological Crisis." *Jason W. Moore*. June 2014. 1–38. PDF. Web. 18 April 2015.

Morris, William. *News from Nowhere; or, an Epoch of Rest, Being Some Chapters from a Utopian Romance*. Ed. Stephen Arata. Peterborough, ON: Broadview Press, 2003. Print.

Nadir, Christine. "Utopian Studies, Environmental Literature, and the Legacy of an Idea: Educating Desire in Miguel Abensour and Ursula K. Le Guin." *Utopian Studies* 21.1 (2010): 24–56. JSTOR. Web. 1 August 2015.

Nersessian, Anahid. *Utopia, Limited: Romanticism and Adjustment*. Cambridge, MA: Harvard University Press, 2015. Print.

Nixon, Rob. *Slow Violence and the Environmentalism of the Poor*. Cambridge, MA: Harvard University Press, 2011. Print.

Orwell, George. *Nineteen Eighty-Four*. London: Penguin, 2000. Print.

Prettyman, Gib. "Daoism, Ecology, and World Reduction in Le Guin's Utopian Fictions." *Green Planets: Ecology and Science Fiction*. Ed. Gerry Canavan and Kim Stanley Robinson. Middletown, CT: Wesleyan University Press, 2014. 56–76. Print.

Sargent, Lyman Tower. *Utopia: A Very Short Introduction*. Oxford: Oxford University Press, 2010. Print.

Sedgwick, Eve Kosofsky. *Epistemologies of the Closet*. Rev. ed. Berkeley: University of California Press, 2008. Print.

Serres, Michel. *Malfeasance: Appropriation Through Pollution?* Trans. Anne-Marie Feenberg-Dibon. Stanford: Stanford University Press, 2010. Print.

Soper, Kate. "Alternative Hedonism, Cultural Theory and the Role of Aesthetic Revisioning." *Cultural Studies* 22.5 (2008): 567–587. Taylor & Francis Online. Web. 1 August 2015.

——. *What is Nature?: Culture, Politics and the Non-Human.* Oxford: Blackwell, 1995. Print.

Stillman, Peter G. "*The Dispossessed* as Ecological Political Theory." *The New Utopian Politics of Ursula K. Le Guin's* The Dispossessed. Lanham, MD: Lexington Books, 2005. 55–73. Print.

Williams, Jeffrey L. "The New Modesty in Literary Criticism." *The Chronicle of Higher Education.* 5 January 2015. Web. 18 April 2015.

Part II

POSTHUMANISM AND MULTISPECIES COMMUNITIES

10

RENAISSANCE SELFHOOD AND SHAKESPEARE'S COMEDY OF THE COMMONS

Robert N. Watson

Re-defining the borders of the human may be the most important task for twenty-first-century environmentalism; and the high road to a posthuman culture—one that acknowledges our enmeshments and denies our exceptionalism—may circle back through the pre-modern. During the Renaissance, the boundaries of the individual person and the boundaries of the human species tended to rise and, more often, fall, in tandem. Shakespeare's tragedies predominantly warn of the costs of a hubristic investment in the self and its insularity; the comedies depict the redemptive force of accepting one's place in nature (and nature's place within oneself); and the late career tragicomedies reconcile human aspirations with the functions of the biosphere.

To the extent that environmentalism depends on an ecological view of life, comedy tends to be greener than tragedy (Meeker 59). The standard moves of the comic genre remind us that we are all flawed and fungible mortal creatures in the same leaky boat. As several ecocritics have recently observed, environmentalist rhetoric must move past the time-worn and spirit-wearing genres of elegy and jeremiad, past even pastoral and its utopian variants, toward laughter and life. But the warnings of tragedy are important—maybe all the more so because, in its greatest instances, that genre does not dismiss the values of the tragic hero in any facile way, only exposes the costs. And to the extent that our environmental crisis involves a large-scale tragedy of the commons, the likeliest way to turn that tragedy comic would be to show—as Shakespeare continues to do—how profoundly communal our lives really are. The selfishness of human beings may be incurable, but expanding our definition of the self may provide some protection against the ravages of the disease.

Practitioners of environmental humanities can show that the intermeshed roles of persons in the *polis* resemble the roles of human beings in the biosphere. The radical constructivist position can be environmentally destructive, because the more we see the human self as entirely a product of an arbitrary human cultural order, the harder it becomes to see it as part of a natural network. Furthermore, the currently fashionable forms of critique that provide an accusatory deconstruction of the subtle mechanisms of social control imply that a fully liberated self would be ideal. That ideal corresponds to the fantasy of a human being wonderfully exempt from the burdens of participation with nature—exempt, that is, from all of the compromises (taints and restraints) imposed by our interaction with fellow life-forms, as if those compromises brought no corresponding blessings.

In recent decades, discussions of "the body" have featured persistently in the top echelons of literary study. The prospect of finally grounding abstract and abstruse critique in something tangible and fundamental without falling into reductive Marxist materialist determinism was irresistible, much as history of "the book" has become a popular respite from the airy speculations of both literary and political readers of "texts." The drawbacks from an ecocritical viewpoint are manifest in the definite article so regularly appended: "the body" is a revealing term for what are nearly always studies of human bodies, and it reinforces an assumption that the unit for study is self-evidently an individual as framed by the outline of skin. My intention is not to erase such bodies, but instead to call attention to the way their outlines are blurred by both nature and culture.

Human selfhood in the Renaissance

Jacob Burckhardt's classic study characterizes the Renaissance as a period of growing individualism. Norbert Elias's argument that *homo clausus* emerges in that same period defines such personal insularity as a rejection of inter-human sociability. My goal here is to extend that argument about the costly illusion of being "separate from all other people" to the no less costly illusion of being separate from all other forms of life. This extension would parallel the New Historicist revision of Burckhardt that shows Renaissance persons as permeable to interpellation (to adopt Althusser's term) by their ideological environments.

The revival of Classical culture that gave the Renaissance its name transmitted Stoic ideals of a centered and self-contained self that was fundamentally indifferent or immune, thanks to human rationality and free will, to the vicissitudes of the world. This legacy is especially strong in the dramatic tradition, in the form of the gloriously—if often tragically—insular Senecan hero (Braden), although a contrary literary legacy in Old and New Comedy laughed away the dignity of that posture. Pythagoras's theories of reincarnation pushed the connection with other species to an extreme, but one so extremely alien to mainstream Christianity that, except in the vegetarian tracts of a few radicals, it appears mostly as parody in the Renaissance. Although the *Metamorphoses* of Ovid (evidently Shakespeare's favorite writer) eroded the outlines of the body and the division of the species, the predominant philosophical traditions supported human exceptionalism, with the mind as the core of personhood. As John Donne would write in "The Funeral," "the sinewy thread my brain lets fall/Through every part/Can tie those parts, and make me one of all."

Christianity locates the essential self in the soul, but the late Renaissance schism between Catholicism and Protestantism disrupted traditional ideas about the boundaries of identity and humanity. Battles over the nature of Christ's presence in the Eucharist, and over the way those substances enter and transform the soul of the recipient, unsettled assumptions about the nature of selfhood. The essays of Michel de Montaigne, a skeptical Catholic, refute the idea of a stable and unified human being.

Protestantism understood the soul and the spiritual life as far more individualized, private, and invisible than Catholicism did—an invention of interiority not unlike that sometimes attributed to Shakespeare (and, in particular, to a Wittenberg student named Hamlet). Yet that Protestant self was also far less in control of itself: Calvinist predestination deprived the soul of any self-determination, and the corresponding emphasis on the corrupt residue of the old Adam within deprived us of any integrity. Any will toward Grace had to be imposed by God. Any self-contained individual was therefore a cursed individual, and the species was outcast from the gates of Eden.

Even the optimistic strains of Reformation theology that considered Eden retrievable tended to dissolve the self. The spiritualist heirs of the Heresy of the Free Spirit sought an ecstatic emptying of selfhood—a doctrine that was reportedly accompanied, among seventeenth-century radical Protestant groups such as the Familists, with doctrines of universal love that mingled bodies freely and welcomed nonhuman creatures into the blissful afterlife. Furthermore, radical early Calvinism in England enabled the modern capacity for protective fellow-feeling with other creatures (Watson, "Protestant").

Galen's ancient theory of bodily humors, which regained preeminence in the Early Modern period, integrated both the material and the psychological aspects of identity with local forces such as climate and diet. Purgative blood-letting was a signature practice. "Highly permeable and thus subject to—even composed of—its environment, the humoral body suggests a material embeddedness of self and surround" (Selleck 149–150). Across the centuries and cultures of the Renaissance, the human self was frequently imagined, often in conjunction with Galen's system, as containing the whole world or universe in miniature. This trope of the microcosm implied—in ways that might be valuable to recover and update in our Anthropocene era—that human beings are partly constituted of other creatures and are parts of a planetary superorganism. In this sense, my argument is a lemma to Ursula Heise's argument for an eco-cosmopolitanism; my premise is that the human self is itself a cosmos, and one whose local signature (each person) and global signature (the human species) are as deeply intertwined as Heise shows place and planet to be in the environmentalist project. This model, which runs from as far back as Democritus, up through Pico della Mirandola and Paracelsus in the Renaissance, and, after some reconceptualization, on to Leibniz, offers the same fractal structure that occurs in the graphing of some algorithms and in the pattern of plants and other natural phenomena. Seeing parts of the whole as mirrors of the whole comprised yet another Renaissance challenge to simple definitions of identity.

Lately the mainstream media have been rife with stories about how many other life-forms inhabit and even enable our bodies: biota variously coating microclimates of our skin and protecting our guts, while mitochondria generate most of our energy. The science would have amazed Renaissance writers—the compound microscope was not invented until the 1590s, and the microscopy of life appears to commence with the work of Robert Hooke and van Leuwenhoek in the 1660s—but the basic concept would have surprised very few of them. They had long understood a human being as a tiny universe containing multitudes, as we were part of the multitudes the universe contained. Even Francis Bacon—now widely decried for promoting science as a tool for human exploitation of nature, and explicitly resistant to mystical ideas about microcosms—understood the human as "the most extremely compounded" of all bodies, built as it is omnivorously (Bacon 117). Invoking that traditionally revered and Christian-inflected conception may help convince conservative minds that we must share the world rather than consume it, and must protect its diversity rather than erase everything that does not fit our idea—a dream turned nightmare—of human dominion.

Human selfhood in Shakespeare

Shakespeare's *Midsummer Night's Dream* is haunted by nightmares about irrational forces controlling humanity from without and within, to the extent that characters become beasts, plants, and other presumptively subhuman creatures (Watson, "*Midsummer*"). Yet the play also dreams gloriously about humanity being adored and protected by supernatural forces lurking in the wilderness that allow us not only to survive, but also to be our best selves. *Midsummer* does not endorse any naïve assumption that interactions across the boundary of the human self are always benign: the passions that run dangerously wild in the wilderness must be brought back to the city and controlled by its institutions. But the comic renewal of life requires a looser grip on human identity than high human civilization in its rational-ist mode, embodied here by the skeptical Duke Theseus, normally permits. This symbiotic vision anticipates recent discoveries of unseen forces that drive our appetites and aversions. The interweaving culminates in the wonder-filled synesthesia of Bottom the weaver's dream-narrative.

Where Bottom's dream ends—with the body parts unfathomably intertwined—is also where Coriolanus's nightmare begins. Arguably the greatest tragedy ever written about delusory pride in impermeable human selfhood (though there are revealingly many), Shakespeare's *Coriolanus* begins with the sympathetic aristocrat Menenius telling Rome's starving plebeians to accept their role as the working limbs to which the noble stomach sends strength. The idea of a community of bread, set up by the play's opening debate about the sharing of grain, turns up with remarkable persistence—as "company," many times, along with "accompany," "accompanied," "companion," "companions," and "companionship"—in speeches by the play's compromisers and conciliators. The word "common" appears far more often here than elsewhere in Shakespeare, becoming the unnamed antagonist of a tragedy of uncommonness. If "trans-" is the prefix that haunts *Midsummer*, in *Coriolanus* it is surely "com-" and its variants. Coriolanus's rejection of Rome's common people—a rejection that focuses on both their creatureliness and their interchangeability, and thereby on their appetite for individual and collective survival—seems valid even as it proves fatal. In the hero's tragic war against "the beast/With many heads" (IV.i.1–2), the anti-ecological boundaries guarding the human individual and the human species stand or fall together. Coriolanus's embarrassment about his wounds, and his determination to "stand/As if a man were author of himself/And knew no other kin" (V.iii.33–37), destroying with purging fires anyone who threatens to compromise or com-plicate his martial definition of himself and the Roman body-politic, offer a fascinating limit case to the classical project of selfhood.

Variants of Coriolanus's disease are epidemic even in Shakespeare's comedies, among those (such as the proud Malvolio in *Twelfth Night*) mocked as enemies of life's appetites. The comic genre—as modern sit-coms demonstrate—loves to humiliate those who take on some fixed identity to set themselves apart and above the group. The young men aspiring to be pure scholars in *Love's Labor's Lost* suffer a similar fate, dragged embarrassingly but finally appealingly back into the business of mortal bodies.

In tragedy, however, that disease turns deadly. Caught between the Roman model of Stoic isolation and the Egyptian allure of shared sensual pleasure, the hero of Shakespeare's *Antony and Cleopatra* loses any grasp on the outlines of his identity: "Here I am Antony,/Yet cannot hold this visible shape" (IV.xiv.1–14). He has lost the "classical" body—a model of ideal, orderly, stable, smooth-outlined, self-enclosed form—as opposed to the "grotesque" porous and protrusive body that emerges in the Renaissance (Bakhtin). As his old compatriots warned, Antony's Roman greatness melts away under Egyptian influences that go with flood, fertility, appetite, love, a more fluid aesthetic, and Cleopatra's own "infinite variety." By making this melting seem simultaneously horrifying and wonderful—a version of the feminine sublime—Shakespeare makes his audiences confront their own biases about selfhood.

Shakespeare's tragic figures seem oversupplied with the death-drive posited by Freud: the destructive urge to escape or erase the organic. A nightmare-ridden Richard III cannot cease striving "Until my misshap'd trunk that bears this head/Be round impaled with a glorious crown" (*Henry VI, Part 3*, III.ii.170–171). Macbeth loses kinship, food, and the cyclical "season of all natures" in his quest to be "perfect,/Whole as the marble" (III.iv.20–21), and ironically the result is a persistently implied war between the parts of his own body. Hamlet seems so disgusted with worms, food, sexuality, and kinship that the play's final poisonings look like a fulfillment of his perverse dark fantasy of the world: a whole chronicle of morbid adolescent disgust, vindicated. Othello's proud fantasies of purity collide with the shaming of his black skin, the shaming of his aging body, and the shaming of sexuality itself, leading him to murder his innocent, adoring wife.

Shakespeare's tragicomic *Winter's Tale*—my abbreviated proof of method here—undoes the unhappy ending of sexual jealousy in *Othello* by teaching King Leontes to surrender his proudly insulated selfhood and accept natural life, in all its flaws and transience. This may seem banal as a moral, and facile in reducing Shakespeare's complexity to a pedantic binary implicitly endorsing ecological and communitarian political views widespread among modern academic humanists. But the binary does reflect something important about the plays in their own right and about the environmental work they can do now.

The Winter's Tale begins by explaining that two kings, Leontes of Sicilia and Polixenes of Bohemia, were "as twinn'd lambs" in their boyhood, but have since become markedly separate and sophisticated people (1.i.67, 23–32). Leontes recalls his courtship of Queen Hermione in terms that evoke a flower opening in spring: "Three crabbed months had sour'd themselves to death,/Ere I could make thee open thy white hand,/[And] clap thyself my love; then didst thou utter,/'I am yours for ever'" (I.ii.101–105). But now Leontes wants the ritual vow of eternity to erase the seasonality of life. Understanding human life as fleshly and cyclical is intolerable. His horrified compulsion to imagine his son as "neat" in the sense of bovine, rather than in the sense of purely civilized, contrasts with his counterpart Polixenes affectionately calling his own son "My parasite" who "with his varying childness cures in me/Thoughts that would thick my blood" (I.ii.122–127, 168, 170–171). Leontes, in contrast to this healthy purgation, compares his discovery of his pregnant wife's supposed infidelity to a man who discovers a spider in a cup from which he has been drinking, and therefore "cracks his gorge, his sides,/With violent hefts," trying to vomit out the toxic otherness of mortal life (II.i.44–5).

The first speech manifesting Leontes's madness notoriously defies coherent paraphrase, but its insistent theme is pernicious interaction and mysterious penetration of the self, signaled by the same co- and com- prefixes that disgust Coriolanus: within eight lines Leontes says, "stabs the centre ... Communicat'st ... thou co-active art,/And fellow'st nothing ... co-join with something ... beyond commission ... to the infection of my brains" (I.ii.138–145). The same theme resounds through Leontes's jealous observations of his wife, who becomes an overly penetrable fowl or fish: "How she holds up the neb! the bill to him! ... she has been sluic'd in 's absence/And his pond fish'd by his next neighbor ... other men have gates and those gates open'd,/As mine, against their will" (I.ii.183–184, 194–195, 197–198). His conclusion emphasizes astrological and sexual penetrations as corrupting humanity:

> It is a bawdy planet, that will strike
> Where 'tis predominant; and 'tis pow'rful—think it—
> From east, west, north and south. Be it concluded,
> No barricado for a belly. Know't,
> It will let in and out the enemy,
> With bag and baggage.
>
> (I.ii.201–206)

Even love-making seems adversarial to Leontes, since it mixes bodies. A man frantically seeking exemption from the ordinary interplay of life, but not quite aware of doing so, could easily perceive his children as evidence of something gone shamefully wrong. A deep horror at adulterations might translate, across levels of the psyche, into an inchoate suspicion of adultery.

When Polixenes flees this deadly jealousy with Leontes's trusted advisor Camillo, Leontes again focuses on the violation of both the boundary of his city and the boundary of his son's body, a body that is inevitably, and to Leontes horribly, intermingled with his wife's body: "How came the posterns/So easily open?" He then immediately turns to and against Hermione as a corrupter of their son: "Give me the boy. I am glad you did not nurse him. Though he does bear some signs of me, yet you/Have too much blood in him" (II.i.52–58). The then-conventional notion that women were leakier vessels than men (Paster 23–63) is unameliorated here by the benign function of lactation, which both Macbeth's wife and Coriolanus's mother disdain in order to support the project of perfected masculine identity. Hermione insists she has been "continent," and demands to know why she has been "barr'd, like one infectious" from the "first-fruits of my body" (III.ii.34, 97–98), but the answer surely lies in Leontes's indistinct phobia about the interplay of bodies, which he now evidently sees as a contagious degradation of the category of human.

Leontes promptly erects new barriers, which the play unrelentingly exposes as contrary to nature. He locks Hermione away in prison: "So have we thought it good/From our free person she should be confin'd" (II.i.193–194). That "confinement" is ironically juxtaposed with the other meaning of the word, the release of Hermione's new-born daughter: "This child was prisoner to the womb, and is/By law and process of great Nature thence/Freed and enfranchis'd" (II.ii.57–59).

By this time, however, Leontes is no longer satisfied with merely imprisoning these reminders of mortal kinship: "say that she were gone,/Given to the fire, a moi'ty of my rest/Might come to me again" (II.iii.7–9). This fantasy combines Coriolanus's vengeful determination to incinerate the disgusting common bodies—the compromising mess— of his native city with Macbeth's discovery that, by cutting himself off from nature, he has forfeited the regenerative surrender of identity called sleep (by which the *Midsummer* night-fairies repair the human race). Hermione's waiting-woman Paulina offers to cure Leontes's insomnia by showing this "unnatural lord" his new daughter as a reminder of the "good goddess Nature, which hast made it/So like to him that got it" (II.iii.113, 104–105): his flesh and blood. But Leontes thrice renews his determination to purge the evidence with fire (II.iii.114, 155–157).

Life in Leontes's Sicilia then closes down to a lifeless winter many years long; and in earthier Bohemia the hyper-civilized fantasy exits pursued by a bear, as Sicilia's emissaries are swallowed up by ferocious nature (III.iii.95–101). But the flow of blood is life as well as death. The crafty Autolycus sings that "the red blood reigns in the winter's pale" (IV.iii.4), which certainly seems to link the return of spring to the return of life's sustaining fluid to the "pale"—the enclosure—of pallid life.

At a sheep-shearing festival in Bohemia, the tension between civilized self-containment and mutualist permeability takes the form of a traditional debate about art versus nature that collapses into paradox. Characteristically, this play seeks a redemptive middle ground between the incontinent mixing of bodies and the chilly isolation of bodies: the offer (consolatory but hardly sufficient in *King Lear*, a fatal surrender of autonomy for Coriolanus) to take another person's hand, which occurs at least ten times during this play. Distributing flowers to the guests at this seasonal feast, Perdita—assumed by all to be a farm lass, but actually the abandoned child of Leontes and Hermione—implies a warning to the maidens by giving them "The marigold, that goes to bed w' th'sun,/And with him rises weeping" (IV.iv.105–106), balanced against a *carpe diem* warning that forbidding sexual penetration has its own risks, since "winter's pale" is always coming back around: "pale primroses,/That die unmarried" (IV.iv.122–123). A less lovely version of that balance emerges in Autolycus's ballads: one about a pregnant woman who "long'd to eat adders' heads and toads carbonadoed" (IV.iv.262–265), another about "a woman who was turn'd into a cold fish for she would not exchange flesh with one that lov'd her" (IV.iv.279–280).

A penitent Leontes welcomes Polixenes's son and his unrecognized daughter back into his kingdom, but his fantasy of unnatural Sicilian purity persists: "The blessed gods/Purge all infection from our air whilest you/Do climate here!" (V.i.168–170). And at first he mistakes the next generation for his own, and wants his wife's statue to look like her younger self, as if she could be wife eternal, as he had fantasized himself "boy eternal." But soon he is "ready to leap out of himself for joy of his found daughter" (V.ii.49–50).

To test whether Leontes really has overcome his fantasies of civilized transcendence, Paulina offers him a convergence between art and nature that obliges him to weigh their relative value: a statue of Hermione by a sculptor "who, had he himself eternity and could put breath into his work, would beguile Nature of her custom" (V.ii.97–103). The gentlemen plan to "with our company piece the rejoicing" (V.ii.107–108); the shared-bread word so important in *Coriolanus* is back, along with the idea of the individual as a "piece"—a word that appears more often in *The Winter's Tale* than in any of Shakespeare's other works.

Even more encouraging is Leontes's excitement at seeing Hermione's "natural posture," even though it is "piercing to my soul. O, thus she stood,/Even with such life of majesty (warm life,/As now it coldly stands), when first I woo'd her!" (V.iii.34–36). As Iago makes Othello feel that his embraces with the supposedly unfaithful Desdemona have only "begrim'd" them both, and that he could now love her only as a cold alabaster statue of herself (III.iii.386–388, V.ii.1–19), Paulina tests Leontes further by warning him that kissing Hermione's statue will "mar" her and "stain" him, as he seems to have felt in retrospect about the conception of their daughter (V.iii.80–83). When Leontes nonetheless consents to let the statue "take you by the hand" (V.iii.89), Paulina tells the supposed statue, "Dear life redeems you" (V.iii.103), and tells Leontes, "Nay, present your hand." Leontes's reply signals his complete acceptance of ordinary companionate life: "O, she's warm!/If this be magic, let it be an art/Lawful as eating" (V.iii.107–111).

Overriding his rationality, and with it his earlier revulsion from drink, Leontes indulges his thirst to believe that Hermione lives,

> For this affliction has a taste as sweet
> As any cordial comfort. Still methinks
> There is an air comes from her. What fine chisel
> Could ever yet cut breath? Let no man mock me,
> For I will kiss her.
>
> (V.iii.76–80)

We may recall here Vasari's comment that, in Michelangelo's *Pietà*, "the fine tracery of pulses and veins are all so wonderful that it staggers belief that the hand of an artist could have executed this inspired and admirable work so perfectly" (Bartlett 288). Yet every living creature is such a masterpiece, beyond the highest achievements of high culture. That is what King Lear understands, tragically too late, when he surrenders his over-investment in his personal sovereignty and says that a single breath from his martyred daughter Cordelia would be a "chance which does redeem all sorrows/That ever I have felt" (V.iii.266–268).

At this point the *Winter's Tale*'s audience must pass a parallel test—a test that many sophisticated commentators have, in my opinion, failed. Despite the play's title, this is not a fairy tale; or if it is, it comes true every spring. The living Hermione is not a singular, transcendent miracle of art, but instead the far too easily overlooked daily miracle of life. The living Perdita who has grown to look so much like her mother did in that time of courtship is surely no less wonder-worthy a creation than her mother returned to the world in guise of a statue—and Paulina rejects Leontes's complaint that the statue looks older than Hermione did when he last saw her alive (V.iii.23–32). The loss of Hermione's old self is a gain. Paulina repeatedly warns her audience not to mistake what she has done for magic, and Shakespeare slips in a remark that Paulina "hath privately twice or thrice a day, ever since the death of Hermione, visited that remov'd house" (V.ii.105–107) where the statue appears, presumably bringing meals. Flesh and its exchanges must not be mistaken for marble.

Shakespeare goes to remarkable lengths to alert his Christian audience that Leontes seeks exemption from the mortal legacy of the Fall that would otherwise make him inherently corrupt and corruptible, doomed to death and vermiculation (Watson, *Shakespeare* 227–279). But by keeping himself so proudly locked up against the world of creaturely appetites (a sophisticated Coriolanus in that regard), Leontes locks in the intolerable toxins of his own mortality. His quest to make Sicilia a timeless Edenic garden is the theological aspect of an error that has an ecological aspect also. He experiences connection to other lives as a degrading contagion, whether understood in the theology of original sin or the biology of procreation. When Leontes recalls Hermione's lily-like acceptance of his hand, she replies, "'Tis Grace indeed" (I.ii.105). Shakespeare thus uses the convenient metaphor of Christianity—from its beginning, a rebuttal to the Roman model of insular manhood and its violent pride—to sacralize his claims for the permeable self and its comic answer to mortality. The very last line of *The Winter's Tale* tells us that its lives are no longer "dissever'd" (V.iii.153–155).

Leontes's Italian-Renaissance-humanist fantasy of a city or civilization so advanced that it excluded all creaturely contamination and hence produced immortality anticipates the fundamental error that threatens the protagonist of Shakespeare's next (and final complete) play, *The Tempest*. The chilly classical idea of *fama* again proves to be no substitute for the warmth of family, any more than a statue should be preferred to a living spouse.

Who better than Shakespeare to propose a comedy of the commons? He was supremely open to the experience of multiple conflicting characters and their internal conflicts, and supremely successful at weaving them together into plays that communicate across cultures and generations. In that Globe where Jaques explicated the seven parts all men play on the world's stage, and Hamlet exclaimed, "What a piece of work is a man," audiences could hardly escape pondering, among other epistemological questions, what a puzzle identity actually is. It becomes especially puzzling in a setting where parts and pieces are the sections of any body, but are also the various characters' speeches that together make a up a wonderfully harmonious imitation of life. Shakespeare's unsurpassed ability to remind people how much they have in common with people in distant centuries and continents refutes a narrow presentism, just as the plays themselves refute a narrow individualism. To understand life as—like Shakespeare's last plays—a series of tragicomedies of separation and reunion, rather than a doomed struggle of the human self toward its own immortality, is a giant step toward an ecological sensibility.

References

Bacon, Francis. *The Advancement of Learning*. Oxford: Benediction Classics, 2008. Print.

Bakhtin, Mikhail. *Rabelais and His World*. Trans. Helene Iswolsky. Bloomington, IN: Indiana University Press, 1984. Print.

Bartlett, Kenneth R. *The Civilization of the Italian Renaissance: A Sourcebook*. Toronto: University of Toronto Press, 2011. Print.

Braden, Gordon. *Renaissance Tragedy and the Senecan Tradition: Anger's Privilege*. New Haven, CT: Yale University Press, 1985. Print.

Elias, Norbert. *The Civilising Process. The History of Manners: Sociogenetic and Psychogenetic Investigations*. Trans. Edmund Jephcott. Oxford: Blackwell, 1978. Print.

Heise, Ursula K. *Sense of Place and Sense of Planet: The Environmental Imagination of the Global*. New York: Oxford University Press, 2008. Print.

Meeker, Joseph W. *The Comedy of Survival: Literary Ecology and a Play Ethic*. Tuscon, AZ University of Arizona Press, 1997. Print.

Paster, Gail Kern. *The Body Embarrassed: Drama and the Disciplines of Shame in Early Modern England*. Ithaca, NY: Cornell University Press, 1993. Print.

Selleck, Nancy. "Donne's Body." *Studies in English Literature* 41 (2001): 147–174. Print.

Shakespeare, William. *The Riverside Shakespeare*. Ed. G. B. Evans. Boston, MA: Houghton Mifflin, 1974. Print.

Watson, Robert N. "*Midsummer Night's Dream* and the Ecology of Human Being." *Ecocritical Shakespeare*. Ed. Lynne Bruckner and Dan Brayton. Farnham, Surrey: Ashgate, 2011. 35–56. Print.

———. "Protestant Animals." *English Literary History* 81 (2011): 1111–1148. Print.

———. *Shakespeare and the Hazards of Ambition*. Cambridge, MA: Harvard University Press, 1984. Print.

11

MULTISPECIES EPIDEMIOLOGY AND THE VIRAL SUBJECT

Genese Marie Sodikoff

In August 2014, an acute outbreak of the bubonic plague hit Madagascar. It quickly spread from the northern town of Amparafaravola along roadways and southwest into the capital, Antananararivo. Many international readers were no doubt surprised that this medieval blight still existed and had the potential to balloon into an epidemic. Online accounts of Madagascar's plague outbreak depicted grisly pictures of dead rats and heaps of uncollected trash in the capital, connecting the dots for readers between rodents, urban squalor, and lethal disease. I was interested in the outbreak not only for humanitarian reasons, having done anthropological work in Madagascar over the past twenty years, but also because I had embarked on a new project on zoonosis and land degradation around a giant industrial mine along the island's rainforest escarpment. I am now collecting stories about two diseases, the bubonic plague, transmitted by rats and vectored by fleas, and rabies, spread most commonly by wild animals to pet dogs and cats, who in turn infect people. These stories tell of viral subjects caught in a downward spiral of bodily and ecological health.

Zoonosis—from *zoo* ("of animals") and *nosos* ("disease")—is the type of infectious disease in which a virus, bacterium, parasite, fungus, or prion crosses from an animal species to humans, causing illness. For modern humans and extinct hominin species, zoonosis is as old as time and may be the origin of many of our contemporary diseases from measles and tuberculosis to the common cold. They may or may not depend on vectors, such as insects or bats, which transmit pathogens from animals to humans. Anthropozoonosis describes the opposite pathway—that is, disease that spills over from humans to nonhuman animals.

Certain zoonoses have occasioned close public scrutiny and alarm, such as HIV/AIDS in 1981, the 1997 Hong Kong outbreak of avian bird flu, and the 2009 swine flu in the United States. Outbreaks of novel viruses continue to capture the attention of readers around the world: Middle Eastern Respiratory Syndrome, transmitted to people via camels; Sudden Acute Respiratory Syndrome in China, likely spread through the consumption of bats or civets; and Ebola in West Africa, thought to be vectored by fruit bats. The Centers for Disease Control in the United States vigilantly scans the globe for the next "big one," a disease outbreak that could devastate huge swaths of the human population, as the bubonic plague did centuries ago. Beyond the sensational outbreaks, chronic zoonoses affect our everyday lives, and some cases are increasing. In the United States alone, people live with the lurking threat of Lyme disease, salmonella, rabies, West Nile Virus, and a host of others.

As the decades of the Anthropocene march forward and the climate warms, we are witnessing a proliferation of microbial life forms pathogenic to humans, particularly in tropical latitudes (Jones et al.). The subject of virality and examinations of viral subjects must take into account the economic activities and ecological changes that foster zoonotic conditions. As people around the planet invade ever-diminishing primary forests and wetlands to eke out a livelihood, they settle beside habitats already overcrowded with wildlife populations, and therefore more pathogenic. People's encroachment on remnant habitats, coupled with the alternately life-threatening or microbe-incubating effects of a warming climate, pushes wild animals, insects, and pathogens into new areas. These contact zones between built settlements and wilderness borderlands are risky spaces, the hotspots of spillover disease. Diseases often lurk within reservoir hosts after outbreaks dissipate. A reservoir host is a "living organism that carries the pathogen, harbors it chronically, while suffering little or no illness," as David Quammen explains in his book about zoonoses, *Spillover* (23). Primary habitats, such as rainforests, are home to a variety of reservoir hosts. As forests erode, and contact between people and hosts intensifies, the viruses carried by hosts sometimes become hyperactive, "chatty," infecting people but not yet reaching widespread transmission (Lebarbenchon et al.; Barrett 84). Scientists discover "viral chatter" where people hunt and eat bushmeat, where wild species invade and root around in human dwellings, and where mosquitos and other insect vectors find watery havens in open spaces.

For anthropology, identifying the risky encounters among humans, animals, and insects, and tracing the pathways of zoonotic disease in specific places call for interdisciplinary spillover: a mingling of epidemiology and multispecies ethnography, as well as medical and environmental anthropology. The study of the "viral subject" must engage the social sciences and life sciences if the goal is not only to understand but also to contain pathogenesis and improve the quality of macroscopic life. My interest in interdisciplinarity and a multiscalar analysis—from pathogen to cultural cosmology—is inspired by an environmental humanities approach that reflects on the natural history of human selfhood in a time of mass extinction (see Thomas). In examining infectious diseases that breach taxonomic boundaries and challenge the anthropocentric conception of the subject or the self, I also draw on posthumanist perspectives within the environmental humanities. Zoonosis unsettles the human subject because it entails an unwholesome intermingling of mammals, insects, and microscopic pathogens. The idea that persons are intrinsically and exclusively human is revealed to be factually incorrect, and therefore able to yield only partial insights into the rapidly changing conditions of life at present.

Zoonosis emerges out of multispecies interactions inside individual bodies and out there in the external world. Merrill Singer holds that medical anthropology "must be concerned with interactions among multiple life forms" not merely as they affect life for humans but also, comprehensively, for ecosystems (1282). The "syndemic" denotes the interface between two or more diseases and the ways in which the actions and adaptive responses of organisms mutually shape disease courses. He asserts that zoonoses, in particular, demand that the concept address ecological context and unstable taxonomic categories. Singer recommends an ecosyndemic approach that encompasses "intersections and blurred species boundaries" within changing environmental conditions (1282).

Anthropologists Eben Kirksey, Craig Schuetze, and Stefan Helmreich coined the term "multispecies ethnography" to denote the new genre of writing that foregrounds nonhuman others and elevates their status to beings of interest rather than mere backdrops of the human theater. They assert that a new generation of ethnographers is exploring "how 'the human' has been formed and transformed amid encounters with multiple species of plants, animals, fungi, and microbes" (2). Laura A. Ogden, Billy Hall, and Kimiko Tanita describe the posthumanist orientation of multispecies ethnography, as anthropologists seek to expand our understanding of species-being, agency, and world. They see multispecies ethnography as "ethnographic research and writing that is attuned to life's emergence within a shifting assemblage of agentive beings," including "biophysical entities as well as the magical ways objects animate life itself" (6; see also Kohn; Kirksey).

Multispecies ethnographers generally concur with the premise that humans and nonhuman others "co-become" over time, their existences intertwined (see Derrida; Latour; Haraway; Descola). The thought that humans, nonhumans, and matter come into existence interactively and dynamically implies that the boundaries of personal and species identities are porous and selves are mutable. Such a view has roots in the ethnology of personhood in the West and parts of the rest of the world, such as India and Melanesia. Whereas the Western concept of personhood has historically been defined by individualism and individuality, the self being "compressed within the physical body" (Fowler 16), elsewhere it has been described as contingent and relational, shifting through different social transactions. A person's identity is contextual and relational; it hinges on the meaning of exchanges with other humans, objects, animals, or plants. Nonhuman entities may also be ascribed personhood. Persons in this rendering are infinitely "partible," more aptly named "dividuals" than individuals, as Marilyn Strathern has argued concerning personhood in Melanesia.

Posthumanism adds new meaning to the concept of partible personhood. Interactions between humans and nonhuman species by which microbes are transferred from one species body to another further trouble the Western ontological view of hermetically sealed species' bodies and fully human agency. Moreover, as genomic science advances, we discover that our human selfhood includes microbial communities—a microbiome within—that may be pathogenic or benign (Turnbaugh et al. 804; Singer). Animals and plants are biomolecular networks of hosts and associated microbes, or "holobionts" (Bordenstein and Theis; Lincoln). Julia Adeney Thomas points out that with the completion of the Human Microbiome Project in 2012 by the National Institutes of Health, scientists have discovered that microorganisms within the human body outnumber human cells by ten to one. Microbes, she argues, are "inseparably 'us,' more responsible than 'we' are for 'our' existence by most calculations on this micro level" (1594). Our specific microbiome may in fact influence our moods, behavior, and even personality, as Jaroslav Flegr controversially asserts with regard to the parasite *Toxoplasma gondii*, transmitted through cats and able to exert a degree of puppet-mastery over its host (Flegr; Lotterman). We begin to suspect that the human self is chimeric and epiphenomenal.

Who is a mind colonized by pathogens transported by insects and mammals? Who controls a rabid body, a febrile mind? Zoonosis reveals the vulnerability and porosity of species boundaries. The dividual here is more than the person forged, temporarily, by a particular social exchange; it also encompasses transactions among other life forms in a rapidly changing landscape.

Mutable, partible personhood, tied to multiple agencies and multispecies encounters, renders the subjectivities of zoonotic disease. As exotic pathogens jump the xenographic barrier to infect human bodies, they induce febrile states, local rumors, and, as is often the case, causal narratives deprived of scientific information. For my new project in Madagascar, I am interested in the dynamic of multiple variables that keep disease in motion and comprise the stories of persons whose bodies are colonized by zoonosis as they navigate the terrain, the broken healthcare systems, and the effects of state and corporate practices that degrade rural and urban environments. Malagasy people also intersect the movements of wild, displaced animals, hungry rodents, rabid dogs, and vectoring insects, and they engage in forms of caretaking for infected individuals that often jeopardize their own health.

The viral subjects of my study include people, animals, and microbes at the interfaces of rainforest fragments and industrial mines along the eastern escarpment of the island. Landscape change in both rural and urban areas and the rising incidence of zoonotic diseases in certain places influence how people and animals, both domestic and wild, interact and move over space. The perceived deterioration of countrysides and cities in Madagascar has generated critique that combines quasi-scientific explanation of disease outbreaks with a postcolonial sensibility. Malagasy people liken the toxic effects and criminal atmosphere of the mining and logging boom by multinational companies to colonialist land grabbing and wealth extraction.

The ethnography of zoonosis in Madagascar

The case of rabies in Madagascar illustrates the ways etiological beliefs, material resource scarcities, human–animal contact, and the changing landscape assemble to create recognizable viral subjects, including sick people, hot topics, mad dogs, disease hosts, and fantastical creatures. Since the early 2000s in Madagascar, and especially after a coup d'état in 2009, a boom in mining and logging has been devastating rainforest remnants. The quality of life for most Malagasy is becoming more precarious as the island's infrastructure and services for health, waste collection, and sanitation deteriorate.

The town of Moramanga lies a couple of hours east of the capital, on a well-traveled national route. It is a large town that has grown rapidly since the early 2000s due to an adjacent nickel and cobalt mine called Ambatovy, which employs hundreds of residents. In Moramanga, as elsewhere on the island, rabies is endemic and spread primarily by stray dogs. The local doctor at the hospital in charge of vaccinating people against rabies once they have been bitten claims to see about four cases of suspected rabies infection per day. The high incidence results from the growing population of mean dogs. Well-appointed homes on the road leading to the entrance of Ambatovy's open pit mine are guarded by aggressive dogs, so some residents associate the worsening rabies problem with the mine.

The stories I collected in June 2015 in Moramanga and Amparafaravola—the ground zero of the 2014 plague outbreak—recount varied entanglements of people, animals, insects, and microbes in conditions of social inequality and rapidly changing ecologies and climate. Folk etiologies, the explanations everyday people offer for the causes of disease, suggest in these cases people's implicit political critiques of defunct governmental services and collusions between the state and the private sector. The larger project aims for an ecosyndemic approach to rabies and the plague; an outbreak of the more lethal pneumonic variety of the plague occurred in the Moramanga area in August 2014. These stories thematize virality in unexpected ways. I offer an example of one viral subject, the *kelibetratra* ("little big chest"), described to me as a species of wild dog whose native habitat is the rainforest.

Dr. Paul (a pseudonym) is the Malagasy physician in charge of the rabies ward at the Moramanga hospital. He authorized me to sit in the hospital room to await rabies victims in need of inoculation and to hear their accounts of how they were bitten. I did not have to wait long one June weekday when a boy about four years old walked in, crying, with his father, his ankle red and oozing from an infected bite he got a couple of weeks earlier. Dr. Paul and a nurse jotted some information into the logbook. Had the dog been rabid? It wasn't certain. The father had not seen the dog again. The vaccine was a precaution. Cases of human rabies in Moramanga had not occurred in recent memory, but a man in the tourist town of Andasibe recently died of rabies, and people recalled one other case nearby.

The nurse rolled up the boy's sleeve as he recoiled from the needle. She swabbed his upper arm and injected him with the thin syringe. Tears streamed down his face. When it was over, he grew calm. He would have to return a total of four times to complete the course of vaccine. The treatment was free to patients. But vaccinations for animals, needed annually, were exorbitant for most residents, so the virus was never tamped. It circulated in roving dog packs in town, or at the more distant forest-village boundary, where cattle, pet dogs, and cats encounter a variety of potentially rabid mammals.

Dr. Paul explained that cases of human rabies were probably more frequent than they knew, but people in the more remote villages, a day's walk from the hospital, did not receive information about what to do if bitten by a dog or bat. If they fall ill and die, people often assume the cause was the fever (*tazo*), malaria. It is not uncommon for bats to fly into people's homes in Moramanga sometimes, flapping about and causing pandemonium. Yet they can also be silent, stealthy biters and vectors of rabies. Two more bite victims walked into the simple, sunlit room that morning. Each patient had a story about a mean dog, assumed rabid. The medical staff could not be blamed for excessive vigilance against rabies; it runs rampant in town.

The next day, I accompanied a colleague from the Pasteur Institute of Madagascar to track down the local veterinarian in town, Madame Marthe (also a pseudonym), who ran a small shop in the midst of the labyrinthine marketplace, where she sold animal medications. It is her role to follow up on dog bite episodes. She explained that the usual process was to instruct a biting dog's owner to leash the animal in a courtyard for several days and to observe its behavior. Did the dog eat? Did it avoid drinking water? Was it uncoordinated, anxious, raging, and aggressive? Was the mouth paralyzed, and did it froth? Did it lose control of its faculties? As Madame Marthe posed such questions to dog owners, she was essentially asking whether their animals were no longer themselves. The rabies virus robs the sense of self from a body, and replaces it with a frothing madness that benefits the virus as it hurls its population toward other bodies via saliva. Unlike most viruses that travel through the bloodstream, rabies courses through the nervous system toward the brain. Once there, it "works slowly, diligently, fatally to warp the mind, suppressing the rational and stimulating the animal" (Wasik and Murphy 3). Madame Marthe would make daily rounds to any animal suspected of having rabies in order to make her own observations. If she determined the dog was infected, it was euthanized: she would have someone decapitate the dog. She would wrap the head carefully, taking care not to touch the blood or brain matter, and ship it to the Pasteur Institute in the capital where scientists analyzed the brains.

Madame Marthe complained about the lack of resources. The marketplace was not electrified, so she could not preserve vaccines in her refrigerator. It was hopeless. She was constantly beseeching the electric company to resolve the problem, to no avail. How could they control rabies if they could not refrigerate the vaccines to inoculate the dogs? "When I sent in the heads over the last year, something like fourteen out of sixteen came back positive for rabies," Marthe told me excitedly. I had heard people say, including Dr. Paul, that the source of the virus was this wild dog, the *kelibetratra*. "Yes!" she confirmed and went on to describe the elusive animal to me. I had never heard of a species of wild dog in

Madagascar, and I was intrigued. Would my project involve a cryptozoological dimension, camping out in the rainforest with a team, waiting for the appearance of the "little big chest"? Multispecies ethnography entails risks. Does one trace the movements of stray dogs to see where they congregate at night, where they come into contact with human passersby?

"The *kelibetratra* has short arms and legs." She began to sketch a shape on a scrap of paper. "Like a pitbull with short limbs. But a big chest, big for its size." She patted her chest with the palm of her hand. "It comes into town at night and gets into dog fights. The next morning, you see a dead dog. Or a wounded dog. Then that dog becomes rabid (*romotra*)." No one has seen a *kelibetratra*. She then corrects herself: "Once! Once I saw one at night. A breed I had never seen before here." The more people I asked about the *kelibetratra*, the more people confirmed its existence, though none had ever spotted one personally.

I returned to the hospital to ask Dr. Paul more about this creature. He also agreed that the *kelibetratra* was the origin of the rabies virus in these parts. "It comes out of the forest and attacks domestic dogs and cats." He worried that the situation appeared to be aggravated by the shrinking forest. As the forest eroded, wild animals that had been in its depths were coming out into the open, into villages, with greater frequency. I searched online and in scholarly articles for the vernacular name of this species and discovered a 2007 article in French from a Malagasy political commentator in *Le Tribune*, a Malagasy daily newspaper: "Recall that last year, the population of Mahajanga lived in terror due to disinformation on the existence of an extraordinary animal commonly called 'Kelibetratra' who does not hesitate to kill. In fact, it is only rumors" (Guy). From another newspaper, *Tana News*, in a 2012 article entitled, "Satanism and Bandits in the South," a commentator offers his analysis: "In our national history ... when our nationalist ancestors wanted to get rid of the colonial powers, they mobilized rumors of the Mpakafo [heart thieves] and Bibiolona [creatures that are part human, part beast]. When the Ratsiraka regime seemed too oppressive, we invented the Kelibetratra" (Raharizatovo).

The figure of the *kelibetratra* had gone viral in the popular imagination, providing a compelling narrative around Moramanga about how forest erosion from agriculture and mining scared dangerous, disease-ridden animals, formerly hidden in the deep forest, out into the borderlands. In this region, the *kelibetratra* narrative sounds plausible, which explains why the local physician and veterinarian believed in its existence. Certainly rainforest erosion has prompted an uptick of the flight of bats into villages and towns, and has possibly exacerbated the spread of rabies within rainforest fragments, as species populations are increasingly overcrowded. Rainforest edges comprise the virally chatty spaces where novel zoonoses emerge, and long-established zoonoses, spread through species movements and interspecies contact, remain active and virulent.

Moramanga residents' desire to locate the origin of the rabies epidemic in the hinterlands, where it allegedly creeps into town on the stumpy legs of the *kelibetratra*, or is relayed from *kelibetratra* to roving strays, echoes assertions I have long heard in Madagascar about the rural–urban divide. In rural villages of the east coast, villagers and town residents would typically attribute behaviors and traits they considered bumpkinish or "primitive" (*sovazy*, from the French, *sauvage*) to more distant settlements, particularly those that lay closer to the rainforest edge. In this case, the rabies virus, which turns already menacing guard dogs into contagious killers, imports untamed wildness into the courtyards and well-trafficked roads of peri-urban Moramanga, adding to local criticisms of the Ambatovy mine.

Yet the *kelibetratra* is not confined to the rural periphery. It is also reputed to stalk neighborhoods of the city, Antananarivo, far from the rainforest, where it attacks pet dogs that sleep outside to protect people's homes, according to a longtime colleague of mine who resides in a wealthy neighborhood near the airport. He claimed that sometimes one can hear the sounds of howls and snarls echo in densely packed quarters at night. But as in Moramanga, the creature inevitably disappears into the night, unseen. The *kelibetratra* is like the werewolf, a *"bibiolona,"* the subject of nightmares and delirium. The narrative's scientific plausibility weakens in the city, where the animal, a "ghost of colonialism," seems a manifestation of a collective anxiety, and maybe, as the political commentators suggest, a product of sublimated anger against oppressive and corrupt state leadership (Good 518).

Ethnographic methods reveal cultural beliefs about pathways of disease that may elude capture by epidemiologists whose data determine who gets infected and by what means, as well as the paths over which a microbe may travel. By the same token, epidemiology and the other sciences recruited into solving multispecies disease outbreaks guide the anthropologist's investigations, pointing the way toward hotspots and routes of transmission and contagion. It is also essential to grasp how pathogens jump the species divide and reproduce over space and time, exploiting errors in human judgment and weaknesses in the political and social structures of places like Madagascar.

As a fantastical creature that explains the spread of rabies while offering tacit political critique, the *kelibetratra* is both a febrile projection of the collective imagination, a subject in its own right, and a figment that lends insight into subjectivity in contemporary Madagascar. Medical anthropologists, such as Byron Good, have argued that the term "subjectivity" entails an awareness of the genealogy of the subject, its roots in colonial subjugation, and its emergence out of evolving forms of exclusion, violence, and modes of governance (Good et al. 3). With "viral subjectivity," I aim to add a posthumanist perspective to the contemporary anthropological treatments of subjectivity, broadly defined by João Biehl et al. (5) as the "inner life processes" shaped by historical and geographical circumstances and refracted "through potent political, technological, psychological, and linguistic registers." Viral subjectivity emphasizes the beyond- and more-than-human dimensions of human affective states and worldviews that evolve within particular places, at particular times. Through the methods of multispecies ethnography and epidemiology, anthropologists can identify viral subjects, tracing how disease vectors and risky interactions conjoin humans and other animals in common suffering, and chronicling the practices, behaviors, and words of those affected by or working with zoonosis. Such efforts illuminate the multiple forms of life that constitute subjectivity in the fast-changing habitats of the late Anthropocene.

References

Barrett, Ron. "Avian Influenza and the Third Epidemiological Transition." *Plagues and Epidemics: Infected Spaces Past and Present.* Ed. D. Ann Herring and Alan C. Swedlund. New York: Berg, 2010. 81–94. Print.

Biehl, João, Byron Good, and Arthur Kleinman, eds. *Subjectivity: Ethnographic Investigations.* Berkeley: University of California Press, 2007. Print.

Bordenstein, Seth R. and Kevin R. Theis. "Host Biology in Light of the Microbiome: Ten Principles of Holobionts and Hologenomes." *PLoS Biology* 13.8 e1002226. 18 August 2015. Web. 15 October 2015.

Derrida, Jacques. "And Say the Animal Responded?" *Zoontologies: The Question of the Animal.* Ed. Cary Wolfe. Minneapolis: University of Minnesota Press, 2004. 121–146. Print.

Descola, Philippe. *The Ecology of Others.* Trans. Geneviève Godbout and Benjamin P. Luley. Chicago: Prickly Paradigm Press, 2013. Print.

Flegr, Jaroslav. 2013. "Influence of Latent *Toxoplasma* Infection on Human Personality, Physiology and Morphology: Pros and Cons of the *Toxoplasma*–Human Model in Studying the Manipulation Hypothesis." *Journal of Experimental Biology* 216 (2013): 127–133. Pub Med. Web. 15 October 2015.

Fowler, Chris. *The Archaeology of Personhood: An Anthropological Approach.* London: Routledge, 2004. Print.

Good, Byron J. "Theorizing the 'Subject' of Medical and Psychiatric Anthropology." *Journal of the Royal Anthropological Institute* (N.S.) 18 (2012): 515–535. Web. 18 December 2015.

Good, B., M. Good, S. Hyde, and S. Pinto. "Postcolonial Disorders: Reflections on Subjectivity in the Contemporary World." *Postcolonial Disorders.* Ed. M. Good, S. Hyde, S. Pinto, and B. Good. Berkeley: University of California Press, 2008. 1–40. Print.

Guy, M. Dolsain. "Des actes de destabilization." *La Tribune Madagascar.* 13 July 2007. Web. 1 July 2015.

Haraway, Donna. *When Species Meet.* Minneapolis: University of Minnesota Press, 2007. Print.

Jones, K. E., N. G. Patel, M. A. Levy, A. Storeygard, D. Balk, J. L. Gittleman, and P. Daszak. "Global Trends in Emerging Infectious Diseases." *Nature* 451 (2008): 990–993. Print.

Kirksey, Eben. *Emergent Ecologies.* Durham, NC: Duke University Presss, 2015. Print.

Kirksey, Eben, Craig Schuetze, and Stefan Helmreich. "Tactics of Multispecies Ethnography." *The Multispecies Salon.* Ed. Eben Kirksey. Durham: Duke University Press, 2014. 1–24. Print.

Kohn, Eduardo. *How Forests Think: Toward an Anthropology Beyond the Human.* Berkeley: University of California Press, 2013. Print.

Latour, Bruno. *Reassembling the Social: An Introduction to Actor Network Theory.* Oxford: Oxford University Press, 2007. Print.

Lebarbenchon, C., R. Poulin, M. Gauthier-Clerc, and F. Thomas. "Parasitological Consequences of Overcrowding in Protected Areas." *RCE EcoHealth* 3 (2006): 303–307. Print.

Lincoln, Martha. "A Holobiont Manifesto: Microbiology, the Culture Concept, and Post-Human Utopias in the Anthropocene." *Parasites and Human Evolution Symposium.* 26 October 2015. Rutgers University, New Brunswick. Presentation.

Lotterman, Charlie. "Entangled Epistemologies, Entangled Sexualities: Toxoplasma Gondii and the Human Subject." *Parasites and Human Evolution Symposium.* 26 October 2015. Rutgers University, New Brunswick. Presentation.

Ogden, Laura, Billy Hall, and Kimiko Tanita. "Animals, Plants, People, and Things: A Review of Multispecies Ethnography." *Environment and Society: Advances in Research* 4 (2013): 5–24. Print.

Quammen, David. *Spillover: Animal Infections and the Next Human Pandemic.* New York: Norton, 2012. Print.

Raharizatovo, Gilbert. "Satanisme et dahalo dans le Sud." *Tana News.* 14 June 2012. Web. 10 July 2015.

Singer, Merrill. "Zoonotic Ecosyndemics and Multispecies Ethnography." *Anthropological Quarterly* 87 (2014): 1279–1309. Print.

Strathern, Marilyn. *The Gender of the Gift: Problems with Women and Problems with Society in Melanesia.* Berkeley: University of California Press, 1990. Print.

Thomas, Julia Adeney. "History and Biology in the Anthropocene: Problems of Scale, Problems of Value." *American Historical Review* 119 (2014): 1587–1607. Web. 1 December 2015.

Turnbaugh, Peter J., Ruth E. Ley, Micah Hamady, Claire Fraser-Liggett, Rob Knight, and Jeffrey I. Gordon. "The Human Microbiome Project: Exploring the Microbial Part of Ourselves in a Changing World." *Nature* 449 (2007): 804–810. Print.

Wasik, Bill and Monica Murphy. *Rabid: A Cultural History of the World's Most Diabolical Virus.* New York: Penguin, 2012. Print.

12

ENCOUNTERING A MORE-THAN-HUMAN WORLD
Ethos and the arts of witness

Deborah Bird Rose and Thom van Dooren

This chapter is a celebration of, and a call for the cultivation of attentiveness to, the diverse living beings and forms of liveliness that constitute our world.[1] Paying attention in this way is not simply an epistemic project; rather, it is about the difficult work of learning to live well with others in this challenging time that is increasingly coming to be known as the Anthropocene. Our approach is rooted in an ethical practice of "becoming-witness" which seeks to explore and respond to others in the fullness of their particular "ethos," or way of life. This is an inherently interdisciplinary environmental humanities approach that brings the humanities into conversation with the natural sciences and a broad range of other ways of knowing the world, to nourish the connectivities and possibilities that these dialogues produce for people and the wider environment, and to turn them toward the cultivation of better futures.

Albatrosses

As we walked amongst nesting albatrosses on a rocky headland on the island of Kaua'i, we were brought sharply into an awareness of the particularity of this unique way of life. These are birds who spend the vast majority of their lives on the wing, soaring just above the ocean's shifting surface. And yet, these birds are tied to the land too, required to return each year to isolated places like the one where we now stood, to breed and raise chicks. It was December when we visited, so most of the nesting birds that we encountered were males taking their first long shift incubating newly laid eggs. To some observers this might have looked like the early stages of breeding, but in reality the laying of an egg is the culmination of years of effort, of courting and of relationship building. Long before albatrosses are physically ready to breed, they start returning to the colony where they hatched to find a mate. They sing and they dance, weaving and bobbing, responding to each other's sounds and movements on land and in the air. After years of these kinds of interactions, a reliable pair bond is formed.

Successfully raising albatross chicks takes months and months of effort, and is a task that definitely requires two parents. Initially, birds will take turns incubating the egg or guarding the young chick, while the other heads to sea to feed. Once the chick is big enough to be left unattended, both birds will take to the air covering thousands of kilometers in search of the fish, squid, and other tasty morsels required to sate a hungry, growing chick. If one bird dies or abandons the effort, the other will often stick with it as long as he or she can, but will eventually be forced to give up on the egg or chick. In some cases, parents have been known to stay on eggs right up to the point of their own dehydration and starvation, waiting for their mate to return.

As we walked amongst albatrosses, we were drawn into an appreciation of this dedication and reminded of the time and the work that goes into bringing forth the next generation, and so holding this way of life in the world. The nesting albatrosses did not move as we walked close by; some quietly clacked their beaks in the direction of their strange visitors, but for the most part they ignored us. As colonial sea birds, albatrosses have historically bred on islands free of predators. For the millions of years that they and their ancestors have been traversing the oceans of this blue planet, terrestrial mammals like ourselves simply have not been a relevant feature of their environment. We either did not exist, or had not yet come into contact with them. But everything has now changed. As albatross populations around the world plummet—as a result of long-line fishing, but also a range of new threats at breeding sites—this superbly adapted way of life is being brought to the edge of extinction.

Ethos

Our encounter with albatross on that warm day was an opening into their particular "ethos" (plural *ēthea*). The word "ethos" is not widely used these days. It comes from old Greek, where it meant things like character or way of life, but also custom, and customary practices and places. Today it retains a place in some anthropological discussions of aesthetics, poetics, and performance, and can be used to describe the spirit or character of a time—a zeitgeist, or an overarching set of values and practices—a style. Clifford Geertz's definition provides a good baseline for our discussion: "A people's ethos is the tone, character, and quality of their life, its moral and aesthetic style and mood; it is the underlying attitude toward themselves and their world that life reflects" (Geertz 127). Our principal point of departure from Geertz is the recognition that it is not only humans who are known by their ethos. In making this point, we reclaim and redo an older, broader, definition of the term: Homer, for example, wrote of the ethos of horses—their habits and habitats.

From this perspective, an ethos is what makes a group or "kind" distinct; we know (or think we know) that given individuals are *a kind* because they are distinct. This distinctiveness takes many different, but interwoven, biocultural forms. For example, Dominique Lestel et al. note that "there will always be a cognitive or behavioral style that will characterize chimpanzees as chimpanzees and distinguish them as much from gorillas as from elephants and humans" (Lestel et al., "Etho-Ethnology" 171). These behavioral and cognitive forms of distinctiveness are a central part of what constitute the *ēthea* of these animal beings, but they are far from all of it. Chimpanzees are, of course, also distinct from gorillas in a range of other ways that we often label as morphological, physiological, and ecological; that is, in terms of characteristics like size, development, and diet. While there are certainly many

121

similarities between these *ēthea*—many characteristics that chimpanzees and gorillas share with each other that they do not have in common with an elephant or an oak tree—each is nonetheless a distinctive way of life. It is this distinctiveness that our focus on ethos aims to draw out. The goal of doing so is not to reify difference, or to insist on a realist notion of species (or the straightforward correlation of *ēthea* with species), but rather to pay attention to what makes life forms unique. Doing so seriously requires us to ask, again and again, how might we recognize and respond to these differences, and with what consequences for whom?

This broad use of the term "ethos" is a deliberate effort to refuse a clean distinction between the biological and the cultural. An ethos is not a cultural or behavioral element in isolation from a biological form. An ethos is an *embodied* way of life: a way of reproducing, of forming social groups, of advertising to pollinators, of swimming upstream to spawn, for example. It is all of this and more, everything that together constitutes a distinctive "way of being."

Much of what is distinctive about a kind is inherited from others. Some of this inheritance takes the form of genetic material passed between parent(s) and offspring, but it is much more than this in terms of both what is inherited and whom it comes from. Diverse inheritances—sometimes labeled under categories like genetic, epigenetic, behavioral, and symbolic (Jablonka and Lamb)—form the foundation of complex developmental processes in which it makes no sense to ask "how much" is culture and how much biology, how much nature and how much nurture, how much innate and how much environmental (Oyama et al.; Keller). Rather we become with and through diverse others, ancestors and contemporaries, human and nonhuman (van Dooren, "Spectral Crows in Hawai'i"; Rose, "Multispecies Knots"). To be is to become with others (Haraway); it is to inherit, inhabit, and to some extent remake a distinctive style of being in the world. In this context, *ēthea* cannot be formed and sustained in isolation: each is a style or way of being *with others*, of connecting and consequently of sharing "meaning, interests and affects" (Lestel et al., "Phenomenology" 155), as well as flesh, fluids, genetic materials, and much more.

And so it is clear that ethos arises through the interplay of sameness and difference. If there were no differences, there would be no background and foreground, no pattern to distinguish, no figure to become meaningful. Ethos involves knowledge, sense of self and other, discernment between kinds. And yet this recognition is always also partial and halting. An ethos is not an essence. *Ēthea* are emergent, performative co-becomings, never uniform, isolated or fixed, bleeding into and co-shaping one another, and yet somehow maintaining their distinctive uniqueness.

Of course uniqueness in this context can be and is gauged in different ways. In the examples given above we have used (common) species names to describe distinct *ēthea*, but there is no reason to assume that what "we" (who?) recognize as the borders of a species will neatly map onto what it is that characterizes a way of life as meaningfully distinct. In this context our approach aims to problematize a notion of "species" "as a grounding [often assumed and realist] concept for articulating biological difference and similarity" (Kirksey and Helmreich 563). We might think, for example, of the killer whales (*Orcinus orca*) off the West coast of North America, now recognized as two distinct "cultures" on the basis of their different dialects, feeding habits, and other learned behaviors (Keim). Sometimes "subspecies" designations are used to mark this kind of "internal" distinctiveness, especially if it takes genetic or morphological form, but species designations are a complex mess with a range of ways of making cuts and understanding what is at stake in them (Mayr; Kirksey; Amundson; Wilkins). In short, we should not be surprised if within, or across, those units that we call "species," we find a range of possibly distinct *ēthea* (as with the diverse forms of life found amongst those we call *Homo sapiens*). The task is to ask what constitutes a *meaningful* distinction, a question that can only ever be asked in the context of a given line of questioning or possible relationship: how might paying attention to differences "within" or "between" open up new possibilities?

In disrupting any assumed correlation between *ēthea* and species, we practice curiosity about the diverse ways in which others, human and nonhuman, might recognize kinds (their own and others). Vinciane Despret, for example, urges us to consider how it is that lions understand lion-ness, how they decide who is and is not a lion (Despret 126). Eben Kirksey reminds us about how wasps align themselves with particular "kinds" of fig trees. As their key pollinators, the way in which wasps identify appropriate figs matters for the future of these species, allowing pollination, but sometimes also "cross"-pollination between what might be thought about as different species of fig tree. But perhaps they are not different species from a wasp's perspective, and so will not be able to be for much longer from other perspectives either, if cross-pollination continues (Kirksey). In these and many other ways, living beings are making distinctions between kinds—for example, as potential collaborators, competitors, predators, or mates—and not always as "we" would. The diverse modes of recognition and distinction at work here are in some cases conscious, but might just as easily be cellular or molecular: can this flower be pollinated by that pollen, does this immune system attack or ignore that "foreigner"? A range of processes at various scales come together in developmental "intra-action" (Barad) to shape the distinctive modes through which embodied ways of life inhabit a world of diverse beings encountered as particular kinds of others. And so, while (some) humans might be the only ones talking about "species" and "biodiversity," or "cultural diversity" for that matter, we are certainly not the only ones living in and communicating about worlds in ways that pay critical attention to the diversity of living forms.

Mistletoe

We are participants in the more-than-human world, a life-world that is communicative through and through. Almost all of the communication has nothing to do with us. As Martin Burd put it in an article on the language of color between birds and flowers, we are "eavesdroppers on the visual conversations in which they are engaged" (Burd).

Mistletoes are great communicators, and they are amongst the greatest givers on Earth. It seems an odd thing to say of a parasite, but mistletoes are almost improbable in their dazzling beauty and generosity. In the ecological world, the improbable always has a story, the story almost always entices us, and the enticement is part of wider systems of connectivity.

The category "mistletoe" includes numerous families and a very large number of species, some of which are found on all continents except Antarctica, and are creatures of influence wherever they live (Watson). In Australia, mistletoe is seen as an indicator of landscape health because of its capacity to create food and habitat for a large diversity of creatures including birds, butterflies, and other insects, as well as possums and koalas. The plants play "a key role in nutrient dynamics, affecting litter and soil composition" (Watson and Herring 1). An overabundance of mistletoe signals wider biodiversity problems, but in balanced abundance it is a positive benefit to both flora and fauna diversity (Watson and Herring; Watson).

The foremost gift from mistletoes to others is nutrition. The bright, showy flowers attract pollinating birds and insects. The nectar is not only high in sugars, but also fats. Some of the Australian *Loranthaceae* produce nectar containing droplets of pure fat. Insects are in and out of mistletoes, and insect-eating birds get the benefit. The berries, too, are highly visible, abundant, and full of nutrition. Worldwide, many "folivores" eat the high-nutrient leaves: deer, camels, rhinoceroses, gorillas, and possums, amongst many others. Mistletoes maintain their gifts throughout the year, and during some parts of the year theirs is the only food available for some creatures.

Mistletoes offer a steady rain of litter, including their own nutritious leaves. Soil in a mistletoe area is high in nutrients, and all the plants get the benefit, including, of course, the host tree, helping it to pump up the resources that mistletoes tap into.

At the same time that mistletoes are great givers, they are also highly dependent. They must have a host; without the host there is no mistletoe. They must be pollinated, and their seeds must be dispersed. And they must be browsed. The main thing that keeps these remarkably successful creatures in check is that other creatures munch on them.

Mistletoe gifts work in both directions: both the offering and the return. Interdependence, or symbiotic mutualism, is the key to it all. The story of mistletoe mutualisms is all about entanglements of interdependencies, nutrient cycles, and seductions. It tells of an ethos of giving that goes around, and comes back, producing entanglements that are veritable orgies of seductive gifts. These are not gifts in the manner imagined by Derrida, where the perfect gift would be so untouched by self-interest that it would not even be discernable as a gift (Derrida, *Given Time*; *The Gift of Death*). No, mistletoes' gifts are their competitive edge, and they announce themselves ostentatiously, abundantly, and promiscuously. Everyone around them benefits.

Becoming-witness

From this grounding in an attentiveness to ethos, we understand ethics as an openness to others in the material reality of their own lives: noisy, fleshy, exuberant creatures with their multitude of interdependencies and precarities, their great range of calls, their care and their abundance along with their suffering and grief. Within entangled worlds of mutual becoming, attentiveness is necessarily a complex mode of participation. Our practice of "becoming-witness" is a mode of responding to others that exceeds rational calculation, one that arrives through encounter, recognition, and an ongoing curiosity. This practice brings emerging approaches in multispecies studies (e.g., Kirksey and Helmreich; Lestel et al., "Etho-Ethnology"; Lestel et al., "Phenomenology"; Despret) into dialogue with an ethics that arrives from outside the self (or, in this case, the ethos) and that exceeds rational calculation (Levinas and Kearney; Hatley), to produce grounded and situated practices of curiosity and care (Puig de la Bellacasa; van Dooren, "Care"; Haraway; Rose, *Wild Dog Dreaming*).

This approach works against the "reductive stance" (Plumwood 177) which in Western thought over several centuries, at least, has abandoned or consigned nonhumans to oblivion. One of the great terms for this arena of rejection is "social death" (Card), a socially constructed power relation wherein the lives of some humans and most nonhumans are deemed to be either useful to the powerful or superfluous, their meanings (if any) irrelevant, their deaths and destruction non-events except, perhaps, as property loss. Those who are socially dead are tossed into an ethical dead-zone from which no voice is heard because the makers of social death have already decided not to hear or take notice. Consigning creatures to social death sets up conditions of abandonment in which others become standing reserve, deemed to be surplus to requirements (Chrulew), machine-like (Crist), finding their lives and bodies taken over for experiments or sacrificed in pursuit of "higher purposes." Or they just get in the way and "we" decide "we" cannot afford to take the steps that would enable "them" to live sustainably and meaningfully.

At the current time such events take place all around us, and as a result ethical questions take a host of different forms: What kinds of response does mass species extinction demand? Can the killing of many millions of plants and animals on the grounds that they are "introduced" species really be acceptable? What about the industrial killing processes of contemporary "meat production"?

In all of these contexts we are called to witness; we are called not to abandon others and, more positively, we are called to engage others in the *meaningfulness* of their lives. The arts of becoming-witness include both attention to others and expression of that experience: to stand *as* witness and actively to *bear* witness. Underlying these arts is a fundamental disposition toward openness to the calls of others. In the body of collaborative work on extinction that we are developing, we have explored the practice of telling stories with the aim of drawing audiences into others' lives in new and consequential ways. At the heart of these stories is an effort to expose and explore something of the ethos of disappearing species and the others they share their worlds with, something of their particular modes of being in the world.[2] In this work, attentiveness to ethos is integral to ethics.

While responses to others are often grounded in immediacy, genuine responsibility, wherever possible, requires a curious attentiveness that exceeds the given moment so that we might better understand the other in order to make an *appropriate* response. It was precisely this absence of curiosity that Haraway criticized Derrida for in his discussion of his cat. As she notes, "caring means becoming subject to the unsettling obligation of curiosity, which requires knowing more at the end of the day than at the beginning" (36). An ethical response takes an interest in what matters to another rather than reading one's own positioning on to them; it takes account of the diverse ways in which others make and live their worlds. For example, as we have explored in other work, how does the particular fidelity of flying foxes and little penguins to their roosting or nesting places—their intimate and abiding tie—alter the nature of our obligation not to expel them or wantonly destroy these places (van Dooren and Rose, "Storied-Places in a Multispecies City")? Knowing more matters, not least because it draws us into new understandings, relationships, and responsibilities.

But to witness is also to participate in the world in its *relational* becoming. In this context, our curiosity about and for others must be definitively expansive, perhaps even explosive. Can our awakening response draw us into this particular creature's story, and on into encounters with the many others who are bound up in relationships of nourishment, care, meaning-making, and more? Within a broader "ecology of selves" (Kohn), can our awareness become recursive, circling back to the many people who are today unequally exposed to violence and suffering through their entangled relations with diverse nonhumans? Can our awareness make the relational leaps that hold trees, pollinators, soils, climates, mistletoes, and many others within a domain in which ethics are mutual and life-enhancing?

Much of our own work has brought us into situations where some fateful combination of ignorance, negligence, selfishness, and deeply conflicting values are spelling disaster for individuals and species. Our arts of witness involve turning-toward rather than away. The turning-toward is a material and semiotic kinesis that seeks engagement with the world of life. Confronting atrocity, we are acutely aware of the prevalent slippage between explanation and justification; our approach refuses to reduce deathwork to a rational calculation. We are in the most difficult place of witness, where to say nothing is abhorrent, and where to say anything is already to risk reductions. And yet we "stay with the trouble" (Haraway) because the trouble has seized us, and we cannot turn our backs. An ethics of refusal grounds this work: it is the refusal, following Levinas, to abandon others (Levinas and Kearney). Put in the positive, it embraces the ethical call others make upon us in the meaningfulness of their lives and deaths.

One of the great challenges for the environmental humanities is that of developing modes of response that hold on to a curious attentiveness to the lives and deaths of others in a world in which so much is slipping away. This will, of necessity, be deeply interdisciplinary and expansively intercultural work, unsettling limited notions of culture; work that is open to complexity and diversity, and to modes of giving an account that do not need to settle neatly into singular narratives. In dark times, the arts of witnessing are a concerted effort against the force of social death, against the reduction of others, and toward modes of knowing and inhabiting lively, more-than-human worlds.

Notes

1 This chapter draws on our previously published article, Thom van Dooren and Deborah Bird Rose, "Lively Ethography: Storying Animist Worlds." Some sections of this chapter are reproduced from this work.
2 The two short vignettes included in this chapter—on albatross and mistletoe—give only the most cursory sense of this effort. For fuller examples see Rose, "Judas Work"; Rose, "Monk Seals at the Edge"; Rose, "Multispecies Knots"; van Dooren and Rose, "Lively Ethnography"; van Dooren, *Flight Ways*; van Dooren, "Spectral Crows in Hawai'i"; van Dooren, "Vultures and Their People in India."

References

Amundson, Ron. *The Changing Role of the Embryo in Evolutionary Thought: Roots of Evo-Devo*. New York: Cambridge University Press, 2005. Print.

Barad, Karen. *Meeting the Universe Halfway: Quantum Physics and the Entanglement of Matter and Meaning*. Durham, NC and London: Duke University Press, 2007. Print.

Burd, Martin. "Colourful Language – It's How Aussie Birds and Flowers 'Speak.'" *The Conversation*. 2014. Web. 21 August 2015.

Card, Claudia. "Genocide and Social Death." *Hypatia* 18.1 (2013): 63–79. Project Muse. Web. 21 August 2015.

Chrulew, Matt. "Managing Love and Death at the Zoo: The Biopolitics of Endangered Species Preservation." *Australian Humanities Review* 50 (2011): 137–157. Web. 21 August 2015.

Crist, Eileen. *Images of Animals: Anthropomorphism and Animal Mind*. Philadelphia: Temple University Press, 1999. Print.

Derrida, Jacques. *Given Time: I. Counterfeit Money*. Trans. Peggy Kamuf. Chicago: University of Chicago Press, 1992. Print.

———. *The Gift of Death.* Trans. David Wills. Chicago and London: University of Chicago Press, 1995. Print.

Despret, Vinciane. "The Becomings of Subjectivity in Animal Worlds." *Subjectivity* 23.1 (2008): 123–139. Palgrave Journals. Web. 21 August 2015.

Geertz, Clifford. "Ethos, World View, and the Analysis of Sacred Symbols." *The Interpretation of Cultures.* New York: Basic Books, 1996. Print.

Haraway, Donna. *When Species Meet.* Minneapolis: University of Minnesota Press, 2008. Print.

Hatley, James. *Suffering Witness: The Quandary of Responsibility after the Irreparable.* Albany: State University of New York Press, 2000. Print.

Jablonka, Eva and Marion J. Lamb. *Evolution in Four Dimensions: Genetic, Epigenetic, Behavioral, and Symbolic Variation in the History of Life.* Cambridge, MA: MIT Press, 2005. Print.

Keim, Brandon. "Hidden Whale Culture Could Be Critical to Species Survival." *Wired.* 24 June 2009. Web. 21 August 2015.

Keller, Evelyn Fox. *The Mirage of a Space between Nature and Nurture.* Durham, NC: Duke University Press, 2010. Print.

Kirksey, Eben. "Species: A Paxiographic Study." *Journal of the Royal Anthropological Institute.* 21.4 (2015). Wiley Online Library. Web. 8 January 2016.

Kirksey, S. Eben and Stefan Helmreich. "The Emergence of Multispecies Ethnography." *Cultural Anthropology* 25.4 (2010). AnthroSource. Web. 21 August 2015.

Kohn, Eduardo. *How Forests Think: Toward an Anthropology Beyond the Human.* Berkeley: University of California Press, 2013. Print.

Lestel, Dominique, Florence Brunois, and Florence Gaunet. "Etho-Ethnology and Ethno-Ethology." *Social Science Information* 45 (2006): 155–177. Sage Journals. Web. 21 August 2015.

Lestel, Dominique, Jeffrey Bussolini, and Matt Chrulew. "The Phenomenology of Animal Life." *Environmental Humanities* 5 (2014): 125–148. Web. 21 August 2015.

Levinas, Emmanuel and Richard Kearney. "Dialogue with Emmanuel Levinas." *Face to Face with Levinas.* Ed. Richard Cohen. Albany State University of New York Press, 1986. 13–33. Print.

Mayr, Ernst. "What Is a Species, and What Is Not?" *Philosophy of Science* 63 (1996): 262–277. Print.

Oyama, Susan, Paul E. Griffiths, and Russell D. Gray. *Cycles of Contingency: Developmental Systems and Evolution.* Cambridge, MA: MIT Press, 2001. Print.

Plumwood, Val. *Environmental Culture: The Ecological Crisis of Reason.* London: Routledge, 2002. Print.

Puig de la Bellacasa, Maria. "'Nothing Comes without Its World': Thinking with Care." *The Sociological Review* 60.2 (2012): 197–216. Wiley Online Library. Web. 21 August 2015.

Rose, Deborah Bird. "Judas Work: Four Modes of Sorrow." *Environmental Philosophy* 5.2 (2008): 51–66. Print.

———. "Monk Seals at the Edge: Blessings in a Time of Peril." *Extinction Studies: Stories of Time, Death and Generations.* Ed. Deborah Bird Rose, Thom van Dooren, and Matthew Chrulew. New York: Columbia University Press, forthcoming.

———. "Multispecies Knots of Ethical Time." *Environmental Philosophy* 9.1 (2012): 127–140. Print.

———. *Wild Dog Dreaming: Love and Extinction.* Charlottesville: University of Virginia Press, 2011. Print.

van Dooren, Thom. "Care: In the Living Lexicon for the Environmental Humanities." *Environmental Humanities* 5 (2014). Web. 21 August 2015.

———. *Flight Ways: Life and Loss at the Edge of Extinction.* New York: Columbia University Press, 2014. Print.

———. "Spectral Crows in Hawai'i: Conservation and the Work of Inheritance." *Extinction Studies: Stories of Time, Death and Generations.* Ed. Deborah Bird Rose, Thom van Dooren, and Matthew Chrulew. New York: Columbia University Press, forthcoming.

———. "Vultures and Their People in India: Equity and Entanglement in a Time of Extinctions." *Australian Humanities Review* 50 (2011): 45–61. Web. 21 August 2015.

van Dooren, Thom and Deborah Bird Rose. "Lively Ethography: Storying Animist Worlds." *Environmental Humanities* 8.1 (2016). Web. 1 March 2016.

——. "Storied-Places in a Multispecies City." *Humanimalia* 3.2 (2012): 1–27. Web. 21 August 2015.

Watson, D. M. "Mistletoe-a Keystone Resource in Forests and Woodlands Worldwide." *Annual Review of Ecology and Systematics* 32 (2001): 219–249. Print.

Watson, D. M. and M. Herring. "Of Mistletoe and Mechanisms—Drivers of Declining Biodiversity in Remnant Woodlands." *Veg Futures 06: The Conference in the Field* (2006): 1–7. Print.

Wilkins, John S. *Species: A History of the Idea.* Berkeley: University of California Press, 2009. Print.

13

LOVING THE NATIVE

Invasive species and the cultural politics of flourishing

Jessica R. Cattelino

George Washington's revolutionary spirit led him to remove non-American plants from his front garden at Mt. Vernon and secure ornamental specimens from across the new United States to adorn his estate. As historian Andrea Wulf explains in *Founding Gardeners* (22–33), his was the first ornamental garden in the United States to be planted with and for native species, in an act of independence. Certainly, his act was patriotic, but it was more: it was also a settler-colonial act, one that anchored him more directly to this land. Invasive species management and the embrace of native species are affective projects that operate distinctly in settler-colonial societies like the United States, past and present. No place better illustrates this than the Florida Everglades.

After a century of reclamation projects reduced the Everglades by half in size, Congress in 2000 authorized the Comprehensive Everglades Restoration Program (CERP), which, at an estimated $13.5 billion, is the world's largest ecological restoration project. I am writing an ethnography that examines how the diverse residents of a 20 by 40 mile region in the northwest Everglades value water and land, and how these practices enact political belonging.[1] Building out from the Everglades, I develop an argument at the intersection of scholarship on settler colonialism (e.g., Wolfe; Simpson) and on the cultural politics of nature (e.g., Moore et al.); namely, that nature and political belonging are co-produced in settler societies in specific, patterned ways that have broad and problematic consequences for both sociocultural and ecological flourishing.

One domain in which we can see how the cultural politics of nature operate distinctly in settler societies is the management of invasive species and the promotion of native species. Scholars including Banu Subramanian, Amelia Moore, and Jean and John Comaroff have identified in the discourse and practice of invasive species management a pervasive socio-political nativism against immigrants and aliens.[2] Subramaniam briefly notes that the "native" in such discourse is the white settler, which she considers an "irony" (34, 36). In a settler society, however, the "irony" by which white settlers are figured as "native" is constitutive, and invasive species management is contradictory in ways that such analyses cannot fully address.[3] For example, nonindigenous occupants of this land must reckon with (or overcome by disregarding) their own arrival and the structural conditions of colonial invasion that so fundamentally shape American ecology and polity alike. Indigeneity becomes, as indigenous critical theorist Jodi Byrd writes, a "transit" through which other relational forms pass and become enacted. In a recent paper, Mastnak et al. have defended native plant advocacy as "botanical decolonization" by emphasizing that botanical changes have been a method and legacy of settler colonialism. While settlers' domesticated species did play a role in colonization, it is imperative not to draw an analogy between indigenous peoples and native species or between settlers and invasive species. Instead, the challenge is to understand the nonanalogical processes—too often obscured by the technique of the analogy, in fact—through which indigeneity and nature are co-produced in settler societies. The following are glimpses at three such processes: the problem of categorization; disturbance and equilibrium; and loving native species.

Category problems

In his keynote address to the 2012 Everglades Coalition conference, which is the major annual forum on the state of the Everglades, US Senator Bill Nelson (D-FL) emphasized the destruction wrought by invasive species upon his beloved state (Fieldnotes, January 8, 2012). Nelson won applause in our standing-room-only audience by touting his efforts, realized just days thereafter, to pass a federal ban on Burmese python importation. Nelson's speech tapped into a dominant discourse and practice on the part of environmentalists, land and water managers, and policymakers that promotes native species and seeks control of invasive ones.

During the early 2010s, the specter of Burmese pythons (*Python molurus bivittatus*) overtaking the Florida Everglades caught Americans' attention and took the Internet by storm.[4] Pythons are generalist apex predators, with a startling result: a decline of approximately 90 percent in nonhuman mammal populations upon establishment in the Everglades (Dorcas et al.; Harvey). Blame for the snakes' spread is often heaped on irresponsible pet owners for setting them loose in the Everglades, though research suggests that Hurricane Andrew was responsible for a significant portion of the population when in 1992 it blew the live contents of pet import warehouses from the suburbs into the swamps. More broadly, as anthropologist Laura Ogden has written, the pythons reanimate a long-standing American fascination with snakes in the Everglades and underscore the ability of snakes to "mesmerize, to halt people in their tracks" (83).

Invasive species can wreak havoc. Often lacking predators, they enter an ecological zone and frequently out-compete native species for food and habitat. Invasive species shape the earth: in the yards and playgrounds of the city of Clewiston (population seven thousand) and the Seminole Big Cypress Reservation (population six hundred), where my ethnographic research is focused, red fire ants build mounds. According to a state park biologist, fire ants may contribute to the grasshopper sparrow's endangerment on the prairies north of Lake Okeechobee (fieldnotes, December 31, 2011). Invasive species alter the air: long-time residents attribute a perceived increase in human allergies to invasive plants and to exotic orange blossoms in citrus groves, in the center of Florida's citrus industry. They fill the water: canals brim with invasive tilapia, lionfish, and oscars, and with surface-dwelling water hyacinth and water lettuce that block light to species below. They change the horizon: Australian pines first planted as wind breaks now overtake woodlands. They create disappearances: as Seminole Tribal Council Representative Mondo Tiger recounted, quail vanished from the Big Cypress Reservation after residents began to keep cats as pets. Invasive species account for approximately 30 percent of all uncultivated plants in Florida (South Florida Water Management District and US Army Corps of Engineers). Nationally, invasive species management costs more than $120 billion annually (Dorcas et al.).

Invasive species pose category problems. The line between "exotic" species (those out of their places of origin) and invasive ones (ditto, but with the added definition of causing harm) is thin, thoroughly blurred, and socially defined. What counts as harm often differs for a water or wildlife manager, a farmer or an angler. The task, then, is not just to break down or destabilize categories but rather to analyze what sustains them, and with what political and economic effects.

Take oranges, for example, in their multiplicity. Cultivated orange trees originated in Asia and are often said to have arrived with Christopher Columbus. They are widely valued as Florida heritage (see, for example, "Citrus Industry History"). Seminole Ahfachkee school pupils on Ah-Tah-Thi-Ki Museum tours learn that citrus trees mark the campsites of their nineteenth-century war heroes (e.g., Abiaka, also known as Sam Jones) who fought against the United States and resisted removal to Indian Territory (now Oklahoma). Meanwhile, the Asian psyllid (*Diaphorina citri*) that carries the citrus greening disease currently threatening Florida's multi-billion-dollar citrus industry is treated as an invasive (Harmon). The spacetime of invasion tracks national borders and colonial timelines, not only with oranges but also with other species. In fact, Florida state law and everyday practice associate nativeness with being here before colonists and existing inside United States territorial borders. Native nature is pre-historical and nationalized statutorily in ways that render the settler state its rightful inheritor and steward.

Assessments of a species' harm and benefit can change over time. The first time I drove to Clewiston and up the levee to Lake Okeechobee, I saw a band of tall, dead trees along the rim canal. These were melaleuca trees (*Melaleuca quinquenervia*) that had become victims of a state eradication program. Americans initially imported melaleucas from Australia as an ornamental. Then, on the theory that the thirsty specimens could be used to drain "useless swamps," twentieth-century land managers dropped seeds from airplanes, and the US Army Corps of Engineers in 1940–1941 planted them along Lake Okeechobee's levees for stabilization (Carter-Finn et al.; South Florida Water Management District). In just a few decades, melaleucas took over approximately 20 percent of South Florida's "natural" lands. As Everglades reclamation gave way to restoration, melaleucas were listed federally as "noxious weeds" and by Florida as "weeds" (making it illegal under federal and state laws to possess or sell them). Under CERP and other policies, coordinated state and federal agencies have devoted substantial resources (over $35 million to date) to regional melaleuca eradication, with notable successes. Some Clewiston residents, unaware of this, bemoan the loss of the tall trees atop the levee, under which they had enjoyed shady picnics and lake breezes on steamy summer afternoons.

Invasives can restore. Out on the same lake, an endangered indicator species, the snail kite, has rebounded recently, thanks to an invasive snail they eat that is making its way into the Everglades. Snail kites had declined because reclamation destroyed the pond apple forest along the southern lakeshore that previously provided habitat for their prey, the apple snail. Pond apple restoration is limited by agricultural development. Within such constraints, invasives can restore. Meanwhile, this same pond apple species was brought to Australia as graft material but escaped an orchard there and now is taking over wetlands and subject to federal eradication projects. In Australia, the pond apple endangers a valued native species: the melaleuca tree.

For some, the *agent* of arrival matters to the category of species. While volunteering at a tour of the J-7 ranch during a nearly all-white birding festival, I gazed up at my favorite swallowtail kites circling above their nest and then gingerly asked the enthusiastic tour leader if, in her view, it is acceptable for birders to have favorite birds. She grinned and admitted that most birders do, so I asked her to name hers. After noting her love of many species, she named the (threatened) Florida scrub jay and then emphasized that she liked them all—so long as they were native. What made a species native, I asked? That it came from here. I asked about the cattle egret (*Bubulcus ibis*), a common wading bird that is believed to have arrived in North America only in the mid-1900s, having established itself decades earlier to the south after crossing the Atlantic, perhaps on a hurricane. She considered the cattle egret native because it was not brought by humans (Fieldnotes, March 31, 2012). Humans, or in any case some humans, stand outside nature, as disturbance. Such an ideology of unpeopled wilderness developed in the United States as part of the settler-colonial project, with disastrous consequences for indigenous land claims and ecologies. Categories do things and sustain structures.

Disturbance and equilibrium

Invasive species, human activity, hurricanes, drought, and fires generally are included among the "disturbances" to the Everglades ecosystem. Invasive species management rests in part on theories of the ecosystem that privilege ecological equilibrium and view change as disturbance. The term *ecosystem*, coined in 1935, came to dominate ecological theory over subsequent decades (Golley). In the ecosystems view, ecological communities are complex systems whose structure and function can be examined with synchronic analysis. The historical period that produced ecosystems theory also produced structural functionalism in anthropology, an approach that sought explanation for social organization in the function of maintaining social structure.[5] Like ecosystems theory, this approach was rightly criticized for an inability to explain historical change and a tendency to interpret change as disruption or dysfunction, at least for colonized, indigenous, and racialized peoples. Thus, the problems of "disturbances" to culture and to ecosystems share roots in mid-twentieth-century thought. In a settler society, as many scholars have shown (e.g., Barker; Clifford), this means that change among indigenous peoples signals cultural loss, inauthenticity, and loss of sovereignty. Change to an ecosystem, meanwhile, degrades it as such. These are not simply analogies or historical curiosities: they are logics that link indigeneity with nature. Together, prevalent ecological and cultural views foreclose historical change in the Everglades and other ecosystems *as such* and, similarly, for Seminoles and other indigenous peoples *as such*. It should come as no surprise that Michael Grunwald's generally helpful history of the Everglades introduces indigenous people first in a section on "native species" (20).

Equilibrium is fantasy, and for that reason maintaining a "native" state requires active and ongoing invasive species management. Without creatively destructive management, invasive species would take over Florida's wetlands. To the untrained eye, invasive species' eradication and maintenance look like environmental destruction: large yellow bulldozers tear down Brazilian pepper trees, US Army Corps of Engineers boats spray herbicides over water lettuce mats in Lake Okeechobee, and controlled burns clear pastures of invasive grasses. Invasive species management can produce additional damage. The Seminole Tribe of Florida's Environmental Resource Management Department (ERMD) is responsible for invasive species eradication at Big Cypress, but the chemical treatments they use also affect valued plants. Plants sprayed with herbicides cannot be used to prepare medicine. Staff members, most of whom are not Seminole, wish to avoid spraying medicine plants and have requested that tribal citizens identify their location to prevent harm. However, norms that protect culturally sensitive knowledge limit disclosure. Invasive species management itself disturbs.

Invasive and exotic species force us to ask what we mean by diversity. All too often in environmental discourse, biodiversity and cultural diversity are laminated onto one another, generally in efforts to preserve indigenous peoples, cultures, and languages in order to maintain human diversity as a good that gains its force through direct or indirect reference to biodiversity. The perils of such discourse are (at least) threefold. First, it contributes to the long-standing and consequential problem of collapsing indigenous peoples into nature. Second, it trades on the specter of extinction that no doubt raises awareness but also participates in the project of indigenous vanishing that is endemic to settler colonialism. And third, it often (though not intrinsically) associates indigeneity with stasis in ways that devalue indigenous cultural change as cultural loss.

Loving native species

In South Florida and across the United States, invasive species eradication is coupled with the promotion of native species, and this is an affective project. At one local Hendry-Glades Audubon Society meeting, the audience soaked up county extension agent Gene McAvoy's presentation on the advantages of planting native species that require less fertilizer and water. But they appeared taken aback when he pointed out that the most extensive cultivated plant in the United States is lawn grass, which accounts for higher concentrations of pollution per square foot than the agricultural crops more often fingered as polluters in Everglades debates (Fieldnotes, March 12, 2012). The big water quality issue in the Everglades is phosphorus, and "Big Sugar" has been the primary target of environmentalist activism around water quality. Native species promotion grows and native nurseries are on the rise; there are annual conferences promoting native plants in South Florida. Like the birding tour guide who loved native species, one becomes emotionally attuned to the native-invasive distinction.

I grew to appreciate native species by contrast with invasive species and even found myself adjusting emotional attachments to flora and fauna upon learning their status. As the summer of 2012 approached, for example, I delighted as each night fell in watching shiny green tree frogs with suctioned feet make nocturnal appearances on the exterior of our rental house's glass patio door. The then-toddler squealed with pleasure when the frogs' curious-looking heads and long legs appeared at the edge of the glass. One morning, we found a frog tucked snugly into the depression behind the handle of our car door. Subtropical living! When I conveyed our enthusiasm to my neighbor and landlord, she looked pleased but then, with an almost embarrassed expression, explained that these were probably Cuban tree frogs, an invasive species that contributed to the population decline of Florida's native frogs. Should I have turned on those invasive critters that had seemed so much to be of the place? Felt less of a thrill when they poked their shiny eyes around the corner at dusk? I just couldn't. But the categories of native and invasive come with their own emotional force: a scientist at an environmental organization recalled being taken aback when a neighbor who was new to birdwatching asked him to identify an invasive species in order that the neighbor could learn which birds to hate.

Some scholars, as mentioned, understand the embrace of native species in the United States to be a decolonizing project. This is because botanical colonization—notably, the cultivation of non-native species—was a method of settler colonialism (Mastnak et al.). Such analyses are helpful but overlook the contradictory structure of nativism in a settler society on the part of all but indigenous peoples. Settlers' embrace of native species anchors *us* in this land, as ours. Recall George Washington's garden. Wulf's otherwise delightful discussion of the founding fathers' gardening politics misses a critical piece of the puzzle: the founders' horticultural severing of America from Europe staked a distinctly settler-colonial claim by gathering native species and caring for them as their own, as their national patrimony. Such acts of power not only displace indigenous peoples' histories and futures but also depoliticize those very acts of dispossession. In recent years, the US Army Corps of Engineers and other agencies have created public education materials that depict invasive species in the Everglades on Old West-style wanted posters. The frontier lives and shifts.

Emma Marris is an environmental journalist who, in *Rambunctious Garden: Saving Nature in a Post-Wild World*, presents a provocative case for jettisoning the long-held environmentalist goal of preserving pristine nature in favor of promoting post-wild nature under human management. In the course of doing so, she asks readers to learn to love exotic species. Marris, moreover, observes that the tendency to view natural landscapes as devoid of humans is especially prominent in places like the United States and Australia. She does not ask why or how pristine nature might be linked to the sociopolitical formation of these nations.

This is where an analysis of settler colonialism comes in. After all, settler colonists had and have every reason to see the land as uninhabited and unchanged, as *terra nullius*. Enlightenment thinkers developed theories of property through the foil of the New World indigene who purportedly failed to cultivate land and thereby held no property right (Tully). It is worth noting that Washington had another garden of mostly exotics, planted for the purpose of improving agricultural practices in the new United States (Wulf 32–33). Still, even as settler societies embrace native species as a kind of patrimony, there is room for a more unsettling politics and practice. Native species do not only anchor settlers in the land or foster Creole exceptionalism. They also have the capacity to facilitate future flourishing, insofar as the institutions that govern them are guided by anti-colonial and anti-nativist practice. More on that below.

Back at the Everglades Coalition conference, Senator Nelson spoke broadly about environmental restoration. In a voice thick with longing, Nelson asked us to imagine Florida as it was almost five hundred years ago, when the "explorer" Ponce de León landed his ship. Picturing that moment, Nelson delivered his rallying cry: "and that's what we're all here today for." We, the listeners, joined him on that ship, admiring that which we and our Spanish hogs were about to invade. As Senator Nelson's speech simultaneously battled invasive species, identified with (invasive) Spanish colonizers, and rallied his audience to restore the Everglades to a moment of naturalness just prior to European conquest, his metaphors of native and nonnative remained in play in, for him, loving contradiction.

On flourishing

In invasive species management, and more generally on the terrain of nature, settlers often narrate and enact a kind of nation-building and nation-sustaining nativeness. One conclusion is that the cultural politics of nature reinforce settler colonialism as a logic, and no doubt that is often the case.

Yet, perhaps we should not focus exclusively on logics and contradictions internal to settler colonialism. Doing so while ignoring the specificity and power of indigeneity risks reproducing the very eliminatory logic—this time by writing indigenous peoples out of the story—that settler-colonial studies aim to upend. Indigenous sovereignty matters to nature. For example, indigenous water rights are a constitutive part of the Everglades, from the Seminole water compact's impact on water allocations and restoration to the activities of the Tribal Historic Preservation Office, which intervenes whenever restoration or development unearths significant objects. What is more, indigeneity is built into US law and public culture even when indigenous peoples are neither the subjects nor the objects of engagement, from water rights doctrine to ecosystems restoration modeling, and from settler narratives of nature's past and future to what wilderness looks and feels like in the most personal and embodied ways.

In invasive species management, some claims open up processes of categorization, disturbance, and loving. For example, Seminole natural resource management staff increasingly respond to community input that promotes native species along with selected nonnative ones (e.g., guava) that offer culinary, aesthetic, sentimental, economic, and other forms of value—that contribute to flourishing. Farmers in the drained Everglades manage invasive species through the cultivation of exotic ones, like oranges and sugarcane, and they argue for a form of flourishing that does not sweep people off the landscape in a wilderness model of restoration.

I am thinking toward an account of ecological and sociocultural flourishing, on the conviction that it is only by unsettling nature that the Everglades can be "saved" in some yet-to-be-determined way. The challenge is to simultaneously to open up categories like native and invasive, delinking them from colonial timelines and dispossessions, and to train attention on governance and the justice imperatives of institutions (e.g., science, law) that sustain the present order of things in the Everglades and other settler-colonial contexts. How a decolonized invasive species management agenda would enact the politics of indigeneity is unforeseeable. Minimally, it would insist on the agency, governance, and scientific participation of indigenous peoples, whose long-term experience with invasive and native species and whose sovereign authority over environmental governance on their territories should inform and delimit non-Indian management practices. Such an approach would bring indigenous dispossession and sovereignty into the environmental accounting of human harm and human value, whether in assessing the value of a species' ecosystems services or preparing cost-benefit analyses for a project's environmental impact statement. And that's a start.

Notes

1 Fieldwork in 2012 was funded by fellowships from the National Science Foundation (#1122727) and the Wenner-Gren Foundation. Additional research was funded by the Howard Foundation.
2 Subramaniam writes: "The point of my analysis is not to suggest that we are not losing native species, nor that we should allow plants and animals to flow freely across habitats in the name of modernity or globalization. Instead it is to suggest that we are living in a cultural moment where the anxieties of globalization are feeding nationalisms through xenophobia. The battle against exotic and alien plants is a symptom of a campaign that misplaces and displaces anxieties about economic, social, political, and cultural changes onto outsiders and foreigners" (34).

3 By settler society, I refer especially (if not only) to the liberal democratic settler states of the former British Empire with indigenous minorities: Australia, Canada, Aotearoa/New Zealand, the United States. The historian Patrick Wolfe differentiates settler colonialism's target of land dispossession from the expropriation of labor in dependent colonies. Thinking in terms of settler society integrates indigenous and non-indigenous lives, while sustaining attention to power, by attending to the ways in which all of our conditions are structured by the legal, historical, cultural, and economic formations that are characteristic of settler societies.

4 I first learned of the reptiles in the early 2000s upon hearing that a concerned non-Seminole employee had circulated an illustrated informational email to all the Seminole Tribe of Florida staff, sparking eye-rolling and more serious disapproval in light of prohibitions that regulate Seminole interactions with snakes. A few years later, tribal citizens and employees recounted when another non-Seminole employee captured a python on the Brighton Reservation and preserved it for research purposes in an administration building refrigerator. The relevant department director soon fielded a call at his distant coastal home that required him to ensure that the snake and the refrigerator itself were removed from the premises by the time the building opened the next morning.

5 Animal ecologist Charles S. Elton is an interesting example here. In his important 1927 book *Animal Ecology*, he theorized ecological niches and compared the loss of a species (a badger) to the loss of a role (vicar) in society (64). Tellingly, Elton moved on to study invasive species. Often, structural functionalism is considered organismic. It is also ecological (cf. Jax, 78–81).

References

Barker, Joanne. *Native Acts: Law, Recognition, and Cultural Authenticity*. Durham, NC: Duke University Press, 2011. Print.

Byrd, Jodi A. *The Transit of Empire: Indigenous Critiques of Colonialism*. Minneapolis: University of Minnesota Press, 2011. Print.

Carter-Finn, Katherine et al. *The History and Economics of Melaleuca Management in South Florida*. Gainesville, FL: Institute of Food and Agricultural Sciences, University of Florida, 2006. Web. 19 Sept. 2013.

"Citrus Industry History." Florida Citrus Mutual. Lakeland, FL. 2012. Web. 20 October 2013.

Clifford, James. *The Predicament of Culture: Twentieth-Century Ethnography, Literature, and Art*. Cambridge, MA: Harvard University Press, 1988. Print.

Comaroff Jean and John L. Comaroff. "Naturing the Nation: Aliens, Apocalypse and the Postcolonial State." *Journal of Southern African Studies* 27.3 (2001): 627–651. JSTOR. Web. 1 October 2015.

Dorcas, Michael E. et al. "Severe Mammal Declines Coincide with Proliferation of Invasive Burmese Pythons in Everglades National Park." *Proceedings of the National Academy of Sciences* 109.7 (2012): 2418–2422. Web. 1 October 2015.

Elton, Charles S. *Animal Ecology*. 1927. Chicago: University of Chicago Press, 2001. Print.

Golley, Frank B. *A History of the Ecosystem Concept in Ecology: More than the Sum of the Parts*. New Haven: Yale University Press, 1993. Print.

Grunwald, Michael. *The Swamp: The Everglades, Florida, and the Politics of Paradise*. New York: Simon and Schuster, 2006. Print.

Harmon, Amy. "A Race to Save the Orange by Altering Its DNA." *New York Times* 27 July 2013. *NYTimes.com*. Web. 25 Oct. 2013.

Harvey, Rebecca G. et al. *Burmese Pythons in South Florida: Scientific Support for Invasive Species Management*. Gainesville: Institute of Food and Agricultural Sciences, University of Florida, 2008. Web. 11 Sept. 2013.

Jax, Kurt. *Ecosystem Functioning*. Cambridge: Cambridge University Press, 2010. Print.

Marris, Emma. *Rambunctious Garden: Saving Nature in a Post-wild World*. New York: Bloomsbury, 2011. Print.

Mastnak, Tomaz, Julia Elyachar, and Tom Boellstorff. "Botanical Decolonization: Rethinking Native Plants." *Environment and Planning D: Society and Space* 32.2 (2014): 363–380. Sage Journals. Web. 1 October 2015.

Moore, Amelia. "The Aquatic Invaders: Marine Management Figuring Fishermen, Fisheries, and Lionfish in The Bahamas." *Cultural Anthropology* 27.4 (2012): 667–688. AnthroSource. Web. 1 October 2015.

Moore, Donald S., Jake Kosek, and Anand Pandian, eds. *Race, Nature, and the Politics of Difference.* Durham, NC: Duke University Press, 2003. Print.

Ogden, Laura. *Swamplife: People, Gators, and Mangroves Entangled in the Everglades.* Minneapolis: University of Minnesota Press, 2011. Print.

Simpson, Audra. *Mohawk Interruptus.* Durham, NC: Duke University Press, 2014. Print.

South Florida Water Management District. "Plants Behaving Badly: Melaleuca." 2004. Flyer in author's collection.

South Florida Water Management District and US. Army Corps of Engineers. "Melaleuca Eradication and Other Exotic Plants." 2014. PDF. Web. 25 January 2016.

Subramaniam, Banu. "The Aliens Have Landed! Reflections on the Rhetoric of Biological Invasions." *Meridians: Feminism, Race, Transnationalism* 2.1 (2001): 26–40. Project Muse. Web. 1 October 2015.

Tully, James. "Aboriginal Property and Western Theory." *Social Philosophy & Policy* 11.2 (1994): 153–180. Print.

Wolfe, Patrick. "Settler Colonialism and the Elimination of the Native." *Journal of Genocide Research* 8 (2006): 387–409. Print.

Wulf, Andrea. *Founding Gardeners: The Revolutionary Generation, Nature, and the Shaping of the American Nation.* New York: Vintage Books, 2012. Print.

14

ARTIFACTS AND HABITATS

Dolly Jørgensen

Humans see distinctions between *artifacts*, which are constructed by human hands with human ingenuity, and *nature*, which we tend to think of as somehow not made by humans even if we acknowledge that little nature is left untouched by humans. An artifact at its core is related to the word *artificial*, meaning made by human hands through art or craft. The word has a long history going back to the classical Latin *artificialis* and is most often used to represent the opposite of *natural*, a word which then implies not manmade. These distinctions play into how scholars in the humanities approach environmental topics, which tend to position *artifact* as something that modifies (often negatively) *nature*. However, I would like to propose that, for nonhumans, artifacts are part of their habitat. While artifacts may not be natural, they are part of nature.

When historians represent an artifact as artificial or natural, they most often focus on the production of the item in question, but not necessarily on the artifact's value or use as habitat. *Habitat* is an unabashedly scientific term. It comes from the Latin *habitare*, meaning to inhabit or live, but it is now defined more specifically as "the locality in which a plant or animal naturally grows or lives" ("habitat, n.," *OED Online*). The use of *naturally* as a part of the technical definition of habitat is noteworthy, because it implies that habitat cannot be artificial. Yet I would argue that an exclusion of the artificial as habitat may be misguided if we consider life from the nonhuman point of view.

In the environmental humanities, we need to shift focus from the artifact's (human) producer to its (nonhuman) user. In the field of history of technology, traditionally the producer of technology was of primary interest, with stories tending to focus on makers and inventors like Ford or Edison. But a shift has been underway over the last thirty years to focus on the user instead of just the producer, with books such as *More Work for Mother* and *How Users Matter* (Cowan; Oudshoorn and Pinch). In these newer histories, how end users react to, incorporate, and modify the technologies is just as important as the original invention and design. I would like to suggest that environmental humanities scholars also need to take a "user" turn when thinking about the meeting of technology and environment. The two cases of artificial reefs and bird nesting boxes provide an opportunity to think through the implications of such a turn.

Artificial reefs

In 1996, the oil company Chevron removed four offshore oil structures off the coast of California. The platforms, named Hazel, Hilda, Hope, and Heidi (or 4-H platforms for short), had reached the end of their productive life for the extraction of offshore underground petroleum. But before they could be removed, a debate ensued about whether these technological artifacts served another kind of productive life.

In the three decades that the structures had been standing in the water, aquatic life had taken up residence on the steel beams and accumulated deposits of shells at the bottom of the structure. The initial colonization had happened quickly. A study published in 1964 of the ecosystems around Hazel (installed 1958) and Hilda (installed 1960) already revealed the presence of fourteen fish species managed under the Pacific Groundfish Fishery Management Plan, including several threatened rockfish species (see Helvey for a summary of scientific studies of fish around the California platforms through 2000). A later study showed vibrant communities of mussels, crabs, sea cucumbers, sea stars, and numerous fish on the mounds (Love et al.). Recreational fishermen had known that fishing around the platforms was particularly productive, so the sportfishing organization United Anglers of Southern California asked Chevron to leave the structures in place as fish habitat instead of removing them (Pattison). Facing issues of permit compliance and liability, Chevron opted to take them out (Frumkes 272–273). The removal of the 4-H platforms sparked a legislative debate about whether or not California should permit offshore structures to stay in place as artificial reefs. I have previously categorized this controversy as a difference of enactment of nature: one side saw manmade structures as a natural way to increase fish stocks, and the other believed nature would only exist if the technological structures were removed from the water (Jørgensen, "Environmentalists").

In the history of these platforms-cum-reefs, one distinctly human idea looms large: artificiality. The categorizations of technological, artificial, and natural invoked in debates like the California rigs-to-reefs issue are human constructs. If we shift the focus from the humans to the nonhumans, a different view emerges. The fisheries scientist Milton Love once remarked about the use of offshore oil structures as reefs in California: "In a lot of cases, fish don't care. A rock or an oil platform is all the same to them" (quoted in McEntee). There is, of course, always the question whether or not the two really are the same. The main opponent of rigs-to-reefs legislation in California, Linda Krop of the Environmental Defense Center, frequently questioned the quality of the habitat provided by offshore structures via comparison: "There are birds on telephone lines and sea gulls in landfills. Just the presence of animals does not mean that it's providing habitat" (quoted in Tran). To decide which view is correct, we need to consider how nonhumans use the technological artifact.

Much of the debate about whether or not artificial reefs are acceptable revolves around how *natural* the materials used in reef construction are. Artificial reefs have been made with a great variety of materials since the end of WWII, including rocks and boulders, concrete, sunken ships, rubber tires, old automobile bodies, and much more. Artificial reefs purposefully constructed on land and then placed into the sea can be made of steel, concrete, or even ashes from cremated human remains. During the course of drafting guidelines on the construction of artificial reefs in the late 1990s, the North Sea international treaty organization OSPAR had to decide what types of artificial reef materials would be acceptable for future use (Jørgensen, "OSPAR's Exclusion"). The national representatives from the UK and Norway wanted a guideline that permitted structures which had originally not been designed as artificial reefs to be repurposed as reefs. Other delegations, including Germany and Spain, wanted to ban the use of post-consumer materials in artificial reefs. In the end, OSPAR's *Guidelines on Artificial Reefs in Relation to Living Marine Resources* (Agreement 1999–1913), adopted in June 1999, stated that no waste material should be used to construct reefs, in essence restricting artificial reefs to virgin materials. The upshot of the guidance document is that artificial reefs must mimic nature in composition as well as form.

A similar desire to make an artificial island as natural as possible is apparent in the construction of Dubai's Jumeriah Palm Island. In the mid-1990s, the Sheikh of Dubai decided to build a manmade island in the Arabian Gulf off Dubai's coast to increase the availability of beachfront property. The design team drew an innovative—and many would argue overtly unnatural—island: a 6-kilometer long palm tree with seventeen fronds. Despite its unnatural shape, the developer Nakheel touted its construction of "natural rock" to encourage reef development (Nakheel).

The process required to construct a huge artificial island from these "natural" materials of rock and sand was anything but natural. The crescent-shaped breakwater section of the island, which rises to a height of 4 meters above low tide level, is a highly engineered construction of multiple layers: a small hill of sand at its base, a water-permeable geotextile to keep the sand in place, a lower protective layer of smaller rocks, and two layers of rocks weighing as much as 6 tons each. The 5.5 million cubic tons of rock for this base were acquired from sixteen quarries across the United Arab Emirates. The island itself has a rock foundation and consists of over ninety-two million cubic meters of sand, which was dredged from the Arabian Gulf sea floor six nautical miles out at sea ("Impossible Island"). In promotional material from the developer, the artificial is said to make the natural: "using natural rock effectively has actually stimulated the propagation of a complex marine ecosystem, creating a natural haven for divers and fish alike" (Nakeel). The developer downplays the technological and artifactual to emphasize that natural materials make a "natural" island.

In artificial reefs and artificial islands, the artificial and natural are played against each other. The concepts are relational: they are defined by the presence, absence, and extent of the other. Of course, who is making the judgment about naturalness or artificiality matters, whether it is an environmental activist, a scientist, an engineer, or even an animal. From the nonhuman point of view, the human distinctions between artificial and natural are meaningless. For barnacles, oysters, red snappers, and crabs, what matters is whether the reef provides good shelter and food—they have no interest in whether it was made by human technology.

Bird nest boxes

Sometime in the early 1800s, John James Audubon installed a large box on a pole near his house as a home for purple martins, a North American migratory swallow species.[1] The martins used the nest box for several years to raise young. One year, Audubon decided to invite bluebirds to nest in the area as well, so he put up several smaller boxes. The birds had other ideas: "The Martins arrived in the spring, and imaging these smaller apartments more agreeable than their own mansion, took possession of them, after forcing the lovely Bluebirds from their abode" (117). In spite of repeated attempts by the bluebirds to reacquire their human-built home, the martin stood its ground. Audubon decided to act:

> I thought fit to interfere, mounted the tree on the trunk of which the Blue-bird's box was fastened, caught the Martin, and clipped his tail with scissors, in the hope that such mortifying punishment might prove effectual in inducing him to remove

to his own tenement. No such thing; for no sooner had I launched him into the air, than he at once rushed back to the box.

(118)

Audubon recaught the bird and clipped its wings, which still did not entice the martin to give up. Exasperated, Audubon "seized him in anger, and disposed of him in such a way that he never returned to the neighbourhood" (118).

Audubon's designer mindset is apparent in this exchange. He had designed the bird houses as artifacts with a particular purpose—the big one was for martins, the small ones were for bluebirds. The martins, however, did not care about his designs. They saw the boxes as potential habitat and picked the ones that suited them. From the martin point of view, these boxes were part and parcel of the environment—good places to set up a nest and raise young—no more, no less.

Setting up bird nest boxes for purple martins was nothing new in Audubon's day. He noted in his text from 1831 that "the erection of such houses is a general practice" (118). This held true in urban areas ("all our cities are furnished with houses for the reception of these birds"), small towns ("almost every country tavern has a Martin box on the upper part of its sign-board"), and in the countryside where Audubon noted that both Native Americans and Southern black slaves hollowed out calabash gourds and stuck them on sticks as martin homes (119). The illustration he made of the purple martin for the *Birds of North America* shows the birds with a manmade gourd nesting box hung on a branch. Later treatises on how to make bird boxes continued to recommend the gourd houses in the southern states, although wooden boxes were recommended as longer-lasting for New England (Scudder 31). Purple martins appear to have rapidly transitioned to life in human-constructed homes. A study from 1974 could confidently claim that the purple martin "now nests almost exclusively in houses provided by man" (Jackson and Tate 435).

In the early twentieth century, bird boxes became a regular part of avian conservation efforts. "Systematic feeding and housing of the birds" was understood as a way to increase the wild bird population, which had drastically decreased from the previous century (Scudder 9). As one author noted in 1919, cultivation and clearance may be signs of human progress, but they had deleterious effects on the bird population: "If we cut the dead wood from our wood lots, parks, and groves; clean out, sterilize, and fill rotting spots in limb and trunk with concrete, we deprive many birds of nesting facilities" (Taverner 119). According to Taverner, the solution was not to abandon these destructive practices but to make up for them, as "bird boxes will largely compensate for natural cavities" (119). Humans wanted to replace lost natural bird habitats with manmade habitats.

Humans may be able to make bird boxes, but they cannot force the birds to live there. The designs, however, can be customized to the nesting habits of particular birds. In the early twentieth century, books like *Conservation of Our Wild Birds* and *How to Attract and Protect Wild Birds* pointed out different designs that met the needs of desirable bird species. Catering to the birds' habits meant building boxes of particular sizes, placing the entrance holes in the right places, and hanging them at the right heights. Purple martins nest in groups, unlike most other birds targeted by bird houses, so their houses could be designed for multiple nests in the same structure. Their designs may have been particularly influenced by the dovecote model which had been common since at least the Middle Ages for raising semi-wild pigeons. Design elements in combination were intended to make the birds adopt the artifact as habitat.

141

Within the ecological sciences community, criticisms of studies that acquired data from birds using artificial nesting boxes surfaced in 1989. Anders Pape Møller argued that nest box studies introduced two "experimental artefacts"—safety from predators and a reduction of parasite populations—because of the way researchers built and maintained the boxes. A reply published three years later argued that "although the structures themselves may be artificial, the ways in which birds take advantage of the opportunities provided by nest boxes are not" (Koenig et al. 305). Although Møller had argued that because the structures were not natural, the behaviors would be unnatural, one can ask, like the authors of the response paper, "What is natural?" To use their example, barn swallows build their homes in barns built by humans—does that make everything those swallows do unnatural? Bird species like the barn swallow and purple martin have been good at making human-constructed artifacts into their habitats, and their adaptability to artifacts as habitats may be keeping them from extinction in the face of declining numbers of tree cavities.

The above history of purple martins and nest boxes is only a teaser; a full history of the humble bird box has yet to be written, and should be. While environmental historians Chris Smout and Tom Dunlap have suggested studying how bird conservation manifested itself tangibly, object-centered environmental histories are rare. Even where nest boxes are central to an environmental history, they seem to be taken for granted. For example, while Etienne Benson shows how nest boxes and food provision were critical to the success of urban squirrel introduction in the United States, he does not investigate the particular technological choices that created those habitats. An environmental history of nest boxes would want to answer some questions about the artifact and its habitat: Where did the bird box designs come from? Why did people at different moments in time want to encourage birds to settle near them? How did these technological artifacts become habitats for birds?

In conclusion, by thinking about the birds as users of the nest boxes, or fish and mollusks as users of the offshore oil platforms, we can shift the focal point for decisions about the naturalness of artifacts. Nonhuman users and their values come to the fore in stories that have been typically seen from the perspective of human makers. The definitional work of deciding what a nest box or platform is and how it can and should be used as habitat is not an exclusively human endeavor. Rather than understanding *habitat* in a scientific sense with a "naturalness" qualifier, we should look to the definition of the much older word *habitation*: "the act of dwelling in or inhabiting as a place of residence" ("habitation, n.," *OED Online*). When nonhumans use artifacts as their dwelling places, those artifacts are habitats.

Within the history of technology, the shift of focus to users has resulted in entirely different stories of innovation and technological development, changing the very core of the field. Likewise, I believe, looking at the users of technological artifacts will change the definitions of "environment" and "naturalness" relevant for environmental humanities scholarship. Environmental humanists need to embrace the perspectives of the nonhuman users of our human-constructed artifacts, realizing that human distinctions between artificial and natural, artifact and habitat, may not hold true from a more-than-human viewpoint.

Note

1 Smout's blog post "Birds and Squirrels as History" brought Audubon's drawing to my attention.

References

Audubon, John James. *Ornithological Biography, or An Account of the Habits of the Birds of the United States of America; Accompanied by Descriptions of the Objects Represented in the Work Entitled The Birds of America, and Interspersed with Delineations of American Scenery and Manners*. Vol. 1. Philadelphia: Judah Dobson, 1831. Print.

Benson, Etienne. "The Urbanization of the Eastern Gray Squirrel in the United States." *Journal of American History* 100 (2013): 691–710. Print.

Cowan, Ruth Schwartz. *More Work for Mother: The Ironies of Household Technology from the Open Hearth to the Microwave*. New York: Basic Books, 1983. Print.

Dulap, Thomas R. "Thinking with Birds." *The Edges of Environmental History: Honouring Jane Carruthers*. Ed. Christof Mauch and Libby Robin. Munich: Rachel Carson Center for Environment and Society, 2014. 25–29. Print.

Frumkes, D. R. "The Status of the California Rigs-to-Reefs Programme and the Need to Limit Consumptive Fishing Activities." *ICES Journal of Marine Science* 59 (2002): S272–S276. Print.

"habitat, n." *OED Online*. Oxford University Press, June 2015. Web. 4 August 2015.

"habitation, n." *OED Online*. Oxford University Press, June 2015. Web. 4 August 2015.

Helvey, Mark. "Are Southern California Oil and Gas Platforms Essential Fish Habitat?" *ICES Journal of Marine Science* 59 (2002): S266–S271. Print.

"Impossible Island: Dubai Palm Island." *Megastructures*. National Geographic. 13 September 2005. Television.

Jackson, Jerome A. and James Tate, Jr. "An Analysis of Nest Box Use by Purple Martins, House Sparrows, and Starlings in Eastern North America." *The Wilson Bulletin* 86.4 (1974): 435–449. Print.

Jørgensen, Dolly. "Environmentalists on Both Sides: Enactments in the California Rigs-to-Reefs Debate." *New Natures: Joining Environmental History and Science and Technology Studies*. Ed. Dolly Jørgensen, Finn Arne Jørgensen, and Sara B. Pritchard. Pittsburgh: University of Pittsburgh Press, 2013. 51–68. Print.

——. "OSPAR's Exclusion of Rigs-to-reefs in the North Sea." *Ocean & Coastal Management* 58 (2012): 57–61. Print.

Koenig, Walter D., Patricia Adair Gowaty, and Janis Dickinson. "Boxes, Barns, and Bridges: Confounding Factors or Exceptional Opportunities in Ecological Studies?" *Oikos* 63 (1992): 305–308. Print.

Love, Milton S., J. Caselle, and L. Snook. "Fish Assemblages on Mussel Mounds Surrounding Seven Oil Platforms in the Santa Barbara Channel and Santa Maria Basin." *Bulletin of Marine Science* 65 (1999): 497–513. Print.

McEntee, Marni. "Pt. Mugu May Get Artificial Reefs." *Los Angeles Daily News* 7 June 1996 Valley edition: N4. Print.

Møller, Anders Pape. "Parasites, Predators and Nest Boxes: Facts and Artefacts in Nest Box Studies of Birds?" *Oikos* 56 (1989): 421–423. Print.

Nakheel. "Artificial Reefs." Nakheel PJSC. Dubai, United Arab Emirates. Web. 3 August 2015.

Oudshoorn, Nelly and Trevor Pinch, eds. *How Users Matter: The Co-construction of Users and Technology*. Cambridge, MA: MIT Press, 2003. Print.

Pattison, Kermit. "Conservationists Envision Offshore Oil Rigs as Artificial Reefs," *Los Angeles Daily News* 25 October 1994 Conejo Valley edition: TO4. Print.

Scudder, Bradford A. *Conservation of Our Wild Birds*. Boston: Massachusetts Fish and Game Protective Association, 1916. Print.

Smout, Chris. "Birds and Squirrels as History." *Environmental Histories: Local Places, Global Processes*. 4 April 2011. Web. 4 August 2015.

Taverner, P. A. "Bird-houses and Their Occupants." *The Ottawa Naturalist* 32. 7 (1919): 119–126. Print.

Tran, Cathy. "A Platform for Debate: Conference Discusses the Fate of Unused Oil Platforms and Their Surrounding Sea Life." *Orange County Register* 31 March 2007. Web. 4 August 2015.

15

INTERSPECIES DIPLOMACY IN ANTHROPOCENIC WATERS

Performing an ocean-oriented ontology

Una Chaudhuri

"Yes, animal, what a word!" Jacques Derrida famously exclaimed, calling out the totalization of animal life by a self-serving and arrogant humanist tradition (32). Today, the concept of the Anthropocene suggests an inverse scandal, and evokes the equivalent of another exclamation: "*Anthropos*—what a prefix!" many seem to be saying, objecting to the term's potential elision of national and economic differences, its tendency to shift ecological responsibility away from the West's carbon-based industrial capitalism and onto a totalized and featureless humanity (see, for example, Haraway and Kenney 259). Yet, undifferentiated though they are, both the Anthropocene and "the animal" are galvanizing concepts for two fields, the environmental humanities and animal studies respectively, whose subjects involve urgent realities that can make certain differences seem less relevant. When terrestrial species are undergoing something called "the Sixth Extinction," and when CO_2 levels are over 400 parts per million, "big picture" constructs like "the animal" and "the Anthropocene" can be valuable not just for political activation, but also for fresh thinking about the species and ecospheres that climate change is plunging into a state of emergency.

The art practice I discuss in this chapter responds to this emergency in a way that balances species-specificity with global ecological awareness. Its intense focus on a single species is motivated by a concern for—rather than disinterest in—the fate of all species. And it understands that fate in a way that is best described as "Anthropocenic," to mark the vital need, at this rapidly unfolding juncture, for concerted and universal human attention to the environment, no matter what the historical causes of—and specific culprits behind—the current crisis.

The Dolphin Dance Project was initiated in 2009 by New York-based dancer and choreographer Chisa Hidaka and her partner, filmmaker Benjamin Harley. The artists describe the project as bringing together "human dancers and wild dolphins to co-create underwater dances in the open ocean" (Dolphin Dance Project). The dances are filmed, most recently in 3D, with the intention of offering viewers the "extraordinary experience of participating in an intimate movement-based conversation with completely wild animals, in whose eyes we recognize shared intelligence, creativity, and rich, meaningful emotional lives." In pursuit of this inspiring goal, the Dolphin Dance Project inevitably encounters many of the ethical and ideological challenges that beset even the most benign and well intentioned of interactions between humans and any of the other animals. Indeed, for some people, the project arouses anxiety and distrust—even outright anger, as was the case with one of my friends, whose immediate comment upon hearing the mere name of the project was: "Leave those dolphins the fuck alone!" This friend is an accomplished artist, a lifelong birdwatcher, highly conversant with the field of animal studies, and deeply committed to safeguarding wild species. I do not dismiss his instinctive reaction; on the contrary, I believe it is based upon the very values that motivate the Dolphin Dance Project itself: urgent concern for increasingly endangered animal species, respect for wild habitats, and unwillingness to excuse or further tolerate the long-standing and deeply exploitative practices of animal exhibition and entertainment.

I interpret the Dolphin Dance Project as a complex corrective and alternative to the shameful history of marine exhibition, which results in such media sensations—and personal tragedies—as the death of the veteran Sea World trainer Dawn Brancheau, and whose systemic violence is documented in recent films like *Blackfish* and *The Cove*. The latter shows how tourist practices like the many "Swim with Dolphins" programs that flourish all over the world are implicated in practices like the annual dolphin massacre in Taiji, Japan. Without question, such interspecies violence deserves universal condemnation and legal opposition. Yet it is also important to seriously consider the fact that marine animal displays and interactive programs draw thousands of spectators and participants worldwide. Clearly, they promise a kind of interspecies experience that many people crave, and they afford powerful affective rewards. To disregard this reality and focus exclusively on the harm that results may be to miss an important opportunity to discover new and better answers to the question posed in that fertile early site of animal studies, John Berger's classic essay "Why Look at Animals?"

The consideration of human–animal encounter, interaction, and representation which that essay initiated, especially its bold claim that modernity had replaced actual animals with their effigies, continues to be both highly contested and enormously generative. My friend's vehement reaction to the Dolphin Dance Project, reflecting a growing cultural awareness of the price other species pay for our curiosity, implies a revision of Berger's question along the lines of "Why Not *Stop* Looking at Animals?"—a reflexively ironic idea coming from a birdwatcher. But that irony arises from the fact that Berger's catchy title has distorted his deeper argument, which was not primarily about looking but rather about encountering animals, about *exchanging* looks with the animal, about *being with* the animal. What Berger bemoans about modernity's animal practices is that they substitute looking at for being with, installing a regime of alienated visuality where once there was embodied co-presence.

If the interspecies fascination that fosters animal shows and touristic animal encounters is not just an idle, self-indulgent curiosity but something else, perhaps a deep-rooted and nature-affirming need to be better connected to the earthly realities that our so-called civilization deprives us of, then these practices may be sites for fostering a much needed biophilia, along with the enhanced ecological and interspecies consciousness that the Anthropocene demands. I read the Dolphin Dance Project as a thoughtful, painstaking attempt to navigate the conundrum arising, on the one hand, from a vital need for interspecies encounter and, on the other, from the potential damage that the fulfillment of this need inflicts on the nonhuman species involved. The effort is undertaken in a spirit of remediation and reparation, with deliberate attention to egregious past errors and a desire to revise the values from which they arose. The Dolphin Dance Project's work answers a different revision of Berger's question: "How might we—how should we—be with animals? Especially now, in the Anthropocene, when their lives and habitats are so threatened by the activities of our species?"

The Dolphin Dance Project combines two art forms—dance and film—to provide a rich, though indirect, experience of interspecies encounter. The "actuality deficit" that the project deliberately incurs—the fact that while the dancers enjoy the thrilling *presence* of actual dolphins, their audiences do not—is carefully compensated for by several features of the project, which work to heighten one aspect of the traditional artist-to-audience relationship: that of messenger between worlds. This aspect of the Dolphin Dance Project would seem to affiliate it with such paradigms as the shamanism, or neo-shamanism, that many contemporary artists have become interested in, but the discourse framing the project points in a different direction, toward a possible "interspecies diplomacy."

I am well aware that the concept of diplomacy carries many negative associations and unpalatable connotations. In a recent interview with Bruno Latour, Heather Davis expresses her hesitation about diplomacy as a political modality, saying that it "seems already to presume two, or more, opposing sides. And the diplomat ... is slippery, not quite trustworthy" (Latour 51). Latour agrees, adding "it is someone who betrays" (51). But then he offers this provocative, and counter-intuitive elaboration: the diplomat, he says, "betrays those who have sent him or her precisely because he or she modifies their values. He or she sees that the official attachment is not one to be ready to die for." By thus refusing to grant preeminence to the values of the group he or she represents, the diplomat "introduces a margin and a space to manoeuver ... So, to say that there is a horizon of diplomacy is to say we have to state our agreement or disagreement" (51). The "horizon of diplomacy" produced by the Dolphin Dance Project emerges from acknowledging species similarity and difference rather than ideological agreement or disagreement; it is shaped by the dialectical character of the discursive framework within which the work is presented to audiences: a framework that carefully balances showing and telling, presence and absence, doing and not doing, knowing and not knowing.

I should note that the artists themselves do not use the term "diplomacy," nor do they make any explicit claim about engaging in interspecies diplomacy. However, their website does have a section entitled "Dolphin Etiquette," where their key principles are articulated. Each of the sentences in the following excerpt exemplifies crucial elements of what I interpret as an interspecies diplomacy:

> We only work with wild dolphins in the open ocean, on the dolphins' terms. We never feed dolphins, nor attempt to coerce or train them in any way. As a rule, the dolphins approach us out of their own curiosity. We do our best to be well-informed about the most current scientific research on the natural behavior of dolphins so that we can interact with them in ways that are safe and appropriate for humans and dolphins. We understand that we are visitors in the dolphins' environment and we never attempt to interact with dolphins if they are resting or feeding or show any signs of disinterest or annoyance.
>
> (Dolphin Dance Project)

To take the last sentence first: understanding the human as visitor effectively reverses the fraudulent claim—ubiquitous in zoo publications and websites—that animals in zoos and aquariums are "ambassadors" from the wild. The Dolphin Project's principle of carefully prepared and "well-informed" visitation counters the practices of violent territorial intrusion that make the institutions of the zoo and the aquarium possible. If there are any ambassadors here, it is the human dancers, and this role is highlighted through the discourse that frames and accompanies the presentation of the filmed dances to audiences. The contextualizing discourse—on the project's website, within the films themselves, and in discussions following screenings—implicitly theorizes the project's ideological and imaginative operations.

Whenever the artists present the project, they always mention two "rules" they have imposed on themselves: one, the dancers never touch the dolphins; and two, the project team never discloses the location of their dances. Each of these rules, fairly simple and straightforward at first glance, is in fact one of the pillars of an ethos of restraint that the Dolphin Dance Project is invested in. Each rule also revises certain pervasive assumptions about the ocean and its inhabitants that are inaccurate or outdated, as well as ecologically dangerous.

The first rule—no touching—can perhaps most readily be unpacked with the help of Susan Davis's analysis of the Sea World slogan and advertisement entitled "Touch the Magic." Davis shows how the ad's designation of the ocean world and its inhabitants as an otherworldly realm of enchantment and wonder is crucially linked to the promise of tactile contact, "the fantastic wish for a total merging with [a] wild nature" that has long been constructed as a distant spectacle (211). The possibility of touching something also brings with it the temptation or fantasy of "grasping" it—physically, emotionally, intellectually. The Dolphin Project rejects this hubristic fantasy, holding itself to an ethos of self-restraint that is also a statement of respect for its animal partners as well as an acknowledgment of the limits of human knowledge.

The Dolphin Project's injunction against touching the animals—even when the dolphins initiate physical contact, as they sometimes do—is especially interesting in light of the fact that the project leaders and dances are rooted in the dance form known as "contact improvisation," which relies heavily on physical contact as a form of information and communication, and where the patterns of the choreographies emerge not only from touch but from full body contact. In the context of their work with the dolphins, the dancers derive this information from the movement and behavior of the animals, as well as from a different kind of contact: eye contact, about which the artists often speak in stirring terms. In one of the films, Hidaka says:

> When you're eye to eye with a dolphin, you really see a person, regarding you. They're going, "Wow, there's a real person there, or there's a real dolphin there." That's a real paradigm shift. Because we're so used to thinking of ourselves as being separate and different from everything else. You know? There's humans, and then there's animals? Well, all of a sudden, you realize: there are these persons in the ocean.

This description of the experience and meaning of the interspecies gaze explicates—better than any other I have come across—Berger's enigmatic formulation of its unique value. Discussing the intersubjective drama that unfolds when the human being and the animal look at each other across the "narrow abyss of incomprehension" he made famous, Berger writes:

> [W]hen he is being seen by the animal he is *being seen as his surroundings are seen by him*. His recognition of this is what makes the look of the animal familiar. And yet, the animal is distinct and can never be confused with man. Thus a power is

ascribed to the animal, comparable to human power but never coinciding with it. The animal has secrets, which, unlike the secrets of caves, mountains, seas, are specifically addressed to man.

(3; italics mine).

The interspecies gaze is, for Berger, the recognition that there are other subjectivities from whose perspective the human is an "other"; in Hidaka's more straightforward terms, it is the recognition that members of other species are "persons" too.

The remainder of Berger's article tracks the transformation of the animal from interacting equal to inert spectacle, and inaugurates a great suspicion—ongoing in contemporary animal studies—about the sense of sight and regimes of visuality. The Dolphin Dance Project intervenes in that history by locating its use of eyesight within a revised sensorium that arises from a mindful engagement with the alien world and the inhospitable element which the animals inhabit, and which the dancers enter as visitors. Both the new kind of interspecies gaze the project achieves and the revised sensorium that makes it possible are part of an ocean-oriented ontology that we can begin to glimpse through the second basic rule of the project, keeping the performance sites secret.

The refusal to disclose where the dolphin dances take place is motivated first, of course, by a desire to keep dolphin habitats commercially undisturbed, free of the intrusions from the "Swim with Dolphins"-style of tourist entertainments. But it has another meaning as well, a symbolic resonance that corrects current (mis)alignments between oceanic realities and human knowledge. Through a kind of negative mimesis, it performs the ocean's essential unknowability, its resistance to easy mapping. It stands as a rebuke, for example, to the kind of assumptions that came into view in 2014, when a lost Malaysia Airlines flight proved impossible to locate. The expression of public outrage and incredulity that accompanied the inability of authorities to find the plane, in spite of a massive international effort, suggested that things like GPS and Google Earth have done some serious cognitive damage in that they have convinced us that we can go anywhere and see anything on this planet.

While this may be close to the truth for terrestrial areas, it is nowhere near the reality of our current command over the ocean, which is feeble. A vast array of statistics and facts could be marshaled to convey the current state of ocean knowledge; I prefer, however, to offer an image drawn from a recent book of popular science writing, James Nestor's account of the sport of free diving: "If you compare the ocean to a human body, the current exploration of the ocean is the equivalent of snapping a photograph of a finger to figure out how our bodies work" (9). Another statistic confirms this limit now and for the future: the US budget for space exploration (NASA) is 150 times larger than that for ocean exploration (NOAA, the National Oceanic and Atmospheric Administration) (Conathan). This deficit is also responsible for a widespread and profound misunderstanding about the ocean as an earthly environment. Were maritime exploration further along, people would presumably be better informed about how thoroughly alien the oceanic environment is to human life and human physiology, a fact that Nestor's body-oriented perspective on the ocean emphasizes:

At sea level, we are ourselves. Blood flows from the heart to the organs and extremities. The lungs take in air and expel carbon dioxide. Synapses in the brain fire at a frequency of around eight cycles per second. ... *At sixty feet down, we are not quite ourselves*. The heart beats at half its normal rate. Blood starts rushing from the extremities toward the more critical areas of the body's core. The lungs shrink to a third of their usual size. The senses numb, and synapses slow. The brain enters a heavily meditative state. ... At *three hundred feet, we are profoundly changed.* The pressure at these depths is ten times that of the surface. The organs collapse. The heart beats at a quarter of its normal rate, slower than the rate of a person in a coma. Senses disappear.

(7, emphasis added)

A space beyond the normal realm of the human senses, the ocean has also long been a space beyond human thought, almost beyond human imagining. The names given to its deepest layers signal our sense of fearful otherness: the second deepest layer, the abyssal zone, is etymologically linked to the mind-boggling concept of bottomlessness, while the hadal zone is named after the Greek underworld of everlasting darkness. These sunless depths are currently enjoying a moment in the media spotlight, thanks to James Cameron's dive to the Marianas Trench, but for centuries they were relegated to the status of a void. As Philip Steinberg has argued, the purported emptiness of the ocean was ideologically useful, a convenient untruth, because it allowed the ocean to be regarded as an "empty transportation *surface,* beyond the space of social relations" (113). This construction was well suited to serve the interests of industrial capitalism, enabling the "free" flow of goods that has, in our own time of globalization, reached its apotheosis in the commercial revolution known as containerization, a massive system of ocean freight using gigantic standardized "modal" steel boxes.

While sporadic efforts of maverick oceanographers occasionally challenged the view of the ocean as a featureless void, it was only late in the last century that a new awareness began to dawn of the vast biological treasures of marine life, as undersea expeditions dredged up not only the lost treasure they were seeking but also thousands of strange new species: "giant worms and slugs, spindly crabs and prawns, delicate sponges and sea lilies" (Broad 37). Today, with marine science still in its infancy, the ocean faces devastation from a host of human activities, including overfishing, trawling, deep-sea mining, oil spills, and toxic dumping. Ocean acidification and warming sea temperatures all but guarantee that countless species of marine animals, plants, and organisms will be rendered extinct before they are even discovered by science—a paradox on which Stacy Alaimo reflects in her essay in this volume. This combination of historical oblivion, current fascination, and overarching threat is the fraught context for artists like the Dolphin Dance Project, and a growing number of others, who seek to perform in, about, or for the ocean today.

Interspecies diplomacy in this context begins with an acknowledgment of the ocean's alien ontology, of the vastly different sensoria of the species native to it, and their profoundly divergent modes of phenomenal experience. Differences between oceanic and terrestrial realms range from obvious ones like the elemental (water instead of air) to perplexing ones like the cartographic (the volume and mobility of seawaters thwart land-mapping techniques suited to static, two-dimensional surfaces). These differences also require us to shift or expand our modes of knowledge production to include forms of embodied experience:

Whilst rationalists "turn away from the waves to admire the wave-born" ... and romantics revel in the ocean's alterity ... those who actually engage the ocean, like sailors and, perhaps even more profoundly, surfers and swimmers, become one with the waves as the waves become one with them, in a blend of complementarity and opposition.

(Steinberg and Peters 4)

The Dolphin Dance Project makes oceanic alterity an integral part of its interspecies diplomacy, embracing the limitations that it imposes upon (and reveals in) human physiology: compared to dolphins, humans are lousy water-dancers! The project also welcomes the ways in which this alterity can shape a mental state with enhanced interactive potential for the artists. This mental state is intimately connected to a key feature of ocean ontology, humans' inability to breathe underwater. To hold our breath for any length of time is a conscious act that can be improved with practice and technique, and one that affects our physical state.

For the dancers to be able to hold their breath and remain underwater long enough to engage the dolphins in dance sequences requires, among other things, that they calm their minds, because the active brain uses more oxygen than other organs. Mindfully measured breathing is a key to this calming:

> A focus on breath is core to our practice, whether we are training or actually dancing with dolphins. We hold our breath for a minute or more repetitively, with little rest in between, and this means we need to have excellent control of our breath. We need to breathe deeply and well between dives, and remain relaxed while holding our breath during the dives. We need to be keenly aware of our oxygen deficit and carbon dioxide build up so we can extend our diving times and depths safely.
>
> (Hidaka)

The mental state that results profoundly shapes the experience of interspecies encounter, especially of the interspecies gaze. When speaking of this, Hidaka is understandably cautious; the project prizes its alignment with marine science and the artists are wary of being dismissed as New Age kooks. Choosing her words carefully, Hidaka allows that "I know this kind of focus on breath is common in various meditation practices, but I wouldn't really call what we do 'meditative' … The breath work does affect our mental, physical and emotional state. I would describe it as a feeling of 'openness.'"

Hidaka's further observations on this state of being link sensation and affect, physicality and emotionality, linkages that in turn foster a new space for ethical interspecies encounter: "The 'open' state is also a relaxed one, and our relaxed demeanor and body posture expresses our trust (and lack of anxiety or aggression). *A sense of trust* definitely seems to be noticed and reciprocated by our dolphin partners." In this account, experience flows in both directions, inward and outward: the dancer's relaxed body (a relaxation achieved through the special breathing practices) makes her feel calm and safe, but it also communicates beyond the dancer, to the dolphins, whose demeanor in turn encourages a creative physicality and a heightened spirit of collaboration. Like the medium in which the work occurs, the dance practice involves circulations and flows, mobilizing a dynamic reciprocity of perceptions and emotions ranging from excitement, curiosity, and fascination to enjoyment and love:

> The engagement with dolphins is very social. Sometimes dolphins look at us wide-eyed as if excited or curious. Sometimes they cast a sleepy, half-open eye towards us. Sometimes the regard expresses deep interest in us. In an "open" state we can respond with our full range of emotional and kinesthetic expressions. Sometimes we are also joyful or fascinated. Sometimes we feel a crushing tenderness, or even love. And if we are clear in our expression, our dolphin partners easily respond

to us—to what we express—just as we respond to them: through eye contact and shared movement. Our meeting turns into a true exchange—a movement-based "conversation"—a dance.

(Hidaka)

To conclude, I return to Latour's provocative formulation about diplomacy to ask: What values does this "movement-based conversation" modify? How does it "betray" the ways that humans have previously related to animals? As I read the project, that "betrayal" happens in two areas, which we can identify, very broadly, as knowledge production and artistic process.

The words "mysterious" and "unfamiliar" recur in Hidaka's discussions of the work, often closely connected to "openness." She notes, for instance, that "the 'open' state helps me accept the unfamiliar without rushing to 'figure it out' or give it an overly anthropomorphic meaning. I don't have to feel anxious about what I don't know" (Hidaka). In place of the inquiring mind of scientific investigation, the calm mind of the relaxed dancer affords a tolerance for the unknown. This acceptance of the unknown and unfamiliar is, for Hidaka, linked to necessary changes in the regimes of knowledge underlying existing relationships to wild animals. According to her, the connection we need to feel in order to care enough to protect animals and their habitats has to leave room for nonhumans "to have needs and desires we cannot fully know. Without allowing for that, we start to imagine—wrongly— that 'caring' for dolphins in captivity on completely human terms could be ok for them. We start to think that we can 'manage' the complex ecology of their ocean habitats with our puny knowledge" (Hidaka). The interspecies diplomacy of the project, then, includes reca-librating the relative roles of scientific certainty and epistemological humility. Balancing the known with the unknown, accepting unfamiliarity while cultivating alertness, produces a result—"I can remain alert and connected to my partner, continuing to move fluidly together"—that resonates with Latour's description of the "horizon of diplomacy" as a space where peers learn what it is to be together.

Paradoxically, the embrace of the unknown makes the ocean feel, says Hidaka, like "less of a 'void,' and exposes humans as less 'exceptional' than we might believe." This felt rejec-tion of human exceptionalism, is, I would suggest, the bedrock principle and fundamental achievement of the project and the cornerstone of its interspecies diplomacy. It is both cause and effect of the artists' willingness to cede control to the other species, even in an area that most artists find hard to let go off: aesthetic control. To use Latour's term again, the artists "betray" the traditional centrality and agency of the artist, making the animals their guides and teachers, allowing the animals' movements to determine the dancers' movements and hence the works' choreography.

By having their choreography emerge from mirroring the animals' movements, the Dolphin Dance Project firmly rejects the principle underlying the long history of animal performances; namely, to make animals imitate *human* movement and behavior. Unlike the orcas at Sea World, the dolphins in this project are not required to wave hello, or clap their "hands," or kiss their trainers on cue. Indeed, they are not *required* to do anything. On the contrary, great care is taken to ensure they will and can do only what they choose to do: either engage creatively with the human dancers, or swim away.

The imagery that emerges from this project—imagery enhanced by brilliant cinematogra-phy, complex sound design, skillful editing, and the latest video technologies—is gorgeous and stirring to behold. Yet it is never allowed to remain just that. The Anthropocenic inter-vention that is the project's deepest goal—and the one I have tried to track and theorize

151

here—is achieved through the careful and extensive discursive framing. Its aim of articulating a new mode of interspecies attention and communication that is particularly attuned to the entanglement of "nature" and "culture" that characterizes the Anthropocene is lodged in the discourse—the carefully introduced film screenings, the conversation with the artists. This heavy mediation is crucial to its successful functioning as Anthropocenic diplomacy, as is its species-specificity. Focusing their attention, research, training, and artistry on one species and one challengingly alien environment, the artists of the Dolphin Dance Project bring us thought-provoking, heart-stirring, and politically promising glimpses of how to look at—and be with—the other animals with whom we share this fragile, endangered world.

References

Berger, John. "Why Look at Animals?" *About Looking*. London: Bloomsbury, 1980. 1–26. Print.

Blackfish. Dir. Gabriella Cowperthwaite. Manny O Productions, 2013. Film.

Broad, William J. *The Universe Below: Discovering the Secrets of the Deep Sea*. New York: Touchstone, 1997. Print.

Conathan, Michael. "Rockets Top Submarines: Space Exploration Dollars Dwarf Ocean Spending." *American Progress* 18 June 2013. Web. 30 Nov. 2015.

The Cove. Dir. Louie Psihoyos. Participant Media, 2009. Film.

Davis, Susan. "Touch the Magic." *Uncommon Ground: Rethinking the Human Place in Nature*. Ed. William Cronon. New York: Norton, 1996. 204–231. Print.

Derrida, Jacques. *The Animal That Therefore I Am*. Trans. David Wills. New York: Fordham University Press, 2008. Print.

The Dolphin Dance Project. "Dancing with Dolphins." 2010. Web. 30 November 2014.

Haraway, Donna and Martha Kenney. "Anthropocene, Capitalocene, Chthulhocene: Donna Haraway in Conversation with Martha Kenney." *Art in the Anthropocene: Encounters Among Aesthetics, Politics, Environments and Epistemologies*. Ed. Heather Davis and Etienne Turpin. London: Open Humanities Press, 2014. 255–270. Print.

Hidaka, Chisa. Personal communication. 15 May 2015.

Latour, Bruno. "Diplomacy in the Face of Gaia: Bruno Latour in Conversation with Heather Davis." *Art in the Anthropocene: Encounters Among Aesthetics, Politics, Environments and Epistemologies*. Ed. Heather Davis and Etienne Turpin. London: Open Humanities Press, 2014. 43–56. Print.

Nestor, James. *Deep: Freediving, Renegade Science, and What the Ocean Tells Us about Ourselves*. New York: Houghton Mifflin Harcourt, 2014. Print.

Steinberg, Philip E. *The Social Construction of the Ocean*. Cambridge: Cambridge University Press, 2001. Print.

Steinberg, Philip E. and Kimberley Peters. "Wet Ontologies, Fluid Spaces: Giving Depth to Volume through Oceanic Thinking." *Environment and Planning D: Society and Space* 33 (2015): 247–264. Web. 30 November 2015.

16

THE ANTHROPOCENE AT SEA

Temporality, paradox, compression

Stacy Alaimo

The concept of the Anthropocene compels us to think the temporal and spatial reach of the human in vast, and predominantly geological, dimensions. But what would it mean to take the Anthropocene out to sea? What problematics, what figurations, what epistemologies would that generate? The seas have long enticed us to imagine immensities of breadth, depth, and volume, as well as unfathomable zones of darkness, pressure, and cold. The ocean, suspected by Charles Darwin and others of harboring hardy "living fossils" that have endured major extinction events by eluding terrestrial time, would seem to have its own temporality (Dobbs 159). Paradoxically, the deep sea has been imagined both as a haven for such "living fossils" and as a lifeless zone. The depths of the oceans, for example, were long considered "azoic," devoid of life, due to the formidable darkness, pressure, and cold. Now, even the most extreme areas such as those around deep sea vents are known to be teeming with life—life based on chemosynthesis, not photosynthesis, life apart from the solar temporalities of day and night. While recent science echoes the earlier sense that time crawls across the abyssal planes, as deep sea species live and die at a slower pace, that temporality actually makes the fish less, rather than more, impervious to harm: "Deep sea fish are highly vulnerable to disturbance because of their late maturation, extreme longevity, low fecundity and slow growth" (Devine 29). Similarly, manganese nodules, the target of deep sea mining, grow "extraordinarily slowly, only a few millimeters per million years … a thousand times less than the sedimentation rate" (Koslow 164). In the 1980s Lauren Mullineaux discovered their "fragile fauna," "a unique ecological community contained within the universe of the nodules" (Koslow 163). In 2015, a team of scientists note that "fragile habitat structures and extremely slow recovery rates leave deep sea communities vulnerable to physical disturbances such as those caused by mining" (Wedding et al. 144). Protecting deep sea ecosystems, or even attempting to decrease the extent of devastation, requires reckoning with abyssal temporalities.

As the environmental humanities grapples with the expansiveness of geological time, marine science suggests an eerie temporal compression. Callum Roberts, a marine scientist and popular author, notes that "with an ever accelerating tide of human impact, the oceans have changed more in the last thirty years than in all of human history before" (3). A multitude of significant alterations crowd together in brief time frames, within ever more empty seas. The massive Census of Marine Life, and other research in the deep and open waters of the oceans, has revealed thousands of new species, in a paradoxical moment marked by both astonishing discoveries and a multitude of extinctions. Indeed, the paradox stings with the realization that much of the technology enabling discovery is due to the very industries—deep sea mining, deep sea fishing and trawling—that decimate species and habitats. The compressed temporality of the Anthropocene ocean is evinced by the title of one of the most comprehensive reports on the state of the deep seas, "Man and the Last Great Wilderness: Human Impact on the Deep Seas" (Ramirez-Llodra et al). The grand title harkens back to eras in which the majestic wilderness beckoned "Man," and yet the subtitle collapses the elevated sense of both "man" and "Nature" into a prosaic account of anthropogenic effects. While the title's barely veiled wish for a wilderness beyond human destruction is dashed by the subtitle, and more importantly, by the devastating data in the article itself, the term "wilderness" indicates that the deep seas remain still relatively uncharted in the twenty-first century.

Even as weird and wonderful creatures are discovered in the depths and delivered to computer screens and coffee table books, the deep seas offer an epistemological provocation, as little is known about most of these photogenic fauna, their habitats, or their ecosystems. Whereas the ability to stand back and map vast terrains—the already emblematic epistemological stance of the Anthropocene—places Man outside of that which he surveys, the compressed time of the Anthropocene seas puts us under pressure, weighs us down with the recognition that even as human impacts may be colossal, human understanding of marine ecologies and species, especially those of the deep seas, is miniscule. Indeed, the not-yet-nor-never-to-be-discovered marine species that (must) have become extinct due to anthropogenic causes—which elude capture by human knowledge systems but nonetheless cannot elude the unintended effects of human actions—would be apt icons for the Anthropocene seas. Icons without images, names, or lineages. Countless creatures who will never appear within human frames, vanishing silently, yet capable of signifying—if we imagine them to do so—how human knowledge is not adequate to account for, nor certainly to ameliorate, the enormity of the effects of a geological epoch distinguished by anthropogenic consequences. Taking the Anthropocene to sea elicits tempestuous speculations as well as curious attachments to creatures that are hypothetical or conjectural. In *Before the Law: Humans and Other Animals in a Biopolitical Frame*, Cary Wolfe argues that we cannot avoid "discrimination, selection, self-reference, and exclusion," and must grapple with the "differences between different forms of life," such as "bonobos vs. sunflowers," since "all cannot be welcomed, nor all at once" (103). He contends, "We *must* choose, and by definition we *cannot* choose everyone and everything at once. But this is precisely what ensures that *in the future we will have been wrong*" (103). But within the compressed time of the Anthropocene seas, which are not entirely encompassed by biopolitical frames, we have been and continue to be inadequate for the task of even beginning to respond to how diverse marine species have vanished and will continue to vanish due to anthropogenic causes. It will not have been a matter of realizing the ways we "will have been wrong," as Wolfe puts it so trenchantly, but of speculating about how we will have been insensible.

Although the deep seas do seem a world apart, when scientific and popular rhetorics cast distant depths and abyssal creatures as "alien," they imaginatively remove them from the planet, from the terrain of human concern, and even from reality.[1] While the sense of wonder may encourage a paradoxical amalgamation of ethical concern and epistemological restraint, it may also spark the detached awe of the spectator and deflect responsibility, as what is alien dwells beyond the domain of earthly concern.[2] While the rhetoric of "alien seas" still predominates, unfortunately, some scientists devise positions where wonder, epistemological humility, and material interconnection coexist. Michael A. Rex and and Ron J. Etter, in *Deep Sea Biodiversity: Pattern and Scale*, insist that the deep sea is not a realm apart from the rest of the planet, and yet it remains a site for wondrous discoveries: "Far from being the isolated, stable environment it was long assumed to be, the deep sea is now understood to be a dynamic and integral part of the global biosphere. Since most of the deep sea remains unexplored, we can hardly guess what wonders exist there" (x). While the shift from the deep seas as already interconnected with more familiar dimensions of the biosphere to the deep seas as unexplored realms of wonder may seem contradictory, not least in the move from dry statement to affect-saturated promise of marvels, their position makes sense. They refrain from casting the depths as alien or segregated in terms of global systems, even as they emphasize that humans know little about these zones. For example, there are "still no reliable estimates of total deep-sea biodiversity" and "deep-sea ecology has not been incorporated into mainstream ecology" (243, x).

Tony Koslow's position is similar, in that he calls the deep sea "a source of scientific wonder" and "the last great human frontier on this planet," but underscores that this does not mean the depths are pristine, as "no portion of the deep sea is today unaffected by human activities" (1, 3). Rex et al. foster cautious epistemologies of wonderment by emphasizing how much is unknown, thus casting the deep seas as neither alien nor easily mastered. Indeed, the abyssal and benthic zones indicate the enormity of the effects of intentional and unintentional human actions, as those actions reverberate all the way to the sea floor. And yet the abyssal and benthic zones undermine anthropocentricism, given that human access to and knowledge of these areas is peripheral. Thus, the depths encourage us to think the Anthropocene in a profoundly nonanthropocentric manner.

It is nonetheless possible, of course, to reconstitute delusions of human mastery, omniscience, and centrality. The popular science book, *The Extreme Life of the Sea*, by Stephen R. Palumbi and Anthony R. Palumbi, promises in its preface to give "a simple sense of guiltless wonder about how wonderful the ocean's life actually is" (vii). They admit that humans "have done sufficient damage to every habitat;" thus the "bell of doom" can always be sounded. Dramatizing the lives of various extreme "characters" of the seas throughout the bulk of its chapters, the book nevertheless concludes by sounding the bell of doom in the end. Given the state of ocean ecologies, how could it not? And yet the wonder evoked by the extraordinary creatures detailed in the previous 157 pages fails to preclude the most ham-fisted assertions of human ownership and agency in the conclusion, with its subtitle, "The future oceans—and what lives in them—are ours to choose" (158). Even worse, perhaps, the final page asserts: "The ocean itself is our single greatest tool when properly harnessed and leveraged. That tool sits ready, and we have a good idea how to use it" (178). True—many things could be done, and should be done, to mitigate some of the threats to ocean ecologies. Individuals could have some impact if they shrunk their carbon footprint, stopped consuming seafood, and avoided plastics. Governments and international bodies should create many more marine-protected areas, and enforce stricter policies regarding fishing, trawling, mining, and dumping. But in the face of overwhelming and even inalterable damages to the ocean—caused by radioactive, chemical, plastic, sonic, and other pollution; wasteful and destructive industrial fishing; trawling; mining; warming waters; and acidification—the authors' call to action, a crude individualist scene of "choice" and "tools," fails to account for epistemological challenges, nonhuman agencies, and the incommensurability of scale.

They reduce the swirling, interactive, altered, and emergent substances, forces, geochemistries, ecosystems, and creatures of the ocean to something like a hammer on a workbench—distinct, apparent, crafted for human use, secreting no surprises.

Ocean acidification, widely dubbed the "evil twin" of climate change, caught the scientific community unaware.[3] The rather chipper title of a 2009 Nature.com blog on climate, "Ocean Acidification Disorients Fish, Riles up Scientists," by Anna Barnett, reports on the 2008 UN conference "The Ocean in a High CO_2 World," by concluding with a preliminary and modest call for policymakers "to realize that ocean acidification is not a peripheral issue." Unlike atmospheric climate change, which is a complex phenomenon, difficult to demonstrate, involving many different sorts of scientific research and mapping systems; ocean acidification, caused by more CO_2 absorption in the seas, is a rather simple phenomenon. But while the process of acidification is straightforward, the enormous diversity of its effects on marine species and ecologies will probably never be known. As marine scientist J E. N. Veron states, "What is still far from clear is exactly how these effects [of acidification] will manifest themselves: which will come first, which species will be most susceptible, and what second-order effects there will be for other dependent components of affected marine webs" (218). As marine science has not even identified or, most likely, discovered all the species in the seas, it would be impossible to map out a comprehensive understanding of how acidification will affect each of the different species and then to determine the wider ecological effects within the webs of interrelations. Despite the uncertainty, however, the predictions are dire. Veron states, in his rather restrained and academically rigorous history of the Great Barrier Reef, "This account of acidification may seem like a science-fiction horror story, but there is little evidence of fiction either in the science on which it is based or in the simplified interpretation I have given here. A continued business-as usual scenario of CO_2 production will ultimately result in destruction of marine life on a colossal scale" (219). He concludes that acidification is the "most serious (if least well understood) of all predicted environmental changes on earth" (220).

Calcifying organisms may be most vulnerable to acidification. "Increased acidity makes life more difficult for species that absorb carbonate from the water to build their shells and skeletons, such as snails and corals. As acidity levels rise, those shells and skeletons begin to dissolve" (Ogden 323). In November 2012, the British Anarctic Survey put out a press release entitled "First Evidence of Ocean Acidification Affecting Live Marine Creatures in the Southern Ocean," describing the work of Dr. Nina Bednaršek and her team, who discovered that the shells of pteropods, or marine snails, were being dissolved by upwelling, more acidic waters. The tiny creatures, a crucial part of the food web and an important carbon sink, become more vulnerable without their shells. The effects of acidification on the pteropod have been rather widely publicized, perhaps because the mechanisms involved, which dissolve the shell of these indispensable creatures, are simple to understand and compelling to visualize (Alaimo, "Your Shell on Acid"). But the figure of the pteropod dissolving may also be read in ways that emphasize multispecies and transcorporeal perspectives on the Anthropocene—no species is safe, removed, or protected from the biological, chemical, and physical alterations of the planet. Marine creatures are submerged in, and literally the stuff of, the world that has been rapidly transformed by human and nonhuman agencies. The very shell that would protect them from predators is no match for the altered alkalinity of the waters. The scene of the dissolve is paradigmatic of the sixth great extinction, a microcosmic glimpse of fragile lives disappearing, one tiny creature that vanishes as so many others have and will have done.

While the pteropod, as it dissolves, attests to a kind of epistemological solidity—the processes involved are simple, visible, and can even be reproduced as a classroom activity ("Understanding Ocean Acidification")—the range of effects of ocean acidification on other species are quite varied, and often more elusive, subtle, and complicated. Veron's small list of types of species that will be vulnerable to acidification include phytoplankton of the Southern Ocean, which would affect krill, "the linchpin of ... all Southern Ocean food webs"; deep sea corals; coralline algae, which consolidates reefs, and other corals and reef organisms "followed by a wide range of reef-dependent biota" (218–219). Gershwin, in her book about jellyfish, lists krill that are hatched deformed; mammals and fish with acidosis; clownfish and damselfish that lose the olfactory clues they need for homing and avoiding predators; sperm motility and fertilization disruption in coral and sea cucumber; slow development in scallops and sea urchins; and reproductive problems in sea urchins and prawns (324–235). The International Geosphere-Biosphere Program's 2013 report, "Ocean Acidification: Summary For Policymakers, Third Symposium on the Ocean in a High CO_2 World," includes a clear chart describing the ecosystem roles, the global economic value, vulnerability, and percent of species affected, of five different groups, molluscs, echinoderms, crustaceans, finfish, and corals (9). The orderly infographics convey more certainty than is warranted, perhaps, given the paucity of research pertaining to different species. (A text insert appearing on a global map, in another section of the same report, for example, explains that how coccolithophores, an important carbon sink, "respond to ocean acidification is an area of intense investigation," since some species seem to tolerate it and some do not.) Strangely, mammals, such as cetaceans, do not appear on the IGBP chart, perhaps because to list them so coldly, on a chart entitled "commercially and ecologically important organisms," would be jarring, given their popularity as charismatic megafauna and their status as highly intelligent, social creatures. The inclusion of marine mammals could also disrupt the placid "objectivity" of the document by invoking the impassioned legal, cultural, and political issues surrounding the hunting and capture of dolphins and whales. For finfish, however, the chart noting their estimated global value at $65 billion (US) summarizes their vulnerability to acidification as follows: "Indirect effects due to changes in prey and loss of habitats such as corals likely. Possibly some direct effects on behaviour, fitness and larval survival." The orderly chart belies the uncertainty of the data it conveys.

Whereas the terrestrial Anthropocene is characterized by the solid structures that cover far too much of the earth's land, the Anthropocene seas will, it is feared, be occupied by the very softest, gelatinous creatures, who proliferate in conditions that other sea creatures cannot survive. Jan Zalasiewicz and Mark Williams in *Ocean Worlds: The Story of the Seas on Earth and Other Planets*, predict that in the future, the "diverse, beautiful ecological systems still dominated by coral reefs and fish will be replaced by 'slime-rock' systems dominated by algal and microbial mats and jellyfish," citing Daniel Pauly's term "Myxocene," an epoch named for slime (191). Marine scientists, popular science writers, and even the NSF on their "Jellyfish Gone Wild" site warn that jellies are taking over the seas. "When jellyfish run wild, they may jam thousands of square miles with their pulsating, gelatinous bodies. In recent years massive blooms of stinging jellyfish and jellyfish like creatures have overrun some of the world's most important fisheries and tourist destinations—even transforming large swaths of them into veritable jellytoriums. The result, major damage to fisheries, fish farms, seabed mining operations, desalination plants, and ships." Although it has long been known that jellies flourish in polluted waters, the innocuous subtitle of the "Jellyfish Gone Wild" site—with the phrase "environmental change"—obfuscates human responsibility for upsetting ocean ecologies and provoking the jellyfish blooms. Instead, the site depicts the jellies themselves, with their "pulsating, gelatinous bodies," as monstrous. It is not clear when the site was created, but it does not mention acidification. Jellyfish expert Lisa-ann Gershwin states that some research has shown jellyfish prosper in more acidic waters, while other research shows no effect, and more needs to be known. She concludes, however, by reminding us that ancient seas, which were more acidic, were "dominated by flora and fauna similar to those that today's seas appear to be shifting toward," basically, "lots of jellyfish" (344).

The Anthropocene seas will be paradoxical, anachronistic zones of terribly compressed temporality where, it is feared, the future will move backwards, into a time when the oceans were devoid of whales, dolphins, fish, coral reefs, and a multitude of other species, but jellyfish (and algae) proliferated. The preference for species diversity and abundance makes many of us wish the oceans were what they were hundreds of years ago (Roberts), but there is little enthusiasm for a future in which the oceans become like the ancient acidic seas, characterized as "'slime-rock' systems." Such epochal preferences may seem baseless. Yet the rapid anthropogenic alterations and extinctions that would bring us (back) to gelatinous futures are a matter of concern for environmental ethics and politics. It is one thing for creatures to have not yet evolved and another for humans to have killed them off.

Even as the concept of the Anthropocene is quite recent—neither scientists nor humanists have had much time to consider its implications—the geological timescale of the Anthropocene is mammoth. The paradoxical moment of thinking something "new," which nonetheless has already been present, and indeed, which extends back into the murky reaches of time, to multiple, diffuse, and rather arbitrarily designated origins, adds another dimension to the epistemological anxieties of the early twenty-first century. Swirling natural/cultural forces threaten human and nonhuman lives, yet elude human knowledge practices, be they expert predictions of weather events or the improvisational everyday practices of transcorporeal subjects in risk society (Alaimo, *Bodily Natures*). The paradoxical moment of the Anthropocene, which was only just recently announced and yet entails the reconsideration of a vast swath of history as having been happening under its banner, submerges human mapping of geological time, as what would be assumed to be the background for the history of industrialized human "progress"—geological, "natural," time as well as the presumed bedrock of terrestrial "nature"—is engulfed by the accumulating effects of human actions. The comforting idea that the solid, permanent, physical world can serve as the backdrop for the exploits of the human has never seemed more delusional. Despite the challenge that the Anthropocene poses for humanists to scale up to the immensity of the geological epoch, the Anthropocene may also feel rather cramped, as the vast world becomes compressed within the effects of one species, our own.

Even as the Anthropocene enters academic and popular parlance as a new concept, many theories in the environmental humanities, science studies, and feminist science studies have long held that the conceptual divide between nature and culture cannot be maintained. The idea that we can no longer think in terms of a nature separate from humanity is hardly new. And yet the designation of a geological epoch signaling humans have left an indelible mark upon the planet does emphasize the enormity of the unintended effects of human activity and may be invoked as a call to action. But by emphasizing the scale of human impact and conceptualizing the human as a transhistorical, transcultural force, the human is set off from the rest of the planet, even as the notion of the Anthropocene would seem to insist that there is, in fact, no chasm between the human and what was once called "nature." For example, both popular visual depictions of the Anthropocene and academic accounts such as those of Dipesh Chakrabarty widen the gulf between "man" and earth by depicting human perspectives as transcendent and human force as abstract, floating above the solid, geological expanse of the planet (Alaimo, "Your Shell on Acid").[4] Moreover, the already iconic visual renderings of the Anthropocene depict the results of human force on transformed terrestrial places, substances, and materials, but living, threatened, and vanished species are not visible. Claire Colebrook writes "The very eye that has opened up a world to the human species, has also allowed the human species to fold the world around its own, increasingly myopic, point of view" (59). This myopic perspective has been translated into the actual production of a material world hostile to other living creatures. But the iconic depictions of the Anthropocene, which feature cityscapes, superhighways, the planet lit up at night, and other aerial shots, take things even further, rendering other species as having already vanished from a planet that looks much like Burtinski's *Manufactured Landscapes* writ large.

We've traveled a long way from our Darwinian origins, it seems, as smooth structures pave over the tangled bank, and the human as a corporeal being evolved from and with other creatures becomes an absent cause. More generative figurations of the Anthropocene, if that is a term environmentalists continue to think with, would consider the human as transcorporeal rather than incorporeal and would foreground a multitude of living species, considering their struggles to survive in rapidly transforming places. But it is also essential to think the Anthropocene from the vast seas, which have been profoundly altered by human activity in ways that are accidental, even as they are exploitative—oil spills and invisible shifts in alkalinity rather than superhighways or suburban sprawl. Thinking the Anthropocene at sea means venturing beyond the safe harbor of human structures and systems of habitation that wrap the human in himself, toward a sense of what it means for myriad creatures to dwell within rapidly changing waters and ecosystems. Figurations of the Anthropocene that insist on the presence—and the enigmatic, disquieting absence—of other species may help us navigate unknown futures while keeping nonhuman lives in mind.

The phrase "to put pressure" on a concept or theory has become all too common in academic parlance. It is fitting, however, to think of the pelagic and abyssal seas—so rarely thought—as zones that compress, transform, unmoor, or render paradoxical the scales of time and distance, the taxonomies, and other conceptual navigation systems that work on land. Taking the Anthropocene out to sea will no doubt result in troubling the concept in ways that this brief foray has not begun to imagine.

Notes

1 Take, for example, James Cameron's film, *Aliens of the Deep*, a documentary about deep sea exploration that repeatedly supplants the seas with the planets. The creatures featured in Cameron's film *The Abyss* end up being an alien civilization, not marine life. See Alaimo, "Feminist Science Studies: Aesthetics and Entanglement in the Deep Sea" and "Dispersing Disaster: The Deepwater Horizon, Ocean Conservation, and the Immateriality of Aliens." While I agree with Stefan Helmreich's argument in his magnificent book, *Alien Ocean*, that the "figure of the alien materializes … when uncertainty overtakes scientific confidence about how to fit newly described life forms into existing classifications or taxonomies" (16), I would add that many figurations of marine species as alien or the seas as outer space are fantasies induced by environmental guilt or even the anachronistic desire for a geography beyond the human, a wish for a timespace before the Anthropocene.

2 For a different critique of the dynamics of wonder within environmental concern, see Heather Houser, who argues in her analysis of Richard Powers's work, that wonder, which oscillates "between familiarity and strangeness," "can tip over into projection and paranoia, relations that divert energies away from ethical involvement" (28).
3 Callum Roberts notes that Joanie Kleypas, a coral reef expert, suddenly realized during a climate change meeting in 1998 that "coral reefs would, by the end of the twenty-first century, be bathed in water corrosive enough to destroy them" (107). Lesley Evans Ogden notes the term "ocean acidification" was coined in 2003.
4 In his essay "Brute Force," for example, Chakrabarty posits, "to say that humans have become a 'geophysical force' on this planet is to get out of the subject/object dichotomy altogether. A force is neither a subject nor an object. It is simply the capacity to do things" (2). Conceptualizing the human species as an abstract agency removes us from the scene, which, in my view, replicates the long tradition of disembodied Western, masculinist rationality.

References

Alaimo, Stacy. *Bodily Natures: Science, Environment, and the Material Self*. Bloomington: Indiana University Press, 2010. Print.
——. "Dispersing Disaster: The Deepwater Horizon, Ocean Conservation, and the Immateriality of Aliens." *American Environments: Climate-Cultures-Catastrophe*. Ed. Christof Mauch and Sylvia Mayer. Heidelberg, Germany: Winter, 2012. 177–192. Print.
——. "Feminist Science Studies: Aesthetics and Entanglement in the Deep Sea." *Oxford Handbook of Ecocriticism*. Ed. Greg Garrard. London: Oxford University Press, 2014. 188–204. Print.
——. "Your Shell on Acid: Material Immersion, Anthropocene Dissolves." *Anthropocene Feminisms*. Ed. Richard Grusin and John C. Blum. Minneapolis: University of Minnesota Press, forthcoming.
Barnett, Anna. "Ocean Acidification Disorients Fish, Riles up Scientists." *Nature.com Blog*. 3 February 2009. Web. 15 January 2015.
British Antarctic Survey. "Press Release: First Evidence of Ocean Acidification Affecting Live Marine Creatures in the Southern Ocean." UK Natural Environment Resource Council. 25 November 2012. Web. 15 January 2015.
Chakrabarty, Dipesh. "Brute Force." *Eurozine*. 7 October 2010. Web. 15 January 2015.
Colebrook, Claire. "Framing the End of the Species: Images without Bodies." *Symploke* 21 (2013): 51–63. Print.
Devine, Jennifer A., Krista D. Baker, and Richard L. Haedrich. "Deep Sea Fishes Qualify as Endangered." *Nature* 439.5 (January 2006): 29. Print.
Dobbs, David. *Reef Madness: Charles Darwin, Alexander Agassiz, and the Meaning of Coral*. New York: Random House, 2005. Print.
Gershwin, Lisa-ann. *Stung! On Jellyfish Blooms and the Future of the Ocean*. Chicago: University of Chicago Press, 2013. Print.
Helmreich, Stefan. *Alien Ocean: Anthropological Voyages in Microbial Seas*. Berkeley: University of California Press, 2009. Print.
Houser, Heather. *Ecosickness in Contemporary U.S. Fiction: Environment and Affect*. New York: Columbia University Press, 2014. Print.
IGBP, IOC, SCOR. "Ocean Acidification Summary for Policymakers—Third Symposium on the Ocean in a High-CO_2 World." International Geosphere-Biosphere Programme. Stockholm, Sweden. 2013. Web. 15 January 2015.
Jellyfish Gone Wild. National Science Foundation. Web. 15 January 2015.
Koslow, Tony. *The Silent Deep: The Discovery, Ecology, and Conservation of the Deep Sea*. Chicago: University of Chicago Press, 2009. Print.
Ogden, Leslie Evans. "Marine Life on Acid." *BioScience* 63 (2013): 322–238. Print.

Palumbi, Stephen R. and Anthony R. Palumbi. *The Extreme Life of the Sea*. Princeton: Princeton University Press, 2014. Print.

Ramirez-Llodra, Eva, Paul A. Tyler, Maria C. Baker et al. "Man and the Last Great Wilderness: Human Impact on the Deep Sea." *PLoS ONE* 6.8 (2011): e22588. Web. 2 September 2016.

Rex, Michael A. and Ron J. Etter. *Deep-Sea Biodiversity: Pattern and Scale*. Cambridge, MA: Harvard University Press, 2010. Print.

Roberts, Callum. *The Ocean of Life: The Fate of Man and the Sea*. New York: Penguin, 2012. Print.

"Understanding Ocean Acidification: Hands On Activities." *Channel Islands National Marine Sanctuary*. Web. 24 November 2015.

Veron, J. E. N. *A Reef in Time: The Great Barrier Reef from Beginning to End*. Cambridge: Harvard University Press, 2008. Print.

Wedding, L. M. et. al. "Managing Mining of the Deep Seabed." *Science* 349.6244 (10 July 2015): 144–145. Web. 1 December 2015.

Wolfe, Cary. *Before the Law: Humans and Other Animals in a Biopolitical Frame*. Chicago: University of Chicago Press, 2013. Print.

Zalasiewicz, Jan and Mark Williams. *Ocean Worlds: The Story of Seas on Earth and Other Planets*. Oxford: Oxford University Press, 2014. Print.

Part III
INEQUALITY AND ENVIRONMENTAL JUSTICE

17

TURNING OVER
A NEW LEAF

Fanonian humanism and environmental justice

Jennifer Wenzel

Humanism … must excavate the silences, the world of memory, of itinerant, barely surviving groups, the places of exclusion and invisibility, the kind of testimony that doesn't make it onto the reports but which more and more is about whether an overexploited environment, sustainable small economies and small nations, and marginalized peoples outside as well as inside the maw of the metropolitan center can survive the grinding down and flattening out and displacement that are such prominent features of globalization.
—Edward Said, *Humanism and Democratic Criticism*

Humanism is just a way of saying that everybody's right to self-creation matters.
—Richard Pithouse, "That the Tool Never Possess the Man"

The emergence of the environmental humanities is at once exciting and perplexing: exciting for its intellectual dynamism and renewed sense of urgency and relevance; perplexing for its re-mobilization of concepts that have come under pressure or erasure, most notably *the human* itself. In other words, "environmental humanities" is something of an oxymoron, as the recent posthuman turn runs up against the etymological anthropocentrism of the humanities and the geneaologies of humanism that are their epistemological foundation. These contradictions—between traditions of thinking the human and the proliferating interest in the more-than or other-than human—are too deep and disruptive to be resolved merely by affixing a *post-*.

From the *Sputnik* panic of the mid-twentieth century to the STEM fever of today, talk of the diminishment of the humanities in academic institutions and public life is nothing new. What is new—and newly paradoxical—is the disjuncture between the various external pressures to *scale back* the humanities and the call to *scale up* our understanding of humans as a species, increasingly seen as having altered the physical processes of the planet. Yet to many ears, Anthropocene species talk is a troubling new universalism that disregards the highly uneven roles that different groups of humans have played in the transformation of the planet, and the uneven distribution of risk and resilience in confronting this human-made world. Newfound interest in geological stratification threatens to displace attention to social stratification.

Debate over the universalizing thrust of Anthropocene discourse is the most recent example of a recurrent tension within environmentalism about the role of intra-human conflict in the relationship between humans and nonhuman nature. Mainstream environmental movements in the northern hemisphere have assumed a normative status that tends to discount other versions of environmental concern, including the "environmentalism of the poor," "popular environmentalism," movements for "environmental justice," or "livelihood" or "liberation ecology." Taken together, these environmentalisms insist upon the indivisibility of the social and the ecological: social inequality inflects relationships between humans and the environment. Running counter to mainstream assumptions that poor people do not care about the environment, these approaches—often in the form of social movements defending life and livelihood—recognize that "the poor sell [their health or natural resources] cheap, not out of choice but out of lack of power" (Martínez-Alier 30). They recognize that the socially marginalized (the poor, the racialized, the colonized) tend to end up on the losing end of conflicts over environmental benefits and burdens: "the unequal incidence of environmental harm gives birth to environmental movements of the poor" (54). These environmentalisms "demand contemporary social justice among humans" (11). Recognizing the many kinds of struggles theorized as environmentalisms of the poor, Joan Martínez-Alier has written that "The world environmental justice movement started long ago on a hundred dates and in a hundred places all over the world" (172).

One time and place for the emergence of this environmentalism-from-below is Frantz Fanon's *Les damnés de la terre* (1961), translated as *The Wretched of the Earth* (1963). In what follows, I show how Fanon's anatomy of the struggle for Third World national liberation is an instructive text for thinking about the relationship between the environmentalism of the poor and the environmental humanities, precisely because it is concerned equally with the roles of *nature* and *humanism* under colonialism and the roles they might play in an emancipated future. In Fanon's view, nature was a crucial terrain for material exploitation and psychic subjugation, and humanism was the handmaiden of a false universalism whose actual constituents were a lucky few on good days in Europe. But Fanon's dialectic of decolonization shows how nature and humanism can also be the terrain and telos of liberation; *The Wretched of the Earth* suggests that the environmentalism of the poor is also a humanism of the poor.

In this chapter I take the lexical root *human*, common to humanism, the environmental humanities, and the Anthropocene, as a rubric for teasing out what this environmentalism-and-humanism of the poor means for the environmental humanities.[1] Both environmentalism and humanism have been criticized, from different angles and with different emphases, for being ethnocentric and anthropocentric. Although challenges to cultural universalism or species exceptionalism reveal contests over what it means to be human, the referent of *human* may not be the same. Humanism's human ideal (and the forms of life excluded from it) may not be the same thing as the human perspective decentered by a multispecies approach. While some communities struggle not to be folded into "nature" (i.e., they are not quite recognized as human), others seek to dissolve the destructive privileging of the human in a more-than-human world. Thinking through these contradictions, and the forms of recognition, privilege, and exclusion at work in them, might mean that instead of just one more occasion to apologize for ethnocentrism and anthropocentrism, our efforts could be an opportunity to forge what Aimé Césaire called in *Discourse on Colonialism* a "true humanism – a humanism made to the measure of the world"; and, we might say, an environmental humanities made to the measure of the planet (56).

166

The first aspect of Fanon's analysis that merits attention is his description of colonialism as the imposition of a species divide. This act of violence, in which the native is made other (and less) than human, is most legible in *The Wretched of the Earth* not in the moment of dehumanization but in the self-liberating rejection of it (which, I show later, amounts to a radical humanism). As he announces in his first paragraph, Fanon sees decolonization as "quite simply the replacing of a certain 'species' of men by another 'species' of men" (in French, "*espèce*"). Decolonization is the moment when the native refuses the "allusion to the animal world" in the colonizer's account of him, "for he knows that he is not an animal; and it is precisely at the moment he realizes his humanity that he begins to sharpen the weapons with which he will secure its victory" (35, 42). In his anxious preface to Fanon's text, Jean-Paul Sartre refines Fanon's notion of colonialism as dehumanization or bestialization by adding a third term between the human/animal divide. In the first sentence of the book, we read: "Not so very long ago, the earth numbered two thousand million inhabitants: five hundred million men, and one thousand five hundred million natives" (7). Sartre posits *the native* as a colonial creation, neither human nor nonhuman animal. This liminal species is not hybrid or chimera, but instead half-degraded: no longer human but not quite reduced to beast; the careful result of economic cunning. Sartre explains: "For when you domesticate a member of our own species, you reduce his output, and however little you may give him, a farmyard man finishes by costing more than he brings in. For this reason, the settlers are obliged to stop the breaking-in halfway: the result, neither man nor animal, is the native" (16).

The obvious objection to this metaphorics of colonial bestialization (even its half-measures) is its *a priori* diminishment of the category of animal as the devalued Other of the human. How could such thinking serve as a foundation for environmentalism of any stripe? Even as Fanon and Sartre rail against the armed ethnocentrism of colonialism, they reify man's subjugation of beast and the specious sense of superiority that underwrites it. But for me, this is precisely the point. (Their humanist language is also stubbornly masculinist—as Fanon says about Europe, he and Sartre "are never done talking of Man"—a fact I want to register historically and move beyond, in their own dialectical spirit.) Fanon's emphasis on the native recognizing himself as a nonanimal human might imply a troubling anthropocentrism, but I would argue that Fanon's vivid account of colonization-as-bestialization offers a prescient reminder of how a species divide has been *deployed* historically to cast some humans as sub- (rather than non-) human:

> serious consideration of the status of animal seems to be fundamentally compromised by the human, often western, deployment of animals and the animalistic to destroy or marginalize other human societies. ... Human individuals and cultures at various times have been and are treated "like animals" by dominant groups. ... The history of human oppression of other humans is replete with instances of animal metaphors and animal categorizations frequently deployed to justify exploitation and objectification, slaughter and enslavement.
>
> (Huggan and Tiffin 135)

If *The Wretched of the Earth* is a manifesto for decolonization, one could say that Fanon puts the question of the human on its agenda from the outset, and in so doing reveals the dark intersections between ethnocentrism and anthropocentrism.

In an essay that later appeared in *Slow Violence and the Environmentalism of the Poor*, Rob Nixon identified four oppositions that obstruct conversations between postcolonialism and mainstream American environmental thought: hybridity vs. purity, displacement vs. place, transnational vs. national frames of analysis, and history vs. timeless transcendence ("Environmentalism and Postcolonialism"). I would argue that Fanon's *The Wretched of the Earth* reveals an additional schism to add to Nixon's list: that between postcolonialism's attention to histories of dehumanization, on the one hand, and environmentalism's critique of anthropocentrism and its conception of humans-as-animals in a more-than-human world, on the other. The category of the human has been wielded as a double-edged sword against those included in Sartre's census of "natives." As we have seen, in one stroke of the sword *les damnés de la terre* are excluded from the Sartrean brotherhood of man, their subjugation and enslavement justified in the name of a species divide: "since none may enslave, rob, or kill his fellow man without committing a crime," Sartre writes, colonial occupiers "lay down the principle that the native is not one of our fellow men" (15). Yet that sword cuts just as deeply when wielded in the opposite direction by an equally ethnocentric "antihuman environmentalism" (Nixon, *Slow Violence* 5), often in the form of national parks or nature preserves, that narrowly defends the interests of charismatic nonhuman animal species against the needs of marginalized humans (often formerly Sartre's "natives") who share their habitats and whose lives and livelihoods also depend upon access to them. In a very literal sense, the environmentalism of the poor must also be a humanism of the poor, a movement that reclaims the category of the human and demonstrates the ethnocentrism sometimes lurking behind pious rejections of anthropocentrism.

Fanon's second contribution to the environmentalism of the poor is his account of colonialism as the capture of natural resources. We might say that the species divide is the *ideological* fiction that colonialism uses to justify the *material* expropriation of the natural wealth of the colonized world, including enslaved human labor. This process entails a massive ecological debt. Behind the veneer of European affluence, Fanon finds a history of theft and an explanation for the Third World's "geography of hunger":

> This European opulence is literally scandalous, for it has been founded on slavery, it has been nourished with the blood of slaves and it comes directly from the soil and from the subsoil of that underdeveloped world. ... Colonialism and imperialism have not paid their score when they withdraw their flags and their police forces from our territories. ... The wealth of the imperial countries is our wealth too. ... Europe has stuffed herself inordinately with the gold and raw materials of the colonial countries. ... From all these continents ... there has flowed out for centuries toward that same Europe diamonds and oil, silk and cotton, wood and exotic products. Europe is literally the creation of the Third World.
>
> (96, 101–102)

Fanon tallies the European theft of nature and charts the cartography of underdevelopment: the natural wealth of the Third World made possible the development of Europe. Drawing a link between captive labor and expropriated natural resources as twinned objects of colonial exploitation, Fanon identifies a *continuing* structural inequality in what Fernando Coronil called the "international division of nature," an understudied corollary to the Marxian international division of labor (29).

As with colonialism's imposition of a species divide, Fanon theorizes decolonization as a dialectical reversal through which formerly colonized peoples forge newly sovereign nation-states and assert resource sovereignty: the right to dispose freely of the natural resources within their national territories.[2] Here too, one might question whether Fanon's account of underdevelopment, resource theft, and the decolonizing assertion of resource sovereignty can underwrite a properly *environmental* analysis, since Fanon's thinking about "the soil and mineral resources, the rivers" (100) remains firmly enmeshed within a *resource logic*, in which nature is understood as natural resource, disposed for human use and subject to human control. For Fanon, the question in resource sovereignty is which humans have the right to dispose over nature, rather than the "sovereignty" of nature itself. Thus, in Fanon's vision of national liberation, nature remains (in one crucial sense) colonized: subject to epistemological capture as the Other of the human.

On both of these questions, species divides and nature-as-resource, I would argue that Fanon's attention to the history of colonialism helps to limn faultlines within environmentalism. Fanon shows who repeatedly ends up on the losing end of European conceptions of humans and nature, who pays for the "externalities" of these categories and their mobilization. He demonstrates the necessity of the perspective that we now call the "environmentalism of the poor," attentive to the role of political power and social inequality in the disposition (conceptual and material) of nature. And he epitomizes what Joan Martínez-Alier observes: that many marginalized peoples' struggles over nature, for justice, have been waged in an idiom that is not explicitly environmental but nonetheless engages the concerns of environmentalism (viii, 54, 62).

Within the rhetorical and dramatic arc of *The Wretched of the Earth*, humanism and nature play similar roles: each is each weaponized under colonialism, and then reclaimed and reconfigured through the dialectical reversal of decolonization. Admittedly, one might say that within Fanon's logic, this process holds true for everything under (and including) the sun: whatever colonialism uses for its own ends, "natives" decolonizing themselves must seize, invert, and make their own. But nature and humanism are not just any old things in Fanon's account of colonialism: they bear a shadow interrelationship in which ethnocentrism and anthropocentrism are unpredictably intertwined. Untangling these relationships is one task of the environmental humanities, from whose perspective Fanon's drama of decolonization appears incomplete: further dialectical turns are necessary to reckon with the persistent anthropocentrism (and masculinism) of Fanonian humanism and the persistent ethnocentrism of mainstream environmentalism. The environmentalism of the poor shows why this should be so, and how the environmental humanities might avoid replicating the kinds of colonizing moves (involving both nature and humanism) that Fanon so excoriated. Picking up the mantle of Fanonian humanism, Edward W. Said named this relation *critique*—the critical embrace and emancipatory emendation of an imperfect inheritance, a task and an orientation that the environmental humanities would do well to adopt.

Fanon's scathing indictment of European humanism is justly famous. In the searing conclusion to *The Wretched of the Earth*, Fanon exhorts:

> Come then, comrades. ... Let us ... [l]eave this Europe where they are never done talking of Man, yet murder men everywhere they find them, at the corner of every one of their own streets, in all the corners of the globe. For centuries they have stifled almost the whole of humanity in the name of a so-called spiritual experience. ... today we know with what sufferings humanity has paid for every one of their triumphs of the mind.
>
> (311–312)

For Fanon, the European humanist tradition amounts to little more than an alibi for genocide, "a succession of negations of Man, and an avalanche of murders" (312). It is for this reason that Fanon remarks, "when the native hears a speech about Western culture he pulls out his knife—or at least he makes sure it is within reach" (43).

This indictment of European humanism (putatively universal, but in Fanon's account murderously ethnocentric) is relevant to the environmental humanities for two reasons. First, I would argue that Fanon builds his case against the hollowness of European humanism in part through his analysis of colonialism and nature that I discussed above: the ideological exclusion of the native from the category *Man* and the material effects of that exclusion (the "geography of hunger," the theft of nature). European humanism is propped up by concepts of nature that mask the dehumanization of *les damnés* and the ravagement of ecosystems and lifeworlds. As Sartre writes in his preface, "With us there is nothing more consistent than a racist humanism since the European has only been able to become a man through creating slaves and monsters" (26). While Sartre (prefatorily) echoes Fanon—peering at the lies and violence that lie beneath the "strip tease of our humanism," and acknowledging of the native-recreating-himself-as-man that "we were men at his expense" (24)—Fanon tallies up the bill, enumerating the costs of European humanism paid in deaths, dehumanization, and the theft of nature.

But the second reason to linger over Fanon's indictment of humanism—introduced in his first chapter, reprised in his conclusion—is to mark its dialectical reversal in the name of something truly remarkable, a newly and truly *universal* humanism. The conclusion of *The Wretched of the Earth* is searing but also soaring, as in the book's final lines: "For Europe, for ourselves and for humanity, comrades, we must turn over a new leaf, we must work out new concepts, and try to set afoot a new man" (315–316). This is not the Fanon of apocalyptic, even genocidal anticolonial violence, nor even the Fanon who places his faith in the nation and nationalism, but instead the visionary of a new, and newly universal, humanism. Ato Sekyi-Otu has argued that *The Wretched of the Earth* must be read as a dramatic arc: the narrative transformations within the text perform diegetically the dialectic of decolonization that Fanon is theorizing. This approach can make sense of the trajectory of humanism in the text. Fanon opens by describing decolonization as genocide and by disavowing Europe's fraudulent universalism; he closes with a universalizing gesture to forge a new humanism "for Europe, for ourselves and for humanity." I have discussed Fanon's account of the species divide and natural resources in similar terms, with decolonization as dialectical reversal. These smaller reversals take their place within the overarching framework of a Fanonian humanism that radically expands the constituency of the "universal." This humanism is a kind of telos within and toward which the dialectic of decolonization unfolds. This is another way of saying that Fanon's environmentalism of the poor is also a humanism of the poor.

In the US academy, Edward W. Said has been the most influential proponent of a humanism recognizable as Fanonian. Fanon and Said articulate an epistemological critique of European ethnocentrism and of colonialism's nexus of knowledge, power, and violence, yet they seek to forge, as if for the first time, a humanism worthy of the name. In his last book, *Humanism and Democratic Criticism,* Said holds humanism to account for having been instrumentalized by identitarian programs and murderous pogroms (77). Yet for Said (as for Fanon), humanism still might be something other than the pillage committed in its name. "It is possible to be critical of humanism in the name of humanism," Said writes; "schooled in its abuses by the experience of Eurocentrism and empire, one could fashion a different kind of humanism" (10–11). Rejecting notions of humanism as the purview of a priestly elite, Said affirms the democratic possibilities of humanist self-knowledge and self-critique: "to understand humanism at all, for us as citizens of this particular republic, is to understand it as democratic, open to all classes and backgrounds" (21). Said's is a humanism-against-hierarchy: a practice of reading and a practice of citizenship. Said's ramification of Fanonian humanism is well suited to the commitment of the environmental humanities to speak to contemporary crises and injustices at multiple scales, and to engage and mobilize broader publics.

The double gesture I have identified in Fanon and Said—bringing murderous exclusions into view, while reaching toward a newly inclusive universality—can provide a model for the environmental humanities in the Anthropocene as it considers both its humanist inheritance and the claims made by the environmentalism of the poor. The process of opening-out entailed in Fanon's dialectic of decolonization finds its counterpart in Said's articulation of critique, which lies "at the very heart of humanism ... as a form of democratic freedom and as a continuous practice of questioning ... open to, rather than in denial of, the constituent historical realities" of colonialism and globalization (48). Calling to account the abuses of humanism—in the colonies or in the Pentagon—is part of the ever-unfolding realization of humanism. "Come, then, comrades," Fanon urges, let us "set afoot a new man" (316); Said, too, calls us to humanism as a "process of unending disclosure, discovery, self-criticism, and liberation" (21).

This process, I argue, can allow the environmental humanities to confront even Fanon's and Said's own blindnesses and to grapple further still with the problem of the human in relation to several distinct versions of otherness. Richard Pithouse argues that Fanon's "reverence and respect for human (self and world making) creativity ... has an extraordinarily persuasive power ... [that] can create, in the reader, the subjectivities that generate an emotional identification with what is human ... [and can] inspire action aimed at realizing a more human world" (127). We, in turn, might consider what it would mean to radically expand this identification, to realize a more more-than-human world.

There is more than a passing resemblance between humanism as I have described it here and Joan Martínez-Alier's account of the environmentalism of the poor. His analysis is grounded in the social science disciplines of environmental economics and political ecology, which are concerned with ecological distribution conflicts, that is, conflicts over the "social, spatial, and intertemporal patterns of access to the benefits obtainable from natural resources and from the environment as a life support system, including its 'cleaning up' properties," and over the unevenly distributed burdens of environmental risk and harm (73). Crucial to the environmental humanities is the implication that the ecological *distribution* conflicts that give rise to the environmentalisms of the poor are also conflicts of ecological *valuation*: contests over language, ideology, and frameworks of interpretation. These questions recur throughout *The Environmentalism of the Poor*: "*Who has the power to impose particular languages of valuation ... Who has the capacity to simplify complexity, ruling some points of view out of order? ...* Who has the power to privilege one analytical point of view (the economic, the social, the environmental) on a chosen time-space scale?" (viii, ix, 161; emphasis in original). In any environmental conflict, multiple temporalities and logics and languages of value are at stake (money and the market being only one among them).[3] The question is always who has the power to make their language stick; the answer is usually those with the money. These questions of values, representation, and interpretation are ones that humanists are well suited to consider.

I want to close by invoking Fanonian/Saidian humanism as a method and interpretive stance from which to offer two provocations. First, I certainly would not want to erase the insurgent histories of specific social movements for environmental justice, or to disregard the importance of power and perspective. Still, I also dream of a world where the "environmentalism of the poor" would be known simply as environmentalism. The point is perhaps more legible if I say that I dream of a humanism of the poor that is known simply as humanism. That is to say, at work in such movements are a fundamental materialist realism (attention to flows of matter, energy, and waste, at multiple spatial and temporal scales) and an eye toward the workings of power and injustice that would be salutary for the environmentalism of everyone, not just the poor.[4]

Failing that—still awaiting an imagined future turn of the dialectic when a new human has been indelibly inscribed on the new leaf that Fanon urges—my second provocation would be to resolve the terminological debates about the relationship between the environmentalism of the poor (a rubric which generally refers to movements in the global South, focused on class stratification and the disempowerment of a majority population) and the Environmental Justice movement in the United States (which grew out of the civil rights movement and is concerned primarily with environmental racism and the toxic burdens borne unevenly by racialized minorities). Again, there is analytical and strategic value in not losing sight of specific histories of marginalization and terrains of struggle by collapsing different kinds of movements into each other. Nonetheless, what if we decided to recognize such movements as shared struggles under the banner of environmental justice, in a generic, lower-case sense? Beyond expanding the human constituency of solidarity, another advantage is that such an approach could facilitate the further dialectical turn against anthropocentrism that I describe above. This version of environmental justice could account for a more-than-human world inextricably bound up in struggles among differently empowered humans: materially, epistemologically, and politically. That is to say, as Martínez-Alier acknowledges, "environmental justice" signifies both the US tradition of social movements against environmental racism, as well as Rawlsian philosophical and ethical inquiry into "the allocation of environmental benefits among people including future generations, and between people and other sentient beings" (168). Inspired by Fanon and Said, I am dreaming of an approach to environmental justice—and environmental humanities—that calibrates these different strands, bringing not only Sartre's "natives," *but also nonhumans and the future*, into the fold of the most universal universalism yet. An environmentalism-and-humanism of the poor, along the lines of *buen vivir*, or the "tree of tomorrow" imagined by Zapatista leader Subcommandante Marcos.[5] Perhaps that constituency will go by the name of the *posthuman*, including, in a literal sense, those who come after us in the future world we humans (and others) will have made—a Saidian notion. Yet a painstaking conceptual and political working-through of the stakes of recognizing fellow creatures would be necessary to forge that constituency: it is no mere matter of affixing yet another *post-* and continuing with business as usual.

Notes

1 One might discuss similar issues around the rubric of cosmopolitanism, but I am interested in how the concept of the human links and troubles these discourses.
2 See Wenzel, "Reading Fanon Reading Nature" for further discussion of Fanon's engagement with the mid-twentieth century effort to assert in international law the principle of Permanent

Sovereignty over Natural Resources, and the discrepant economic and ecological implications of that endeavor.

3 Other frameworks of valuation include "the ecological value of ecosystems, the respect for sacredness, the urgency of livelihood, the dignity of human life, the demand for environmental security, the need for food security, the defence of cultural identity, of old languages and of indigenous territorial rights, the aesthetic value of landscapes, the injustice of exceeding one's own environmental space, the challenge to the caste system, and the value of human rights" (Martínez-Alier 149).

4 To say a bit more about materialist realism, the environmentalism of the poor, in Martínez-Alier's account, recognizes that "economic growth unfortunately means increased environmental impacts, and it emphasizes geographical displacement of sources [of natural wealth] and sinks [for waste]" (10). Why should it take the environmentalism of the poor to establish in economic analysis the notion that the poor "sell cheaply" their health and their resources not because they want to, as free agents in the marketplace, but because they have little choice?

5 Marcos describes a "Fourth World War" between neoliberalism and humanity. In the face of this war, Marcos envisions the somewhat quixotic planting of a slow-growing "tree of tomorrow," which will become "a place with democracy, liberty, and justice" ("Marcos on Memory and Reality" 293).

References

Césaire, Aimé. *Discourse on Colonialism*. 1955. Trans. Joan Pinkham. New York and London: Monthly Review Press, 1972. Print.

Coronil, Fernando. *The Magical State: Nature, Money, and Modernity in Venezuela*. Chicago: University of Chicago Press, 1997. Print.

Fanon, Frantz. *The Wretched of the Earth*. Trans. Constance Farrington. New York: Grove Press, 1968. Print.

Huggan, Graham and Helen Tiffin. *Postcolonial Ecocriticism: Literature, Animals, Environment*. London: Routledge, 2010. Print.

Marcos, Subcommandante. "The Fourth World War Has Begun." *The Zapatista Reader*. Ed. Tom Hayden. New York: Nation Books, 2002. 270–285. Print.

——. "Marcos on Memory and Reality." *The Zapatista Reader*. Ed. Tom Hayden. New York: Nation Books, 2002. 286–296. Print.

Martínez-Alier, Joan. *The Environmentalism of the Poor*. Cheltenham: Edward Elgar, 2002. Print.

Nixon, Rob. "Environmentalism and Postcolonialism." *Postcolonial Studies and Beyond*. Ed. Ania Loomba et al. Durham, NC: Duke University Press, 2005. 233–251. Print.

——. *Slow Violence and the Environmentalism of the Poor*. Cambridge, MA: Harvard University Press, 2011. Print.

Pithouse, Richard. "'That the Tool Never Possess the Man: Taking Fanon's Humanism Seriously." *Politikon: South African Journal of Political Studies* 30.1 (2003): 107–131. Print.

Said, Edward W. *Humanism and Democratic Criticism*. New York: Columbia University Press, 2004. Print.

Sartre, Jean-Paul. "Preface." *The Wretched of the Earth*. By Frantz Fanon. Trans. Constance Farrington. New York: Grove Press, 1968. 7–31. Print.

Sekyi-Otu, Ato. *Fanon's Dialectic of Experience*. Cambridge, MA: Harvard University Press, 1996. Print.

Wenzel, Jennifer. "Reading Fanon Reading Nature." *What Postcolonial Theory Doesn't Say*. Ed. Ziad Elmarsafy, Anna Bernard, and Stuart Murray. New York: Routledge, 2015. 185–201. Print.

18

ACTION-RESEARCH AND ENVIRONMENTAL JUSTICE

Lessons from Guatemala's Chixoy Dam

Barbara Rose Johnston

Environmental justice

The term suggests an idealized goal, a hopeful outcome of political struggle to secure acknowledgment that injustices have occurred, responsible parties will apologize, and remedial action will occur to repair human and environmental harm. It is a catchall phrase deployed by those who study as well those who fight for transformative change. Environmental justice struggles seek to achieve fundamental alterations in the architecture of power by confronting that which is denied—life-threatening conditions that are the inequitable result of corruption and dysfunctional governance. In essence, environmental justice demands governance that, first and foremost, prioritizes an environmental humanity.

In this chapter I explore the struggle and consequences of environmental justice through the lens of one case, the Chixoy Dam development and related human rights abuse in Guatemala. My perspective reflects what I have learned as an anthropologist whose action-research focuses on environment, human rights, and social justice. This case first came to my attention in 1999 when I was asked by the World Commission on Dams (WCD) to review complaints about hydrodevelopment involving forced eviction, failures to provide promised compensation, and other human rights abuses. I produced a "Reparations and the Right to Remedy" briefing to guide the WCD that included a summation of the Chixoy Dam experience.

Instituto Nacional de Electrificación (INDE), an independent public utility run by Guatemala's military leaders, submitted designs for the Chixoy Dam that were reviewed and approved by the World Bank, Inter-American Development Bank, and other international financiers. Construction began in 1978 without notifying the local population; without legal acquisition of land supporting the construction works, dam, hydroelectric generation facility, reservoir, or farms needed to support resettled communities; without a comprehensive census of affected peoples or plan to address compensation, resettlement, and alternative livelihoods for some 3,445 mostly Mayan residents whose homes would soon be under water (Johnston et al., *Chixoy Dam Legacy Issues Study*; Johnston, "Chixoy Dam Legacies"). Once built, the Chixoy Dam and hydroelectric works became the primary source of electricity for the Guatemalan nation, and the associated highways and transmission lines allowed access and the means to exploit resources in a previously isolated mineral- and gas-rich region. Its development displaced thousands, including the river basin Maya Achí community of Río Negro, whose communal title to indigenous lands dates back to the 1800s. Before construction, the fabric of life was tightly woven across a culturescape of ancient pyramids, fields and pastures, villages and hamlets, maintained by trade, familial ties, cultural beliefs, and historical relationships.

During and after construction, life was hellacious. In Río Negro, government military forces evicted people at gunpoint, with violence and massacre. Pregnant women were raped; sometimes bellies were sliced open before death. Some women were taken to military bases to serve as sex slaves. Some children were captured by soldiers and taken home to work as household slaves. Homes and fields were burned. From the military government's point of view, civilian complaints and protests were evidence of insurgent influence and the Army declared that dam-affected villages were subversive "resistant communities." Civil rights were suspended and state-sponsored violence in the region escalated. On March 4, 1980, the village of Río Negro suffered a massacre. Survivors later filed a complaint with the Inter-American Commission on Human Rights which, in its October 1981 report on human rights in Guatemala, noted that dam construction in the area precipitated the events that led to the massacre. In the months that followed that March 1980 massacre, the Río Negro community suffered three more massacres, along with the residents of villages that survivors fled to. In 1999, when survivors shared their experiences with the WCD, Río Negro survivors told a tale of continuing pain and struggle: surviving "development" meant a life of hardship and sorrow made all the more difficult by extreme poverty and continued oppression.

In 2003, I was asked to advise Chixoy Dam-affected community leaders and their advocates on how, in a land where there is no viable judiciary, they might plead their case for justice. As I delved further into this history and its consequences, I came to understand that this village and the experiences of surrounding communities embody all that is worst about the corruption and structural violence accompanying internationally financed hydrodevelopment. Financed by the World Bank (WB), Inter-American Development Bank (IADB) and others, development loans were the sole source of foreign aid to a nation ruled by a military dictatorship engaged in systematic state-sponsored destruction of Mayan peoples, a fact demonstrated by the United Nations-sponsored Guatemalan truth commission report *Memory of Silence* (Commission for Historical Clarification).

The essential lesson here is that understanding environmental injustice requires stripping away and making visible the collusion and power of economic interests. Exposing the underbelly of the beast can generate outrage and the political will to provide some measure of remedial attention; yet, given the powerful forces involved, such actions can also prompt a dangerous backlash. The stories below illustrate this fundamental tension between action and reaction, a tension that drives societal change in the struggle to secure justice.

July 2003: encountering power, tasting terror

After months of preparation, I took a flight from California to join advocates from International Rivers, Reform the World Bank Italy, and Rights Action in Guatemala. Arriving on a Tuesday night we immediately set to work on our agenda. Wednesday, we had appointments with government, WB, and IADB officials that hopefully would include access to institutional archives. Thursday, we would meet with lawyers, massacre survivors, and other human rights advocates at Rights Action's office in the morning, and then travel several hours north to the resettlement village of Pacux for a Friday meeting with representatives of fourteen dam-affected communities.

Wednesday was a productive blur. Thursday morning meetings were intense: briefings on the status of exhumation, forensic analysis, and gathering of witness testimony on indigenous community massacres. These lawyers were working to build cases in anticipation of a time when the rule of law might be present, when the state might actually prosecute culpable parties for their crimes against humanity. Many of these crimes occurred during the reign of dictators Fernando Romeo Lucas García and Efraín Ríos Montt, whose sole source of funding for these military governments came from international humanitarian loans to build the Chixoy Dam.

Our meeting was interrupted by phone calls; a son and then a daughter called to let their father know they were okay. They reported that masked men armed with sticks, stones, and guns were marching in the streets, burning tires, and shutting down traffic. Tuning in to the radio, we heard that a journalist covering this riot, Hector Ramírez, was killed when he fled the mob. Police officials nationwide had failed to report to work. Other organizations reported break-ins and intimidating phone calls: today is the day when all the threats will be carried out with no fear of reprisal.

The *jueves negro* (Black Thursday) riots were a show of force. Thousands of men were bused into the city, purportedly paid, armed with guns and gasoline, and directed to wreak havoc outside the homes and business of the judiciary, progressive media, and politicians who opposed the presidential bid of former dictator General Efraín Ríos Montt, an elected member of Congress whose candidacy was temporarily derailed by a Constitutional Court ruling (Wise 151–152).

By lunchtime a state of emergency had been declared and the soundscape on our shady sidestreet had changed from birdsong and occasional traffic to the drone of helicopters and the report of gunfire. Should we keep to schedule and attend Friday's reparation meeting in Pacux? Should we travel out of the city on a day when men with guns ruled the roads, to meet with people whose efforts to secure justice placed them in the crosshairs of this political conflict? The consensus was to leave these decisions up to Mayan community leaders. They had seen their relatives die; they had received death threats. They, more than anyone on this day of impunity, risked everything.

Their decision was to proceed. With this news, I was handed the phone and told to call home. Tell my family I loved them, say my goodbyes, but not mention the events of the day, our plans to leave, or where we would be in days to come. Our phone lines were not secure. Calls made, we piled into a dark-windowed van and took the highway out of the city.

Hours later we stopped at a mountain-pass restaurant adjacent to a biosphere reserve that typically hosted busloads of tourists attempting to catch a glimpse of the national bird, the Quetzal. This stop was planned as a safe way to rendezvous with a journalist whose life had been threatened because three weeks earlier he had reported that Ríos Montt, while giving a campaign speech in the Rabinal town plaza, had been hit in the head with a rock. The rock was thrown by angry villagers whose funeral procession had been disrupted by a rally for the man they saw as responsible for the death of their relatives. Those villagers were displaced survivors from Río Negro, now living in the military-guarded compound known as Pacux. Radio broadcast of this encounter between massacre survivors and the former dictator prompted global press coverage that cast Ríos Montt's campaign for the presidency in a negative light. The journalist was fleeing the country and my companions, the activists facilitating the Chixoy reparations meeting, had agreed to give him a ride to a safe transfer point.

The biosphere reserve parking lot was empty; the state of emergency had put a halt to tourist traffic. We entered the restaurant. Seeing no journalist or patrons of any sort, we opted to sit and wait. We ordered food, watched staff make a whispered call, and attempted to project a "we are just tourists!" attitude. As we settled in to our coffee and snacks, a man walked in with nickel-plated guns holstered on his hips. He passed our table to sit with his back to the wall. For the next hour and more we ignored his stare. We chattered in English and acted out our part: loud, happy, carefree, ignorant tourists.

As afternoon light waned we grabbed binoculars, paid the bill, and strolled down the bird path behind the restaurant. Attention diverted, one member of our group slipped through the bushes to the parking lot, hopped in the van, coasted in neutral to the road before starting the engine and heading out to search for the missing journalist. Evening found us clinging to shadows waiting for their return. As darkness fell, our van rolled down the hill and into the parking lot, engine and headlights off. We ran, jumped in, and took off just as a truck full of men with guns pulled in.

They followed us. We huddled down in the seats in case they started shooting. And then we hit the slow traffic and warning lights indicating a roadblock ahead. We slowed, keenly aware of that truck several cars back. We crept through the blockade; with the press of head-lights we could see that highway repairs were underway. Peeking out the back window, we saw the movement of heavy equipment behind our car. We were waved through; ours was the last car before traffic was completely halted.

We were giddy with the exuberant feeling of relief. Instead of traveling on to our planned stop for the night, we took a random road, stopped in a town, grabbed a meal, and secured rooms at a motel. I brushed my teeth, put my pajamas on, stared at the bare lightbulb, and attempted an innocuous yet coded entry in my notebook.

A truckload of men pulled into the parking lot, stopping to talk at the motel office. I announced their arrival to my roommate, whose savvy reply was: "Our door is made of steel. The walls are thick. And the journalist is at opposite end of the motel on the bottom floor. Go to sleep." Night turned to day and on we went to the resettlement village of Pacux.

That terror-filled day taught me so much. The luxury of a life led not knowing, until now, such fear. The absolute inadequacy of my intellectualized understanding of this place and case. Written descriptions of a horrific history cannot convey the tension and pragmatic resolve that characterize the life of a survivor determined to seek justice in a land where impunity reigns.

Forging an environmental justice movement

To enter Pacux we drove through a military compound, past an armed sentry gate, down the singular road to the community hall of a village surrounded by chain link and razor-wire fencing. Other "model villages," as these guarded compounds are called, were built throughout Guatemala's highlands according to a US military design originally crafted to contain and control civilian populations in Southeast Asia. In Guatemala they are home to indigenous massacre survivors who, once rounded up, were housed in a guarded compound, "re-educated," and used as slave labor. Most model villages were decommissioned after the 1996 Peace Accords. In 2003, the only one left in the country was the prototype built to house Río Negro massacre survivors, Pacux.

Representatives of fourteen villages attended this meeting, some eighty people. With the help of translators I outlined my suggestions. My role: develop a participatory strategy for dam-affected villagers and their advocates to collaborate in a study that would document the past and bring clarity to the issues of consequential damages, meaningful remedy, and how—through reparation—they might build a more peaceful and sustainable future. Their role: help document community histories, identify needs, and begin to articulate their visions for that hopeful future. After a period of deliberation, community representatives lined up and signed an *Acta* to formally codify their intent. They would participate, help shape categories of concern, and select young men and women from each of the villages to help collect and record information. Accompanied by a linguist, these citizen scientists would venture out to near and distant homes, walking eight to twelve hours to meet with people who survived the violence of the 1970s and 1980s. They would interview families about household resources and conditions before and after the dam and, with digital cameras, would capture images to illustrate conditions, problems, and attributes that define their current way of life (Johnston, "Notes"; "Reparations").

The most difficult part of this meeting was the psychosocial process that each individual went through as all shared the stories from this gathering of communities affected by the dam. Handing out paper and colored markers, I asked each community to draw a map of their village before the dam, the journey taken to survive, evictions and violence, and where and how they live now. An hour later, maps covered the walls and we moved from map to map to hear each story. This was the first collective airing of wartime experience. I witnessed a hugely traumatized population redefine themselves as they listened to near and distant neighbors, strangers, friends, and enemies. The chaos of violence in the region was not random; it was opportunistic, choreographed by distant actors bent on controlling and silencing the indigenous communities in the Chixoy basin. An internationally financed dam provided the money and means to install military control over the region. The need to depopulate the region prompted episodes of intimidation and violence that, as the military searched for survivors, led to massacre after massacre. By the end of the day the struggle of Río Negro survivors to demand accountability was clearly evolving from a community-specific set of complaints to a regionwide struggle of dam-affected communities.

Environmental injustice: study findings

The reports that I produced in collaboration with the dam-affected communities not only chronicled INDE's development of Chixoy Dam, the March 1980 massacre, and subsequent state-sponsored violence, but also investigated the international funding mechanisms that allowed the development and associated persecution. Major financing for Guatemala's Chixoy Dam was twice provided by the World Bank and three times by the Inter-American Development Bank. Yet despite five loan review and contract negotiations requiring the government of Guatemala to provide proof of title to land, and despite language obligating the banks to insure proof of title had been legally secured before releasing loan funds, there was no compliance. Financing was provided for resettlement, yet when construction was near completion communities were still living in the basin. An independent audit concluded that funds earmarked for resettlement were used for other purposes, which prompted $20 million for resettlement to be added as an emergency line item to IADB's 1981 loan (Partridge, "Comparative Analysis"; "Recommendations"). Funds were transferred shortly before a tunnel carrying water to the hydroelectric plant collapsed and electrical generation ceased. Extensive repairs were needed to reroute water around soft and porous karst deposits; a comprehensive geological assessment had never been done. INDE reallocated resettlement funding for repairs and for site security operations, including death squad actions targeting Río Negro and communities that sheltered Río Negro massacre survivors. When dam repairs were finally completed in 1983, ten communities in the Chixoy River Basin had been destroyed by massacres. In Río Negro alone, some 444 of the 791 original inhabitants had been killed.

While resettlement villages were eventually built, plans made in the 1970s with community input were discarded and the militarized compound of Pacux built in its place. The Río Negro massacre site was exhumed in 1992. Attention to this case occurred in the lead-up to the 1996 Peace Accords with the World Bank brokering a compensation agreement with the incoming Guatemalan government. This agreement was never fully honored and was grossly inadequate to meet the needs of indigenous communities who suffered the loss of lands, resources, livelihoods, and lives.

How to translate such a history into a compelling case for reparation, especially when the rule of law is so fragile? How can you try such a case when potentially culpable parties include international entities that cannot be tried in a court of law, such as the World Bank?

Our strategy: produce an independent report accepted by all as a legitimate accounting of events and damages, rather than a politically motivated, activist report. To do this, we sought support from foundations with no vested interest in the outcome. We simultaneously published all material in Spanish and English. We secured scientific association sponsorship from the American Association for the Advancement of Science, American Anthropological Association, and the Society for Applied Anthropology, and submitted all work to rigorous peer review. We distributed research findings as broadly as possible to influence the court of public opinion and encourage powerful actors to pressure responsible parties. And we created or expanded rights-protective space by establishing relationships with international human rights bodies; every time a representative from these entities met with the Guatemalan government, the status of Chixoy reparations was on the agenda.

Action and reaction

Before the analysis was completed, our research was already having a demonstrable impact. One concern in the inquiry was the current status of legal title; was it ever finally secured for the lands needed to build the dam and generate electricity? To address this question we hired a land title expert who reviewed the Guatemala Registry and Central American Cadastral and produced a report (Johnston et al. vol. 5). Much of the land under the Chixoy Dam, its reservoir, and the hydroelectric works was still the legal property of indigenous communities. One upstream community was still paying taxes on submerged lands to allow their continued use of cropland on the hillsides. News of these findings spread rapidly through the region and, despite advice to delay until the full study was completed and published, community outrage prompted action.

On September 7, 2004, five busloads of people from the displaced communities traveled to the dam, secured entry, occupied it, and threatened to turn off the nation's primary source of electricity unless the government committed itself to a negotiated reparations process. After three days of tense negotiations, an agreement between community leaders, INDE, and the government was signed by the head of Guatemala's national human rights commission.

The following week INDE filed charges against the seven indigenous signatories and an eighth member of the community, a worker who was charged with allowing entry. They were arrested, charged with crimes against the state, and went through years of court hearings supported by pro bono legal counsel before charges were dismissed.

The completion of the *Chixoy Dam Legacy Issues Study* in 2005, delivery to responsible and sponsoring parties, and press conferences in Guatemala and the US set the stage for a new phase in this struggle. In 2006, the government of Guatemala reconfirmed its commitment to reparations following the 2004 dam protest, though with no concrete outcome. In 2008, a new government of Guatemala agreed to participate in a verification and reparation negotiation with affected community, WB, and IADB representatives in a process facilitated by the Inter-American Commission on Human Rights (IACHR). Our study was formally adopted as one of three key evidentiary documents and a definitive statement of consequential damages.

The elusive experience of success

April 18, 2010. While I was delivering the Earth Day speech at California State University Sacramento, my cell phone rang. I looked down and saw the call was from International Rivers activist Monti Aguirre. I put my lecture and the audience on hold: "Hang on, I really gotta take this call." The verification of damages and reparation negotiation was concluded with a legally binding accord signed by President Álvaro Colom. I celebrated the moment with the Earth Day gathering: sometimes a measure of justice can actually be achieved.

Yet, months went by and the Reparations Plan was still held as a confidential document pending an *Acuerdo Gubernativo*—an executive action from the president ordering his government to implement the plan. Anticipating an *acuerdo*, in December 2010 Guatemala's Congress passed a budget that included the first of fifteen annual payments to implement reparation. Shortly after, I received a copy of the confidential agreement. Major details included compensation for material and nonmaterial damages and losses totaling US$154.5 million, construction of 191 homes in Pacux, repair of 254 homes in other communities, improvement of roads, water and sewage systems, provision of electricity, and an environmental restoration and management plan for the Chixoy Basin including reforestation with native plants and management of river flow to support water quality, basin ecology, and community livelihoods. And the president of Guatemala will personally present an apology.

Eleven months after the government signed on to the plan? Still no *acuerdo*. In March 2011, President Colom announced his decision to reject the signed Reparation Agreement and instead accept INDE's 2006 statement that all obligations had been met years ago. News coverage was nonexistent—the world's attention was on the earthquake, tsunami, and nuclear meltdown in Japan, and Guatemala's press was singularly focused on internal dramas: the filing of presidential divorce papers that, in theory, might allow Colom's wife to run in the next presidential election.

Accountability and justice

In 2012, human rights violations accompanying the development of Guatemala's Chixoy Dam were considered in the IACHR court findings in *Río Negro Massacres v. Guatemala*.

In this judgment, hydroelectric development of the Chixoy river basin is the background event that led to the Río Negro massacres and an inhibiting factor to achieving full reparation. Construction of the dam and its reservoir destroyed indigenous sacred sites, and inundation permanently prevented the return of the Río Negro communities to their ancestral lands. Living conditions in resettlement villages did not allow inhabitants to resume traditional economic activities, basic health, education, electricity and water needs have not been fully met, and these conditions have caused the disintegration of the social structure and cultural and spiritual life of the community. Observing that the massacres of the community of Río Negro took place within a systematic context of grave and massive human rights violations, the Court found that in addition to historical damages "the surviving victims of the Río Negro massacres experience deep suffering and pain ... which fell within a state policy of 'scorched earth' intended to fully destroy the community" (IACHR 8). Compensation and apology were ordered.

When Efraín Ríos Montt's genocide trial began in January 2013, President Otto Pérez Molina had taken no action to implement the IACHR judgment in the Río Negro massacre case, rejected the validity of the 2010 Chixoy Dam reparations agreement, and rejected the 1999 UN finding of genocide. The Ríos Montt genocide trial, the May 7, 2013 verdict that a former president (and dictator) of Guatemala was guilty, and the subsequent political manipulations that resulted in an annulment of conviction ten days later were the subject of intense global media coverage. Significant attention was given to behind-the-scenes actors, including questions of how genocide was financed (Chixoy loans), and whether the United States has any obligation given its role as the major contributor to these loans (Deardon). This public scrutiny prompted US senators to review key documents and propose corrective action in the Senate Foreign Appropriations Bill. The Chixoy study and supporting documents were requested, and I helped prepare staff briefings.

In January President Obama signed the 2014 Congressional Appropriations Bill that included language requiring international financial institutions (IFIs) to ensure the negotiated reparation agreement between the Guatemalan government and communities affected by Chixoy Dam is fully implemented. It also instructed US representatives at IFIs to vote against any loan, grant, strategy, or policy of such institution to support the construction of any large hydroelectric dam; and it required IFIs to undertake independent outside evaluations of all its lending to ensure that each institution responds to the recommendations of its accountability mechanisms and provides just compensation and other appropriate redress to those who suffer violations of human rights (Consolidated Appropriations Act 504, 549–550).

This legislation generated foreign intervention complaints from Guatemalan president Otto Pérez Molina (Reeves). Months later, after civil society submitted letters of concern and dam-affected community representatives traveled to Washington DC to update US legislators, World Bank Board members, and US Treasury representatives, the World Bank Executive Board withheld its vote on a major loan financing operations of the Guatemalan government. The political will to proceed with the reparation plan in Guatemala was quickly resuscitated.

In October 2014, Guatemalan president Otto Pérez Molina apologized in person to dam-affected communities for the massacres, violence, and varied failures associated with development of the Chixoy Dam. His apology was entered into the US Congressional record. In November 2014, a $155 million commitment for social and economic development was adopted by Guatemala's Congress and codified as a line item in the national budget over the next fifteen years. In October 2015, with former President Molina incarcerated on corruption charges and the interim president hospitalized for heart problems, interim Vice President Juan Alfonso Cifuentes Soria hand-delivered the first monetary compensation to survivors of the Río Negro massacre. Recognizing the huge distance between a political promise and on-the-ground outcomes, US Congress continues to include language in annual appropriations restricting aid to Guatemala pending evidence of Chixoy Dam reparations.

Environmental justice?

When the reparations study began in 2003, there was no rights-protective forum for dam-affected communities. Financiers, with loans paid back in full in 1996, saw no remaining obligation and Guatemalan politicians rejected the validity of complaints. Thus, the primary goal of the study was to force such a mechanism to come into being. This push generated a backlash that, in turn, generated a more expansive struggle. The eventual outcome of the Chixoy case demonstrates that in a land of impunity, with collaborative and participatory effort, rights-protective space can be created and a measure of accountability can be achieved. Dam-affected community representatives have, thanks to the reparation agreement, an increasingly powerful voice in determining the shape and form of remedial social and economic development in their region.

However, there is much evidence to suggest that this environmental justice outcome may be the relatively small price paid to achieve more substantive gains in the larger game of plunder. Case in point: at the same time national and international attention was focused on this historical case, hydrodevelopment was financed and approved upstream in the Chixoy Basin through a sadly familiar rights-abusive process, as a recent Oxfam summation of the Santa Rita Dam project illustrates ("Suffering of Others"). Proposed in 2008 and approved by the Guatemalan government in 2010, Santa Rita dam development has stimulated protest and conflict in the region. Financed with investments from private and public equity funds (including European development banks and the World Bank's International Financial Corporation), plans were finalized and construction commenced in violation of national law requiring the free and prior consent of indigenous communities. When preliminary site work began in 2012, communities protested, Guatemalan president Pérez Molina declared them subversive, suspended civil rights, and sent some 1,500 national police in to protect development interests. By 2014, when the UN Clean Development Mechanism accepted the request for carbon credit registration for the Santa Rita Dam, twenty-two indigenous communities had expressed their opposition and were subject to targeted violence, forced evictions, burning of property and crops, and other abuses. As of this writing, five indigenous leaders who opposed the Santa Rita dam and two young children of another leader have been assassinated. Once completed, the Santa Rita Dam will generate carbon credits tradable under the European Union's emissions trading system. Much of the energy produced will be exported to Mexico and the United States. The dam and its reservoir will displace thousands and adversely impact an estimated two hundred thousand people.

Thanks to cell phone and Internet access, indigenous communities in Guatemala today are keenly aware that, despite the official apology and plan to repair harms from Chixoy development, rights-abusive development continues with the same callous disregard for the rule of law. The experience of Chixoy communities has served as a model for other communities fighting in defense of their rights. In the Santa Rita case, international organizations have collaborated with community members to document abuses, file complaints, and demand their rights. The use of state-sponsored violence to insure institutionalized plunder is what drove thousands of rural workers into Guatemala City streets in early February 2016 to call for an end to development projects that displace communities and exploit natural resources. Reaction to this protest was swift. A few days later, Vice-Minister of Sustainable Development Roberto Velasquez announced that the prior administrations' moratorium on new mining licenses, which was prompted by an IACHR ruling, will not be maintained by the Ministry of Energy and Mining. Instead, this agency will speed up the process of granting requests for licenses because such projects "help to reduce the high levels of poverty within the country" (Guatemala Human Rights Center).

The tensions that drive environmental justice struggles in Guatemala, and indeed throughout the world, continue to morph and expand. Years ago, Eric Wolf reminded us that when society fails to address or reflect the needs of its citizens, a window for substantive transformation is opened where old orders of society can be discarded and new centers of power, action, and lines of cleavage can emerge (92). Fear of this drives backlash and repression when human rights violations are exposed and demands for accountability capture the public imagination. Backlash is inevitable, yet backlash also prompts further progressive change. The push for justice requires tackling environmental, social, political, and economic inequities head-on, resulting in actions that reinvigorate or reinforce the meaning, power, and integrity of the community.

References

Commission for Historical Clarification (Comisión para el Esclarecimiento Histórico). *Memory of Silence: The Guatemalan Truth Commission Report*. 1999. PDF. Web. 9 March 2016.

Congressional Record 12 November 2014: S5947–S5948. *Congress.gov*. Web. 9 February 2016.

Consolidated Appropriations Act, 2014. Public Law No. 113–176. 17 January 2014. *Congress.gov*. Web. 16 February 2016.

Deardon, Nick. "Guatemala's Chixoy Dam: Where Development and Terror Intersect." *Guardian* 10 December 2012. Web. 15 January 2015.

Guatemala Human Rights Center (GHRC). "Guatemala News Update, February 6–12." Monitoring Guatemala: GHRC's Human Rights Blog. 12 February 2016. Web. 15 February 2016.

Inter-American Commission on Human Rights (ICHRC). *Report of March 4, 1980 Massacre in the Dam-affected Village of Río Negro*. October 1981. Print.

Inter-American Court on Human Rights (IACHR). *Case of the Río Negro Massacres v. Guatemala: Official Summary Issued by the Inter-American Court*. 4 September 2012. PDF. Web. 18 February 2016.

Johnston, Barbara Rose. "Chixoy Dam Legacies: The Struggle to Secure Reparation, and the Right to Remedy in Guatemala." *Water Alternatives* 3.2 (June 2010): 341–361. Web. 18 February 2016.

———. "Notes from the Field: Guatemalan Community Struggles with Chixoy Dam Legacy of Violence, Forced Resettlement." *American Association for the Advancement of Science Report on Science and Human Rights* 23.3 (Fall/Winter 2003): 1–4. Print.

———. "Reparations and the Right to Remedy." *Thematic Review 1.3: Displacement, Resettlement, Reparations, and Development*. Cape Town, South Africa: World Commission on Dams, 2000. Web. 18 February 2016.

———. "Reparations for Dam Displaced Communities? Report from the Chixoy Dam Legacy Issues Meeting, July 23, 2003, Pacux, Guatemala." *Capitalism, Nature, Socialism* 15.1 (2004): 113–119. Print.

Johnston, Barbara Rose et al. *Chixoy Dam Legacy Issues Study*. Vols 1–5. Santa Cruz, CA: Center for Political Ecology, 2005. Web. 9 February 2016.

Partridge, William L. "Comparative Analysis of BID Experience with Resettlement Based on Evaluations of the Arenal and Chixoy Projects." Report to the Banco Interamericano de Desarrollo. December 1983. Print.

———. "Recommendations for the Human Resettlement and Community Reconstruction Components of the Chixoy Project." Report to the World Bank. June 1984. Print.

Reeves, Benjamin. "Guatemalan President Spars with US over Chixoy Dam Reparations." *The Tico Times* 11 February 2014. Web. 9 February 2016.

"The Suffering of Others: The Human Cost of the International Financial Corporation's Lending through Financial Intermediaries." Oxfam Issue Briefing, April 2015: 11–13. PDF. Web. 11 December 2015.

Wise, Michael B. "Judicial Review and Its Politicization in Central America: Guatemala, Costa Rica, and Constitutional Limits on Presidential Candidates." *Santa Clara Journal of International Law* 7.2 (2010): 145–180. Web. 9 February 2016.

Wolf, Eric. "Distinguished Lecture: Facing Power—Old Insights, New Questions." *American Anthropologist* 92 (1992): 586–96. Print.

19

FARMING AS A SPECULATIVE ACTIVITY

The ecological basis of farmers' suicides in India

Akhil Gupta

In the summer of 2015, newspapers in Karnataka were once again full of news of farmers' suicides. Every day, there were reports of more farmers in the state who had committed suicide ("BJP Disputes Govt Data on Farmer Suicides" 5; Singh and Dadhich). Indeed, largely because of the persistence of the pioneering, award-winning journalist P. Sainath, who first drew attention to the phenomenon in the mid-1990s, the story of farmers' suicides has been constantly in the news. However, headlines have not led scholars to carefully examine the evidence. Farmers' suicides have become, instead, an arbitrary signifier used by different interest groups to push their own political agendas. As I examine below, the careless and biased use of "farmers' suicides" as a sign has in fact prevented an understanding of the causes of this epidemic, and thereby done little to ameliorate the tragedy.

The phenomenon of farmers' suicides brings together ecological, social, and existential problems that evade analysis from the perspective of any one discipline. It requires the convergence and interpenetration of scientific, social scientific, and humanistic knowledge and insights, precisely the terrain of the environmental humanities. Although economists have been at the forefront of scholars attempting to explain the causes of farmers' suicides, their interpretations are too "experience distant" and sociologically thin to yield a careful and complex understanding of the issue. I introduce the idea of a "speculative climate" to reference the conjuncture of environmental change and neoliberalism in producing precarity. I argue that the financialization of farming through seed and input loans and the increasing exposure to risk in output markets because of global fluctuations in prices is producing conditions that encourage indebtedness among small and marginal cultivators (Ataulla and Anand 2; Krishna Kumar and Veeraraghav 11). And the one thing we know without a doubt about farmers' suicides is that it is the inability to pay back loans that leads farmers to commit suicide.

Outline of the problem

On August 31, 2012, the *Indian Express* reported that the minister of state for agriculture had stated, in response to a question in parliament, that 290,740 farmers had committed suicide in India during the period from 1995 to 2011 ("2.90 Lakh Farmers"). On average,

Table 19.1 Farmer Suicides in Selected States and All India 1997–2007 (from Gruère and Sengupta 319).

State	1997	1998	1999	2000	2001	2002	2003	2004	2005	2006	2007
Maharashtra	1,917	2,409	2,423	3,022	3,536	3,695	3,836	4,147	3,926	4,453	4,238
Andhra Pradesh	1,097	1,813	1,974	1,525	1,509	1,896	1,800	2,666	2,490	2,607	1,797
Karnataka	1,832	1,883	2,379	2,630	2,505	2,340	2,678	1,963	1,883	1,720	2,135
Madhya Pradesh	2,390	2,278	2,654	2,660	2,824	2,578	2,511	3,033	2,660	1,375	1,263
Gujarat	565	653	500	661	594	570	581	523	615	487	317
Other states	5,821	6,979	6,152	6,105	5,447	6,892	5,758	5,909	5,557	6,418	6,882
All India	13,622	16,015	16,082	16,603	16,415	17,971	17,164	18,241	17,131	17,060	17,131

therefore, eighteen thousand farmers have committed suicide every year for the last sixteen years (Gruère and Sengupta 319).

Most of these suicides have happened in six out of the 640 districts in India (Sainath 3; see Table 19.1). The paradox of these suicide "hotspots" is that they are located in wealthy states such as Maharashtra, Seemandhra, Karnataka, and Kerala. These states have higher rates of growth and better indices of social development than the rest of the country (Vasavi 39). Together with Telengana and Madhya Pradesh, these six states account for between half and two-thirds of all farmers' suicides in the country.

The districts and the states where reported suicides have been highest are also highly correlated with the use of genetically modified seeds. Vandana Shiva and others have therefore labeled genetically modified seeds, and Bt cotton in particular, as "seeds of suicide" (Shiva et al.; Shiva and Jalees).

Looking at the data and the extensive literature on this subject, however, has made me wary of such formulas and formulations. The old saw that "correlation should not be mistaken for causation" holds especially true for this tragic phenomenon. The only thing that is clear about "farmers' suicides" as a phenomenon is that it escapes easy generalizations. But before I come to the question of how farmers' suicides are to be explained, let me delve a little deeper into the issue.

Let us first look at the sources of data. In India, suicide is classified as a criminal offence. Vasavi points out the colonial origins for this: committing suicide was seen by the British colonial state as a sign of disaffection with its policies, and a statement about its inability to provide order (23). Thus, the data on suicides come from the National Crime Records Bureau (NCRB), which is part of the Home Ministry, and are posted online (http://ncrb. nic.in/adsi/main.htm). Since policing is a "state subject," that is, under the jurisdiction of individual regional states, the national data are an aggregate of data collected from each state (this fact is important, as we shall see later).

Farmers' suicides have been connected to globalization and to the introduction of genetically modified crops in particular. This gives us two points in time, after which we should have expected changes in the number of suicides by farmers. The first point is 1991, when the neoliberal "reforms" that opened up the Indian economy to the pressures of globalization were introduced. The second point is 2002, when Bt cotton was first introduced commercially.

If you look at the data on the number of farmers committing suicide, there is clearly an upward trend from 1998 onwards. The NCRB reports do not list farmers as a separate category before that time. However, there is an upward trend in the number of suicides from all

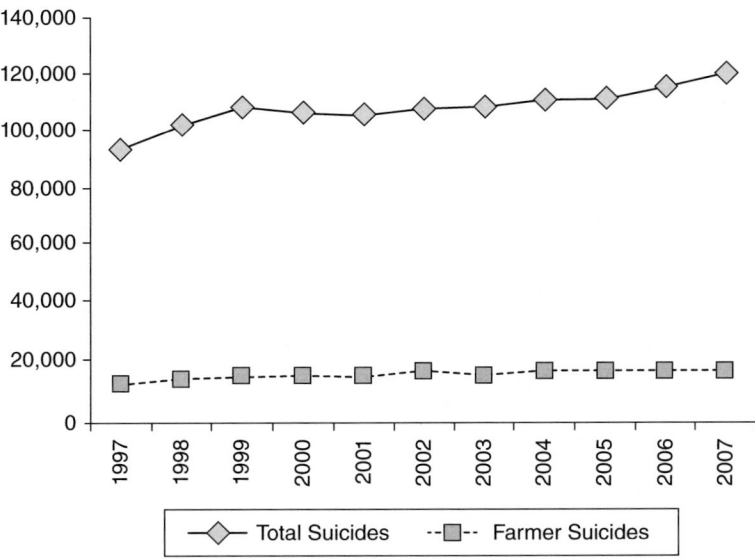

Figure 19.1 Farmer and total suicides in India, 1997–2007 (from Gruère and Sengupta 320).

causes, and farmers' share of total suicides has actually dropped since 1997 (see Figure 19.1). Second, if you look for spikes in the number of farmers' suicides following 1991 and/or 2002, there is a noticeable absence of any sharp changes following the advent of neoliberal globalization or the introduction of Bt cotton.

If you take the statistics at face value, then, not much has changed. Yet analyzing the statistics carefully is useful only if the underlying data are reliable. As I document in *Red Tape* (Gupta), anybody who has done work in rural India knows that statistical data on rural populations are highly unreliable. Government officials, including the police, know very little about what goes on in villages, and, depending on the wealth and connections of the offending party, homicides are regularly classified as suicides and, depending on the sensitivity of local authorities, from headmen to district officials, suicides are often classified as occurring due to "natural causes." Autopsies are unheard of, and thus there is no way to discern whether a death by natural causes is actually a suicide, and since bodies are cremated within a few hours of death, physical evidence of foul play in a homicide can never be recovered.

There is another, more important reason to suspect these statistics, and that is tied to the complex forms in which the state has responded to farmers' suicides. Some states, like Gujarat, which is one of the centers of Bt cotton production, report very few suicides as compared to other Bt cotton-growing areas like Maharashtra and Andhra Pradesh. The image-conscious chief minister of Gujarat, Narendra Modi, who is now India's prime minister, is reported to have said that the reason why Gujarat does not have farmers' suicides is that "Gujarat's farmers are not like those in other states. Our farmers drive Maruti cars" (Vasavi 12). However, official denial is also found in other states like Madhya Pradesh, despite the fact that it is one of the leading states for farmers' suicides according to the NCRB (Vasavi 12). In order to suppress data on farmers' suicides, the government of Gujarat has refused to offer compensation to families of suicide victims, and suppressed data on the extent of the phenomenon. What little data has emerged has been due to activists using the Right to Information (RTI) Act to extract information from a state government that is reluctant to part with those data (Vasavi 15–16).

One of the most important interventions made by different levels of the government is to compensate the families of victims with a cash payment (usually about INR 50,000, or US$ 800), and to "write off" the loans made to the victim. This helps pull the remaining members of the family from out of crushing debt, and provides them with some resources to begin a new life. However, as the number of "farmers' suicides" multiplied, state governments ran up against budgetary problems, and orders were discreetly issued to local police to stop recording peasant deaths as "farmers' suicides." Also, there was some concern that the payments were giving incentives to heavily indebted farmers to actually commit suicide. Thus, the statistics on "farmers' suicides" reflect not just a technique for the enumeration of life, but also complex processes for its regulation and control, of biopower and biopolitics (Münster, "Farmers' Suicides as Public Death").

Seeds of death?

In India, farmers' suicides have most often been associated with the cultivation of cotton. The main exception to this story is Wayanad district in the southwestern state of Kerala, where farmers' suicides have been associated with the cultivation of other cash crops, such as pepper, ginger, vanilla, and rubber (Münster, "Farmers' Suicides and the State in India"). Because of the link between cotton production and farmers' suicides, new, transgenic varieties of cotton seeds, called Bt cotton, have been termed "seeds of death" by prominent activists such as Vandana Shiva (see Shiva et al.). It is estimated that close to 90 percent of the cotton grown in India is now Bt cotton. This fact has led to the conclusion that Bt cotton cultivation is responsible for farmers' suicides. Is it possible, however, that the correlation between Bt cotton cultivation and farmers' suicides hides a larger structural issue of risk and vulnerability that extends well beyond Bt cotton? I will tentatively label the structural conditions of precarity with the term "speculative climate."

There are good reasons to doubt the narrative of "seeds of suicide" used for Bt cotton. I pointed out above that the recorded suicide rate did not go up after the introduction of Bt cotton. Some suicide hotspots such as Wayanad are areas where Bt cotton has never been grown. Therefore, there must be other reasons why farmers in those suicide hotspots are taking their lives in large numbers. Finally, even in areas where a crop such as Bt cotton is grown, the vast majority of farmers do not commit suicide. There seems to be a particular conjunction of ecological and financial circumstances that leads some farmers to be so heavily indebted that their only recourse is to take their own lives. As Münster argues ("Farmers' Suicides as Public Death" 1591–1595), even in suicide-prone districts, farmers rarely kill themselves as a form of political protest, but "farmers' suicides" have been seized upon by various groups to construct their own narratives that oppose, among other things, globalization, genetically modified crops, the Indian state, and the party in power in the regional state.

Ecological roots

Farmers' suicides have been linked to environmental degradation in two ways. Suicide hotspots are places marked by the monocropping of cash crops. If we look beyond Bt cotton, there does appear to be some truth in the claim that both monocropping and cash crops may have contributed to the crisis in agriculture for different reasons.

The ecological dangers of monocropping are well recognized. Growing the same crop season after season depletes the soil of nutrients essential for the crop, and makes the soil less fertile. It also encourages pests to flourish since they find a continuous source of food. Finally, it can lead to the pollution of the soil and of groundwater by encouraging the use of chemical fertilizers and pesticides. In many cases, monocropping can also lead to soil erosion, by encouraging the planting of crops that would not grow there without the heavy use of groundwater and fertilizers.

The story of Bt cotton intersects with ecological precarity in two ways, best illustrated by the cotton-growing district of Warangal, another suicide hotspot, in the state of Telengana (formerly Andhra Pradesh). Cotton does poorly on rain-fed land with thin soils, and these are the conditions found in Warangal. The entire region was suffering from poor rainfall since 2001; in 2004, the rains failed spectacularly, and plunged farmers, clinging on for the previous years, deeper into debt and despair. The biggest difference between Bt cotton and the traditional varieties is that the seed for Bt cotton is more expensive. However, because it is resistant to the bollworm, Bt cotton requires less expenditure on pesticides, so the outlay in growing the crop is potentially smaller (Herring 150). Farmers are thus growing a crop poorly suited to agro-ecological conditions, and the fragile environmental backdrop is worsened by monocropping.

The second, essential, component of ecological precarity is the direct result of climate change, namely greater variability in the quantity and timing of rainfall. India as a whole is considered highly water-stressed, and all the regions that are suicide hotspots are severely water-stressed, with the exception of Wayanad in Kerala. Most Indian agriculture is rain-fed, and this leaves it especially vulnerable to fluctuations caused by climate change ("As Indian Ocean Warms" 19; "Unseasonal Rains" 5; Prabhu 7). A farmer deep in debt hopes that the monsoons will be better next year. If the rains are good, and cotton prices are high, then even a farmer in Warangal can expect to emerge from debt the following year.

There is a long history of ecological precarity in Telengana, and in the drought-prone areas of the Deccan more generally, centered on cotton and peanuts. In the late nineteenth and early twentieth centuries, the colonial government expanded the network of railways and road transport, thereby better integrating the drylands of the Deccan into the world market. Then, too, cotton production by smallholders predominated, and the region was racked by famines and regular crop failures (Washbrook 131). Not infrequently, this resulted in uprisings against the burdensome tax demands of the colonial state. For example, the so-called "Deccan Riots" of 1875 were partially motivated by the heavy indebtedness of smallholding agriculturalists. There is thus a long history to the immiserization of the Deccan peasant, which revolves around cotton, and which connects ecological precarity to debt peonage and insertion into global markets.

Markets and stratification

Perhaps the only thing that studies of farmers' suicides agree upon is that the victims were heavily indebted. Is there a pattern about *who* is likely to go into debt and *why*? Suicides are relatively uncommon among the richest farmers *and* among the landless. If one follows the classical method of classifying the peasantry into the six categories of absentee landlords,

1. absentee landlords: lease out land to tenants; paid cash or share of output by tenant

2. rich peasants: farm themselves; hire in labor; grow food for family but sell most of their output

3. middle peasants: use own labor and some hired labor; grow food for own consumption and a little for sale

4. poor peasants: use own labor and hire out own labor;grow food for own consumption and purchase food from market

5. the landless: hire out labor, purchase food from market

6. non-cultivating artisans: sell products of labor

Figure 19.2 List of peasant classes.

rich peasants, middle peasants, poor peasants, the landless, and noncultivating artisans (see Figure 19.2), then it appears that the suicides have affected mostly the poor peasant (No. 4 in Figure 19.2) and middle peasant (No. 3) groups, and to a lesser extent, rich peasants (No. 2).

Why have poor peasants ended up as the group with the highest suicide rates? (The category of "poor peasants" includes small tenants, people who lease land from an absentee landowner or a rich peasant.) It is a complicated story, which is clearly different in different parts of the country, but the general outlines go something like this: when the first Green Revolution (input-intensive, industrial agriculture) was adopted in the late 1960s and 1970s, mostly in the wheat-growing areas of north India that had secure and perennial sources of irrigation, large landholders, rich peasants, and middle peasants were best positioned to benefit from it. As this model of industrial agriculture spread to other parts of India, and especially to semi-arid areas like eastern Maharashtra, Telengana, and northern Karnataka, larger landholders took advantage of intensive methods of farming because they had the resources to invest in tubewells, and to purchase seeds, inorganic fertilizer, and pesticides. The Indian state supported intensive agriculture by subsidizing electricity (and hence water from tubewells) and by providing price supports for output. When economic liberalization was announced in 1991, the top tiers of the peasantry, who had benefitted most from new industrial farming techniques, diversified their economic activities, sometimes moved to small towns and cities, and invested in industrial enterprises. Their children obtained service-sector jobs, and these families eventually abandoned farming. Their lands became available for hire or purchase, and many former artisan castes, who earlier depended on making goods for farming, but whose occupations were undermined by the opening up of markets, moved into farming either as tenants or as marginal and small farmers. These groups were largely new to farming.

It was in this context that new seeds, of which Bt cotton is an important part, were introduced. Cash crops offered the prospect of quick returns when prices of commodities were high. But not all cash crops had price supports, and crops like cotton, pepper, ginger, coffee, and vanilla, which were well integrated into global markets, experienced enormous price fluctuations depending on worldwide production. Unlike those with large- and medium-sized farms, who had the land area to spread their risk by cultivating some food crops and some cash crops, small and marginal farmers cultivated only cash crops on the little land they owned. The new crops demanded much larger cash expenditures up front for seeds, water, and fertilizer before any money could be made by selling the output. One of the most important properties of GM seeds is that they cannot be replicated by the farmer himself, but that a new stock of seeds has to be purchased from the patent-holding corporation every year. Since most marginal and small farmers do not have such cash reserves, they took loans at high interest rates from seed companies, moneylenders, and relatives. The new seeds also demanded different, new knowledge about appropriate farming techniques, which small farmers did not possess. For example, many farmers who planted Bt cotton continued to waste money on expensive pesticides, not realizing that they did not need to do that anymore.

South-central India is dominated by rain-fed agriculture, and the variability and quantity of the monsoons has begun to fluctuate wildly with climate change. If the crop failed due to late or unseasonal rain, or if the world price for the commodity that they were farming collapsed, small and marginal farmers were left deep in debt, with no prospect of repaying those loans (Krishna Kumar and Veeraraghav 11; Prabhu 7). One common feature of farmers' suicides has been their degree of indebtedness, which has been consistently found to be several times that of other farming households.

The paradox of farmers' suicides lies in their selectivity. The image of peasant communities as places where social mechanisms ensured that prosperity and hardship were shared by everyone, even if such sharing was not equal, is squarely refuted by the phenomenon of farmers' suicides. Village India has always been highly unequal, but it is now riven with divisions that make it especially difficult to imagine villages as communities. Not only did an earlier round of the Green Revolution create sharp class divisions among cultivators, but it also created new conditions of risk for different sections of the peasantry, and thereby deep differences in the possibilities for upward mobility. With the same seeds, and the same technology, some farmers are able to prosper while others are exposed to enormous downside risks. It is not as if the crop is wiped out in whole villages or regions; rather, some small and medium farmers are affected, while rich farmers are able to leverage the technology for further income gains. In the same village and in the same agroclimactic zone, farmers are exposed to very different risks because of their ability to access water, high-quality seed, or markets. Thus, risks are differentiated by class and knowledge rather than by region.

A speculative climate?

What is striking about farmers' suicides in India today is that they emerge from historical conditions in which it appears that little has changed. For instance, during colonial rule, cotton farmers in the Deccan were exposed to the busts and booms of global commodity markets, exposed to periods of drought resulting in severe famines and want, and had their lands usurped by moneylenders who took advantage of their indebtedness. Similar conditions, marked by incorporation into global markets, indebtedness, and climactic fluctuations characterize the present. Although much attention has been devoted to Bt cotton, it is not genetically modified crops that are causing the current crisis, but a certain conjunction between ecological conditions, finance, and risk that I want to label "speculative climate."

In using the term *speculative climate*, I want to underline not only that climate change has made the conditions for agriculture more uncertain, but also that such uncertainty is occurring in a context where social life is being lived in a climate of speculation. Not since Lenin and Hilferding first defined imperialism as the highest stage of capitalism have we seen finance capital play such a dominant role in global capitalism. Farming has increasingly become a speculative activity because farmers have to take out production loans to start the cycle of production. They are then gambling on the weather and on global commodity prices: if either of those variables turns against farmers, they fall into debt. If that pattern recurs, they fall further into debt, until they reach a point of no return.

It is not surprising that those with small- and medium-sized farms are more likely to engage in such risky behavior, betting on small probabilities that they will make a fortune, and that is why there is a preponderance of suicides among them. Poor peasants see richer peasants and urban people profiting from a neoliberal economy. They aspire to it, and it lures them into risky endeavors, which in turn delivers them not to a middle-class life, but instead exposes them to social death, the loss of face, and, eventually, to death. Neoliberal economic policies have accentuated this precarity, and many suicide victims were reported to be afraid of "losing face," of "having one's nose cut," and of being publicly humiliated by debt collectors (Vasavi 127–128). The term "precarity" has been used to denote the existential condition and the daily experience of downward mobility amongst the middle classes in the global North (Allison). If that term has any currency in the global South, it is surely in the lure of aspirational economies and their eventual failure: not so much the experience of downward mobility, but the dashing of the hopes of upward mobility.

References

"2.90 Lakh Farmers Committed Suicide During 1995–2011: Govt." *Indian Express* 31 August 2013. Print.

Allison, Anne. *Precarious Japan*. Durham: Duke University Press, 2013. Print.

"As Indian Ocean Warms Up, Agri and Fin Sectors Sweat." *Economic Times* 17 June 2015: 19. Bangalore Edition. Print.

Ataulla, Naheed and Anand J. "How Loan Sharks Pull Poor Farmers into a Debt Trap." *The Times of India* 27 July 2015: 2. Bangalore Edition. Print.

"BJP Disputes Govt Data on Farmer Suicides, Says 221 Deaths in 7 Months." *Deccan Herald* July 30, 2015: 5. Bangalore Edition. Print.

Gruère, Guillaume and Debdatta Sengupta. "Bt Cotton and Farmer Suicides in India: An Evidence-Based Assessment." *Journal of Development Studies* 47.2 (2011): 316–337. Print.

Gupta, Akhil. *Red Tape: Bureaucracy, Structural Violence, and Poverty in India*. Durham, NC: Duke University Press, 2012. Print.

Herring, Ronald J. "Whose Numbers Count? Probing Discrepant Evidence on Transgenic Cotton in the Warangal District of India." *International Journal of Multiple Research Approaches* 2.2 (2008): 145–159. Print.

Krishna Kumar, R. and T. M. Veeraraghav. "50 Suicides in 15 Days." *The Hindu* 19 July 2015: 11. Bangalore Edition. Print.

Münster, Daniel. "Farmers' Suicides and the State in India: Conceptual and Ethnographic Notes from Wayanad, Kerala." *Contributions to Indian Sociology* 46.1–2 (2012): 181–208. Sage Journals. Web. 15 October 2015.

——. "Farmers' Suicides as Public Death: Politics, Agency and Statistics in a Suicide-Prone District (South India)." *Modern Asian Studies* 49.5 (2015): 1580–1605. Cambridge Journals. Web. 15 October 2015.

Prabhu, Nagesh. "Spectre of Drought Rubs Salt into Farmers' Wounds." *The Hindu* 12 August 2015: 7. Print.

Sainath, P. "Neoliberalism and India's Farm Crisis." Boulder, CO: Alternative Radio 25 (September 2006). Print.

Shiva, Vandana and Kunwar Jalees. *Farmers Suicides in India*. New Delhi: Research Foundation for Science, Technology and Ecology (RFSTE), 2005. Print.

Shiva, Vandana, Afsar H. Jafri, Ashok Emani, and Manish Pande. *Seeds of Suicide: The Ecological and Human Costs of Globalisation of Agriculture*. New Delhi: Research Foundation for Science, Technology and Ecology (RFSTE), 2002. Print.

Singh, Charan and C. L. Dadhich. "Take Steps to Prevent Farmer Suicides." *Indian Express* 12 August 2015. Web. 15 October 2015.

"Unseasonal Rains Take Juice Out of Sugar Cane Crop: 2 Farmers End Lives, One Makes Attempt." *Deccan Herald* 13 July 2015: 5. Print.

Vasavi, A. R. *Shadow Space: Suicides and the Predicament of Rural India*. Gurgaon: Three Essays Collective, 2012. Print.

Washbrook, David. "The Commercialization of Agriculture in Colonial India: Subsistence and Reproduction in the 'Dry South,' c. 1870–1930." *Modern Asian Studies* 28.1 (1994): 129–164. Print.

20

ECOLOGICAL SECURITY FOR WHOM?

The politics of flood alleviation and urban environmental justice in Jakarta, Indonesia

Helga Leitner, Emma Colven, and Eric Sheppard

Within the rich vein of geographical scholarship in the environmental humanities, significant attention has been paid to the relationship between cities and nature. Such scholars show how, just as humans are part and parcel of the other that we dub nature, even the greyest urban landscape is every bit as bound up with the more-than-human world as any rural landscape, in terms of how it co-evolves locally and globally with nonhuman species and materialities (Braun; Cronon; Gandy; Kaika). This scholarship also has taken up the question of how cultural and political processes intersect, substantially contributing to the interdisciplinary field of urban political ecology which attempts to untangle the economic, political, social, and ecological processes that form contemporary cities (Heynen et al.). With the majority of the world's population now residing in urban settlements, interrogating the relation between cities and the more-than-human world has become increasingly urgent around such questions as urban sustainability, urban ecological security, and climate change (Hodson and Marvin).

For some time, the key trope has been urban sustainability: can cities potentially work with the more-than-human world, offering a less ecologically destructive way of spatially organizing humans than spreading them evenly across Earth (Davis), or are their enormous carbon and ecological footprints evidence that cities and nature are inherently in opposition? On the former side of this ledger, cities are imagined as sites of possibility for sustainable living and green urbanism, often framed by overly blithe claims about the possibility of ecological modernization or sustainable development under capitalism. As sustainability has come to be critiqued for imagining that some kind of long-term balance or equilibrium between humans and nature can be engineered, those representing cities as potentially good for nature have turned to discourses of urban resilience. Resilience thinking emphasizes the ability of cities to "bounce back from climate-related shocks and stresses" (Leichenko 164), while also imagining nature as resilient in the face of urbanization.

On the other side, cities are represented as dystopian threats to both humanity and the more-than-human world, and nature's agency as a threat to urban life. Responding to the intensifying concern for the threats posed to society by anthropogenic climate change and

194

resource constraints, urban ecological security (UES) has become increasingly influential as a framing for environmental governance. UES refers to strategies that seek to "reconfigure cities and their infrastructures in ways that help to secure their ecological and material reproduction" (Hodson and Marvin 193).

Yet who gets to define what counts as ecological security, and for whom? Hodson and Marvin argue that ecological security is essentially an elite agenda of "global" cities, as they develop policies to secure their prosperity against the threats of resource scarcity and climate change. Such policies should then travel to Third World "mega-cities," like Jakarta, where environmental problems seem intransigent. Here, however, we enter a domain of thought dominated by the developmental imaginary and the presupposition that precarious places (Jakarta) and bodies (the urban poor) can only improve by imitating the (capitalist) practices of prosperous bodies and places.

Discourses of urban ecological security focus on risk associated with environmental threats and the capacity of cities to develop resilience in defending against them. Yet this neglects the fact that not all bodies and places are equally affected, nor are they endowed with equal capacities. Recent work on sustainable development has made questions of equity and justice explicit, seeking development that achieves a low ecological footprint, empowers disadvantaged humans, and reduces social inequality. In this chapter, we first examine Jakarta's politics of flood alleviation and how international consultants, government agencies, the World Bank, and political elites are framing hydrological engineering projects as the city's best option for fending off the threat of flooding. Second, we examine the implications of these projects for environmental justice. We argue that the politics of flood alleviation with its emphasis on hydrological engineering is likely to improve the ecological security of elites and middle-class populations, while undermining the social and economic security of the urban poor. Condemned to flood-prone locations along river banks and around retention ponds, these populations face eviction while being blamed for the flooding. This politics also marginalizes grassroots alternatives that focus on working with, rather than controlling, nature.

Jakarta

From the perspective of Jakarta's and Indonesia's political and economic elites (Jakarta being Indonesia's political capital, and far and away its largest city), the city's principal environmental threat is urban flooding compounded by climate change. Located on Java's northwest coast, Jakarta historically has been prone to flooding during its monsoon season from November to February or March. However, Jakarta suffers with increasing frequency from widespread and devastating flood events. The 1996, 2002, and 2007 flood events were the most destructive ever recorded in Jakarta (WHO 2007, cited in Douglass). In 2007, 340,000 people were displaced as waters reached depths of up to four meters (MacKinnon), claiming the lives of eighty people in the greater metropolitan region of Jakarta (Padawangi). Furthermore, with extraordinarily high rates of land subsidence—averaging 7.5cm per year but as high as 20cm a year in some areas (Abidin et al.)—hydrological engineers forecast that 80 percent of northern Jakarta will be below sea level by 2030 (Witteveen+Bos), dramatically heightening the risk of flooding.

Seeking to protect Jakarta against flooding, the Dutch, Japanese, and Indonesian governments have in recent years embarked upon on a number of very ambitious, high-modernist

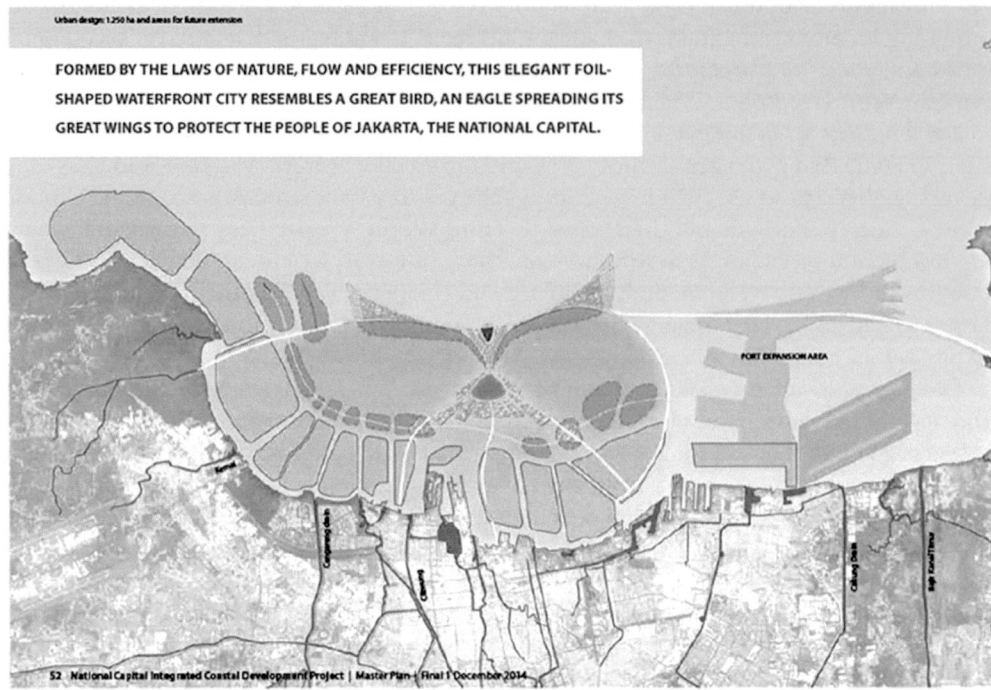

FORMED BY THE LAWS OF NATURE, FLOW AND EFFICIENCY, THIS ELEGANT FOIL-
SHAPED WATERFRONT CITY RESEMBLES A GREAT BIRD, AN EAGLE SPREADING ITS
GREAT WINGS TO PROTECT THE PEOPLE OF JAKARTA, THE NATIONAL CAPITAL.

Figure 20.1 Artistic rendering of the Great Garuda Sea Wall. (Source: NCICD Consortium 52.
Project: NCICD consortium. Urban Design and Images: KuiperCompagnons. Reproduced
with the permission of KuiperCompagnons, Rotterdam, The Netherlands.)

engineering projects, including the World Bank-funded Jakarta Urgent Flood Mitigation
Project (JUFMP) and the "Great Garuda Sea Wall" (GGSW) Project. As indicated by its
name, the JUFMP intends to provide Jakarta with urgent and immediate flood mitigation
by dredging and "normalizing" fifteen priority sections of floodways, drainage canals, and
reservoirs to restore them to their full capacity, thereby reducing the extent and duration
of flooding.

The GGSW project has been slated as one of the world's most ambitious coastal defense
projects currently underway (Stedman). Designed by a consortium of Dutch architecture
and engineering firms under contract with the government of the Netherlands, and in partnership with the government of the Republic of Indonesia, the Dutch master plan proposes
a giant seawall (Figure 20.1). The design suggests extensive land reclamation designed, controversially, to resemble Indonesia's national symbol and mythical bird, the *garuda*, upon
which new integrated towns will be constructed, including residential, commercial and recreational spaces, and technical and social infrastructure. The seawall is designed to close
Jakarta Bay, transforming it into a retention lake that would absorb water from the rivers
and canals, relying on pumps to maintain a lower water level and expel excess outflow into
the sea. The wall is also supposed to protect the city from flooding from the ocean. The
project began in October 2014, with the strengthening of the existing inner sea wall; a final
decision about whether to proceed with the larger project is pending.

Hydrological engineering

These flood alleviation projects invoke a common technological imaginary, in the spirit of hydrological engineering as it emerged in North Atlantic societies, which envisions humankind as developing hard technological solutions to fence human society off from the predations of nature. When water gets out of control from a human perspective, it needs to be separated from society and channeled elsewhere. Under the banner of development, invoked since the 1950s as the solution to Third World impoverishment, large-scale hydrological engineering has become a hallmark of modernity. Rivers were dammed on larger and larger scales across Asia, Africa, and the Americas; dams came to represent technological prowess and modernity, bending nature to the economy's will (Nixon).

Political discourses in Jakarta present the GGSW, JUFMP, and related hydrological engineering projects, which aim to rein in and control nature, as not just the preferred but also the only solutions to protect Jakarta from flooding. While the JUFMP project has been designed to drain the city and channel unruly floodwaters as efficiently as possible into the Java Sea, the GGSW project will involve the construction of a huge artificial dyke that will enclose Jakarta Bay, transforming it from salt- to sweet-water and reducing the impacts of flooding. Both projects draw heavily on Dutch expertise. The Dutch have achieved global recognition as pioneers in technologies of hydrological engineering to tame water in the wild: building dams and sluices, canalizing the movement of water, dredging waterways, digging flood holding ponds, and so on, and their presence shaping water in Jakarta dates back to its early days as Batavia.

Reminiscent of North Holland's *Afsluitdijk* completed in 1932 as part of the Zuiderzee Works—one of the Netherlands largest hydrological projects of the twentieth century—the GGSW project is indicative of the continued influence of Dutch water expertise into the postcolonial present (Disco and Toussaint; Latschan). Presented as a massive dam collecting floodwaters from the rivers and canals and excluding seawater from Jakarta, the project will also protect future reclaimed land. These would not be like agricultural polders, but massive "integrated" residential and commercial real-estate projects where Jakarta's middle class can escape their poorer co-residents and recreate in green spaces and clean waters. The JUFMP project is entirely about dredging, canalization, and storm- and floodwater control.

The rationality and control associated with this European tradition of engineering brings with it two other attitudes toward nature and society: compartmentalization and top-down planning. Rather than seeing humans' relationship with the more-than-human world as one of co-evolution and uncertainty, with unexpected twists and turns, this particular form of hydrological engineering compartmentalizes environmental processes to be tackled by separable modularized projects—this bridge, that dam—that will tame and enroll nature one step at a time. Engineering is also about optimization—using mathematics to design the best possible project given the requirements. In this view, expertise is unquestioned and consultation with those designated as nonexperts at best a distraction from achieving this goal. This reinforces a state- and elite-led top-down philosophy of leaving it to the experts to decide who should be evicted and which mangroves should be cleared.

Transforming threats into opportunities

Political discourses in Jakarta have actively constructed flooding as an imminent and overwhelming threat to the city's safety and prosperity. River dredging and reinforcement projects are justified on the grounds of the urgent need to contain the danger of flooding. The GGSW project is planned to address a longer-term but no less urgent trajectory: the prediction that 80 percent of northern Jakarta will be underwater by 2030. Its Master Plan warns that "the lives of 4.5 million people are at stake," declaring that "there is no time to lose" (NCICD Consortium 25, 33). At the same time, the deployment of hydrological engineering through the GGSW project promises to transform the threat of flooding into an *opportunity*: to generate revenues through real-estate markets. Large-scale reclamation offers a means to extend the city into the ocean, create virgin lands for settlement, and escape the complexities of the existing city, where land is perceived as scarce, expensive, and difficult to assemble in large tracts. It is in this sense that, despite circulating discourses of environmental threat and disaster, Jakarta's coast has been reimagined as the city's "last frontier" (Kusno; Juniardhie/Sekretariat Kabinet Republik Indonesia).

Normalisasi

The term *normalisasi* ("normalization" in Bahasa) circulates widely in Jakartan official and popular discourse as an appellation for the river dredging and canalization initiatives. At one level this references the goal of "returning" polluted and blocked rivers to their "natural" (i.e., pre-Jakarta) state. Notably absent is any acknowledgment that for many centuries these rivers have overflowed and flooded during the rainy season, or that the canals constructed by the Dutch in the seventeenth century were unable to completely resolve the problem of flooding. Dutch colonial settlers eventually abandoned the old city (Kota) for higher ground. But *normalisasi* has come to mean more than this. It means clearing out "illegal" residents, relocating them into "proper" (i.e., North Atlantic style public) housing, an environment in which they are expected to live right and take up "proper" (formal sector) jobs. It is associated with the normalization of public opinion, around a new norm that legitimizes evictions and represents the poor as responsible for their own failure to prosper (Budiari) and for "choosing" to occupy marginal environments (van Voorst). It is also associated with the normalization of nature, putting it anthropocentrically in its place as humans' resource for better living.

Although the GGSW project is not described as *normalisasi*, this project also seeks to redefine what should be considered normal. The master plan presents an offshore wall as the new normal: the rational choice from among three scenarios, rejecting the current norm (a city that floods, increasingly frequently) outright as an irredeemably poor option (see Figure 20.2). Reading across these projects, the new normal is hydrological engineering, conformity with North Atlantic imaginaries of a developed society, and nature at humans' beck and call.

Doing no harm

Hydrological engineering projects also are framed by the discourse of doing no harm. The Resettlement Policy Framework, negotiated and signed in December 2010 by the

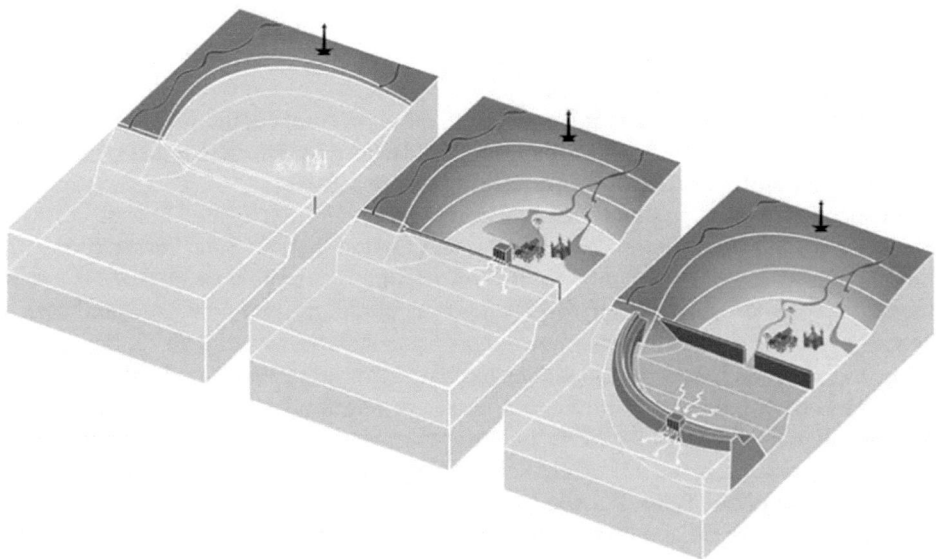

Figure 20.2 "Three principal solutions: a. abandon the National Capital, b. onshore protection by high dykes and onshore waduks and c. offshore sea wall with large offshore waduk." (Source: NCICD Consortium 36. Project: NCICD consortium. Urban Design and Images: KuiperCompagnons. Reproduced with the permission of KuiperCompagnons, Rotterdam, The Netherlands.)

Indonesian national government and the World Bank as the foundation for implementing JUFMP (DKI Jakarta), goes to considerable lengths to ensure that evictions in the name of JUFMP, or under related non-World Bank dredging projects, are minimal in number, and that those evicted receive full "market" compensation for the loss of property and livelihoods and are resettled in housing that is of higher quality. The JUFMP agreement does not include harm to the more-than-human world—evidently river dredging is presumed to be beneficial to nature as it "normalizes" rivers and reduces flooding. The GGSW Project Master Plan considers harm to nature, albeit narrowly: any acreage of mangroves that has to be destroyed to implement the Sea Wall will be replanted nearby.

Doing no harm has been the discursive frame for all the World Bank's resettlement initiatives (International Consortium of Investigative Journalists), but the Bank has been heavily criticized for failing to live up to this principle in practice, a failure acknowledged by its recently appointed director Jim Yong Kim, who wants the Bank to transition from doing no harm to doing good. The principle also can be criticized for not seeking to reduce social and environmental injustice, and for a rudimentary and anthropocentric view of justice to nature (what's good for human ecological sustainability is good for nature).

Environmental (in)justice

In the global North, scholarship on environmental justice has documented the uneven "socio-spatial distribution of environmental hazards" (Reed and George 838), including the disproportionate environmental and health impacts of air toxins, pesticides, incinerators and landfills on communities of color and/or lower-income neighborhoods. In the global South, studies have examined the disproportionate impact of environmental degradation, due to resource extraction or energy production, on poor, rural communities and on indigenous groups (Martínez-Alier), often framing struggles for environmental justice in terms of an "environmentalism of the poor" (Guha and Martínez-Alier).

More recently, a growing literature on climate justice has interrogated how distributional and procedural justice questions articulate with climate change. Such research has highlighted global-scale North–South inequities in the impacts of climate change, the greater vulnerability of the urban poor (who tend to occupy more marginal land and lack the resources necessary for resilience), and unequal participation in climate policy decision-making (O'Brien and Leichenko; Dodman and Satterthwaite; Paavola and Adger). Yet little attention has been paid to environmental justice in relation to adaptation and mitigation strategies and policies of concern here, including the disproportionate distributional impacts of flood mitigation projects on already vulnerable populations and the procedural marginalization of these populations' knowledge and experience in policy debates. In short, such strategies "undermine the stability of some social groups or places" even as they improve the situation for others (Heynen et al. 10).

While the techno-managerial solutions to flood alleviation underway in Jakarta are presented as being in the public interest, their social impact is frequently uneven, raising important questions about social and environmental justice. They disproportionately affect the urban poor, who are the most vulnerable and exposed to environmental threats and also lack the resources necessary for resilience. Poor residents face a triple burden: disproportionately exposed to flooding, particularly vulnerable to eviction as a part of flood mitigation, and blamed in popular discourse for causing flooding. Annually recurring floods of the major rivers criss-crossing Jakarta particularly affect the estimated 3.5 million poor residents who have built their shelters on marginal land in large informal settlements along the city's rivers. For example, Kampung Pulo on the east side of the Ciliwung River and Kampung Bukit Duri on the west side flood every year during the rainy season, damaging the homes and belongings of poor residents along the river, often forcing them to evacuate temporarily.

Since 2013, residents in these kampungs have faced ever-increasing threats of eviction due to river dredging and reinforcement programs. After several years of delay, Governor Basuki "Ahok" Tjahaja Purnama ordered the eviction of Kampung Pulo residents. Early in the morning on August 20, 2015, police, bulldozers, heavy machinery, and construction crews moved in to remove the housing, leveling the area and destroying hundreds of homes within a few days (Figure 20.3). Although the river dredging and reinforcement programs include a plan for resettlement and full compensation for those displaced, Ahok and Jakarta government officials declared that the inhabitants are illegally occupying state land, asserting the government's right to evict them (de facto eminent domain). Ahok has made compensation conditional on residents holding land right certificates and/or an identity card, but many of those residing along the Ciliwung River do not possess such documents. In addition, there has been insufficient provision of housing for resettlement, whether in-situ or on the outskirts of Jakarta. As with previous evictions, such as that of the Pluit reservoir in North Jakarta, where thousands of residents were evicted and had their homes destroyed in order to restore the reservoir to its full capacity,

Figure 20.3 After the eviction. Kampung Pulo, August 2015. Source: Authors.

the newly built housing units are too few to accommodate all those evicted and ill-suited to residents' needs.

Kampung residents living in marginal environments not only experience the brunt of flooding and evictions, but also increasingly have been blamed by political leaders, the media, and the middle and upper classes for contributing to clogged-up rivers and thus to flooding. The current governor of Jakarta has publicly stated that Jakarta floods because the rivers "are surrounded by illegal residences. Therefore the rivers' surroundings cannot properly absorb the overflowing water" (Wardhani, "Governor Ahok Defends"). Residents occupying waterways often internalize these narratives of blame, further shifting responsibility from the state onto vulnerable populations (van Voorst). While depositing waste of all kinds in Jakarta's rivers has reduced the throughflow of water, flooding results from a variety of human and more-than-human factors. These include unregulated deforestation upstream, unchecked urban development in water-catchment areas, poor water management, urban poverty, and lack of adequate municipal waste collection and disposal. Indeed, shopping malls that have been constructed on land once designated for water retention are not subject to the scrutiny faced by the poor (Tri Irawaty, cited in Wardhani, "Governor Ahok Defends"). World Bank and Dutch consultants' reports also highlight the need for an integrated strategic approach that "addresses the root causes of flooding, at the catchment level" (Netherlands Commission for Environmental Assessment 2).

Struggling for environmental justice and engineering with nature

It would be wrong to portray inhabitants of these informal settlements as passive victims, however. They are struggling for social and environmental justice, helped by such nonprofit organizations as Ciliwung Merdeka, the Urban Poor Consortium (UPC), and RUJAK. Their joint initiatives make visible the disjuncture between promises and practices, demand remedy and compensation, and propose alternatives. For example, Ciliwung Merdeka has helped residents of Bukit Duri to document and map the shelters they have constructed, maintained, and improved over time, drawing attention to the historical significance of these settlements and their contributions to the city's economy in ways that justify demands for compensation for evictions. Nonprofits have also been working with residents to develop alternative proposals and plans for social housing.

The government's top-down approach to providing replacement housing, rejecting consultation with affected residents and the alternative housing designs, has resulted in concrete housing blocks that ignore the specific socioeconomic needs of the poor. For example, the majority of inhabitants of these unauthorized kampungs work in the informal sector, frequently operating small home enterprises. Yet the public housing built does not accommodate such enterprises and is prohibitively expensive for those working in the informal sector. Alternative proposals, sensitive to the specific socioeconomic reality and needs of the poor, rarely make it beyond the blueprint stage. At best they receive a hearing; at worst they are ignored or rejected—a violation of procedural justice. This is an ongoing struggle. In September 2015, on the heels of overwhelmingly negative media coverage of the Kampung Pulo evictions, Ciliwung Merdeka proposed an alternative to evictions to Governor Ahok. Designed with the help of community residents, this would be an elevated and vertical kampung village (*kampung susun*) incorporating sustainable design principles and spaces for informal economic activities and sociability (Wardhani, "Ahok Reopens Dialogue"). This in-situ upgrading alternative would negate the need for further evictions. With the next election for governor looming in 2017, Ahok originally agreed to the plans, but it remains to be seen whether they will be realized.

With respect to the GGSW project, grassroots groups criticize it as simply more urban development at the expense of the urban poor and fishing communities, masquerading as a coastal defense strategy (Elyda). Although the master plan refers to the need for socially just resettlement and makes space in its design for fishing communities to be relocated (to the wingtips of the garuda), will these commitments be honored? With the threat of eviction looming as the GGSW project moves forward, questions of procedural justice for affected residents also are increasingly urgent. Yet the space for local communities to participate in project planning and implementation is very limited to date. Public "socialisasi" (information) meetings have been organized by the government, but these often occur outside of the most affected communities, with presentations that sometimes deploy technical language. In response, KIARA (The People's Coalition for Fisheries Justice) and other NGOs have organized their own meetings to inform residents.

Underneath struggles for environmental justice lie questions of how urban life should relate to nature. Calling into question the techno-managerial approach to flood mitigation and the deployment of hydrological engineering, grassroots groups are challenging the emphasis on hard infrastructure and land reclamation. Even as they continue to be excluded from public debates, these groups are articulating alternative, small-scale initiatives to mitigate flooding in Jakarta—engineering with nature. These initiatives include proposals to protect and preserve the mangroves that act as natural barriers to tidal flooding and sea-level rise and to reforest areas in the surrounding mountains to reduce surface runoff into the rivers flowing through Jakarta. Such proposals focus on planning *with* nature rather than against it, in ways that reconfigure human–nature relationships in less binary terms. In articulating such alternative strategies, groups such as KIARA draw on shifting international models for flood mitigation. The recent "ecological turn" in engineering being rolled out across North America and Western Europe is significant in this context (Disco). It had led to recent proposals for "floodable designs" which accommodate floodwaters in Rotterdam (Doesburg), "living breakwaters" to protect Staten Island, NYC, from sea-level rise (Rebuild by Design), and permeable and flexible designs for the New York City coastline such as "oyster-tecture," urban estuaries, and recycled-glass reefs (Bergdoll).

In addition to alternatives that re-engineer the city's infrastructural relationship with nature, other projects seek to "re-engineer" how urban residents relate to the river and nature more generally. For example, rather than imagining and treating rivers as external to human livelihoods—as places to dump trash—alternative initiatives seek to educate and encourage residents to think of the river as potentially vital to their lives and treat it accordingly. One example, frequently referenced but as yet not practiced in Jakarta, is the Code river rehabilitation project in Jogjakarta, in central Java (Wahyu Titiyoga et al.). This grassroots initiative contrasts sharply with the river *normalisasi* initiative in Jakarta. Initiated by Y. B. Mangunwijaya, a Catholic priest, it has been designed to enroll residents living along the river in the rehabilitation process from the very beginning, including collectively cleaning up the river, upgrading the kampung houses in-situ, setting them back from the river, and last but not least reorienting them to face the river. The intent of this redesign is also to reorient residents' understanding of the river as part of their lives rather than a neglected appendage. For the first time kampung residents look toward the river rather than turning their backs on it.

References

Abidin, Hasanuddin Z. et al. "Land Subsidence of Jakarta (Indonesia) and Its Relation with Urban Development." *Natural Hazards* 59.3 (2011): 1753–1771. Print.

Bergdoll, Barry. *Rising Currents: Projects for New York's Waterfront*. New York: Museum of Modern Art, 2011. Print.

Braun, Bruce. "A New Urban Dispositif? Governing Life in the Age of Climate Change." *Environment and Planning D: Society and Space* 32 (2014): 49–64. Print.

Budiari, Indra. "Urban Poor See Gap with Middle Class." *Jakarta Post* 14 September 2015. Web. 20 September 2015.

Cronon, William. *Nature's Metropolis: Chicago and the Great West*. New York: Norton, 1991. Print.

Davis, Mike. "Who Will Build the Ark?" *New Left Review* 61 (2010): 29–46. Print.

Disco, Cornelis. "Remaking 'Nature': The Ecological Turn in Dutch Water Management." *Science, Technology & Human Values* 27 (2002): 206–235. Print.

Disco, Nil and Bert Toussaint. "From Projects To Systems: The Emergence of A National Hydraulic Technocracy, 1900–1970." *Two Centuries of Experience in Water Resource Management: A Dutch-U.S. Retrospective*. Ed. John Lonnquest, Bert Toussaint, Joe Jr. Manous, and Maurits Ersten. Alexandria, VA: Institute for Water Resources, US Army Corps of Engineers, and Rijkswaterstaat, Ministry of Infrastructure and the Environment, 2014. 155–202. PDF. Web. 19 December 2015.

DKI Jakarta. *Jakarta Urgent Flood Mitigation Project (JUFMP) Resettlement Policy Framework*. Jakarta: DKI Jakarta, 2010. Web. 20 August 2015.

Dodman, David and David Satterthwaite. "Institutional Capacity, Climate Change Adaptation and the Urban Poor." *IDS Bulletin* 39.4 (2009): 67–74. Print.

Doesburg, Anthony. "Car Parks and Playgrounds to Help Make Rotterdam 'Climate Proof'." *The Guardian* 11 May 2012. Web. 15 October 2015.

Douglass, Michael. "Globalization, Mega-projects and the Environment: Urban Form and Water in Jakarta." *Environment and Urbanization Asia* 1.1 (2010): 45–65. Print.

Elyda, Corry. "Giant Sea Wall Islet Project Could Cause More Harm Than Good." *Jakarta Post* 28 October 2013. Web. 20 June 2015.

Gandy, Matthew. *The Fabric of Space: Water, Modernity, and the Urban Imagination*. Cambridge, MA: MIT Press, 2014. Print.

Guha, Ramachandra and Joan Martínez-Alier. *Varieties of Environmentalism: Essays North and South*. Delhi: Oxford University Press, 1998. Print.

Heynen, Nik, Maria Kaika and Erik Swyngedouw. "Urban Political Ecology: Politicizing the Production of Urban Natures." *In the Nature of Cities: Urban Political Ecology and the Politics of Urban Metabolism*. Ed. Nik Heynen, Maria Kaika, and Erik Swyngedouw New York: Routledge, 2006. 1–19. Print.

Hodson, Mike and Simon Marvin. "'Urban Ecological Security': A New Urban Paradigm?" *International Journal of Urban and Regional Research* 33 (2009): 193–215. Print.

The International Consortium of International Journalists. "Explore 10 Years of World Bank Resettlement Data." *The International Consortium of Investigative Journalists*. Web. 14 October 2015.

Juniardhie, Widyanto/Sekretariat Kabinet Republik Indonesia. "Waterfront City Jakarta" YouTube. 3 September 2012. Web. 1 December 2015.

Kaika, Maria. *City of Flows: Modernity, Nature, and the City*. London: Routledge, 2005. Print.

Kusno, Abidin. "Runaway City: Jakarta Bay, the Pioneer and the Last Frontier." *Inter-Asia Cultural Studies* 12 (2011): 513–531. Print.

Latschan, Thomas. "'Great Garuda' Wall to Save Jakarta from Rising Sea Level." *Deutsche Welle* 3 November 2014. Web. 15 December 2015.

Leichenko, Robin M. "Climate Change and Urban Resilience." *Current Opinion in Environmental Sustainability* 3.3 (2011): 164–168. Print.

MacKinnon, Ian. "Four-Metre Floodwaters Displace 340,000 in Jakarta." *The Guardian* 5 February 2007. Web. 2 July 2015.

Martínez-Alier, Joan. "The Environmentalism of the Poor." *Geoforum* 54 (2014): 239–241. Print.

NCICD Consortium. "Master Plan for Jakarta." Witteveen+Bos 2015. Web. 1 June 2015.

NCICD Consortium. *National Capital Integrated Coastal Development Master Plan*. Jakarta: Witteveen+Bos, 2014. Print.

Netherlands Commission for Environmental Assessment. *Advice on the Terms of Reference for the EIA for the Jakarta Urgent Flood Mitigation Project*. Utrecht: Netherlands Commission for Environmental Assessment, 2008. Web. 1 September 2015.

Nixon, Rob. "Unimagined Communities: Developmental Refugees, Megadams, and Monumental Modernity." *New Formations* 69 (2010): 62–80. Print.

O'Brien, Karen L. and Robin M. Leichenko. "Double Exposure: Assessing the Impacts of Climate Change within the Context of Economic Globalization." *Global Environmental Change* 10 (2000): 221–232. Print.

Paavola, Jouni and W. Neil Adger. "Fair Adaptation to Climate Change." *Ecological Economics* 56 (2006): 594–609. Print.

Padawangi, Rita. "The Right to Flood-Free Homes: Urban Floods, Spatial Justice and Social Movements in Jakarta, Indonesia." *Global Visions: Risks and Opportunities for the Urban Planet*. Ed. Adriana Gonzalez-Brun, Low Boon Liang, Jürgen Rosemann, and Johannes Widodo. Singapore: National University of Singapore, 2012. 199–211. Print.

Rebuild by Design. "Living Breakwaters." *Rebuild by Design.* 2015. Web. 15 October 2015.

Reed, Maureen G. and George, Colleen. "Where in the World is Environmental Justice?" *Progress in Human Geography* 35 (2011): 835–842. Print.

Stedman, Lis. "Plan Progress for the Great Jakarta Sea Wall." *Water 21: Magazine of the International Water Association* February 2014. Web. 20 June 2015.

van Voorst, Roanne. "The Right to Aid: Perceptions and Practices of Justice in a Flood-Hazard Context in Jakarta, Indonesia." *The Asia Pacific Journal of Anthropology* 15 (2014): 339–356. Print.

Wahyu Titiyoga, G., A. Nur Yasin, M. Rifqy Fadil, and P. Wicaksono. "Cleaning up Code." *Tempo Magazine* (30 August 2015): 54–56. Print.

Wardhani, Dewanti, A. "Governor Ahok Defends Eviction Policy." *Jakarta Post* 24 December 2014. Web. 20 June 2015.

——. "Ahok Reopens Dialogue, Revives 'kampung susun' Idea." *Jakarta Post* 21 September 2015. Web. 22 September 2015.

21

OUR ANCESTORS' DYSTOPIA NOW

Indigenous conservation and the Anthropocene

Kyle Powys Whyte

Conservation in the Anthropocene

The proposed Anthropocene epoch is understood geologically as a time when the collective actions of humans began influencing earth systems in marked, unprecedented ways. Some claim that the Anthropocene could have started in the year 1610 with "colonialism, global trade and coal" (Lewis and Maslin 177). Scientists and environmental ethicists often characterize futures in the Anthropocene as involving climate destabilization that will likely threaten the very existence of certain ecosystems, plants, and animals (Kolbert; Thompson and Bendik-Keymer; Vaidyanathan; Sandler). Ever-expanding human economic activities and consumer lifestyles are major drivers of climate destabilization through their dependence on burning fossil fuels and certain kinds of land-use such as deforestation. Some conservationists argue that we will inevitably have to learn to live with these changes, make careful decisions about conservation priorities, and, in some cases, learn to let go of certain ecosystems and species (Kareiva and Marvier). Yet others in the conservation community take an adamant position that these changes, especially extinctions, are morally dreadful (Vaidyanathan; Cafaro and Primack).

Cafaro and Primack express this latter position clearly when they argue that "Anthropocene proponents" have "selfish and unjust" views that deem the extinction of species morally acceptable as long as humans suffer no harm. Human expansion that extinguishes the "polar bear" and other species ends the value of these natural species as "the primary expressions and repositories of organic nature's order, creativity, and diversity … Every species, like every person, is unique, with its own history and destiny." To destroy species through human expansion is to bring a "valuable and meaningful story to an untimely end" (2).

Conservation views à la Cafaro and Primack can help to explain why some scholars and writers have noted various strains of dystopian thought in Anthropocene discourses (Trexler; Singleton; Weik von Mossner; Johnson et al.). Such views depict futures in which many hundreds of "valuable and meaningful stories" (Cafaro and Primack) are irreversibly extinguished, leaving human societies to reckon with a world marked by greatly limited historical memories, biodiversity, and expressiveness. In an article titled, "Climate Change Is So Dire We Need a New Kind of Science Fiction to Make Sense of It," futures writer

Claire Evans writes that "we need an Anthropocene fiction. Since sci-fi mirrors the present, ecological collapse requires a new dystopian fiction … a form of science fiction that tackles the radical changes of our pressing and strange reality" (Evans).

As a Neshnabé (Potawatomi) and scholar-activist at a US university working on indigenous climate justice, I was initially struck by what seemed to be some similarities between the dystopian Anthropocene views and the views motivating quite a few indigenous projects to conserve and restore native species. Indeed, indigenous peoples have long advocated that the conservation and restoration of native species, the cultivation of first foods, and the maintenance of spiritual practices that require the existence of plants and animals of particular genetic parentage whose lives are woven with ecologically, economically, and culturally significant stories, knowledges, and memories. I wondered whether indigenous peoples share the same dread of species extinction in an Anthropocene dystopian future.

While surface similarities are present, it is perhaps more accurate to say that indigenous conservationists and restorationists tend to focus on sustaining particular plants and animals whose lives are entangled locally—and often over many generations—in ecological, cultural, and economic relationships with human societies and other nonhuman species. We try to learn from, adapt, and put in practice these relationships, ancient as some may be, to address the conservation challenges we face today and in the future, especially the environmental destruction caused by settler colonialism in North America. In this sense, while we may embrace the value of species such as the polar bear—even when we may have never interacted with one—it is also true that we are unlikely to invoke the polar bear in the absence of also invoking the species' significance to particular human and nonhuman communities with whom it has long, local, complex, and unique relationships.

What is more, the environmental impacts of settler colonialism mean that quite a few indigenous peoples in North America are no longer able to relate locally to many of the plants and animals that are significant to them. In the Anthropocene, then, some indigenous peoples already inhabit what our ancestors would have likely characterized as a dystopian future. So we consider the future from what we believe is already a dystopia, as strange as that may sound to some readers. Our conservation and restoration projects are not only about whether to conserve or let go of certain species. Rather, they are about what relationships between humans and certain plants and animals we should focus on in response to the challenges we face, given that we have already lost so many plants and animals that matter to our societies. In this way, indigenous conservation approaches aim at negotiating settler colonialism as a form of human expansion that continues to inflict anthropogenic environmental change on indigenous peoples—most recently under the guise of climate destabilization.

The dystopia of our ancestors

A long-standing environmental advocate and friend, Lee Sprague (Potawatomi/Odawa, Little River Band of Ottawa Indians), always reminds me that Anishinaabek[1] already inhabit what our ancestors would have understood as a dystopian future. Indeed, settler colonial campaigns in the Great Lakes region have already depleted, degraded, or irreversibly damaged the ecosystems, plants, and animals that our ancestors had local living relationships with for hundreds of years and that are the material anchors of our contemporary customs, stories, and ceremonies. Settler colonial campaigns refer to the various global projects of combined military, commercial, and cultural expansion of European, North American, and many other states (e.g. New Zealand). These waves of settlers, such as those forming the US and Canadian states, continue to deploy strategic tools and weapons to establish permanent roots in indigenous territories with the continued hopes of inscribing homelands for their own families and societies in those territories (Lefevre).

I use the term *campaigns* because these waves of settlement are sustained, strategic, and militaristic. These campaigns include both war-like violence and the tactics for suppressing populations that are used alongside belligerence, from assimilative institutions (e.g., boarding schools) to containment practices (e.g., reservations) to the creation of dependency (e.g., commodity foods). As a means of carving out settler homelands from indigenous homelands, waves of settlers harnessed industrial means, from military technologies to large-scale mineral and fossil fuel extraction operations to sweeping, landscape-transforming regimes of commodity agriculture. Industrial settler *states* are the corresponding polities, from federal nation state governments to local municipalities and subnational provincial governments, that create and enforce the laws, policies, and jurisprudence that serve to protect and incubate the homeland-inscribing process from indigenous resistance, refusal, and resurgence in such territories.

The fallout of what I will call "industrial settler campaigns" is that, as indigenous peoples, we continue to exercise political and cultural self-determination even though there are now states such as the United States and Canada that are perceived by most people as being the preeminent sovereigns in the places where indigenous communities live, work, and play. Our degree of success in exercising self-determination is irreversibly coupled with our political relations with states whose constituent people and institutions wield daunting financial, military, and police resources and regulatory and legal enforcement capabilities. Ecosystems have been reshaped to such a degree by settlers and their institutions that it is hard to recognize anything "indigenous" about them. Hence many scholars and activists describe settler colonialism as a structure of oppression that erases indigenous peoples (Lefevre).

It would have been an act of imagining dystopia for our ancestors to consider the erasures we live through today, in which some Anishinaabek are finding it harder to obtain supplies of birch bark, or seeing algal blooms add to factors threatening whitefish populations, or fighting to ensure the legality in the eyes of the industrial settler state of protecting wild rice for harvest. Yet we do not give up by dwelling in a nostalgic past even though we live in our ancestors' dystopia. My friend Deb McGregor (Ojibwe, Whitefish River First Nation) always points out to me that we are really living in just the tiniest sliver of Anishinaabe history. The vast majority of our history precedes the campaigns that have established states such as the United States and Canada. Our conservation and restoration efforts are motivated by how we put dystopia in perspective as just a brief, yet highly disruptive, historical moment for us—at least so far.

This historically brief, highly disruptive moment, "today's dystopia of our ancestors," sounds a lot like what others in the world dread they will face in the future as climate destabilization threatens the existence of species and ecosystems. Yet for many indigenous peoples, the Anthropocene is not experienced as threatening in precisely the same sense because the particular era of settlement I am describing wrongfully forced many of our societies to let go of so many relationships with plants, animals, and ecosystems at a rapid pace. Rather, if there is something different in the Anthropocene for indigenous peoples, it would be just that we are focusing our energies *also* on adapting to another kind of anthropogenic environmental change: climate destabilization. Indeed, in the nineteenth and twentieth centuries, we already suffered other kinds of anthropogenic environmental change at the hands of settlers, including changes associated with deforestation, forced removal and relo-cation, containment on reservations (i.e., loss of mobility), liquidation of our lands into individual private property and subsequent dispossession, and unmitigated pollution and destruction of our lands from extractive industries and commodity agriculture. While all soci-eties alter the environment in which they dwell, anthropogenic environmental change here refers very specifically to how industrial settler campaigns *both* dramatically changed ecosys-tems, such as through deforestation, overharvesting, and pollution, *and* obstructed indigenous peoples' capacities to adapt to the changes, such as through removal and containment on reservations.

Though the climate destabilization described in Anthropocene futures may be a distinct ecological challenge for indigenous peoples, we experience it nonetheless as associated with the repeated patterns of industrial settler tactics that we know all too well. Indeed, settler industrial campaigns paved the way for industrial and capitalist collective actions whose ecological footprint contributes significantly to today's climate destabilization ordeal. For many indigenous peoples, this ordeal involves adapting to sea-level rise, warming waters, and increased severity of droughts. Indigenous peoples are on the frontlines of dealing with these changing conditions, from coastal communities who have to relocate permanently to communities who are losing habitats in their homelands that they need to continue their relationships with the remaining culturally and economically significant plants and animals. What makes many of these communities particularly vulnerable to the local impacts of cli-mate destabilization is their continued intertwinement today with the same industrial settler strategies that already degraded many of their relationships with plants and animals in the past. For example, many indigenous peoples lived as highly mobile, multispecies societies capable of exercising great agency in dealing with environmental changes of the sort we are discussing now. Yet industrial settler containment practices in North America rendered many indigenous peoples immobile, confining them to reservations, treaty harvesting areas, or small islands that provide fewer options for adapting to changes that threaten their land base or shift the habitats of significant plant and animal populations farther away. As in the past, industrial settler campaigns today also obstruct the efforts of indigenous peoples to respond—from legal and diplomatic failures to mitigate dangerous climate change by lowering emissions to the enactment of laws, policies, and bureaucratic institutions that stymie indigenous efforts to adapt within current confines such as reservations (Marino; Whyte; Wildcat).

In all these ways, climate destabilization fits into a larger pattern of a particular kind of anthropogenic environmental change taking place across this brief but disruptive period of settlement. The ecosystems in which we live today are already drastically changed from those to which our ancestors related—a fact which shapes how we approach discussions of Anthropocene futures. Our ways of approaching conservation and restoration, then, are situated at the convergence of deep Anishinaabe history and the vast degradation caused by settler colonial campaigns in such a short time. I think of this junction as our ancestors' dystopia.

Anishinaabe restoration and conservation

Anishinaabek/Neshnabék throughout the Great Lakes region are at the forefront of native species conservation and ecological restoration projects that seek to learn from, adapt, and put into practice local human and nonhuman relationships and stories at the convergence of deep Anishinaabe history and the disruptiveness of industrial settler campaigns. These projects also seek to find ways to reconcile—as much as makes sense—with settler societies so that indigenous and settler conservation can share responsibilities and hold each other accountable. Consider three projects.

Nmé

Nmé (Lake Sturgeon) is the largest and oldest living fish in the Great Lakes basin, sometimes exceeding 100 years in age. Nmé served the Anishinaabek as a substantial source of food, an indicator species for monitoring the environment, and a clan identity, playing a role in ceremonies and stories. However, industrial settler campaigns in the Great Lakes region threatened the stability of nmé populations and the Anishinaabe system through over-harvesting, dams, stocking rivers with nonnative fish species for sport fishing, and pollution. Nmé used to be plentiful, yet are now reduced to less than 1 percent of their historic numbers (Holtgren). Kenny Pheasant, an elder, says, "Decline of the sturgeon has corresponded with decline in sturgeon clan families. Only a few sturgeon clan families are known around here" (Little River Band). The Little River Band of Ottawa Indians sought to restore nmé in the early 2000s.

The Natural Resources Department of the tribe started a cultural context group, composed of a diverse range of tribal members and biologists, which developed goals and objectives for restoration. The cultural context group facilitated "a voice" that "was an amalgamation of cultural, biological, political, and social elements, all being important and often indistinguishable" (Holtgren 135). The goal was to "restore the harmony and connectivity between Nmé and the Anishinaabek and bring them both back to the river … Bringing the sturgeon back to the river was an obvious biological element; however, restoring harmony between sturgeon and people was steeped in the cultural and social realm. Each meeting began with a ceremony, and the conversation was held over a feast" (Holtgren 136). Ultimately, the department created the first streamside rearing facility for protecting young sturgeon before they are released each fall in order to preserve their genetic parentage (Holtgren et al.).

Annually in September, at a public event featuring a pipe ceremony, feast, speeches and education about Ottawa traditions, the Band's sovereignty, and conservation science, attendees use buckets to personally release nmé back to the river. The event can attract hundreds of attendees from all over the watershed. The relationships between people and sturgeon change and become stronger as individuals realize their dependence on nmé and recognize how the deep historical relationship between the fish and Anishinaabek can guide innovate restoration and conservation efforts to improve environmental quality and heal people's relationships in the watershed today. Participants, including many children, begin to feel a sense of responsibility to nmé, developing a lasting connection when holding the fish in an atmosphere framed by Anishinaabe traditions and conservation science. I have talked to attendees who proclaim that they have come to realize through the event that it is people who also depend on nmé. This is especially significant in a watershed where the relationships among people have been strained by settler colonialism. The participants do not necessarily adopt the Anishinaabe way of thinking or living, yet they come to feel a sense of themselves as co-occupants of and relatives in a shared watershed. Success, perhaps, lies in how nmé restoration has changed the relationship between the settler Americans in the Manistee area and the Anishinaabek through situating the project at the convergence of deep Anishinaabe history and industrial settler campaigns (Holtgren et al.).

Manoomin

Wild rice, or *manoomin*, is another important native species for Anishinaabek. Wild rice grows in shallow, clear, and slow-moving waterways and can be harvested in early autumn. After harvesting, manoomin is processed through activities such as drying, parching, hulling, winnowing, and cleaning (i.e., handpicking out any leftover rice husks). Manoomin is rich in vitamins, minerals, and protein contents and is easily stored as a dried good. The origin story tells of how Anishinaabek migrated from the East until they reached the land where food grows on water, or the Great Lakes region. Neighboring groups of US and Canadian citizens and companies engage in activities such as mining, damming, growing commercial paddy rice for mass distribution, and recreational boating. These activities directly affect manoomin and its habitat—especially the interdependency of manoomin and water. Historically, settlers drove prices of manoomin down, making it uneconomical for Anishinaabe to sell it commercially as an alternative to subsistence use (Wallwork). For example, in states such as Minnesota, manoomin abundance has declined by half in the last one hundred years (Andow et al.).

Anishinaabe people are leaders in the conservation of wild rice. The Nibi (water) and Manoomin Symposium, which takes place every two years, brings tribal rice harvesters in the Great Lakes, indigenous scholars, paddy rice growers, representatives from mining companies and state agencies, and university researchers interested in the genetic modification of rice together. Elders share their stories about manoomin and youth share their perspective on how manoomin fits into their futures. Indigenous persons working as scientists in their tribes share the experiences working with elders to understand the deep historical implications of the work they do to study and conserve manoomin. Other indigenous peoples are often invited to share their experiences restoring and conserving other native species, such as taro and maize.

Many at the symposium emphasize the importance of wild rice for the collective well-being of Anishinaabe people. While members of settler society may come to the event understanding manoomin as something of scientific curiosity, or a commodity to sell in a niche market, or a nuisance to the growth of mining, they are instead exposed to the deep historical relationship between Anishinaabek and manoomin and the significance of manoomin for Anishinaabe self-determination today. In the most recent white paper from the symposium, Norman Deschampe, former Minnesota Chippewa Tribal President, speaking of the state of Minnesota and the US federal government, said that "we are of the opinion that the wild rice rights assured by treaty accrue not only to individual grains of rice, but to the very essence of the resource. We were not promised just any wild rice; that promise could be kept by delivering sacks of grain to our members each year. We were promised the rice that grew in the waters of our people, and all the value that rice holds" (Andow et al. 3). All of these actions have formed an important front of manoomin-based conservation that has challenged mining and other polluting activities, research on genetic modification, and the mass distribution of commercial "wild" rice.

Nibi

For Anishinaabek, as exemplified in the previous projects, it is hard to talk about native species conservation without *nibi* (water). Nibi has traditional value for Anishinaabek because it is among the basic elements of Anishinaabe cosmology, as told in the creation story, which frames how community members view their relations to water. Water quality and abundance benefits human and animal health (McGregor, "Honouring"). Anishinaabe scholar Deb McGregor says that:

> We must look at the life that water supports (plants/medicines, animals, people, birds, etc.) and the life that supports water (e.g., the earth, the rain, the fish). Water has a role and a responsibility to fulfill, just as people do. We do not have the right to interfere with water's duties to the rest of Creation. Indigenous knowledge tells us that water is the blood of Mother Earth and that water itself is considered a living entity with just as much right to live as we have.
>
> (McGregor, "Honouring" 37–38)

Water is sacred for Anishinaabek. Though the Great Lakes region has a high proportion of the world's fresh water supply, things have changed in recent times. The waters are now the dumping ground of numerous pollutants. Agricultural runoff, sewage disposal, and contamination from industry, such as mercury, become entangled in the food chain and affect habitats for species important to Anishinaabe, such as fish. Changes in the Great Lakes, including the recent profusion of algal blooms that threatens whitefish populations, are projected to affect the ecological contexts needed for Anishinaabe to harvest first foods.

Anishinaabek are taking collective action to protect nibi. A group of Anishinaabe women began walking around the Great Lakes in the early 2000s, calling their efforts the Mother Earth Water Walk. The purpose is to help people in the basin recognize and re-recognize human relationships to water in its spiritual dimensions instead of seeing water as an inanimate resource. The walks, which take place in the spring, include a water ceremony, feast, and celebration, and the participating grandmothers take turns carrying a water vessel and eagle staff. Similarly, the grassroots women's group Akii Kwe, made up of Anishinaabe women from Walpole Island, "have been diligently trying to protect water in their territory for years. Guided by their traditional responsibilities, they consider it their duty to speak for the water" (McGregor, "Traditional Ecological Knowledge" 107). Akii Kwe members are guided by their knowledge of how to be sensitive to water and care for water, which arises from their living near and attending closely to rivers and lakes (McGregor, "Considerations").

These collective actions by Anishinaabe women are changing decision-making processes in Canada. The Anishinaabek Nation, an indigenous multiparty organization that plays an important role in Canadian politics, created the Women's Water Commission for bringing women's voices into Ontario and Great Lakes water issues. The explicit goal of the commission includes fostering "the traditional role of the Women in caring for water." "Traditional," here, is important not because of pseudo-factual claims about indigenous women's roles, but because it indicates that certain kinds of orientation toward water imply cultural understandings of one's responsibilities to the Earth's living, nonliving, and spiritual beings, as well as criticisms of industrial settler campaigns in the region that have damaged indigenous systems of stewardship. The commission seeks to encourage recognition of traditional responsibilities along with the need to include women as part of the decision-making processes (McGregor, "Considerations"). The Walk too has also spread across North America, becoming a regional form of action that includes more people each year, not just Anishinaabe women alone (McGregor, "Considerations"; Mother Earth Water Walk). Again, these water conservation efforts show how the recent degradation of the Great Lakes looks from a deep Anishinaabe historical perspective.

Nmé, Manoomin, and Nibi

The three restoration and conservation projects just described all take place today in our ancestors' dystopia. In each project, the focus on native species puts in perspective the convergence between deep Anishinaabe stories and histories and the more recent industrial settler degradation of the environment. The projects bring attention to how industrial settler campaigns erase the particular systems of interdependent relationships of humans, nonhumans, and ecosystems that matter to many indigenous peoples, from more historic seasonal rounds to more contemporary structures of self-determination. Participants in the projects learn from, adapt, and put into practice ancient stories and relationships involving humans, nonhuman species, and ecosystems to address today's conservation challenges in the Anthropocene. The value of these local stories and relationships derives from indigenous peoples' knowledge of what it means to survive and flourish in times our ancestors would have likely imagined to be dystopian. They are not based on dread of certain futures; rather, they arise from indigenous perspectives on how to respond to anthropogenic climate destabilization based on having already lived through local losses of species and ecosystems.

It is certainly true that sturgeon or wild rice are not likely to play the same prominent dietary or cultural roles that they once did. Yet the very act of engaging these stories and relationships through restoring and conserving native species brings attention to indigenous perspectives on the harms of settler colonialism and engenders collective actions based on the lessons we can draw from those experiences. These kinds of native species conservation go along with many other forms of adaptation that indigenous peoples do that are often seen as "untraditional," from investing in science education for youth to updating building codes and other tribal infrastructure to engaging in advocacy through direct actions coordinated by social media. It is also true that the water relations of Anishinaabek will not be restored to what they were. But the Mother Earth Water Walk seeks to engender a greater understanding of how the water was and is degraded through industrial settler activities such as pollution and to foster respect for the sorts of responsibilities humans of all nations and heritages must have in order to protect the water now and moving into the future. The projects emphasize that the industrial settler campaigns erase what makes a place ecologically unique in terms of human and nonhuman relations, the ecological history of a place, and the sharing of the environment by different human societies.

These projects acknowledge that before people can work together to grasp the nettles that are conservation problems in the Anthropocene, there needs to be reconciliation among people so that they come to have sufficient appreciation of their different histories and can share responsibilities and be accountable to each other. The sturgeon, wild rice, and water restoration programs feature public events that bring together indigenous people and members of settler society to learn about how humans are entangled with other species and with the environment. These multispecies engagements are not aimed solely at avoiding the physical loss of certain species or ecosystems, but also at building people's appreciation of what it means to share local places in light of how they are implicated in more regional and global forces such as industrial settler colonialism. The conservation of native species, then, is not *only* about restoration for the species' sake in the absence of histories and relationships with other human and nonhuman species, but because the act of conserving *some* native species—the ones that are still around—serves to raise questions about environmental justice and colonialism that too often are marginalized in global discussions of the future.

Note

1 Anishinaabe or Anishinaabek (plural) and Neshnabé and Neshnabék are English spellings that correspond to different accents of the language of Ojibwe, Odawa, and Potawatomi peoples. There are multiple English spelling systems.

References

Andow, David, Theresa Bauer, Mark Belacourt et al. "Wild Rice White Paper: Preserving the Integrity of Manoomin in Minnesota." People Protecting Manoomin: Manoomin Protecting People: A Symposium Bridging Opposing Worldviews. Mahnomen, MN, White Earth Reservation, 25–27 August 2009. Print.

Cafaro, Philip and Richard Primack. "Species Extinction is a Great Moral Wrong." *Biological Conservation* 170 (2014): 1–2. Print.

Evans, Claire J. "Climate Change Is So Dire We Need a New Kind of Science Fiction to Make Sense of It." *The Guardian* 20 August 2015. Web. 10 January 2016.

Holtgren, Marty. "Bringing Us Back to the River." *The Great Lake Sturgeon*. Ed. Nancy Auer and Dave Dempsey. East Lansing: Michigan State University Press, 2013. 133–147. Print.

Holtgren, Marty, Stephanie Ogren, and Kyle Powys Whyte. "Renewing Relatives: Nmé Stewardship in a Shared Watershed." Archive of Hope and Cautionary Tales. North American Observatory, West Cluster, Humanities for the Environment. 1 December 2014. Web. 7 March 2016.

Johnson, Elizabeth, Harlan Morehouse, Simon Dalby, Jessi Lehman, Sara Nelson, Rory Rowan, Stephanie Wakefield, and Kathryn Yusoff. "After the Anthropocene: Politics and Geographic Inquiry for a New Epoch." *Progress in Human Geography* 38.3 (2014): 439–456. Print.

Kareiva, Peter and Michelle Marvier. "What Is Conservation Science?" *BioScience* 62.11 (2012): 962–969. Print.

Kolbert, Elizabeth. "The Anthropocene Debate: Marking Humanity's Impact." *Yale Environment 360*. 17 May 2010. Web. 17 Dec. 2015.

Lefevre, Tate A. "Settler Colonialism." *Oxford Bibliographies in Anthropology*. Ed. J. Jackson. Oxford: Oxford University Press, 2015. 1–26. Print.

Lewis, Simon L. and Mark A. Maslin. "Defining the Anthropocene." *Nature* 519 (12 March 2015): 171–180. Web. 7 March 2016.

Little River Band of Ottawa Indians. *Nmé (Lake Sturgeon): Stewardship Plan for the Big Manistee River and 1836 Reservation*. Special Report 1. Manistee, MI: Natural Resources Department, 2008. Print.

Marino, Elizabeth. "The Long History of Environmental Migration: Assessing Vulnerability Construction and Obstacles to Successful Relocation in Shishmaref, Alaska." *Global Environmental Change* 22 (2012): 374–381. ScienceDirect. Web. 7 March 2016.

McGregor, Deborah. "Honouring Our Relations: An Anishnaabe Perspective on Environmental Justice." *Speaking for Ourselves: Environmental Justice in Canada*. Ed. Julian Agyeman, Peter Cole, and Randolph Haluza-Delay. Vancouver: University of British Columbia Press, 2009. 27–41. Print.

——. "Traditional Ecological Knowledge: An Anishnabe Woman's Perspective." *Atlantis: Critical Studies in Gender, Culture & Social Justice* 29.2 (2005): 103–109. Print.

——. "Traditional Knowledge: Considerations for Protecting Water in Ontario." *The International Indigenous Policy Journal* 3.3 (2012): 11. Print.

"Mother Earth Water Walk." *Mother Earth Water Walk*. Web. 14 April 2015.

Sandler, Ronald. "The Ethics of Reviving Long Extinct Species." *Conservation Biology* 28.2 (2014): 354–360. Print.

Singleton, Benedict. "Anthropocene Nights." *Architectural Design* 82.4 (2012): 66–71. Print.

Thompson, Allen and Jeremy Bendik-Keymer. *Ethical Adaptation to Climate Change: Human Virtues of the Future*. Cambridge, MA: MIT Press, 2012. Print.

Trexler, Adam. *Anthropocene Fictions: The Novel in a Time of Climate Change.* Charlottesville: University of Virginia Press, 2015. Print.

Vaidyanathan, Gayathri. "Can Humans and Nature Coexist?: Conservationists Go to War over Whether Humans are the Measure of Nature's Value." *Scientific American* 10 (November 2014). Web. 1 September 2015.

Wallwork, Deborah, dir. *The Good Life: Mino-Bimadiziwin.* Saint Paul, MN: RedEye Video, 1997. Film.

Weik von Mossner, Alexa. "Science Fiction and the Risks of the Anthropocene: Anticipated Transformations in Dale Pendell's *The Great Bay.*" *Environmental Humanities* 5 (2014): 203–216. Web. 1 September 2015.

Whyte Kyle P. "A Concern about Shifting Interactions between Indigenous and Non-indigenous Parties in US Climate Adaptation Contexts." *Interdisciplinary Environmental Review* 15 (2014): 114–133. Inderscience Online. Web. 7 March 2016.

Wildcat, Dan. *Red Alert! Saving the Planet with Indigenous Knowledge.* Golden, CO: Fulcrum, 2009. Print.

22

COLLECTED THINGS WITH NAMES LIKE MOTHER CORN

Native North American speculative fiction and film

Joni Adamson

Many environmental humanists, seeking to prepare students for unfolding social and environmental crises, have embraced speculative fiction and film (Pérez-Peña). Margaret Atwood, author of the trilogy of novels *MaddAddam* (2013), and one of the best recognized authors of the genre, employs speculative fiction because it allows her to address not just climate change but a wide range of global ecological transformations she prefers to call "*everything change*" (Finn). Atwood playfully imagines genetically altered "bio-beings" living in a plausible near future. All of the biological entities in her novels "already exist, are under construction, or are possible in theory" (see the acknowledgments page for the last book of the trilogy, *MaddAddam*). Other writers of the genre explore the future of food in densely overpopulated, dystopian worlds. Harry Harrison's novel *Make Room! Make Room!* (1966) and its film adaption, *Soylent Green* (1973), imagine a New York City with forty million people. The wealthy still have access to fresh food, while the poor have never tasted a fruit or vegetable, and live on a green wafer mass-produced by the Soylent Corporation. In the iconic last scene, Charlton Heston, playing a detective investigating corruption in the corporation, learns, to his horror, that the wafer is not made from plankton but from recycled human flesh. Similar anxieties about the future are explored in Paolo Bacigalupi's speculative fiction *The Windup Girl* (2009), in which "calorie companies" control the world's food systems by "genehacking" existing seeds, as their agents comb the planet searching for seeds thought to be extinct. The future of food, as these examples show, plays a central role in much speculative fiction of the last fifty years.

Many contemporary Native North American fictions and recent experimental films, including works by Diane Glancy (Cherokee), Leslie Marmon Silko (Laguna), and Nanobah Becker (Navajo/Diné), also explore concerns surrounding control and access to food and genetic resources. In United Nations and other international legal instruments, this notion is called "food sovereignty," meaning the "ability of peoples, families, countries, and communities to control their own food systems," and has been a topic of international concern since the late 1940s (for a fuller history, see Adamson, "Seeking the Corn Mother"

232–234). These novels emerge out of decades of dedicated global organizing not only for human and civil rights, but also for rights to cultural and food sovereignty, as expressed in Article 31 of the United Nations Declaration on the Rights of Indigenous Peoples (UNDRIP): to "maintain, control, protect and develop their intellectual property" of "genetic resources, seeds, medicines, [and] knowledge of the properties of fauna and flora" (for a history of trans-American indigenous political organization, see Adamson, "'Todos Somos Indios!'"). The work of Glancy, Silko, and Becker, which will be the focus of this chapter, features characters innovating adaptive strategies connected to Mother Corn, as a plant, a seed, and a food, in times when "everything has changed." However, this fiction and film is typically not categorized as speculative, but rather as historical fiction because it is judged to be set in the "past." However, as Indigenous Studies professor Grace Dillon (Anishinaabe) argues in her introduction to an anthology of indigenous science fiction, *Walking the Clouds*, indigenous writers, from North America to New Zealand, tend not to set their work in either the past or the future. Rather, many weave stories from traditional culture into narratives set in present or past, but the result is an engagement with forms of "space-time thinking" that have been described as a subgenre of speculative fiction identified as "Native slipstream" (3). Writers of Native slipstream view time as "pasts, presents, and futures that flow together like currents in a navigable stream" (3). Native speculative fictions embrace the "science" in "science fiction" as "indigenous scientific literacies" rather than the modern or futuristic technologies that are typically the focus of science fiction or speculative fiction (Dillon 7–8). Writers of Native speculative fictions raise questions about the ways science fiction, as a genre that emerged "in the mid-nineteenth-century context of evolutionary theory and anthropology," is "profoundly intertwined with colonial ideology" (Dillon 2). Dillon argues that in many cases, nineteenth-century science fiction can be shown to be a direct mapping of imperialist concerns about "competition, adaption, race, and destiny" (Dillon 2). Native speculative fiction, as a subgenre of science fiction, "defies neat categorization" because of its concern with "nonlinear thinking about space-time" and directly confronts these imperialist constructions (Dillon 3).

Dillon explains that "Indigenous futurisms" involve characters determining how communities "affected by colonization" are adapting to the "post-Native Apocalypse" (10). As I explained in *American Indian Literature, Environmental Justice, and Ecocriticism*, indigenous communities have long understood themselves to be living in a post-colonial "apocalypse" and hundreds of years of resistance should push the advent of the movement termed "environmental justice" and said to have originated in the 1980s, back at least to the late fifteenth century, when indigenous uprisings and slave revolts were occurring on every continent (47–53). In indigenous fictions and films, argues Dillon, characters who resist injustices are often imbued with the traits of ancient supernatural beings understood to represent "interconnected relationship among all 'persons'" (7). In Native North American fictions, "persons" are often suggested to be transformational humans, bears, whales, dolphins, or tricksters living in the modern world. These persons—bears, snakes, coyotes, ravens, and other more-than-human entities—occur frequently in Native North American oral traditions and "illustrate a world view in which no sharp lines can be drawn dividing living beings" (Adamson, "Why Bears" 31). "Persons" may sometimes be animal and on other occasions human. They often exhibit mysterious abilities and qualities, for example, a bear's ability to hibernate, or a snake's ability to shed a skin, that suggest they may at once be symbols "of both death and life" ("Why Bears" 31–32). As symbolic anthropologists and cultural theorists from Claude Lévi-Strass to Victor Turner to Donna Haraway to Eduardo Viveiros de Castro have long established, "transformational persons" represent the "peculiar unity of the liminal: that which is neither this nor that and yet both" (Turner 99). They are singled out, "not because they are 'good to eat' or 'good to prohibit' but because they are 'good to think'" (Babcock 167; Adamson, "Why Bears" 32). Donna Haraway takes this formulation even further to say that transformational beings are "collected things" that are not only good to think because of their liminal characteristics, but, like dogs and other species that live in relation to humans, they are "good to live with" (160).

Following the language of Quiché Mayan scribes who translated the Popol Vuh into Spanish, I refer to transformational characters as "seeing instruments" who, like the gods of the Popol Vuh, work as a "complex navigational system for those who wish to see and move beyond the present" (*American Indian Literature* 145; *Popol Vuh* 32, 71). Whether "human or animal in form or name," these seeing instruments behave "like people, though many of their activities are depicted in a spatiotemporal framework of the cosmic, rather than mundane, dimension" (Adamson, "Why Bears" 45n11). When represented in contemporary fiction, they work in a time-space continuum of past, present, future, because they simultaneously comment on the "relic" or ancient story and the new genre of the novel in order to speculate about the present and the future in a rapidly changing world. In short, they work as thinking instruments that allow humans to "see" what might be done in a situation not only to survive, but also to flourish (Adamson, "Why Bears" 36; *Popol Vuh* 236).

With reference to the work of Eduardo Viveiros de Castro and Marisol de la Cadena, I have also explored how transformational entities referred to by Viveiros de Castro as "persons" become, in contemporary indigenous poetry, fiction, and film, an imaginative force for thinking about how the Earth was formed through a series of relations, often predatory and violent, over geologic and biospheric time. These "persons" offer insights into a "multiverse"; rather than a "universe," or one universal world, the world is actually multiple worlds, at multiple scales, from the microscopic to the cosmic (Adamson, "Why Bears" and "Source"). These entities were the subject of a 2015 conference organized by Viveiros de Castro, "*Os Mil Nomes de Gaia*/The Thousand Names of Gaia," where Donna Haraway explained how they name "collected things" that give presence to an entangled myriad of temporalities and spatialities, or "entitites-in-assemblages—including more-than-human, other-than-human, inhuman" (160).

Notions of a "post-Native Apocalypse" also grow out of cosmologies that hold that Mother Earth is an "entity-in-assemblage" who will remember acts of mass violence. This philosophy was drafted into the 2010 Universal Declaration on the Rights of Mother Earth (UDRME) that emerged from the World Peoples' Conference on Climate Change held in Cochabamba, Bolivia. UDRME attributes to indigenous peoples, nations, and organizations an ancestral indigenous "cosmovision" that understands the Earth as a "living being with whom [all persons] have an indivisible and interdependent relationship" (Article 2). Delegates decried "aggression toward Mother Earth" as an assault on "Us," a word that is interpreted to mean all human groups (not just indigenous) *and* other "persons" including "soils, air, forests, rivers, [and] lakes." In an essay in *Nature*, climate scientists Simon L. Lewis and Mark A. Maslin present scientific evidence of human aggression toward other humans and other species suggesting that indigenous cosmologies about Earth's "memory" may not be mere romanticized mythology. They propose the year 1610 CE as the date that Arctic ice core samples show the lowest point in a decades-long decrease in atmospheric carbon dioxide levels. They deduce that this change was caused by the deaths of over fifty million indigenous residents of the Americas in the first century after European contact, as the result of "exposure to diseases carried by Europeans, plus war, enslavement and famine" (Lewis and Maslin 175). The decimation of the indigenous population, which left only an estimated six million survivors on the northern and southern American continents by the mid-seventeenth century, meant a significant decline in farming, fire-burning, and other human activities that affect atmospheric carbon levels.

Lewis and Maslin's study suggests that human aggression may indeed be "remembered" in the stratigraphy of Mother Earth. Other evidence of a "post-Native Apocalypse" (Dillon 10), although Haraway does not use this neologism, can be found in histories of "rapid displacement and reformulation of germ plasm, genomes, cuttings, and all other names and forms of part organisms and of deracinated plants, animals, and people" (Haraway 162n5). Haraway and her frequent collaborator, Anna Tsing, suggest another neologism for a period of transition out of the Holocene—the "Plantationocene," a term that also suggests the memory of massive numbers of deaths due to colonization and the rise of the plantation system and the slave trade (Haraway 162n5). The Holocene, Tsing explains, was a long epoch in which humans and plants co-evolved in ways that gave rise to many innovations, including agriculture, on a scale that still allowed time and space for what she terms "refugia," or refuges that still sustained "reworlding in rich cultural and biological diversity" even during and after times of ecological catastrophe. For over ten thousand years, humans engaged in innovation, hybridization, and development of thousands of varieties of new staple crop seeds, as well as the collection and selection of particular traits of fruits and vegetables. What changed with the transition into the "Plantationocene" was the scale of movement of "material semiotic generativity around the world for capital accumulation and profit" (Haraway 162n5). Tsing adds that boundaries, or "inflection points," that have been crossed have put the ability of the Earth to protect and provide the necessary conditions to maintain diverse species assemblages (with or without people) at risk (Haraway 160; Tsing). This helps to explain why UDRME declares that "persons," including "Pachamama," the "Mother Earth" of indigenous cosmologies in Latin America, deserve rights that mirror human rights. Just as humans have a "right" not to be murdered, according to the Declaration on Human Rights, earth beings and other species assemblages deserve rights and protections, through time and space, "to regenerate … biocapacity and continue … vital cycles" (UDRME, Preamble).

In this chapter, I will focus on speculative fiction and film by Glancy, Silko, and Becker that enriches understanding of characters who are depicted as having the traits of transformational beings or "collected things." In an analysis of one of these transformational being, Mother Corn, I will explore how indigenous "seeing instruments" promote deeper understandings of biodiversity, cultural diversity, and refugia, and why a "cosmopolitical" movement is forming around advocacy of "rights" for seeds and other diverse assemblages critical to the maintenance of food sovereignty for all "persons"—indigenous and nonindigenous, human and nonhuman (see also Adamson, "Seeking" and "'Todos Somos Indios!'"). This analysis of what Native North American novelists and filmmakers wrap tightly in the husk of the "Mother Corn" will also challenge the argument that indigenous understandings of biodiversity promote romanticized notions of "Nature" as "pure" or "balanced," and that indigenous opposition to transgenics is based on their incommensurability with traditional indigenous cultural values. Rather, I will show that indigenous arguments against GMOs are not based on notions of "balanced nature," but on entities such as Mother Corn that are understood as "collected things" in all their complex assemblages and relationships. I will also touch briefly on new environmental humanities approaches, including biosemiotics, multispecies ethnography, and material ecocriticism, and offer insights into indigenous scientific literacies that promote "scenario-imagining" for those who "wish to see and move beyond the present."

Collected things: Mother Corn

To the botanist, corn is maize, or *Zea mays*, but to the indigenous peoples of North America, "Mother Corn" is a descendent of the progenitor that, according to molecular biologists and ecologists, is called "teosinte," a wild grass. In the Nahuatl (Aztecan) language, this word means "mother of corn." This ancient grass can still be found growing near Mexico City at the edges of corn fields where indigenous farmers still credit it today with cross-pollinating and reinvigorating their modern seed (Nabhan 143, 149). Cross-pollination between wild and domesticated varieties of seed offers a vivid example of Tsing's description of the rich "reworlding" capabilities of "refugia" that support resilience and regeneration through vast spans of geological space and time.

Diane Glancy's *Pushing the Bear* (1996) represents the experience, inventiveness, and hard work of indigenous Cherokee peoples whose cultures and agro-ecological knowledges co-evolved throughout the Holocene. They, and other tribal groups in Central and North America, developed thousands of varieties of this ancient grass into grain cultivars, each named for a particular ecological niche: Oaxaca Green, Hopi Blue, Pawnee Black Eagle, Cherokee White Eagle, Iroquois White. Each of these varieties of corn seed was shared and traded along ancient routes that reached from Mexico to the Great Lakes, and very quickly after encounters between Europeans with the first peoples of the Americas, corn was being planted around the world (Adamson, "Seeking"; Kleindienst 228–29). In the "post-Native Apocalypse," however, many of these varieties have been lost because the people they were associated with were removed from the lands where the seeds were developed. Today, many ancient varieties, hidden away by "seed keepers," are being rediscovered and grown so that they may continue to evolve in relationship with their evolving pests and changing weather conditions (see Nabhan on "gene banks" and "seed collection").

Glancy illustrates what is at stake when the relationship between indigenous peoples and the seeds they have developed is put at risk, and why agro-ecological knowledge as a form of cultural expression is so vigorously defended in UNDRIP and UDRME. *Pushing the Bear* tells the story of the Cherokee on the Trail of Tears, from the moment they are forcibly torn from their farms in the 1830s, through the wrenching days of hunger they experience as they are driven to Oklahoma, mourning the deaths of loved ones and lamenting the loss of their seed corn. The novel is set in the nineteenth century, but rather than being a historical novel, it can be categorized as speculative fiction, since, for the main character, Maritole, a young mother, this *is* the "post-Native Apocalypse," and she is forced to face a dramatically changed future, to adapt and to change as she struggles to survive (Dillon 10). At every step along the way, Maritole remembers the place she was born and calls upon stories from the oral tradition as "seeing instruments" that provide insight into the way this character understands the world around her: "The cornstalks were our grandmothers. ... Their voices were the long tassels reading the air" (4).

The introduction of biosemiotics and multispecies ethnography into the environmental humanities facilitates analysis of "long tassels," or corn silk, "reading" the air as an "entity-in-assemblage" that reveals the semiotic abilities and multiversal worlds of plants constantly perceiving and interpreting their environments through chemical gradients or intensities of light. This, in turn, points to their relationship to Pachamama—the entity understood by many indigenous groups in the Americas to mean "Source of Light" (de la Cadena 335, 350), and to corn as one of the "planet-changing" lifeforms, in this case grasses, that tell of the "spread of seed-dispersing plants millions of years before human agriculture" (Haraway 159). Envisioned by Maritole as a very old sentient being, corn as a "grandmother" evokes this thousands-of-years-in-the-making natural history that resulted in a plant that engages in photosynthesis, turning carbon into sugar and water into oxygen. Co-evolving in relationship with human agro-ecological domestication activities, teosinte, a grass, became a "collected thing," Mother Corn. The "silk" and "tassels reading the air" in this passage call upon environmental humanists, in the words of Deborah Bird Rose, to engage in readings that pay more attention to "situated connectivities that bind us into multi-species communities" (87).

A material ecocritical reading of the words of Maritole's dying father further suggests Mother Corn to be a complex "situated connectivity." Far away from his tight log cabin and thriving farm, he fears for the future of his family if they cannot regain the food security their seeds provide. He cries, "Corn! That's what we eat. We can't live without corn. It's our bodies. Our lives" (79). As Acoma Pueblo writer Simon Ortiz reminds his readers, "earth/corn" is regarded as sacred because it regenerates the human body when people eat the plant; when humans die, their bodies return to the soil, where they feed growing plants (346). Scientifically, corn is one of the very few drought-tolerant plants that creates a compound of four carbon atoms, rather than the three created by most other plants (Pollan 21). Corn takes in more of the isotope C13 than other carbon-hungry plants. This "signature" C4 has been found in the flesh and hair of Central American peoples who have consumed corn for hundreds of years. Most modern people today eat so much corn, in the form of processed foods and sodas sweetened with corn syrup, that like the people who are ethnically Central American, they too are becoming "people of the corn" (Pollan 19). It is now quite correct, writes Michael Pollan, to call most North Americans "corn walking" (23).

All of this makes "Mother Corn," a grass domesticated to become one of the world's most important foods, a "person" who embodies multiversal, ethnobotanical, biotechnological, ecological, social, cultural, and culinary relationships to humans.

Inflection points: *Gardens in the Dunes*

Like *Pushing the Bear*, Leslie Marmon Silko's *Gardens in the Dunes* (1999) is set in the nineteenth century, and might therefore be categorized as a historical novel. However, the main character, Indigo, is a young indigenous girl living in the Sonoran Desert on the Arizona/California border who is captured by the US calvary and forcibly taken away from her home to boarding school. For her, it *is* the "post-Native Apocalypse" (Dillon 10). This turns the novel from a purely historical one into a speculative one that explores a question very like the one formulated by Haraway about the "Plantationocene": when "do changes in degree become changes in kind, and what are the effects of bioculturally, biotechnically, biopolitically, historically situated people (not Man) relative to, and combined with, the effects of other species assemblages and other biotic/abiotic forces?" (159). Silko presents two characters with different approaches to biodiversity and agriculture, who offer insights into issues surrounding the multiply scaled "worlds" and "inflection points" that Haraway's question addresses. Indigo is taught by her grandmother to plant polycultural groupings of corn, beans, and squash in traditional rain-irrigated gardens common among Native North American tribal groups in the Southwestern regions of North America. Her adoptive father, Edward Palmer, is depicted as a landed gentleman and botanist involved in secret dealings with the US Bureau of Plant Industry in Washington, DC, and the Kew Botanical Gardens in Britain. He pursues orchids, citron, and grains of all kinds in the practice of "economic botany," a term that emerged in the nineteenth century when Europeans and Americans were collecting plants deemed valuable by the indigenous peoples and ethnic groups they encountered in their colonial endeavors (Mushita and Thompson 27).

After Indigo is captured by territorial authorities and forced into a California Indian Boarding School, then adopted by Edward and his wife, she is taken on a tour of Europe's most famous botanical gardens. In England, she meets her adoptive mother's British relative, Aunt Bronwyn, who has planted corn, beans, squash, and tomatoes in her "kitchen garden." When Indigo sees foods "she knew," she begins to understand that "seeds must be among the greatest travelers of all" (Silko 240, 291). Read in the context of Gary Paul Nabhan's *Where Our Food Comes From*, a study of the fieldwork of Russian botanist Nikolay Vavilov (1897–1943), a founder of modern crop breeding who traveled the world searching for new seeds and novel ways for "humanity to feed itself" (9), Indigo's story begins to mirror his. Like Vavilov, Indigo carefully follows her grandmother's advice to collect, save, and "carry home" as many seeds as possible, plant them, and experiment to see which of them will grow in new conditions and weather (Silko 83). Grandmother is advising Indigo to follow time-tested agro-ecological methods of innovation, the basis of which is free exchange, hybridization, and domestication in order to increase botanical biodiversity.

Edward's mission is not to "feed humanity." He travels with a "list of plant materials desired by his private clients," who are "wealthy collectors" (Silko 129). Without permission, he smuggles seeds and plant materials away from their original cultivators. His activities reveal the links between colonization, imperialism, displaced farmers, and biodiversity loss that have been called "biopiracy" (Mushita and Thompson). They mirror the work not of Vavilov, but Henry Wallace, founder of the American company Hi-Bred Corn. In the late nineteenth century, Wallace and other North American corn breeders collected hundreds of corn seed varieties that had been domesticated all over Central and North America. Promising their customers unlimited yields and progress on an infinite scale, they selected only a few of these "primitive cultivars" as the focus of their research. By the twentieth century, their work resulted in the development of select "advanced or elite" cultivars, patented as private property, and sold for profit. Wallace proudly advanced the theory that hybridized varieties of corn proved that the white man had done more for corn, more quickly, than the indigenous domesticators of the seed over millennia. However, later, he began expressing alarm as he realized that monocropping elite patented cultivars was leading to an erosion of biodiversity, and lost varieties of corn would lead to lost opportunities for cross-pollination and reinvigoration (Fussell 86). By 1956, Wallace was predicting impending disaster and warned researchers to think more carefully about the head-long rush to increase yields and profits. The disaster occurred in 1970, writes Fussell, when genetic uniformity of commercial hybrid corn led to the loss of one-fourth of the US crop, which was discovered to be "susceptible to a new mutant strain of southern corn-leaf blight" (93).

When Indigo returns from Europe with the seeds she has collected, she returns to her grandmother's gardens in the Sonoran desert. There she grows drought-tolerant "Mother

Corn" at the center of a garden in which she plants all her newly collected seeds (Silko 166). As used by Atwood and others, speculative fiction often suggests some degree of extrapolation from current technologies and social structures to forms that do not (yet) exist. In *Gardens in the Dunes*, Silko employs a collected thing, Mother Corn, and other seeds, as a "current technologies" with long ethnobotanical histories to delve into the inflections between epochs and seek deeper understanding of the space, time, and scale it takes for both evolutionary processes and botanical development of genetically stable staple crop seeds and drought-tolerant foods in modern times. When Indigo arrives home, she immediately begins experimenting with seeds that have never been planted in her home region to see which will grow most successfully. Like Haraway and Tsing, Indigo's garden, and her experimentation with new seed "technology," asks whether or not it might be possible to "reconstitute refuges" (Haraway 160). At one time, over three thousand plant species were used by humans for food, but today only one hundred and fifty are cultivated, and of these, only a few, including transgenic corn, rice, and soy, produce half the world's food. Since the late 1800s, not one new, nontransgenic food has been genetically stabilized. Indigo's story, like the story of her historically real counterpart, Vavilov, calls upon modern readers to better understand where their food comes from and why contemporary geneticists and plant breeders are returning to the collection of wild relatives of important modern foods, like ancient teosinte (Mother Corn) and modern varieties of corn, as they plan for drought and disease in view of pervasive ecological change.

Grace Dillon argues that Native American "science fiction" is renewing, recovering, and extending "First Nations peoples' voices and traditions" (2). *Gardens in the Dunes* deepens understanding of the reasons that Article 5 of UNDRIP calls for protection of biodiversity and maintaining landbases where indigenous breeders and farmers, as they have for centuries, are employing seed diversity and a variety of "technologies to adapt with resilience to shifting conditions" (Nabhan 198).

Interrogating the ethics of biotechnologies: *The Sixth World*

In *The Sixth World* (2011), emerging Navajo director and screenwriter Nanobah Becker debuts a fifteen-minute speculative film in which corn is genetically modified to produce oxygen for astronauts to breathe as they travel to Mars. The main character, Tazbah Redhouse, played by Jeneda Benally (Navajo), is a Navajo astronaut on a mission jointly funded by the Navajo Nation and a futuristic world corporation. While Redhouse is still on Earth training for her launch, she dreams about a traditional cornfield that withers and dies. Becker notes that ideas for the film were influenced by Michael Pollan's argument in *The Omnivore's Dilemma* about why genetically altered monocultural agribusiness presents a frightening prospect for the future of global food systems (Estrada). Tazbah's dream raises questions about the risks to community food sovereignty presented by monocropping of patented corn on farms required by corporate patent contracts to use synthetic fertilizers and pesticides—chemicals that are degrading ecosystems and contributing to biodiversity loss.

As foreshadowed by Tazbah's dream, the mission to Mars almost fails when the genetically modified corn that powers the ship's oxygen system fails to thrive. Before take-off, however, Redhouse received two traditional varieties of Navajo corn—yellow and white—given to her by her commanding officer General Bahe, played by Roger Willie (Navajo). With these two traditional seed varieties, she leads her fellow scientist, Dr. Smith, played by Luis Aldana, to save the spacecraft's critical systems. The traditional corn thrives, the oxygen system is rebooted, and Redhouse is able to go into "hibernation" for the remainder of the trip. Given that the film is set on a space ship, and the mission is co-funded by a biotechnology corporation, Becker's film does not imply simplistic opposition to technology. The astronauts carefully keep the traditional corn and the genetically modified corn separated by glass. The glass nods to efforts to avoid genetic contamination between traditional and transgenic corn, since corn is a wind-pollinated plant and the genetic information of transgenic corn could potentially spread though open pollination and contaminate the natural viability of traditional varieties. It is not known if transgenic organisms will actually contribute to the loss of valuable maize traits (McAfee). Most indigenous advocates of food sovereignty "oppose genetic engineering as a strategy for agriculture not because it affronts an idealized 'harmony with Nature' or the 'purity' of traditional crop landraces" but because it introduces the threat that living organisms can be patented and a loss of seed crop biodiversity will follow (McAfee 34–35).

While in her hibernation pod, Tazbah dreams of herself as Changing Woman, a Navajo transformational entity, and "Mother of Corn," and sees herself chanting in traditional dress with a red stripe of paint across her face. In an interview with Gabriel Estrada, Becker notes that she wanted to give Tazbah "a warrior look" and encourage her audience to think of themselves as having political agency (Estrada). As a "collected thing," Changing Woman calls the audience into the long history of cosmopolitical organizing around cultural and ethnobotanical sovereignty. This history has included indigenous groups and nations working not only with the UN Permanent Forum on Indigenous Issues, but also with the United Nations Environmental Programme's Convention on Biological Diversity (CBD) and diverse non-governmental organizations (NGOs) that focus on food sovereignty. Together, representatives of all these groups are entering into alliances with small-scale indigenous farmers, plant breeders, ethnobotanists, microbiologists, progressive food policymakers, and genomic scientists to maintain food sovereignty and seed diversity (Adamson, "Seeking").

In the last scene of the film, Tazbah awakens safely on Mars, and Becker pans across green fields of Navajo corn growing successfully in the red Martian landscape. In the interview with Estrada, Becker explains that in Navajo emergence stories, "Navajos and their earlier spiritual ancestors find their way through several worlds, each time making their way through the sky of each world into the next colored-world." It is said that the present time period is the Fifth World, and so the awakening on Mars signals an emergence into an imagined Sixth World, one that suggests, like Silko's novel, that it might be possible to "reconstitute refuge" and employ seed diversity to adapt to change with resilience.

Environmental humanities scholars and the public have been drawn to speculative fiction not for its focus on the "end of the world" but because it "allows a kind of scenario-imagining" for plausible, survivable futures (Pérez-Peña). Native North American fiction and film should be read as speculative art, replete with biosemiotic, ethnobotanical, and multispecies information that may spur students to imagine, narrate, design, grow, or engineer scenarios for adapting and surviving in what the writers and filmmakers analyzed here are suggesting has long been a "post apocalyptic world" in which indigenous communities, and nonindigenous communities as well, are living, adapting, and innovating in the face of accelerating "everything change."

References

Adamson, Joni. *American Indian literature, Environmental Justice, and Ecocriticism: The Middle Place*. Tucson: University of Arizona Press, 2001. Print.

——. "Seeking the Corn Mother: Transnational Indigenous Community Building and Organizing, Food Sovereignty and Native Literary Studies." *We the Peoples: Indigenous Rights in the Age of the Declaration*. Ed. Elvira Pulitano. New York: Cambridge University Press, 2012. 228–249. Print.

——. "Source of Life: Avatar, Amazonia, and an Ecology of Selves." *Material Ecocriticism*. Ed. Serenella Iovino and Serpil Opperman. Bloomington: University of Indiana Press, 2014. 253–268. Print.

——. "'Todos Somos Indios!': Revolutionary Imagination, Alternative Modernity, and Transnational Organizing in the Work of Silko, Tamez and Anzaldúa." *Journal of Transnational American Studies* (May 2012): 1–26. Print.

——. "Why Bears are Good to Think and Theory Doesn't Have to Be Another Form of Murder: Transformation and Oral Tradition in Louise Erdrich's *Tracks*." *Studies in American Indian Literatures* 4.1 (Spring 1992): 28–48. Print.

Babcock, Barbara. "Why Frogs are Good to Think and Dirt is Good to Reflect On." *Soundings* 58.2 (1975): 167–181. Print.

Becker, Nanobah, dir. *The Sixth World: An Origin Story*. Los Angeles: Future Estates Production, 2012. Film. Web. 15 August 2015.

de la Cadena, Marisol. "Indigenous Cosmopolitics in the Andes: Conceptual Reflections beyond 'Politics,'" *Cultural Anthropology* 25 (2010): 334–370. AnthroSource. Web. 15 August 2015.

Dillon, Grace, ed. *Walking the Clouds: An Anthology of Indigenous Science Fiction*. Tucson: University of Arizona Press, 2012. Print.

Estrada, Gabriel. "Navajo Sci-Fi Film: Matriarchal Visual Sovereignty in Nanobah Becker's the 6th World." *Journal of the American Academy of Religion* 82 (2014): 521–530. Print.

Finn, Ed. "An Interview with Margaret ATwood." *Slate Magazine* 6 February 2015. Web. 15 August 2015.

Fussell, Betty. *The Story of Corn*. Albuquerque: University of New Mexico Press, 1992. Print.

Glancy, Diane. *Pushing the Bear*. New York: Harvest Book, 1996. Print.

Haraway, Donna. "Anthropocene, Capitalocene, Plantationocene, Chthulucene: Making Kin." *Environmental Humanities* 6 (2015): 159–165. Web. 15 August 2015.

Kleindienst, Patricia. *The Earth Knows My Name: Food, Culture and Sustainability in the Gardens of Ethnic Americans*. Boston: Beacon Press, 2006. Print.

Lewis, Simon L. and Mark A. Maslin. "Defining the Anthropocene." *Nature* 519 (12 March 2015): 171–180. Web. 15 August 2015.

McAfee, Kathleen. "Corn Culture and Dangerous DNA: Real and Imagined Consequences of Maize Transgene Flow in Oaxaca." *Journal of Latin American Geography* 2.1 (2003): 18–42. Print.

Mushita, Andrew and Carol B. Thompson. *Biopiracy of Biodiversity: Global Exchange as Enclosure*. Trenton NJ and Asmara, Eritrea: Africa World Press, 2007. Print.

Nabhan, Gary. *Where Our Food Comes From: Retracing Nikolay Vavilov's Quest to End Famine*. Washington, DC: Island Press. 2009. Print.

Ortiz, Simon. *Fight Back: For the Sake of the People, For the Sake of the Land. Woven Stone*. Tucson: University of Arizona Press, 1992. 285–365. Print.

Pérez-Peña, Richard. "College Classes Use Arts to Brace for Climate Change." *New York Times*. 31 March 2014. Web. 4 April 2014.

Pollan, Michael. *The Omnivore's Dilemma: A Natural History of Four Meals*. New York: Penguin, 2006. Print.

Popol Vuh: The Definitive Edition of the Mayan Book of the Dawn of Life and the Glories of Gods and Kings. Trans. Dennis Tedlock. New York: Touchstone, 1985. Print.

Rose, Deborah Bird. "Introduction: Writing in the Anthropocene." *Australian Humanities Review* 49 (2009): 87. Web. 15 August 2015.

Silko, Leslie Marmon. *Gardens in the Dunes*. New York: Simon and Schuster, 1999. Print.

Tsing, Anna. "Feral Biologies." Anthropological Visions of Sustainable Futures. University College London, February 2015. Presentation.

Turner, Victor. "Betwixt and Between: The Liminal Period in *Rites de Passage*." *The Forest of Symbols: Aspects of Ndembu Ritual*. Ithaca: Cornell University Press, 1967. 93–111. Print.

United Nations Declaration on the Rights of Indigenous Peoples. The United Nations Permanent Forum on Indigenous Issues Homepage. Web. 1 June 2009.

"Universal Declaration on the Rights of Mother Earth." World People's Conference on Climate Change and the Rights of Mother Earth. Web. 17 September 2010.

Viveiros de Castro, Eduardo Batalha. "Cosmological Deixis and Amerindian Perspectivism." *Journal of the Royal Anthropological Institute* 4 (1998): 469–488. Print.

23

THE STONE GUESTS

Buen Vivir and popular environmentalisms in the Andes and Amazonia

Jorge Marcone

A crucial question to ask in any serious reflection on our times is undoubtedly: "How are human identities and responsibilities to be articulated when we understand ourselves to be members of multispecies communities that emerge through the entanglements of agential beings?" (Rose 3). The question is, of course, a recognizable item in the agenda of the environmental humanities, although it undermines two basic assumptions for defining the objects of study of the humanities and the social sciences: the opposition between subject and object, by which the properties of the subject, such as agency, intentionality, and communication, are denied to the object; and the split between society and nature, which leads to either explaining society with the laws that rule nature, or conversely, explaining society as a domain autonomous from nature. These issues have come up too, loud and clear, in the blare of popular social movements' battles against actual, perceived, or potential environmental injustices. In these struggles for territorial rights, access to and control of natural resources, and equity regarding environmental services, the fight for legitimizing other ontologies and axiologies regarding the interrelationships between humans and nonhumans has taken its own place in the public spaces where political confrontations and negotiations occur. These beliefs and values are raised by indigenous and nonindigenous peoples at the front line of climate change impacts and of the consequences of extractive development policies enforced by right-wing and left-wing governments alike.

Beyond their battlegrounds, popular environmentalisms show that the goals and aspirations of the environmental humanities belong too to other actors outside academia. In these cases, however, it would be counterproductive to reflect on those ontologies and values separately from the movements and struggles where they originate and, especially, evolve. We must understand the complexity of popular environmentalisms' local, national, and transnational political scenarios; the alliances and exchanges that are demanded by such complexity; the balance between restoration of traditional knowledge and technological and social innovation; the fluidity of their identities as their struggles are carried out; the self-imposed respect for the diversity within and among the movements; and the prominence of unexpected links that are revealed by the study of such complexity. For instance, the ordeal of such movements and their ontologies calls our attention too to a very particular key problem: even if philosophical resolutions, or revolutions, were to be reached regarding posthumanist political ontologies, their viability in countries which rely heavily on extractivism depends on a great effort of imagination and will by nation-states or plurinational states (depending on the country) for figuring out actual alternative economic policies. This plea for considering the historical contexts of popular environmentalisms when interpreting and valuing their ontologies and axiologies should function at least as a warning against reifying popular environmentalisms into just another *indigenous ontology*, where the adjective "indigenous" carries the essentialist meaning of "fixed non-modern," and the ontology has become a sort of stone guest in an academic philosophical debate.

The situation of popular environmentalisms in Latin America can be illuminated with the old Spanish expression of the *convidado de piedra*, or "stone guest"; that is to say somebody who has been invited to a negotiation out of a sense of duty, but who does not participate in the discussion, or whose participation is ignored by the rest. Because of international agreements such as Convention No. 169 on Indigenous and Tribal Peoples by the International Labor Organization, states ask these movements to participate in the decision-making processes regarding development projects. The Convention requires that indigenous and tribal peoples be consulted on issues of policy, governance, and development that affect them in free, prior, and informed participation. The ultimate goal of the consultation is achieving agreement and *appropriate solutions* in an atmosphere of mutual respect and full participation. However, are these popular environmentalisms invited to the negotiation table because their opinion will truly be heard? Or do their hosts actually prefer to treat them as stone guests with no voice, even if they speak?

In any event, it is becoming clear that these stone guests are neither mute nor passive; quite the contrary, they are playing important dramatic roles, just like that of the stone guest figure in the Spanish Golden Age theater of the sixteenth and seventeenth centuries. In several versions of the Don Juan legend, this well-known character invites the statue of a man offended or wronged by Don Juan's actions to a dinner where Don Juan celebrates his triumphs. These stone guests, however, end up being ghosts who drag Don Juan down to death. Although Convention No. 169 does not grant indigenous and tribal peoples a right to veto a governmental initiative, popular environmentalisms in Latin America have nevertheless become stone guests who, unexpectedly, threaten their hosts at the negotiation table. On city streets and rural highways, in mass media and social networks, and in what seems to have become their preferred visual genre, the documentary, the exposure of their struggles casts a shadow over a number of alleged environmentalist initiatives by the states, such as the positioning of governments as environmental champions in the international arena. Popular environmentalisms have also become active stone guests regarding the extractivism of a former ally, the "New Left" of Latin America. Early in the opposition against neoliberal governments, popular environmentalisms were welcomed by the political Left. Now, this "New Left" is stoning them with scorn and repression.

The most influential environmental or ecological politics are coming from various indigenous and nonindigenous popular environmental movements. And yet, study of the discourses, narratives, images, and media that articulate their environmental ethics, political ecologies, and political ontologies regarding the human and the nonhuman is a task still pending for the environmental humanities. Popular environmentalisms, for instance, have produced or collaborated in a significant number of documentaries narrating their resistance to environmental injustices and unfair political ecologies. A few of these documentaries have received attention by scholars in postcolonial and decolonial studies. Indeed, to a considerable extent popular environmentalisms are linked to movements for indigenous autonomy confronting the history of colonial domination and twentieth-century modernization that are, of course, behind crises of pollution, climate change, scarcity of water, food security, loss of biodiversity and, in general, local and planetary environmental degradation. Thus, the consensus is to acknowledge *differentiated responsibilities* in the rise of the Anthropocene. But if the environmental humanities are truly to engage in dialogue with these popular environmentalisms, it is going to be necessary to go further than merely tracing the history of resistance to injustices, or adding environmental charges to the critique of coloniality. As much as the narratives of injustice are necessary for public awareness and indispensable for legal reparation, popular environmentalisms also need a variety of genres and media recounting the alternatives to development that they actually have imagined and delivered. What are the practices of sustainability and community resilience that take place at the same time that the community or movement is fighting injustices and confronting corporations and the national or plurinational states? This is, after all, the specific goal of oppositional politics against economic development. On the other hand, "ecology," "environment," "human and nonhuman," and "the rights and intrinsic values of nature" are increasingly becoming keywords too in decolonial studies, not only because they refer to actual impacts of the colonial structure of power, but mostly because they are keywords in the decolonial movements themselves which are redefining the notion that only the human is the subject of politics.

Buen Vivir and other popular environmentalisms

Since the early 2000s, particularly in Bolivia and Ecuador, *Buen Vivir* or *Vivir Bien* in Spanish refers to a movement toward developing ontological and axiological alternatives to economic development that draw on indigenous traditions as well as on other critiques of development (see my "Filming the Emergence of Popular Environmentalism" for a more detailed introduction to *Buen Vivir*). BV is an emergent network of more-than-environmentalist discourses and social movements inspired by indigenous knowledge about humans, animals, plants, landscapes, and the nurturing of life and community (Fatheuer). Although it is not behind every single instance of popular environmentalism, BV is spreading like wildfire throughout Andean and Amazonian countries, and beyond.[1] Early formulations of BV reacted against development strategies, either due to their negative social and/or environmental impacts, or to their debatable economic effects (Gudynas, "*Buen Vivir*"). Although BV embraces some Western and modern ideas of quality of life, BV's specific and controversial contribution lies in the conviction that well-being is only possible within a community understood in an expanded sense that includes nature. The ontologies of BV have made their way into two new Constitutions in Ecuador (2008) and in Bolivia (2009). Thus, at this point in its history, BV exceeds the definition of the "environmentalism of the poor" originally proposed by Joan Martínez-Alier. BV's power to intervene in state policies, and even to re-define what beings qualify as subjects of politics, goes far beyond the struggle of local communities concerned for the natural resources they need for survival.

The definitions and interpretations of BV depend on cultural, historical, and ecological settings (Gudynas, "*Buen Vivir*"). And that plurality is indeed encouraged within the network. The most well-known approaches to BV are the Ecuadorian concept of *sumak kawsay* (the Quechua phrase for a fulfilling life in a community, together with other persons and other-than-humans) and, in Bolivia, the similar Aymara concept of *suma qamaña*. However, neither concept is supposed to be followed by all other indigenous groups in Latin America. Other indigenous cultures are exploring and building their own versions of BV.

The thinkers and practitioners of BV do not recognize it as a static concept, but as a project that is continually being created. Therefore, BV should not be understood as a restorative nostalgia for a distant precolonial time, but as a reflective unearthing of indigenous cosmologies. It encourages indigenous and nonindigenous communities to reflect on indigenous beliefs and values as alternatives.

But it calls too for exchange and innovation. Other influences in BV identified by Gudynas came from critical positions within modernity ("*Buen Vivir*"), such as the radical environmental postures of deep ecology and other biocentric approaches, and the feminist critique of gender roles and their part in the acceptance of domination over nature. There is an uninhibited comparativism underlying BV's discourses and political alliances, and an explicit concern that such decisive recovery of indigenous cosmologies and values may fall into ethnocentrism regarding the West or modernity. In his most recent account of BV, the Ecuadorean Alberto Acosta, former minister of energy and former president of the Constitutional Assembly, insists on the need to balance bringing indigenous worlds and values to life with avoiding cultural prejudice: "Entonces, sin minimizar este aporte desde los marginados, hay que aceptar que la visión andina no es la única fuente de inspiración para impulsar el Buen Vivir. Incluso desde círculos de la cultura occidental se han levantado y ya desde tiempo atrás muchas voces que podrían estar de alguna manera en sintonía con esta visión indígena y viceversa" (22). ("Therefore, without minimizing this contribution coming from marginal peoples, we must accept that the Andean vision is not the sole source of inspiration for advancing *Buen Vivir*. Many voices have emerged, even from within the circles of Western culture, and for a long time now, which may be in tune in their own way with this indigenous vision, and vice versa" [all translations are mine].)

Since the early 1990s, indigenous cultural and educational policies, grassroots movements, and even political theories of the state in Bolivia, Colombia, Chile, Ecuador, and Peru have drawn from and elaborated on the notion of interculturalism. Its fundamental premise is rejecting the Anglophone notion of multiculturalism, which has allegedly failed to recognize the internal plurality of each culture, and the inescapability and desirability of intercultural dialogue for mutual enrichment, for questioning our own and each other's cultures, and for building new social relationships and living conditions (Walsh 140). Indeed, the goal of interculturalism among indigenous movements is to reach a negotiation where differences can contribute to the development of new forms of coexistence, collaboration, and solidarity, and not merely to achieve recognition, tolerance, or even acceptance of those differences. Interculturalism is not naïve regarding the inequalities that impede this horizontal exchange. But it is by making such exchanges happen that those inequities can be addressed and subverted.

BV is the most elaborate and emblematic expression of the rise of popular alternative political ontologies in Latin America, but it is also the product of an accumulation of struggles. In fact, the emergence of indigeneities in twenty-first-century Latin America, notes Marisol de la Cadena in her "Indigenous Cosmopolitics in the Andes," has come with an increasingly frequent presence of "earth beings" and "earth practices" on political stages in the Andes. "Earth beings" means animals, plants, and landscapes as sentient beings; "earth practices" refer to human interactions with "earth beings," the respect and affect, or nurturing, necessary to maintain the relationship between humans and other-than-humans that sustains life in the Andes (340–341). In an echo of BV's self-conception, de la Cadena clarifies that this excess to the modernist limit between humans and nonhumans in politics is neither residual of "primitive" indigenous cultures nor simply a strategic essentialism for the interpellation and mobilization of indigenous subjectivities (336). Even beyond the influence of BV, "earth beings" and "earth practices" are disrupting the conceptual field of political ontology to the point where they are no longer just a manifestation of "indigenous culture," but a disagreement between worlds.

The transnational, international, and interethnic political strategies and the comparative ecologies of popular environmentalisms conflict with any essentialist way of thinking the indigenous Other. This clarification is very useful at a time when the exclusion of other political ontologies is justified by questioning the authenticity of the indigenous when it does not match essentialist definitions. De la Cadena has proposed a definition of indigeneity that in fact helps elucidate the confusion created by an essentialist standard: "albeit hard to our logic, indigeneity has always been part of modernity and also different, therefore never modernist" (348). Furthermore, argues de la Cadena, the indigenous is an *indigenous-mestizo* aggregate, not only because it is a historical formation coming out of colonization, but also because it is a socionatural formation where identity is not divorced from the other-than-human. The terrible pattern of defining indigenous peoples in an essentialist manner is not due simply to a lack of education in cutting-edge anthropological research. It happens, among governments and the public, in spite of the guidelines offered by those in charge of bringing together indigenous organizations and development projects at the negotiating table, such as the International Labour Organization's Convention No. 169. Although the Convention does not define who indigenous and tribal peoples are, it provides criteria for describing these peoples that certainly do not include the expectation of isolation or an essential difference with the West or modernity. Along with self-identification, other criteria include traditional lifestyles; a culture and way of life different from the other segments of the national population; their own social organization and political institutions; and inhabiting a certain area in historical continuity, or before others "invaded" or came to such area.

Between a rock and a hard place

In 2013, Ecuador abandoned a pioneering conservation plan in the Amazon started six years earlier. The ITT-Yasuní initiative (Ishpingo-Tambococha-Tiputini region of Yasuní) attempted to raise funds from the international community instead of drilling for oil in this corner of the Yasuní National Park, one of the world's most biodiverse regions. The country's energy minister at the time, Alberto Acosta, put together a BV coalition that drew up this proposal in 2007, after drilling firms discovered 796 million barrels of crude under the ITT region. The initiative promised to leave the oil in the ground, preventing more than four hundred million tons of carbon dioxide from going into the atmosphere, if half the $7.2 billion value of the reserve could be raised by the international community by 2023 (Watts). Ecuador's President Rafael Correa, reported Jonathan Watts for *The Guardian*, blamed the failure of the ITT initiative on the lack of foreign support, after a trust fund set up to manage the initiative received only the tiny fraction of $13 million in deposits. Correa has held that the oil drilling would affect less than 1 percent of the park, but the area affected could be twenty or thirty times larger once access roads are factored in (Watts). With new roads for oil inside the Yasuní, new settlers, illegal logging, and over-hunting will inevitably follow.

President Correa has also argued that the money from oil drilling in Yasuní would help alleviate poverty, particularly in the Amazon. Indeed, not all indigenous peoples are opposed to oil drilling, including some of the Waorani living in the Yasuní National Park who aspire to modern education, transport, healthcare, and housing.[2] Others have made a progressive case for drilling, since apparently in Ecuador two-thirds of all energy comes from renewables, and the country has grown its economy at above the regional average while reducing inequality with revenue from the profits of oil exploitation ("Ethnic Conflict"). There are other issues at stake, though. The global economic crisis and Ecuador's default on sovereign debt in 2008–2009 made the country increasingly dependent for finance on China, which will partly be paid back in oil. Additionally, Chinese companies have been directly linked to ITT exploitation (Watts; Hill). Despite polls showing that between 78 percent and 90 percent of Ecuadoreans are opposed to drilling in this sensitive region (Watts), on May 2014, Ecuador's government approved permits for oil drilling in the Yasuní National Park starting in 2016.

This controversy around the ITT-Yasuní initiative shows that, in spite of a good reputation derived from other environmental initiatives, the "progressive" governments of Venezuela, Bolivia, and Ecuador are all highly dependent on exporting hydrocarbons and minerals, and very willing to capture the substantial rents that lie in their nations' subsoils (Wood). This "new extractivism," explains Gudynas, directs part of the surplus generated by the extraction of these resources toward financing social programs and the development of infrastructure, with which the state gains social legitimization ("New Extractivism"). The New Latin American Left in power, concludes Gudynas, shares with neoliberal extractivism the idea of continuing progress based on technology, and has taken the decision to prioritize the modern and capitalist understanding of the interface between nature and society. The conflict between popular environmentalisms and the "New Left" in power—giving this conflict the benefit of the doubt, as Unai Villalba does in "*Buen Vivir* vs. Development"—indicates the inevitable complexity of transitions to policies that could lead to a full BV model for the state. In the meantime, the "New Left" is attacking popular environmentalisms with tactics not too different from the strategies of governments that openly support neoliberal policies. In recent years, Ecuador has seen a process of radicalization on issues of mining, oil exploitation, and the use of water sources. Correa's government has reacted to the demands and increasing mobilization of communities against mineral and oil extraction by criminalizing social protest and persecuting indigenous leaders and students. There is a significant rise of the number of socioenvironmental conflicts in other Andean and Amazonian countries as well.[3] These scenarios invite some fine-tuning of the diagnoses and goals of the environmental humanities as a project. A major argument about the need for the environmental humanities seems to be an awareness of the complexities underlying the "failure" to implement public policies based on scientific and economic approaches and of the inadequacy of environmental organizations in fostering the values needed for building sustainable societies and environmental citizenship (Nye 4). In Latin America, the major "failure" and inadequacy is not the lack of attention to cultural factors due to disciplinary prejudices; rather, it is the deliberate disowning or co-optation of those cultural values that once were embraced by the "New Left."

Buen Vivir and posthumanist political ontologies

Imagining a dialogue between BV and cutting-edge research on new political ontologies and posthumanist thinking is inevitable. BV and popular environmentalisms may benefit from such an intercultural exchange as an opportunity for strengthening their intervention in the public debate in Andean and Amazonian countries. The French anthropologist and sociologist of science Bruno Latour takes as one of his starting points in *The Politics of Nature* (2004) the fact that some nonmodern cultures are conceptually interested in and politically concerned with the relations between humans and nonhumans, not those between the social and the natural order (45). In spite of BV's commitment to indigenous values and ontologies, in the end it does not propose them as an alternative to Western or modern metaphysics, but as the starting point for a collective *experimental metaphysics*. Or as Latour puts it: "start all over and compose the common world bit by bit" (83). In other words, comparisons, dialogues, and even Convention No. 169's prior consultation and participation are not about negotiating an agreement between epistemologies or subjective perspectives on the world; or, even worse, between objective and subjective perspectives on the world (Latour 78). Such consultations should be horizontal, collective, and innovative efforts, not for debating the validity of different perspectives, but for moving from a disagreement to an agreement between worlds.

The close connection between new political ontologies and indigenous cosmologies has been at the core of the influential work done by Brazilian anthropologist Eduardo Viveiros de Castro. In "Exchanging Perspectives: The Transformation of Objects into Subjects in Amerindian Ontologies" (2004), he also criticized the modern assumptions behind anthropology's explanation of the interface between nature and society by contrasting them with Amerindian cosmologies, which he had previously studied in "Cosmological Deixis and Amerindian Perspectivism" (1998). In modernity, we believe that "social relations—that is, contractual or instituted relations between subjects—can only exist internal to human society" ("Exchanging" 481), and that the interface between nature and society is natural, ruled by necessary laws of biology and physics, since bodies and objects are in ecological interaction with other bodies and forces (475, 481). In Amazonian cosmologies, the world is populated by a diversity of human and nonhuman persons, such as gods, spirits, the dead, entities from other cosmic levels, animals, plants, things, and the weather. These persons perceive and grasp reality from different points of view. Humans and nonhumans share the qualities that modernity identifies with subjectivity, such as intentionality, agency, and self-awareness. Because of this ontologicality, humans and nonhumans think of themselves as "humans" ("Cosmological" 470, 472, 477). The interface between them is, to put it in modern terms, "social." But humans and nonhumans do not see the same things due to the differences between their bodies. Perspectivism derives from bodies, not due to their biological or physical constitution, but out of an "assemblage of affects and ways of being" or *habitus* ("Cosmological" 478–480): what the person eats, how it communicates, where it lives, and so on.

The point of Viveiros de Castro's challenge is not to assert, once more, that "the West got everything wrong, positing substances, individuals, separations, and oppositions wherever all other societies/cultures rightly see relations, totalities, connections, and embeddings" ("Exchanging" 483). For Philippe Descola, who has studied the Achuar people inhabiting territories in the Amazonian border between Ecuador and Peru extensively, such a statement would be a legitimate philosophical profession of truth, but it simply inverts the common ethnocentric prejudice and makes non-modern ontologies "a faithful account of the complexity of the experience of beings" (66). The point that Viveiros de Castro finally makes coincides with BV's openness to dialogue but mistrust of anthropology: "the West and the Rest are no longer seen as so different from each other ... so the great polarity now is between anthropology and the real practical/embodied life of everyone, Western or otherwise" ("Exchanging" 483). BV's expectations of intercultural dialogues and its rejection of essentialisms can also be found between the lines in this statement by Viveiros de Castro. It certainly would be a great irony if the advance of new political ontologies had the unintended consequence of overlooking actual public confrontations in this field. It would be an even worse irony if the theoretical debate reduced BV and other popular environmentalism to just ontologies, and thus made us fall into an old mistake: a new version of the "ecological Indian."

BV shows that the epistemologies and ontologies being developed among the indigeneities of the twenty-first century are not merely non-Western or non-modern discourses available to those elaborating the new political ontologies inspired by and responding to the environmental issues of our times. Instead, they are "new" political ontologies that are already intervening in national and international public spheres, not only by resisting very influential powers and interests, but also by imagining sustainable and resilient alternatives that are built on unearthing traditions and on alliances and interculturalism beyond their frontiers. Popular environmentalisms implicitly suggest to the environmental humanities that, at least in this century, indigenous epistemologies and ontologies should not be separated from the social movements they are part of. The bold proposal they are making to the environmental humanities lies less in their notions of sustainability or resilience than in the expectation that the environmental humanities should collaborate interculturally with twenty-first-century indigenous ontologies to build the common world bit by bit.

Notes

1 For instance, in an unforeseen turn of events, the peace talks between the Colombian government and the *Fuerzas Armadas Revolucionarias de Colombia* (FARC, "The Revolutionary Armed Forces of Colombia") in Havana apparently include a text that defines the main goal of the *Reforma Rural Integral* ("Comprehensive Rural Reform") as pursuing the *buen vivir*.
2 The history of the ITT initiative, but also the rise of the suspicion in BV quarters that Correa was not going to keep the oil in the ground if the fund-raising failed, is narrated in the documentary *Yasuní: El Buen Vivir* (2012), directed by Arturo Hortas (Ecuador).
3 The Ecuadorean case evokes an analogous one taking place in Bolivia. It started in 2010, when President Evo Morales decided to pursue a 190–200 mile, $420-million jungle highway funded by Brazil through the Isiboro-Secure Indigenous Territory National Park (TIPNIS), in the eastern lowlands state of Beni. The story of the repression of the indigenous peoples' protest, and the *exposé* of the economic interests behind the highway is told by director Karen Gil in the documentary *Detrás del TIPNIS* (2012).

References

Acosta, Alberto. *El Buen Vivir: Sumak Kawsay, una oportunidad para imaginar otros mundos.* Barcelona: Icaria, 2013. Print.

Cadena, Marisol de la. "Indigenous Cosmopolitics in the Andes: Conceptual Reflections beyond 'Politics.'" *Cultural Anthropology* 25 (2010): 334–370. Print.

Descola, Philippe. *The Ecology of Others.* Chicago: Prickly Paradigm Press, 2013. Print.

Detrás del TIPNIS. Dir. Karen Gil. Bolivia: Sólo Fuego Producciones, 2012. YouTube. Web. 1 August 2015.

"Ethnic Conflict in Ecuador's Yasuní National Park." *Survival News*. Survival. 15 April 2013. Web. 1 August 2015.

Fatheuer, Thomas. *Buen Vivir: A Brief Introduction to Latin America's New Concepts for the Good Life and the Rights of Nature*. Berlin: Heinrich Böll Foundation, 2011. Print.

Gudynas, Eduardo. "*Buen Vivir*: Today's Tomorrow." *Development* 54 (2011): 441–447. Print.

——. "The New Extractivism of the 21st Century: Ten Urgent Theses about Extractivism in Relation to Current South American Progressivism." *Americas Program Report*. Washington, DC: Center for International Policy, 21 January 2010. Web. 1 August 2015.

Hill, David. "Why Ecuador's President Is Misleading the World on Yasuní-ITT." *The Guardian*. 15 October 2013. Online edition. Web. 1 August 2015.

Indigenous and Tribal Peoples Convention, 1989 (No. 169): Convention Concerning Indigenous and Tribal Peoples in Independent Countries. International Labour Organization. Web. 1 August 2015.

Latour, Bruno. *Politics of Nature: How to Bring the Sciences into Democracy*. Cambridge, MA: Harvard University Press, 2004. Print.

Marcone, Jorge. "Filming the Emergence of Popular Environmentalism in Latin America: Postcolonialism and Buen Vivir." *Global Ecologies and the Environmental Humanities: Postcolonial Approaches*. Ed. Elizabeth DeLoughrey, Jill Didur, and Anthony Carrigan. New York: Routledge, 2015. 207–225. Print.

Martínez-Alier, Joan. *The Environmentalism of the Poor: A Study of Ecological Conflicts and Valuation*. Cheltenham: Edward Elgar, 2003. Print.

Nye, David, Linda Rugg, James Fleming, and Robert Emmett. *The Emergence of the Environmental Humanities*. Stockholm, Sweden: MISTRA, The Swedish Foundation for Strategic Environmental Research. 2013. Print.

Rose, Deborah Bird, Thom van Dooren, Matthew Chrulew, Stuart Cooke, Matthew Kearnes, and Emily O'Gorman. "Thinking Through the Environment, Unsettling the Humanities." *Environmental Humanities* 1 (2012): 1–5. Web. 1 August 2015.

Villalba, Unai. "*Buen Vivir* vs. Development: A Paradigm Shift in the Andes?" *Third World Quarterly* 34 (2013): 1427–1442. Print.

Viveiros de Castro, Eduardo. "Cosmological Deixis and Amerindian Perspectivism." *The Journal of the Royal Anthropological Institute* 4 (1998): 469–488. Print.

——. "Exchanging Perspectives: The Transformation of Objects into Subjects in Amerindian Ontologies." *Common Knowledge* 10 (2004): 463–484. Print.

Walsh, Catherine. "Interculturalidad, plurinacionalidad y decolonialidad: Las insurgencies político-epistémicas de refundar el Estado." *Tábula Rasa* 9 (July–December 2008): 131–152. Print.

Watts, Jonathan. 2013. "Ecuador Approves Yasuní National Park Oil Drilling in Amazon Rainforest." *The Guardian*. 16 August 2013. Web. 1 August 2015.

Wood, Rachel Godfrey. "New Left=New Extractivism in Latin America." *International Institute for Environment and Development*. 29 June 2010. Web. 1 August 2015.

Yasuní: El Buen Vivir. Dir. Arturo Hortas. Ecuador: EJOLT (Environmental Justice Organisations, Liabilities and Trade), 2012. Vimeo. Web. 1 August 2015.

Part IV

DECLINE AND RESILIENCE:
Environmental narratives, history, and memory

24

PLAY IT AGAIN, SAM

Decline and finishing in
environmental narratives

Richard White

Over the last half-century, popular environmental literature in the United States has responded to environmental crises ranging from pesticides to global warming with a single narrative. Writers have recycled a pastoral fable of decline and possible redemption far older than any of the crises. George Perkins Marsh launched the modern version of the fable in *Man and Nature*, but it initially coexisted with and drew upon a very different narrative of "finishing." In the twentieth century the fable took its dominant form in Rachel Carson's *Silent Spring,* in many ways the urtext of modern American environmentalism. Much of the environmental writing that followed emulated it.

The Marsh and Carson versions are undeniably similar, but it is their differences that reveal much about modern environmentalism. Marsh and Carson both asserted that once upon a time humans had created a society of comparative harmony and balance, but now the United States (and the world) faced an unprecedented environmental crisis that threatened life as we know it. This crisis was of our own making. We had, in effect, declared war on our own planet. And, if something dramatic did not change, we were doomed. Carson and Marsh, however, differed in how they understood the causes of the crisis and its historical trajectory.

We have grown so accustomed to this declensionist narrative that we tend to miss a competing ameliorative narrative of "finishing" critical to both *Man and Nature* and *Silent Spring*. Finishing granted humans power over the natural world, but they were part of natural processes, and their power was by and large benign. Human labor, in effect, put the final touches on a landscape to create the pastoral: wheat growing in a field, trees in an orchard, and cattle in a pasture.

It is easy to miss finishing in *Man and Nature* because George Perkins Marsh made the book an encyclopedic catalog of destruction. "It is in general true, that the intervention of man has hitherto seemed to insure the final exhaustion, ruin, and desolation of every province of nature which he has reduced to his dominion" (352). The destructive tendencies Marsh saw everywhere caused him, like Charles Darwin, to reject narratives of human progress that stressed God's or Nature's design. Unlike Darwin, however, Marsh denied that humans were part of nature. As he wrote his publisher:

nothing is further from my belief, that man is a "part of nature" or that his action is controlled by what are called the laws of nature, in fact a leading object of the book is to enforce the opposite opinion, and to illustrate that man, so far from being, as Buckle supposes, a soul-less, will-less automaton, is a free moral agent working independently of nature.

(Lowenthal 290)

Human destructiveness proved that humans were unnatural and "not controlled by natural compensations and balances" (Marsh 40–41). Humans forced change and disruption on a nature Marsh conceived of as harmonious, balanced, stable, and interconnected (29). It was only a seeming paradox that the human ability to unbalance and destroy was a sign of mankind's superiority: "Though living in physical nature, he is not of her ... he is of more exalted parentage, and belongs to higher order of existences than those born of her womb and submissive of her dictates" (36).

Superiority did not insure inevitable progress. Most often, it guaranteed declension. "The ravages committed by man subvert the relations and destroy the balance which nature had established between her organized and her inorganic creations; and she avenges herself upon the intruder, by letting loose upon her defaced provinces destructive energies hitherto kept in check by organic forces destined to be his best auxiliaries, but which he has unwisely dispersed and driven from the field of action" (42). Marsh was left with seemingly irreconcilable ideals of a stable nature, gendered as female, and a progressive but destructive society gendered as male.

Marsh resolved this dismal paradox of a progress that seemingly insured collapse by creating a narrative that, while usually declensionist, was hardly hopeless. He turned republican Rome into the tragic hero of his tale. Rome was a society able to achieve social progress without undermining the natural processes that sustained its people. Marsh described a virtuous Roman landscape of plenty as well as of moderation and labor. Gold and silver were "not found in the profusion which has proved so baneful to the industry of lands rich in veins of precious metals" (7). This was not an Eden. This finished landscape was a product of labor. "Every loaf was eaten on the sweat of the brow. All must be earned by toil. But toil was nowhere else rewarded by such generous wages" (8). Only beauty was a pure gift of unaided nature. Marsh created a moral republican landscape that reflected virtue rooted in labor. For all practical purposes it was an idealized version of Marsh's native Vermont transported to the Mediterranean.

In the end, of course, Rome declined and fell and the Mediterranean sank into ruin and depopulation. This happened for two reasons. The first, and the one emphasized in most readings of Marsh, was a "disregard of laws of nature." This is analytically odd since Marsh's presumption was that the ability to disregard natural laws was what made humans human. The second reason seems to me the more critical one: "the causa causarum of the acts and neglects which have basted with sterility and physical decrepitude the noblest half of the empire of the Caesars, is first the brutal and exhausting despotism which Rome herself exercised over the conquered kingdoms and even over her Italian territory, then the hold of temporal and spiritual tyrannies which she left as her dying curse to all her wide dominion" (11). By despotism and spiritual tyranny he meant monarchy and Catholicism. "Man," Marsh concluded, "cannot struggle at once against crushing oppression and the destructive forces of inorganic nature" (11–12).

For Marsh, Republicanism and Protestantism provided the keys for preventing environmental harm. A Protestant republic like the United States could escape the fate of Europe

240

and Asia. In essence, it was the human social order that had to preserve a natural order otherwise unable to stand before human beings.

Marsh, like Rachel Carson, was part of an American tradition of thinking with nature— they never disentangled their social and environmental thinking. Changes in the American social order threatened the natural order and vice versa. Their difference—and it is a critical one—was that the destructive power that Marsh saw in history, Carson saw in a new technology. Carson, like Marsh, made the United States a scene of pastoral tranquility, an equivalent of Rome, but unlike Marsh, she did not dwell on a long human history of environmental destruction. "Only within the moment of time represented by the present century," she wrote, "has one species—man—acquired significant power to alter the nature of his world … During the past quarter century this power has not only increased to one of disturbing magnitude but it has changed in character" (1–2).

Silent Spring begins with a fable. "There was once a town in the heart of America where all life seemed to live in harmony with its surroundings" (1). And then, in the space of a few paragraphs, the fable describes how birds and bees, plants and animals, and finally humans mysteriously sickened and died. The result was a silent spring. "No witchcraft, no enemy action had silenced the rebirth of new life in this stricken world," Carson wrote. "The people had done it themselves" (3).

Carson's town was the equivalent of Marsh's Rome; it was a product of human finishing. She began her account not with a critique of American life but with an expression of the basic goodness and harmony of ordinary life. Where in Marsh the United States was exceptional in providing a possible break in a long and sorry history of destruction, in Carson the contemporary United States was exceptional because it was in the vanguard of a movement heading not away from destruction but toward it (Carson 1–3). Marsh had marked the human condition as one of exile from nature, but Carson assumed humans were part of nature. Where Marsh understood the United States as a possible break in a human history of destruction, Carson approached her contemporary moment as a break with an older history of harmony. She argued that humans had moved outside evolutionary change. Life no longer had the chance to adapt and adjust, and as a result balance was disrupted.

As history and science, this argument had its problems, but as a rhetorical exercise, it was brilliant. The strength of the writing was rooted very much in Carson's own life, her sense of the world, her place in it, and her confidence that her ordinary experiences and concerns resonated with those of millions of other Americans. Carson preached a basic contentment with the finished landscape. She did not set herself up as having a privileged knowledge, nor did she demand that people change their lives or their values. Instead she demanded that they recognize the insidious forces that had persuaded them to dismantle the very world they loved and valued.[1]

Carson was inventing the modern environmentalist narrative, and I want to emphasize three things at the core of the invention. The first, as I have already mentioned, was a reorientation away from problems evident over millennia to new problems of the present and future: the present is a break in the fabric of history; from here on out everything will be different if we do not act now. Second, she reoriented environmentalism away from the concerns with male production that permeated an older conservation ethic typified by Marsh's work toward concerns with consumption that were feminized and often dismissed.

Prepared to be denounced as a mere birdwatcher when she attacked DDT and other dangerous pesticides considered critical to agricultural production, Carson struck first, labeling her opponents as authoritarian and undemocratic:

> Who has decided—who has the *right* to decide—for the countless legions of people who were not consulted that the supreme value is a world without insects, even though it be also a sterile world ungraced by the curving wing of a bird in flight? The decision is that of the authoritarian temporarily entrusted with power; he has made it during a moment of inattention by millions to whom beauty and the ordered world of nature still have a meaning that is deep and imperative.
>
> (127)

The final element in Carson's reinvention of environmentalism was her appropriation of metaphors of war, particularly nuclear war and subversion. Carson had learned during her years of popular science writing that when the debate was on a complicated or esoteric topic, it was best to argue by analogy to something else more familiar and easily understood. The argument becomes to some extent metaphorical, and the side with the best metaphors wins.

Carson likened environmental damage to war. "The question is whether any civilization can wage relentless war on life without destroying itself, and without losing the right to be called civilized" (99). The metaphors of war worked because they were aimed at a public that *lived* in a war metaphor. The Cold War was a metaphor; it was like a war but not actually a war since the two sides never fought against each other. The great fear of that confrontation was that the metaphor would yield the real thing: a nuclear holocaust. There was, however, also a second, lesser fear that the very preparation for such a war would itself destroy and kill. Carson opened *Silent Spring* with a quote from Albert Schweitzer that she included in the book's dedication to him: "Modern man no longer knows how to foresee or to forestall. He will end by destroying the earth."

The language of destroying the earth was in the 1950s and 1960s the language of nuclear war. Carson drew an exact parallel between insecticides and nuclear fallout. In the 1950s fallout from nuclear tests—strontium 90—was found in milk and human bodies, sparking widespread concern. Carson used strontium 90 to build a bridge to insecticides: "Similarly, chemicals sprayed on croplands or forests or gardens lie long in soil, entering into living organisms, passing from one to another in a chain of poisoning and death" (6). They end up in the human body, where they can "alter the very material of heredity upon which the shape of the future depends" (8). Carson carried the association to its logical conclusion: "Along with the possibility of extinction of mankind by nuclear war, the central problem of our age has therefore become the contamination of man's total environment" (8; see Glotfelty for a slightly different interpretation of Carson's Cold War rhetoric).

Once she adopted the metaphors of Cold War and nuclear war, Carson worked them for all they were worth. Cold Warriors claimed the central danger to Americans was subversion from within. They used this incessantly to silence critics of the United States, whether the subject was civil rights, labor laws, or social justice. Carson expected it to be used against her, and it was. Its most famous formulation, repeated endlessly after the book's publication, was attributed to a letter from ex-Secretary of Agriculture Ezra Taft Benson to former President Dwight D. Eisenhower. Benson wondered, "Why a spinster with no children was so concerned about genetics?" He thought she "probably was a communist?" (Lear 429).

But Carson had already played the Cold War card, far less crudely, and far more effectively than Benson. Having attacked the "authoritarian" who uses the power he has achieved against the people themselves, she was already after "the enemy within." We would destroy ourselves. Carson made the pesticide industry and its academic and governmental allies the equivalent of internal subversives. Like popular images of communists, her enemies within

242

were misguided, powerful, and totalitarian. They would brook no opposition. They could, however, only win if we let them. She preempted the rhetoric of the political right and used it against particular kinds of science and scientists, and particular corporations, all of whom had turned the government into their agent. The enemy could be found within government, narrow science, and capitalism. Carson argued that the "rise of pesticides was indicative of an 'era dominated by industry, in which the right to make money, at whatever cost to others is seldom challenged'" (Gottlieb 86). Only reform of the American social, political, and economic order could save the natural order.

Carson's revised narrative, and many of her metaphors and analogies, were recycled again and again over the next fifty years and more. In *The Population Bomb* in 1968, Paul Ehrlich declared that the moment of crisis had arrived. The world had neared its Malthusian limits. He gave society ten years to correct the population problem or face inevitable famine and mass starvation (18–25). The only solution was "extremely fundamental changes in our society … in order to preserve any semblance of the world we know" (156). In *The Closing Circle* (1971), Barry Commoner asked: "Why after millions of years of harmonious co-existence, have the relationships between living things and their earthly surroundings begun to collapse? Where did the fabric of the ecosphere begin to unravel? How far will the process go?" (11). The answer was: "The chief reason for the environmental crisis that has engulfed the United States in recent years is the sweeping transformation of productive technology since World War II" (177). The cure was to retreat back to production based on natural rather than synthetic products. The same narrative of modern crisis and decline showed up in John McPhee's *Encounters with the Archdruid* (1971), when David Brower—the Archdruid—said, "I'm trying to do anything I can to get man back into balance with the environment. He's way out—way out of balance. The land won't last, and we won't" (16–17).

Bill McKibben's *The End of Nature* (1990) pretty much summarized things for those who had not been paying attention for the previous twenty years and brought to center stage the issue that has dominated down to the present: climate change. McKibben struck the familiar notes: "we have done this to ourselves"; everything has changed (45). McKibben, however, resembled Marsh in that he saw human alienation from nature as the human condition. He regarded the separation as a good thing for reasons different from Marsh's. It gave humans something outside of themselves against which they could judge and measure their actions. But with climate change, humans had affected everything, and the idea of nature as something separate vanished. Meaning had been drained from the world (47–48).

The narrative continued in the usual parameters in the 1990s. In *Earth in the Balance* (1992), Al Gore wrote of "humankind's assault on the earth" (25; cf. 27). He argued that "[g]lobal warming, ozone depletion, the loss of living species, deforestation—they all have a common cause: the new relationship between human civilization and the earth's natural balance" (31). And Gore, like Carson, saw the crisis as paralleling nuclear war. He appealed to nuclear arms negotiations as a model for negotiating solutions to environmental problems (34–35).

Gore made the current crisis the critical crisis. He, like Commoner, Ehrlich, and Carson, argued that "this century has witnessed dramatic changes in two key factors," population, and scientific and technological revolutions (30). "[T]he entire relationship between humankind and the earth has been transformed because our civilization is suddenly capable of affecting the entire global environment, not just a particular area" (29–30). Gore did not deny previous dramatic human alterations, but in distinguishing the current crisis he appealed to a logic that would culminate in our current invention of the Anthropocene.

Essentially, Gore denied the power of addition. Changes become globally significant only when they can affect everything, everywhere, all at once. Regional human crises cannot *add up to* a global crisis. The crisis has to be born global.

Academic studies often anticipate popular concerns, but in the case of the narrative of climate change, popular writing has anticipated academic discourse. The Anthropocene is the culmination of the popular narratives that there has been a break in the fabric of time, that we face a crisis unprecedented in human history, that older crises were local or regional but this one is global, and that humans are now the dominant environmental force. I believe fully in the reality of climate change, but I am an Anthropocene skeptic, even though I have the highest regard for the person who has most successfully pushed for the Anthropocene in the humanities and social sciences: Dipesh Chakrabarty. Chakrabarty is far more sophisticated than Bill McKibben, but his stance is not that different. For Chakrabarty, what has collapsed is "the age-old humanist distinction between natural history and human history," and what is at stake is the loss of "our capacity for historical understanding" (201). But I would argue that the declensionist narrative in both its Marsh and Carson versions collapsed the humanist distinction between human and natural history long ago. Marsh asserted the distinction in theory and undermined it in practice. Carson undermined it in theory and practice even as she anticipated the Anthropocene by half a century in claiming a break with the past. I mention the Anthropocene here only to suggest that even in Chakrabarty's sophisticated formulation, it echoes the basic environmental narrative of rupture and declension while oddly claiming that this narrative has always previously maintained a distinction between human and natural history.

The narratives of decline plowed the ground for a very unlikely set of narratives of resilience that began appearing in the 1970s and which now have their parallels in the Anthropocene. Charles Reich, in *The Greening of America* (1970), announced that resilience had begun, a change of consciousness was underway, and that matters were well in hand. E. F. Schumacher in *Small is Beautiful* (1973) made an appeal for personal transformation in which consumption and self-discovery went hand in hand. Reich's own personal Age of Aquarius was "Consciousness III"—the supposed cultural transformation of the United States that began in the mid-1960s. The specific content of Consciousness III was always a little vague. "Consciousness III says, 'I'm glad I'm me,'" Reich wrote approvingly, and at that point the 1970s were surely in trouble (219, 223). Where Reich looked ahead to a new consciousness, however, Schumacher tended to look backward to ancient wisdom, particularly Christianity. Ultimately, however, he, too, regarded the central problem as one of internal transformation. "We can each of us work to put our inner house in order," Schumacher writes in the conclusion of his book. The answers could be found in the "traditional wisdom of mankind" (Schumacher 281). This emphasis on personal transformation—the goals of "Life-style ecology activists," as it got translated into 1970s Sierra Club speak—resonated with the local environmental movements of the 1970s (Flatboe 15). It was a telling turn.

Declensionist narratives are literary sunsets, and they almost inevitably yield their opposites, rhetorical sunrises. Marsh, writing in the 1850s, could incorporate both into a single volume. He so thoroughly documented the decline that relatively few noted this narrative of resilience made possible by the combination of democracy and Protestantism in the United States. The Anthropocene serves a similar function today. It is over, Anthropocenists announce, and we have just begun. A very odd and very interesting form of this appears in the "Ecomodernist Manifesto" from the Breakthrough Institute run by Michael Shellenberger and Ted Nordhaus.

It is easy to dismiss the "Ecomodernist Manifesto" with its *Whole Earth Catalog* optimism, its libertarian and technological fantasies, and its glib totalizing as a fundraising brochure aimed at Silicon Valley. And in part, it clearly is that, but the question is whether it is more. The "Manifesto" imagines a "good Anthropocene"; the authors take Stewart Brand's aphorism (he is on their advisory board) in the *The Last Whole Earth Catalog*—"We are as gods and might as well get good at it" (1)—and assume that we have gotten good enough at it to "mitigate climate change, to spare nature, and to alleviate global poverty." This seems but a technocratic libertarian version of Marsh's republican Protestant hopes, but its goal of "land sparing" is less a version of finishing than a new version of Paul Shephard's old vision of humans clustered in coastal cities largely devoid of nature while the wild interior flourishes (237–278). But if the last several centuries have taught us anything, it is that nature cannot be banished from the most human-dominated environments, and that human influence is—as the Anthropocene has recognized—everywhere.

The most fruitful finishing narrative, the one still largely unwritten, lies in this paradox captured in the ambiguity of John McPhee's wonderful title: *The Control of Nature*.

Note

1 This was not a stance that many environmental leaders have taken before or since. Compare Carson's self-positioning with a nineteenth-century man who also became a twentieth-century environmental icon: Henry David Thoreau. Thoreau's *Walden* also begins with a small town and a rural landscape. But Thoreau set up this landscape in order to satirize it; he wanted his readers to transform it and themselves. Thoreau's book was a relentless demand for change and transformation. Thoreau took a guru-ish stance that both antagonized his fellow citizens and anticipated the style of many modern environmental leaders.

References

Carson, Rachel. *Silent Spring.* 1962. Boston: Houghton Mifflin, 1987. Print.

Chakrabarty, Dipesh. "The Climate of History: Four Theses." *Critical Inquiry* 35 (2009): 197–222. JSTOR. Web. 1 August 2015.

Commoner, Barry. *The Closing Circle: Nature, Man, and Technology.* 1971. New York: Knopf, 1972. Print.

Ehrlich, Paul. *The Population Bomb.* 1968. New York: Ballantine Books, 1971. Print.

Flatboe, Connie. "Environmental Teach-In." *Sierra Club Bulletin* 55 (March 1970): 14–15.

Glotfelty, Cheryll. "Cold War, *Silent Spring*: The Trope of War in Modern Environmentalism." *And No Birds Sing: Rhetorical Analyses of Rachel Carson's Silent Spring.* Ed. Craig Waddell. Carbondale: Southern Illinois University Press, 2000. 157–173. Print.

Gore, Al. *Earth in the Balance: Ecology and the Human Spirit.* New York: Penguin, 1992. Print.

Gottlieb, Robert. *Forcing the Spring: The Transformation of the American Environmental Movement.* Washington, DC: Island Press, 1993. Print.

The Last Whole Earth Catalog. Ed. Stewart Brand. Menlo Park, CA: Portola Institute, 1971. Print.

Lear, Linda. *Rachel Carson: Witness for Nature.* New York: Henry Holt, 1997. Print.

Lowenthal, David. *George Perkins Marsh, Prophet of Conservation.* Seattle: University of Washington Press, 2000. Print.

Marsh, George P. *Man and Nature, or Physical Geography as Modified by Human Action.* 1864. Cambridge, MA: Belknap Press of Harvard University Press, 1988. Print.

McKibben, Bill. *The End of Nature.* New York: Doubleday, 1990. Print.

McPhee, John. *Encounters with the Archdruid: Narratives about a Conservationist and Three of his Natural Enemies*. New York: Farrar, Strauss, and Giroux, 1971. Print.

——. *The Control of Nature*. New York: Farrar, Straus, and Giroux, 1989. Print.

Reich, Charles A. *The Greening of America*. New York: Random House, 1970. Print.

Schumacher, E. R. *Small is Beautiful: Economics as if People Mattered*. New York: Harper and Row, 1973. Print.

Shellenberger, Michael et al. "Ecomodernist Manifesto." *Breakthrough Institute*. 15 April 2015. Web. 1 August 2015.

Shepard, Paul. *The Tender Carnivore and the Sacred Game*. New York: Charles Scribner's Sons, 1973. Print.

25

HUBRIS AND HUMILITY IN ENVIRONMENTAL THOUGHT

Michelle Niemann

This chapter investigates how environmentalists have used the concepts of hubris and humility to define practices both of sustainable living and of conservation. Though hubris and humility arguably do not fit our current environmental situation well, they are still invoked often: what makes these concepts urgent, despite their inadequacies? In the first section, I examine a classic environmentalist debate structured around hubris and humility and, in the second, turn to recent revisions of these concepts, including Michael Shellenberger and Ted Nordhaus's polemics and the Dark Mountain Project's literary journal and blog. While Shellenberger and Nordhaus's reclamation of hubris has had positive effects on mainstream environmentalist rhetoric and politics, I argue that the Dark Mountain Project's eclectic, syncretic efforts to craft new ways of imagining and practicing humility resonate most vitally with the environmental humanities now, at a time when humility seems both impossible and necessary.

The conceptual dyad of hubris and humility has long structured environmentalist thought. In *Silent Spring*, for example, Rachel Carson accused the "practitioners of chemical control" who "hurl [DDT and other pesticides and herbicides] against the fabric of life" of having "no humility before the vast forces with which they tamper" (297). In *Small Is Beautiful*, E. F. Schumacher praised the "non-violence and humility" (159) of those whom he called "home-comers" against the hubris of "the people of the forward stampede" (155). Aldo Leopold, in articulating a "land ethic [that] changes the role of *Homo sapiens* from conqueror of the land-community to plain member and citizen of it" (204), held that the "ability to see the cultural value of wilderness boils down … to a question of intellectual humility" (200). Here I am not as concerned with the language of humility and hubris, however, as with how classic environmentalist arguments swirled around these conceptual poles and how they haunt current debates.

Environmental philosophers have generally been more interested in hubris and humility than ecocritics, but recent scholarship in both fields attempts to define humility positively, not simply as "the dialectical opposite of hubris" (Weinstein 765). Philosopher Ronald Sandler notes that varied "theories of environmental ethics tend to converge" in classifying hubris among environmental vices and humility among environmental virtues (5); he argues that, for people who have such virtues, activities that others would consider sacrifices become pleasures (6–7). Ecocritic Josh A. Weinstein also focuses on humility as a virtue, tracing the long history of humility in Christian, Jewish, and Buddhist traditions (760–770); "attentiveness" is central to his definition of "ecological humility" (766). Christine Gerhardt, via an ecocritical reading of the poetry of Walt Whitman and Emily Dickinson, makes the essential point that "complete self-effacement is not a position from which one can be humble" (226). For her, "environmental humility" thus does not mean replacing anthropo-centrism with ecocentrism, but instead involves both skepticism about "the human ability to achieve a balance" between human and nonhuman interests and the responsibility to nevertheless continue striving to do so, in the recognition that "humble and human share the same root: the earth [*humus*]" (226, 14). Environmentalist practices subtend such prom-ising scholarly redefinitions of humility. Stephanie Mills's "Epicurean simplicity" and Kate Soper's "alternative hedonism" aptly characterize those practices: both concepts emphasize the sensuous, embodied pleasures afforded by everyday efforts to live more sustainably, like the pleasure of biking to work rather than idling in traffic. As we will see, such practices both intertwine with and diverge from the rhetoric of hubris and humility that structures environmentalist thought.

Hubris and humility, progress and decline

Accusations of hubris or wrongful pride are integral to the jeremiad, a narrative genre named for the prophet Jeremiah that warns of imminent apocalypse unless the people reform their hearts and change their ways. Popular environmental jeremiads, such as Rachel Carson's *Silent Spring*, extend hubris and humility from individual character traits to the larger scale of institutions, corporations, and socioeconomic structures. While hubris is usually associated with narratives of progress, humility is linked with narratives of decline. Techno-optimists who look to the future and think the next innovation will solve our environmental prob-lems are accused of hubris by those who advocate humility and the precautionary principle. Those who advocate humility are in turn accused of nostalgia, of rejecting progress, and of seeking to return to an idealized past.

A classic example of this kind of exchange is the debate about space colonies that Stewart Brand and Wendell Berry had back in the 1970s. Brand and Berry had been collaborators in the *Whole Earth Catalog*'s project of disseminating small-scale tools for sustainable living in the late 1960s, but when Brand devoted twenty-six pages of the fall 1975 issue of his *CoEvolution Quarterly* to physicist Gerard O'Neill's proposal to build space colonies, and then printed and rebuked environmentalists' responses to the idea in the Spring 1976 issue, Berry broke with him publicly. Though the word "humility" only appears once in their exchange, the debate was quintessentially structured around the poles of hubris and humility.

Brand, ever the provocateur, no doubt knew that his environmentalist audience would criticize O'Neill's space colonies and seemed to delight in that. O'Neill argued that building land in space from materials in space would get humanity out of Earth's "zero-sum game" of ecological limits and give us access to free materials and abundant solar "energy without guilt" (6–7, 11). By the fall of 1975, O'Neill had received much mainstream journalistic coverage and testified before a subcommittee in the US House of Representatives (Brand 4; O'Neill 10–19). In introducing the comments he solicited, Brand wrote, "I was partially aiming at environmentalists, including me, who have become too predictable of late, too smug, certain, convergent, uninquiring and unimaginative. We have come to love our famous problems (population, inequity, technology, etc.) and would feel meaningless if they went away. That's a lousy design posture" (Brand et al. 5). Clearly, part of Brand's purpose was to question dominant environmentalist narratives of decline.

Questioning decline narratives still motivates Brand's provocations, including his current advocacy and funding of Revive & Restore, a de-extinction project that is attempting to "resurrect" the passenger pigeon through biotechnology. In a 2011 email to E. O. Wilson and George Church proposing the idea, Brand wrote, "The environmental and conservation movements have mired themselves in a tragic view of life. The return of the passenger pigeon could shake them out of it … It could, as they say, advance the story" (qtd. in Rich). But then as now, Brand seems too eager to trade in narratives of decline for the latest progress narrative. Back in 1976, he concluded that even the "failure" of space colonies would be a win, because it "would bring needed whole system humility" (Brand et al. 52).

If Brand took aim at declensionist narratives by championing a big, flashy project that purported to solve the problem of energy supply and change the game on Earth, Berry and the other environmentalists who objected to space colonies were concerned, as Brand accurately put it, "that Space Colonies are merely more of the same—the same old technological whiz-bang and dreary imperialism" (Brand et al. 52). As Berry wrote, O'Neill's proposal "faithfully utters every shibboleth of the cult of progress," promising, like "the Army Corps of Engineers, the strip miners, the Defense Department or any club of boosters," to solve a slew of economic, social, and environmental problems at once (Brand et al. 8). Others who responded with similar concerns included Lewis Mumford, Gary Snyder, Garrett Hardin, David Brower, Schumacher, and Mills (Brand et al.). Historian Andrew Kirk argues that the space colonies debate exposed a divide among environmentalists, even those who supported the *Whole Earth Catalog*'s "access to tools," over "appropriate technology" versus "ecotechnocracy" (173). Indeed, Berry held that Brand's promotion of space colonies was "a betrayal" of those whom he had encouraged to embrace small-scale attempts at sustainable living "in craft shops, in communes, and on small farms" (Berry 10).

But while top-down technocracy and the concentration of money and power in the hands of an elite characterizes the hubris that Berry criticized, what most centrally defines this hubris is the dynamic by which big "solutions" proliferate new, unforeseeable problems, "ramifying ahead of foresight" (Berry 8–9). Berry gives an example:

> The flush toilet is a social-technological solution to a problem: How to get rid of excrement inoffensively. This solution immediately creates two problems (soil depletion and water pollution) which call for solutions (agricultural chemicals and sewage treatment plants) which create many other problems which call for more solutions which create even more problems, and so on and on.
>
> (9)

Berry argues that, because people have never been able to foresee and prevent all such cascading problems, forging ahead regardless commits us to a risky "policy of technological 'progress' as a perpetual bargaining against 'adverse effects'" (9). On a global scale, the major inventions that inaugurated the industrial revolution—the steam engine and later the internal combustion engine—are examples of tremendous accomplishments that also caused new problems. They not only helped effect the shift to our current fossil fuel "energy regime" and thus allowed for an incredible growth in personal freedom and the shake-up of traditional social hierarchies, as Hannes Bergthaller shows in his chapter in this volume, but also led to climate change. As Schumacher put it, "the new problems are not the consequences of incidental failure but of technological success" (30).

To forestall this series "of problems and solutions madly leapfrogging over the top of each other" (Brand et al. 10), Berry defines a countervailing humility that turns on limits, character, and holism. He argues that O'Neill and Brand are merely dressing up in a space suit the frontier morality of stealing resources from elsewhere (8–9). For Berry, only the humble practices of small-scale holism, or what he calls "the disciplines of finitude," can foster cultural respect for limits (9). While both technological and social "solutions" proliferate new problems, Berry argues that "[w]hen the whole is considered, it becomes possible to conceive of solutions of which the standard is not technological and/or social (wealth, power, comfort) but ecological or organic (health)" (9). His example of this is the outdoor toilet in which "we compost the excrement of our household and make it fit to return to the soil" (9), "a solution that causes a ramifying series of solutions" (10).

Berry's organic solutions still appeal to many, as burgeoning local farming and food movements across the United States show (see, e.g., Plumer; USDA Agricultural Marketing Service). By local food movements, I do not mean the privileged eating pork belly in expensive locavore restaurants (though that is one co-opted component of them), but the farmers and "eaters" who are growing, buying, and dining on "mostly plants" (Pollan) in the context of farmers' markets, cooperative stores, and Community Supported Agriculture (CSA) schemes by which farms provide subscribers a weekly box of produce throughout the growing season. What participants in such movements find appealing about Berry's organic solutions is not so much the doctrine of holism as the empowerment that small-scale action offers (White).

Though the energy in local food movements is an essential dimension of today's environmentalism, the kind of humility Berry articulated in the space colonies debate with Brand is in many ways inadequate to our current situation. First of all, the moralism and focus on character inherent in it are a poor fit for problems that are global in scale and systemic in structure. Systemic problems demand systemic changes, though those should not be confused with big-ticket silver bullet solutions. Moreover, humility seems inadequate because now we feel even more caught in "madly leapfrogging" problems and solutions than in the 1970s, as the fact that some environmentalists have come to support nuclear power attests (*Pandora's Promise*). Isolating a moment before humans became embroiled in such devil's bargains is as difficult as dating the start of the Anthropocene for good reason: both rely on unhistorical, prelapsarian thinking, as Richard White argues in his essay in this volume. But that does not temper the demoralizing experience of being entangled in a chain of wagers against "adverse effects." Hubris and humility also tend to flatten the processes of cultural appropriation that make new technologies part of people's everyday lives into simplified pro- or anti-technology positions (Hård and Jamison).

The concepts of hubris and humility are inadequate in part because the dividing line between big technological solutions and small-scale, appropriate technology is less and less clear. The same digital technologies that allow activists to organize also facilitate governmental surveillance; the same devices that proliferate new social fora also enable corporate and bureaucratic data-collection and their corollaries, targeted advertising and the storage of private information vulnerable to theft. The *Whole Earth Catalog* and its early advocacy of personal computing has left Silicon Valley with a green sheen, despite all the environmental and human costs of digital technologies, from toxic e-waste and exploitative labor practices to the energy required to fuel the material infrastructure of the "cloud." Some proposals, like geoengineering as a "solution" to climate change, still seem to fit the old hubris bill, but technologies big and small, biological and digital, are generally a mixed bag, their effects turning more on who uses them for what ends than on any tendencies toward liberation or oppression inherent in the technology itself.

Still, the inadequacy of hubris and humility is not as much about technological paradoxes as about changing experiences and concepts of agency in an era now being called the Anthropocene. Our experience of our own agency as individuals has changed: as Dale Jamieson argues in his essay in this volume, our feelings of agency have become vastly inflated in some realms (we can send a message to the other side of the globe instantly), but diminished in others (try holding anyone accountable for a gas leak, or lead in the tap water). Human agency writ large is suffering a similar inflation-cum-deflation: wow, we can transform the whole planet! Oops, we transformed the whole planet unintentionally and for the worse. Nonhuman agency, too, has changed, and not just for new materialists or object-oriented ontologists. The microbiome, all the good bugs living in our guts and helping us digest and stay healthy, has been getting a lot of media coverage and, in restaurants and on ubiquitous TV cooking shows, vegetables and fruits, in simple preparations or elaborate molecular de- and re-construction, persuade us of their own delectable virtues. Histories of species and materials fascinate us. From academic to popular discourses, these refigurations of the human subject change what hubris and humility mean. Traditional concepts of human humility before Nature, understood as a separate and overweening force, cause cognitive dissonance in our current situation, surrounded as we are by evidence that humans are changing nature on a global scale. When humans collectively alter the planet without meaning to, what is hubris? How can an individual's humility make any difference? And what should we be humble toward?

Revisions of hubris and humility

Despite their increasing inadequacy, hubris and humility are still often invoked in environmental discourse. Some of these invocations reinscribe classic environmentalist uses of these concepts, while others revise hubris, humility, or their rhetorical dynamics in some way. To hear traditional environmentalist anti-hubris arguments amplified, one need only listen to the charged debates, often under the sign of the Anthropocene, about what is being called the new conservation, which includes everything from more interventionist management of wilderness areas to the de-extinction projects Brand promotes. The many scientists and conservationists who oppose such proposals rely on another familiar environmentalist version of humility: leaving the wild alone.

In his introduction to *Keeping the Wild*, a collection of essays critiquing the new conservationists, Tom Butler sees their conservation practices and the "growing chorus of Anthropocene boosterism" as the embodiment of "hubris" (xi). For Butler, hubris comes down to "anthropocentrism" (xi); he argues that environmentalists should not "merely green up a flawed system," but contest that system fundamentally (xiii). Butler's argument here is less nuanced and more Manichean than Berry's: his version of humility does not center on small-scale, holistic practices, but on restriction—"reducing human numbers and economic pressure on the biosphere" (xii)—and especially on the preservation of wilderness. In response, Emma Marris, a chronicler and advocate of some new conservation strategies, has turned the hubris–humility dichotomy upside down. By refusing to consider unorthodox techniques, conservationists are, Marris argues, sacrificing species to their own purist ideas of wilderness. On the contrary, she contends, "To be truly humble is to put other species first, and our relationship with them second."

Marris's rhetorical reversal, whether or not one finds it persuasive, does not yet take us to the heart of what keeps hubris and humility urgent despite their inadequacy. To investigate that, I turn first to Shellenberger and Nordhaus's call to embrace hubris in the wake of the death of environmentalism, and then to the Dark Mountain Project's eclectic set of attempts to reimagine humility. Shellenberger and Nordhaus burst onto the scene in 2004 with their controversial article, "The Death of Environmentalism," a wake-up call about the problems with mainstream environmentalist politics and organizations in the United States, especially their inability to get legislative agendas on climate change passed. Shellenberger and Nordhaus attributed these failures to a lack of political vision and coalition-building. In *Break Through: From the Death of Environmentalism to the Politics of Possibility* (2007), their follow-up book, Nordhaus and Shellenberger argued that environmentalists need a paradigm shift from the "politics of limits" to a positive politics that embraces hubris. They reinterpreted the legislative successes in the 1960s and 1970s, arguing that while environmentalists think they won victories by framing problems in terms of science and appealing to love of nature (21–26), they actually won victories because of postwar prosperity (27–28). Because people will only protect the environment once their material needs are met, Shellenberger and Nordhaus held that environmentalists should promote global prosperity (15–16, 269). They also argued that environmentalists would only succeed in US politics with a positive vision—a vision of American "greatness" and overcoming (241–273)—that links major federal investments in clean energy with job growth.

In this context, Shellenberger and Nordhaus repurpose the Icarus myth, holding that those who warn against hubris "exaggerate the falling and mask the overcoming," and redefine "hubris" as "the aspiration to imagine new realities, create new values, and reach new heights of human possibility" (271–272). Since 2004 and especially since 2008, some mainstream environmental organizations and Democratic politicians in the United States have taken up their provocative "politics of possibility," at least rhetorically, by promoting clean energy investment, high-speed rail lines, improvements to the electricity grid, green jobs, and the like. These changes in mainstream environmentalist rhetoric are important, though none of these proposals has yet been carried out on the scale necessary to make the emissions reductions we need.

While Nordhaus and Shellenberger, in *Break Through*, reduce humility to merely a mirror for hubris by holding that "humility" is "not timidity but *gratitude*" for "our wealth, freedom, and privilege" (272, 250, emphasis in original), in the "Ecomodernist Manifesto" (Shellenberger et al.), they instead use hubris to rescue the traditional environmentalist version of humility that insists on preserving wilderness intact. Shellenberger and Nordhaus co-authored the manifesto with a group of sixteen others, including Stewart Brand, and published it online through their Breakthrough Institute in April 2015. The manifesto is a strange marriage of their previous techno-optimist environmentalism-through-prosperity arguments with the conservationist vision of setting aside "half the earth" for nature. The authors argue that, for a "good, or even great, Anthropocene," we must "decouple" human economic and social systems from nature, so that human prosperity and well-being does not cause environmental harm, and that we can do this precisely by further intensifying agriculture, urbanization, and industrialization (6–7, 16–19).

Scholarship across the environmental humanities shows that we cannot "decouple" human cultures, histories, and societies from nature in any coherent conceptual way or any viable practical way. Even if it is true that accelerating urbanization and the transition away from fossil fuels will reduce per-capita environmental impacts, as the manifesto's authors argue, it hardly follows that such processes will allow us "to re-wild and re-green the Earth" (15). To assert that is to dismiss, among other things, the foreseeable and unforeseeable problems that the transition to nuclear, solar, and other non-fossil-fuel energy sources—which we do indeed need!—will undoubtedly cause.

According to one of the manifesto's co-authors, ecomodernism took off like a lead balloon at its official launch in London in September 2015 (Lynas). Perhaps this should not come as a surprise: the "Ecomodernist Manifesto" recommends "decoupling" humans from nature when the real energy in environmentalist movements is in fact in small-scale efforts to *recouple* humans with nature. *Break Through* anticipates this disconnect. While Shellenberger and Nordhaus argued that environmental organizations, unlike evangelical churches, have failed to build communities that address middle-class needs for belonging and fulfillment in the postindustrial Unites States (191–203), they ironically overlooked kinds of environmentalism that *do* involve building communities through empowering, small-scale action, such as local food movements. These communities are not only demographically diverse, ranging from African-American and Latino urban farming movements (e.g., Allen; *The Garden*) to the rural movements described by Barbara Kingsolver, but also disrupt the usual political divides, bringing socially conservative Christians and secular progressives together to support and operate food cooperatives, farmers' markets, and CSA programs.

The authors of the "Ecomodernist Manifesto" attack such movements to recouple as dangerous, warning that "any large-scale attempt at recoupling human societies to nature" with "the technologies [of] humankind's ancestors" would "result in an unmitigated ecological and human disaster" (16). Other critics have addressed the absurdity of equating current organic agricultural practices with the farming techniques of "humankind's ancestors" (Monbiot; Smaje). Moreover, a "large-scale attempt" to recouple is a rhetorical fiction which implausibly asks us to imagine that troops of back-to-the-landers are the real threat to the environment, when in fact such movements are inherently small-scale and gradualist. The manifesto's authors, rather than investigating the appeal of recoupling, condescend to such goals as irrational "aesthetic" choices that they will magnanimously allow communities to continue making (27).

While Shellenberger and Nordhaus amplified the infamous line with which Brand opened the *Whole Earth Catalog*—"We are as gods and might as well get good at it" (qtd. *Break Through* 271)—that line often appears in Dark Mountain Project writings as a shorthand for the hubris they oppose. Shellenberger and Nordhaus put a premium on technocratic optimism at the expense of both realism about environmental problems and the humble optimism born of small-scale efforts to create more sustainable socioeconomic forms. The Dark Mountain Project shows that environmentalists and postenvironmentalists long not for brittle, greenwash-thin optimism, but for honesty about our environmental circumstances and our emotional responses to them.

The project began when Paul Kingsnorth and Dougald Hine, both disaffected environmental activists and journalists, called for such honesty in the "Uncivilisation Manifesto" that they self-published in the UK in the summer of 2009. It appeared not long after the housing bubble had burst; Kingsnorth and Hine write that, through the financial crisis, "Hubris has been introduced to Nemesis" (I), and that the ultimate "bubble is civilisation" itself (II). The manifesto attacks "the myth of progress," the related myth that humans are separate from nature, and mainstream environmentalism, which they argue has been co-opted by capitalism (II). Because we are "trapped inside a runaway narrative, headed for the worst kind of encounter with reality," they call for likeminded people to join them in crafting new stories—for "Uncivilised writing" and art that faces our environmental situation, rejects hubris, and forges new kinds of humility (III). Since 2009, the Dark Mountain Project has grown into a blog and a literary journal that has published nine thick issues so far, with the latest devoted to "humbling" ("Dark Mountain"). Here I focus on essays by Kingsnorth and Hine, which give just a taste of the eclecticism embodied in the journal and blog.

The Dark Mountain Project revises humility by linking it to a refusal to deny environmental problems—a linkage that is not as intuitive as it might at first seem. In the manifesto, Kingsnorth and Hine point out that outright climate change denial "may serve as a distraction from a far larger form of denial, in its psychoanalytic sense" (II), calling out all of us who "believe" climate change is happening, but do not change how we live. Sociologist Kari Marie Norgaard, in her ethnographic study of a small town in Norway, concluded that the "social organization of denial," rather than lack of information or caring, best explained community members' avoidance of talking about or taking action on climate change (6). But she also found that a "Norwegian sensibility"—constituted by "humility and thrift," "love of the mountains," "leading a simple life," humanitarianism, and environmentalism—not only did not militate against this kind of denial, but also in fact undergirded it, since community members' complicity in climate change threatened their identity (27–29, 88). In other words, Norgaard ties traditional forms of environmentalist humility to the socially organized denial of climate change.

The Dark Mountain Project remakes humility by tying it, instead, to an insistence on facing large-scale environmental problems. It is arguably more difficult to respond to a global threat with humility than with hubris: the scale of the problem seems to call for a proud assertion that we can vanquish it, as Shellenberger and Nordhaus assume. But in turning toward ongoing environmental crises unarmed with techno-optimist self-assurances about "greatness," the Dark Mountain Project makes possible a greater range of emotional response. In Norgaard's analysis, emotions such as guilt and helplessness play a key role in the social organization of denial; Kingsnorth and Hine found that, in the "thousands" of responses they received to the manifesto, "what came through most powerfully was a sense of *relief*" and "the hope that comes from letting go, from no longer having to pretend" ("End" 2). The Dark Mountain Project gives its participants permission to face and accept their feelings of grief, fear, and loss. Matthew Adams, in one of the few scholarly articles on Dark Mountain, argues that the project usefully questions the ways in which narratives enable the social organization of denial.

Though the manifesto was called apocalyptic (e.g., Gray), Kingsnorth and Hine argue that they simply advocate facing the reality that we are entering a time of rapid ecological change and social and economic collapse ("End" 2). The project certainly participates in environmental apocalyptic discourse, but it revises that discourse in significant ways. While apocalyptic environmental narratives have traditionally warned of disaster in order to prompt action to forestall and avoid it (Buell 285, 308), the "Uncivilisation Manifesto" represents catastrophe as already underway. In the editorial opening issue 4 of *Dark Mountain*, the editors call for "post-cautionary tales." They argue that the typical environmental cautionary tale—in other words, the jeremiad—presumes that we can still avert catastrophe, and thus undercuts the urgency of its own message. The editors call instead for post-cautionary tales which "do not seek to avert crisis or radical change, but which acknowledge that we are already living through those things and that we are going to have to deal with the consequences" (Hine et al. 2–3).

Though the Dark Mountain Project contests orthodoxies that would condemn attempts to question progress or techno-optimism as nostalgic, it does not reject technology: the blog is well-designed and active; they are on Twitter. Kingsnorth has a penchant for deep ecology and a more anti-technological bent, as his essay "Dark Ecology," which appeared in *Orion* as well as in *Dark Mountain*, shows. There he reflects on scything and teaching others to scythe, on reading the writings of Ted Kaczynski (the Unabomber) and finding himself agreeing, uncomfortably, with Kaczynski's criticisms of technology, though not his calls for violence. Kingsnorth also argues that "neo-environmentalists," including Stewart Brand, Emma Marris, Michael Shellenberger, and Ted Nordhaus, risk completing environmentalism's sell-out to progress. But he acknowledges that they are right about some key things, including the fact that traditional environmentalist campaigning has "hit a wall" and that "we are not headed, now, toward convivial tools" (15–17, 23).

Dougald Hine takes a different tact. In an essay titled "Remember the Future?," Hine advocates a "historical humility" which acknowledges "the unknowability of the future, its capacity to humble us and take us by surprise, our inability to control it" (270). He argues that "improvisation" is "the deep skill and attitude which we need for the times we're already in" (264) and that we need an Epimethean rather than Promethean gaze—hindsight rather than foresight—not in order to return to the past, but in order to draw on, revise, and improvise upon ways of living and thinking which "become useful again" in new registers (269). In an interview titled "Rehoming Society: Ivan Illich & the Vernacular," Hine resists dismissing commodified forms of the vernacular that appear in industrial capitalism, such as $5 artisanal loaves of bread, as "totally counterfeit," treating them pragmatically as a possible "line of transmission" for certain practices (97–98). In both the essay and the interview, Hine uses a telling image for the relationship to the past that he envisions: that of the improvisational storyteller, who looks backwards rather than forwards because she does not know the end of the story yet. When the storyteller weaves back in a motif or character from earlier in the story, Hine notes, both the storyteller and the audience experience a deep satisfaction. Hine argues that this kind of return, the serendipitous recognition that something from earlier in the story can be of use in stumbling forward, is "not the same thing as a desire to rewind" (105). Hine revises traditional environmentalist humility by changing that which we are humble toward: this is not human humility before Nature, but humility toward the usually spurned knowledge and practices of other humans in the past and present, toward the contingency of history, and even toward collective human activities that have themselves become a historical force.

Humility comes to mean not only a humble attitude toward the unpredictable changes we face, but also the willingness to slow down and act, improvisationally, on the scales available to us. At a time when environmentalists and environmental humanists need ways to think and feel about environmental problems that are not hubristic, techno-optimist proclamations that all will be solved, the Dark Mountain Project's syncretism, eclecticism, and willingness to challenge orthodoxies are salutary. Such new narratives reframe the small-scale practices of humility that some environmentalists have long advocated, enabling us to stitch together intellectually and emotionally coherent ways of living in spite of the continuing rhetorical contests between escapist fantasies of miraculous overcoming, on the one hand, and apocalyptic fears of catastrophe, on the other.

References

Adams, Matthew. "Inaction and Environmental Crisis: Narrative, Defence Mechanisms, and the Social Organisation of Denial." *Psychoanalysis, Culture & Society* 19 (2014): 52–71. Web. 19 April 2016.

Allen, Will, with Charles Wilson. *The Good Food Revolution: Growing Healthy Food, People, and Communities.* New York: Gotham Books, 2012. Print.

Berry, Wendell. "Wendell Berry Angry." *The CoEvolution Quarterly* Summer 1976: 8–11. Print.

Brand, Stewart. "Space Colonies: Summary." *The CoEvolution Quarterly* Spring 1976: 4. Print.

Brand, Stewart et al. "Is Balance Really Possible Where Even Gravity Is Manufactured?: Comments on O'Neill's Space Colonies." *The CoEvolution Quarterly* Spring 1976: 5–53. Print.

Buell, Lawrence. *The Environmental Imagination: Thoreau, Nature Writing, and the Formation of American Culture.* Cambridge, MA: Harvard University Press, 1995. Print.

Butler, Tom. "Introduction: Lives Not Our Own." *Keeping the Wild: Against the Domestication of Earth.* Ed. George Wuerthner, Eileen Crist, and Tom Butler. Washington, DC: Island Press, 2014. Print.

Carson, Rachel. *Silent Spring.* 1962. Boston: Houghton Mifflin, 1994. Print.

"Dark Mountain: Issue 9: The Humbling." Dark Mountain Project blog. 15 April 2016. Web. 19 April 2016.

The Garden. Dir. Scott Hamilton Kennedy. Black Valley Films, 2008. Film.

Gerhardt, Christine. *A Place for Humility: Whitman, Dickinson, and the Natural World.* Iowa City: University of Iowa Press, 2014. Print.

Gray, John. "Uncivilisation: The Dark Mountain Manifesto." Review. *The New Statesman* 10 September 2009. Web. 16 March 2016.

Hård, Mikael and Andrew Jamison. *Hubris and Hybrids: A Cultural History of Technology and Science.* New York: Routledge, 2005. Print.

Hine, Dougald. "Rehoming Society: Ivan Illich & the Vernacular: A Conversation with Sajay Samuel." *Dark Mountain* 3 (Summer 2012): 90–105. Print.

——. "Remember the Future?" *Dark Mountain* 2 (Summer 2011): 260–271. Print.

Hine, Dougald, Nick Hunt, Paul Kingsnorth, and Adrienne Odasso. "Editorial: Post-cautionary Tales." *Dark Mountain* 4 (Summer 2013): 1–3. Print.

Kingsnorth, Paul. "Dark Ecology." *Dark Mountain* 3 (Summer 2012): 7–27. Print.

Kingsnorth, Paul and Dougald Hine. "Editorial: It's the End of the World as We Know It (and We Feel Fine)." *Dark Mountain* 1 (Summer 2010): 1–4. Print.

——. "Uncivilisation: The Dark Mountain Manifesto." Summer 2009. The Dark Mountain Project. Web. 1 August 2015.

Kingsolver, Barbara. *Animal, Vegetable, Miracle: A Year of Food Life.* New York: HarperCollins, 2007. Print.

Kirk, Andrew. *Counterculture Green: The Whole Earth Catalog and American Environmentalism.* Lawrence: University Press of Kansas, 2007. Print.

Leopold, Aldo. *A Sand County Almanac and Sketches Here and There.* New York: Oxford University Press, 1949. Print.

Lynas, Mark. "Ecomodernism Launch was a Screw-up of Impressive Proportions." *The Guardian* 30 September 2015. Web. 15 March 2016.

Marris, Emma. "Handle with Care." *Orion Magazine* April 2015. Web. 15 March 2016.

Mills, Stephanie. *Epicurean Simplicity.* Washington, DC: Island Press, 2002. Print.

Monbiot, George. "Meet the Ecomodernists: Ignorant of History and Paradoxically Old-fashioned." *The Guardian* 24 September 2015. Web. 15 March 2016.

Nordhaus, Ted and Michael Shellenberger. *Break Through: From the Death of Environmentalism to the Politics of Possibility.* Boston, MA: Houghton Mifflin, 2007. Print.

Norgaard, Kari Mari. *Living in Denial: Climate Change, Emotions, and Everyday Life.* Cambridge, MA: MIT Press, 2011. Print.

O'Neill, Gerard. "The High Frontier," "Testimony," and "'Is the Surface of a Planet Really the Right Place for an Expanding Technological Civilization?': Interviewing Gerard O'Neill." *The CoEvolution Quarterly* Fall 1975: 6–28. Print.

Pandora's Promise. Dir. Robert Stone. Impact Partners, 2013. Film.

Plumer, Brad. "After a 70-year Drop, Small Farms Make a (Small) Comeback." *The Washington Post* 12 October 2012. Web. 8 August 2013.

Pollan, Michael. *In Defense of Food: An Eater's Manifesto*. New York: Penguin, 2009. Print.

Rich, Nathaniel. "The Mammoth Cometh." *The New York Times Magazine* 27 February 2014. Web. 27 August 2015.

Sandler, Ronald. "Environmental Virtue Ethics." *The International Encyclopedia of Ethics*. Ed. Hugh LaFollette. Hoboken, NJ: Blackwell, 2013. Web. 1 February 2016.

Schumacher, E. F. *Small Is Beautiful: Economics as if People Mattered*. New York: Harper & Row, 1973. Print.

Shellenberger, Michael and Ted Nordhaus. "The Death of Environmentalism: Global Warming Politics in a Post-Environmental World." 2004. Breakthrough Institute. Web. 1 February 2016.

Shellenberger, Michael et al. "An Ecomodernist Manifesto." April 2015. Breakthrough Institute. Web. 1 February 2016.

Smaje, Chris. "Dark Thoughts on Ecomodernism." The Dark Mountain Project Blog. 12 August 2015. Web. 15 March 2016.

Soper, Kate. "Alternative Hedonism, Cultural Theory and the Role of Aesthetic Revisioning." *Cultural Studies* 22.5 (September 2008): 567–587. JSTOR. Web. 15 January 2013.

USDA Agricultural Marketing Service. "National Count of Farmers Market Directory Listings Graph: 1994–2013." 3 August 2013. Web. 8 August 2013.

Weinstein, Joshua A. "Humility, from the Ground Up: A Radical Approach to Literature and Ecology." *Interdisciplinary Studies in Literature and Environment* 22 (Autumn 2015): 759–777. Web. 19 April 2016.

White, Monica. "D-Town Farm: African American Resistance to Food Insecurity and the Transformation of Detroit." *Environmental Practice* 13.4 (December 2011): 406–417. Print.

26

LOSING PRIMEVAL FORESTS

Degradation narratives in South Asia

Kathleen D. Morrison

It is not only in the modern imagination that forests cast their shadow of primeval antiquity; from the beginning they appeared to our ancestors as archaic, as antecedent to the human world. We gather from mythology that the vast and somber wilderness was there before, like a precondition or matrix of civilization ... the forests were *first*.

—Robert Pogue Harrison (1)

In its rude beginnings, the greater part of every country is covered with wood, which is then a mere incumbrance of no value to the landlord, who would gladly give it to any body for the cutting. As agriculture advances, the woods are partly cleared by the progress of tillage, and partly go to decay in consequence of the increased number of cattle.

—Adam Smith (190)

Some ideas about the past, and the narratives built from them, are so pervasive that they have almost disappeared as analytical objects. These historical narratives are nevertheless powerful, structuring scholarship across a range of disciplines and activism across a range of political positions. This chapter considers some narratives about forests in South Asia, accounts with significant contemporary implications despite claims to represent the distant past. One such sylvan tale is that all of India was once covered in forests, a primeval condition consistently degraded through time, but with an acceleration of loss following the introduction of modernity, colonialism, or Islam. All good stories contain at least an element of truth, and these are no exception. Indeed, one might argue that the power of narratives lies in both their structure and their content—they feel right and they contain some nugget of fact, comfortably confirming what we thought we knew. Because taken-for-granted environmental narratives come to be so on the basis of both evidence and form, I undertake the somewhat perilous business of considering both.

It is telling that narratives of change, many operating as unquestioned truisms or tropes, move rather easily across venues. The notion of a balance of nature, for example, may be invoked by both an environmental historian and an activist. A climate modeler and a policymaker may both believe that significant human impact on India's vegetation began only with British colonization. Poets and archaeologists alike may invoke an ancient past in which forests covered the subcontinent. But in none of these three cases is the dominant narrative entirely correct. Instead, these narratives all elide important exceptions and in some cases entirely misrepresent past trajectories of change.

The stakes of these easy elisions are multiple. First, they impede us from building more accurate pictures of the past and from understanding the ways in which past human and nonhuman action created the world of today, itself a critical goal. Also at stake are the conceptual foundations by which we partition the world—nature and culture, nonhuman and human, stability and instability—and, by extension, our scholarship, imagination, and political action. Coming to terms with the long and diverse histories of landscape anthropogenesis forces us to call into question simple binaries and pushes us toward the development of new forms of inquiry and new intellectual frameworks which meaningfully incorporate analyses of both humans and the natural world.

Both the paleosciences and the historical social sciences and humanities share a commitment to studying the past, but for the most part, these fields have run on parallel, nonintersecting tracks. When there is interaction, it can be highly selective or mediated through secondary accounts. One increasingly encounters the idea in the physical sciences, for example, that significant human impact on the earth is very recent, and that we have now entered a radically new period—the Anthropocene (Zalasiewicz et al.). Such a view, while usefully highlighting the role of human beings as planetary-scale geological agents, is surprisingly blind to the longer-term trajectory of anthropogenic change, which began with the first agriculture around ten thousand years ago. While the scale and pace of human-induced change certainly increased after the industrial revolution, it is necessary to consider evidence from the Holocene as a whole and not simply focus on the recent past (Ramankutty and Foley). The fact of recent and significant change—especially in Western Europe—does not, however, mean that earlier changes were insignificant. The use of dates such as 6,000 years ago or 1850 CE as baselines elides longer-term trajectories of change and magnifies European histories onto the globe (Morrison, "Provincializing").

As a concept, the Anthropocene is unnecessary, since its key feature—humans as agents of global change—is true of much of the Holocene as well. The Anthropocene concept, further, risks falsely implying that the less recent history of human–environment interactions was one of balance and harmony, with humans having minimal environmental impact, a conclusion that cannot be sustained. Instead, we must consider our present dilemma as the culmination of long-term trends extending across much of the Holocene (Malm and Hornborg 2014).

In this chapter, I address one pervasive idea structuring much environmental understanding in South Asia: the myth of the primeval forest, an extensive woodland that "once" covered the subcontinent. This myth is linked in complex ways with other beliefs, including the existence of a universal colonial ecological watershed and the use of the "precolonial" as a homogenized starting point for "baseline thinking" about environmental change (Morrison, "Conceiving Ecology"). These ideas are expressed in the form of basic narratives—stories of primitivity and improvement, nationalism and romanticism, development and stasis—widely shared across science, history, and governance. These narratives are built from basic conceptual elements that can be combined in different ways so that even diametrically opposed perspectives may be constructed from similar assumptions. Calling historical understandings narratives suggests a way forward, one that attends not only to the content of our histories, but also to their structure (Cronon).

Still, content matters. It is not the case that the humanities give us stories and science facts; instead, we share fundamental preoccupations. In much scientific literature, the same narratives of the past that structure environmental activism and history are also operative. The data we generate about the past are never innocent, and we are responsible for how we represent them.

Sylvan tales: forests as locations of cultural value

Every landscape is a work of the mind, and thus any consideration of forest history, from the scientific to the humanistic, contends with rich cultural imaginations of forests. Forests are vexed objects, poorly defined yet meaningfully rich. While ecologists can, and have, provided definitions of forests, none are entirely satisfactory and significant controversy still reigns over how precisely to identify forests, a pressing concern given their status as objects of management, law, and governance (Grainger). As analytical objects, then, forests can be shadowy, even in the natural sciences. This ambiguity often works in ways that invite reliance upon established narratives of change.

Centuries of textual representations in India celebrate locations of cultural value on the landscape, of which the two most prominent are irrigated fields and forests (Morrison and Lycett, "Constructing Nature"). Forests, in Indian literature, are often places apart, home to religious figures and locations for spiritual or tactical retreat as well as potential lairs for bandits and criminals (Guha; Thapar). As literary devices, forests are liminal, providing transition or refuge. Forests, too, were (and are) locations of alterity, containing people even today distinguished as "tribal" in contradistinction to caste society. Conversely, forests have also long provided reserved locations for elite sport.

Considering the "received view" of forest loss in light of such textual valorizations, we can perhaps understand notions of precolonial sustainability. Associations between forests and religious ascetics, for example, lubricate the idea of relationships with nature as essentially spiritual, a welter of cultural valences inflecting stories of progress, stability, decline and, occasionally, redemption.

Constructing narratives: origins, improvement, degradation, and development

In contexts as diverse as literature, science, and activism, one can locate strikingly similar narratives. While these may themselves be at odds—a developmentalist view differs from a romantic-conservation view, for example—all are built from fundamental, taken-for-granted assumptions: they counterpose balance, homeostasis, harmony, and stability to progress, improvement, and civilization, on one hand, or to decline, decay, and degeneration, on the other. Below, I highlight narratives from environmental science and popular activism, accounts invoking common tropes of loss and degradation but also scientific progress and development.

One kind of narrative is progressive and civilizational, indexed in the colonial context especially by the notion of *improvement*. In much of the formerly colonized world, this perspective was a dominant official narrative, seeking to justify engagement in terms of uplifting, improving, or "civilizing" colonized peoples. Discourses of improvement abounded in British India, exemplified by the development of "scientific forestry" in the nineteenth century, with its logic of management, rationality, profit, and planning.

In an inversion of the colonial perspective, some scholars embraced romantic visions of the past, shifting the signs of value in the colonial perspective. Here, negative valuation of a "backward" past is replaced by a golden age in which traditional arrangements exemplify an ecosystem-like state of balance, harmony, and stability (Gadgil and Guha; Shiva). Traditional harmony with nature is interrupted by colonial intervention bringing deforestation, degradation, and a breakdown of traditional forms of ecological management, the latter assumed to be communal, harmonious, and equitable (but see Mosse). Depicting precolonial worlds as unchanging and equilibrial deprives both humans and the natural world of their dynamic histories, inverting colonial discourses without subjecting them to serious challenge (Prasad). Such romantic inversions are still perspectives of the powerful, supporting a simplified and temporally evacuated status quo placing high castes and others at the apex of the society and ecosystem. It is thus surprising that ecological romanticism also holds such fascination for liberal environmental activists.

Ecological romanticism relies fundamentally on declensionist accounts. There is not much action before the fall from grace; a good story requires a crisis and for many, especially in science and environmental activism, the crisis is central. Indeed, the crisis itself—the loss of forests, traditions, innocence—is so centrally imagined that little effort is expended in actually evaluating the state of affairs in the prelapsarian past. This is certainly the case in South Asia, where precolonial pasts are more often imagined than investigated. What was the "original," "natural" state of South Asian vegetation? The subcontinent was a human-occupied landscape long before the end of the Pleistocene, so there is no human-free analogue for paleoecology, no originary period against which to measure later transformations. The search for a prehuman baseline is futile. Within the Holocene, there is a widespread assumption that the trajectory of landscape change has been from forests to open areas, with forests always in decline relative to the growth of human population. Even without evidence of paleoenvironmental conditions, many have assumed that clearing of primeval forest occurred at some time in the past and that "originally" the subcontinent was entirely cloaked in forests.

Progressivist and declensionist narratives are more alike than different, mirror images assembled from a common stock of assumptions (Morrison, "Conceiving Ecology"). Allowing for interstitial moments of stability, these narrative modalities come together in development discourses, both technocratic and activist. Much development discourse partakes of both declensionist and progressivist narratives, postulating, for example, that indigenous people once lived in harmony with nature until modernity or colonialism destroyed both nature and culture. This sorry state of affairs requires, then, Western science to redirect the story into a narrative of progress (O'Brien).

Environmental science, too, shares certain narrative conventions and assumptions with both colonial and romantic perspectives. In ecology, the notion of a *climax vegetation* is related to concepts of plant succession, especially "primary" succession taking place under "natural" conditions (Chazdon; Kingsland). Like models of cultural evolution, succession studies substitute space for time; although vegetation succession refers to a diachronic process, it is often (re)constructed on the basis of observations of a spatial series of extant plant communities rather than by using paleoecological data. Historical evidence is thus subverted to imaginative history (Davis). Concepts of vegetation succession, and especially climax, suggest a beginning and an end, the latter governed by a stable (albeit dynamic) equilibrium. The vegetation climax constitutes the mythical origin point for both progressivist and declensionist discourses; it *is* the state of nature humans will eventually despoil, a separation of humans from the natural world that is itself profoundly nonecological (Morrison, "Conceiving Ecology").

The pervasive frameworks of ecology which call for the construction of human-free nature as analytical baselines can lead to some curious results. Paired with the common assumption that negative human impacts are progressive and accelerating, the result is narratives that may completely reverse chronology in an effort to make "nature" precede "culture." Dhakal and colleagues, for example, in a study of the impact of cardamom cultivation on montane forest ecosystems in Sri Lanka, carefully sampled, identified, and analyzed vegetation assemblages from plots in the Knuckles Forest Reserve in central Sri Lanka, where cultivation has been banned since 1985. At first glance, the analysis seems straightforward. Areas of "cardamom plantation" are compared to areas of "natural forest" in terms of stem density, stand basal area, canopy openness, and soil nutrients.

A closer look at this specific forest reveals a problem, however. As Dhakal and colleagues note, *all* of the forest has been potentially subject to agriculture in the past. Although official records note that 3,000 ha of the 17,500 ha reserve was formerly under cardamom cultivation, this is known to be a vast underestimation. Indeed, they write, "we observed very little natural forest without planted cardamom during three years of fieldwork," explaining that some areas have been under cultivation for more than a century (152). This is borne out by the presence of cardamom in all plots. In order to create a baseline, the older agroforestry plots are, however, simply re-cast as "nature" (*natural forest*), their human histories erased, while more recently abandoned agroforestry plots stand in for degraded "culture" (*cardamom plantations*). The long-term, complex history of planting, fallowing, replanting, and abandoning specific locations is thus re-cast as (older) nature and (younger) culture, the former pristine and the latter degraded, by definition.

This construction of a nature–culture duality out of a mosaic of socionatural spaces creates a curiously inverted chronology for a study of vegetation succession. Dhakal et al. note that "the densities of seedlings, saplings, and small trees in cardamom plantations have not recovered to pre-disturbance values during the 25 years since active management ceased" (157). But the so-called natural forest here is not pre-disturbance at all—if anything, it is *post-disturbance, further along or later in a successional pathway, not earlier*. The natural as an analytical category was first *constructed* by carving away the human contribution to this socionatural landscape, and then set up as necessarily *prior to* human disturbance. But the temporality is inverted here. The plots imagined to represent "nature" are really successionally later than the plots imagined to represent "culture." This inversion leads the authors to overlook the fact that they know the answer to how long "recovery" may take. They note rather wistfully, "The existence of pervasive impacts of cardamom cultivation so long after active management ceased suggests that recovery to pre-disturbance conditions will take a very long time" (159), in fact, if they are right about the history of cultivation, they know precisely how long this process will take. If the oldest cardamom plantings are really around a hundred years old, but with most beginning in the 1960s (152), then it will take 50–110 years for the "degraded" plots to look like the "natural" plots. This is clear when we re-imagine the "pre-disturbance, natural forest" plots as fields left fallow for longer than neighboring locations and not as untrammeled nature.

Stories of forests lost and regained are not limited to scholars, but also underpin environmental activism. Consider an account presented by the Energy and Resources Institute's website for children, Edugreen (2015) on the history of forests in India:

There is enough evidence to show that dense forests once covered India. The changing forest composition and cover can be closely linked to the growth and change of civilizations. Over the years, as man progressed the forest began gradually depleting. The growing population and man's dependence on the forest have been mainly responsible for this.

All ancient texts have some mention of the forest and the activities that were performed in these areas. Forests were revered by the people and a large number of religious ceremonies centred on trees and plants ... Sacred groves were marked around the temples where certain rules and regulations applied.

When Chandra Gupta Maurya came to power ... he realized the importance of the forests and appointed a high officer to look after the forests. Ashoka stated that wild animals and forests should be preserved and protected. He launched programmes to plant trees on a large scale ...

During the Muslim invasions a large number of people had to flee from the attacks and take refuge in the forests. This was the beginning of a phase of migration to the forest. They cleared vast areas of forests to make way for settlements.

During the early part of the British rule, trees were felled without any thought ... The history of modern Indian forestry was a process by which the British gradually appropriated forest resources for revenue generation ... But after some time, the British began to regulate and conserve.

This account raises more issues than can be discussed here, but the narrative structure is clear. Forest loss and precolonial balance somehow co-exist, though appropriately indigenous rulers like the rulers of the Mauryan empire on the Gangetic plain have impeccable conservation credentials. Muslim "invaders" upset the balance, a loss further accelerated under British colonialism. Science and management intervene, however, in developmentalist mode, providing hope for the future.

Does the empirical basis of such narratives matter? Surely it does. Basic standards of veracity aside, there are always multiple interests at stake in any environmental narrative, many with consequences for habitats and livelihoods. One can read of the imminent destruction deforestation poses to the Himalayan slopes in more or less the same terms, and in the same locations, in texts dating hundreds of years apart (Saberwal). This receding horizon of disaster must eventually reduce the credibility of such accounts. We do better to situate contemporary action on solid historical footings rather than assumed progressions, or regressions, of change.

While the human footprint may be evident in areas long under intensive agriculture, it is less obvious that forested uplands are also humanized landscapes, with images of stable precolonial landscapes only moderately affected by human land use quite common (Collins et al.). Modern forests are almost uniformly seen as "remnants" of deforestation, postclimactic (and postclimax) residues of a story beginning in a vast primeval forest (Morrison, "Discourses of the Remnant"). Despite literature on recent forest transitions, or shifts from net loss of woody cover to net increase (Hecht et al.; Rudel et al.), perceptions of unalloyed deforestation continue. The power of deforestation narratives derives in part from the hold of declensionist narratives in general, but also perhaps from fuzziness about prior conditions, vast tree-covered landscapes of the indefinite past looming large. In South Asia, several of the most persistent accounts of forest loss can be found in locations where, as it turns out, there either were *never* extensive forests or where forest expansion is very recent.

263

The Gangetic plain: forests conquered?

The floodplain of the River Ganges in northern India is one of the most densely settled parts of India today. If any part of India seems fully cultural, this is it. But even here, forests loom in the imagination. For many years, archaeologists suggested that agriculture could only have been possible on the densely forested Gangetic plain after the development of iron tools made clearing dense forests and plowing heavy soils possible (Agrawal; but see Lal). The first blow to the heroic account of forests felled by iron axes came from recognition that elsewhere, ancient people handily cut forests with stone axes. The postulated resistance of the Gangetic forests to farming also began to seem less likely as archaeologists began to look for pre-iron settlements—and found them (Allchin 65). Early farmers did settle on the fertile northern floodplains and while talk of heavy soils declined, forest clearing was still presumed.

Gangetic forests show up in scholarship on later time periods, too. Historians discussing the seventeenth and eighteenth centuries also "see" extensive forests on the Gangetic plain, creating a baseline for colonial-era depredations. Forests thus play a role in both the progressivist narrative of an impenetrable forest conquered by humans through technological ingenuity and in a declensionist narrative of rupture and loss. Recent paleoecological studies, however, show definitively that many parts of the Gangetic plain *were never densely forested at all* (Farooqui and Sekhar; Sharma et al.), with open vegetation dominating throughout the Holocene.

The Western Ghats: remnant forests, remnant people

Unlike the Gangetic plain, the Western Ghat Mountains of Southern India have considerable forest cover today. Represented as remote and exotic, the Ghat uplands are invisible to mainstream history in a way that has allowed upland peoples to be represented both as apart from and irrelevant to the main currents of history and as agentive threats to contemporary biodiversity. Indeed, the historical constitution of the Ghats has worked to set aside politics, power, and cultural production in favor of "nature" and difference, constructing both people and places as natural and fragile (Morrison and Lycett, "Constructing Nature"). By the late fifteenth century and into the Early Modern period, Ghat forest products featured in both peninsular and international trade, with spices such as pepper of major economic importance (Morrison and Lycett, "Forest Products"). Many upland peoples, later to be represented by anthropologists as untouched primitives, grew or gathered forest products and became key players in global trade circuits. Thus, even the verdant slopes of the Ghats cannot be seen as culture-free. At higher elevations, discourses of natural forests and human "remnants" give way to narratives of culturally affected forests and forest "remnants" (Morrison, "Discourses of the Remnant").

Scientists have been just as prone as others to structuring narratives using imagined pasts. While colonial accounts of landscape change in the high-elevation Nilgiri Hills blamed local groups for woodland loss (Sutton), ecologists, too, often represent the patchwork of forests and grasslands as products of loss. Open grasslands are often argued to be degraded evergreen Shola forests, newer "culturalized" places formed as products of degraded "natural" forests (e.g. Mohandass and Davidar 20). Grasslands, it is argued, are recent human creations, caused by grazing, burning, and cultivation. Paleoenvironmental studies show, however, that this is *almost exactly the reverse of the actual vegetation change*. Caner et al., among others, demonstrate that grasslands have characterized much of the vegetation above 1,500 meters since the Late Pleistocene ("Spatial Heterogeneity"). Several studies of pollen, soils, and stable carbon isotopes show that much of the Nilgiri Shola forest is of quite recent origin, formed within the last one to two hundred years (Caner et al., "Occurrence"; Caner at al., "Spatial Heterogeneity"; Rajagopalan et al.; Sukumar et al.). Most sampling locations record upland grasslands in the early Holocene, open formations that were slowly afforested starting either at about 2000 BCE or, in other cases, more quickly transformed from grassland to forest during the last several hundred years.

In these high elevation regions, paleoenvironmental data challenge ahistorical accounts, leading the biologists conducting these studies to conclude that, in fact, the patchiness of the Shola forests is *not* a consequence of recent degradation. Even in these newer accounts, however, humans are only ever mentioned as agents of forest destruction. Evidence for very recent forest expansion, not found across all sampling locations, is *never* attributed to human action but to climate change instead. While details of this change require study, the expansion of the Shola forest took place within a cultural landscape, in a region with a complex history of human land use. It might be expected that over the centuries, agriculturalists, foragers, and others have modified vegetation in this biodiversity hotspot in ways easily misrecognized as natural. The asynchrony of vegetation change may not be fully explained by climate change alone. The patchy high-elevation forests of the Nilgiris are likely to have changed in response to both climate and human activity, the latter effecting both woodland expansion and retreat. This does not appear to be a simple story.

Discussion

All ecological concepts are cultural constructs, but because "forest," like "vegetable" or "weed," is so fundamental, its status as a contested scientific category seems surprising. Like weeds and vegetable, forests are imprecise, laden with affect and imagined histories. Forests are, as Harrison writes, *first*. But are they? There is no doubt that significant forest loss was a widespread consequence of agricultural expansion throughout the Holocene, but not all parts of the earth "began," if indeed we can fix an origin point, covered in forests.

In both European and South Asian traditions, forests are special places, revered and feared, standing in for multiple forms of alterity: caste and tribe, cultivated and unsown, domesticated and wild, law and banditry. As such they participate easily in the powerful narratives of change also applied to human history—savagery to civilization, falls from inno-cence, scientific progress. Stories of primeval forests do important cultural work, no matter what their empirical basis may be. These stories build histories, but often do so on the basis of space for time substitutions or "thought experiments" without recourse to historical evidence. Narratives that ignore evidence are, however, vulnerable—hollow accounts easily inverted to serve alternate agendas.

The power of environmental narratives explains, to some extent, the curious resistance of stories of forest loss to the corrosion of empirical evidence. On the Gangetic plain, it took decades for scholars to eschew the story of triumphant iron tool-users cutting back primeval forests, an account even now repeated. In the Nilgiris, "remnant" Shola forests join the west African "forest islands" described by Fairhead and Leach as examples of forest patches (not fragments) possibly formed through human land use. Here, as elsewhere, narratives are consequential. Not only do they inform scientific concepts and practice, but they also guide policy, with the descendants of forest communities, the very people who created these landscapes over the last several thousand years, increasingly evicted from protected areas and blamed for a history of deforestation that never happened. The content of these pasts thus matters and is, in many cases, recoverable, even when critical changes took place long ago.

References

Agrawal, D.P. *The Copper Bronze Age in India*. Delhi: Munshiram, 1971. Print.

Allchin, F.R. *The Archaeology of Early Historic South Asia*. Cambridge: Cambridge University Press, 1995. Print.

Caner, L., Fancois Toutain, Gerard Bourgeon, and Adrien-Jules Herbillon. "Occurrence of Sombric-like Subsurface A Horizons in Some Andic Soils of the Nilgiri Hills (Southern India) and Their Palaeoecological Significance." *Geoderma* 117 (2003): 251–265. Print.

Caner, L., D. L. Seen, Y. Gunnell, B. R. Ramesh, and G. Bourgeon. "Spatial Heterogeneity of Land Cover Response to Climatic Change in the Nilgiri Highlands (Southern India) since the Last Glacial Maximum." *The Holocene* 17.2 (2007): 195–205. Print.

Chazdon, R. *Second Growth: The Promise of Tropical Forest Regeneration in an Age of Deforestation*. Chicago: University of Chicago Press, 2014. Print.

Collins, N. M., J. A. Sayer, and T. C. Whitmore. *The Conservation Atlas of Tropical Forests: Asia and the Pacific*. Cambridge: UNEP-WCMC, 1991. Print.

Cronon, William. "A Place for Stories: Nature, History, and Narrative." *The Journal of American History* 78.4 (1992): 1347–1376. Print.

Davis, M. B. "Ecology and Paleoecology Begin to Merge." *Trends in Ecology and Evolution* 9 (1994): 357–358. Print.

Dhakal, B., M. A. Pinard, and I. A. U. Nimal Gunatilleke. "Impacts of Cardamom Cultivation on Montane Forest Ecosystems in Sri Lanka." *Forest Ecology and Management* 274 (2012): 151–160. Print.

Edugreen. "Forestry: History of Forests in India." The Energy and Resources Institute (TERI), New Delhi. 10 October 2015. Web. 5 December 2015.

Farooqui, A. and B. Sekhar, "Climate Change and Vegetation Succession in Lalitpur Area, Uttar Pradesh (India) During Late Holocene." *Tropical Ecology* 52.1 (2011): 69–77. Print.

Gadgil, Madhav and Ramachandra Guha. *This Fissured Land: An Ecological History of India*. Delhi: Oxford University Press, 1992. Print.

Grainger, Alan. "Pan-Tropical Perspectives on Forest Resurgence." *The Social Lives of Forests: The Past, Present, and Future of Woodland Resurgence*. Ed. Susanna B. Hecht, Kathleen D. Morrison, and Christine Padoch. Chicago: University of Chicago Press, 2014. 84–96. Print.

Guha, S. *Environment and Ethnicity in India*. Cambridge: Cambridge University Press, 1999. Print.

Harrison, Robert Pogue. *Forests: The Shadow of Civilization*. Chicago: University of Chicago Press, 1992. Print.

Hecht, Susanna B., Kathleen D. Morrison, and Christine Padoch. "From Fragmentation to Forest Resurgence: Paradigms, Representations, and Practices." *The Social Lives of Forests: The Past, Present, and Future of Woodland Resurgence*. Ed. Susanna B. Hecht, Kathleen D. Morrison, and Christine Padoch. Chicago: University of Chicago Press, 2014. 1–13. Print.

Kingsland, S. *Modeling Nature: Episodes in the History of Population Ecology*. Chicago: University of Chicago Press, 1995. Print.

Lal, M. "Iron Tools, Forest Clearance, and Urbanization in the Gangetic Plains." *Man and Environment* 10 (1986): 83–90. Print.

Malm, A. and A. Hornborg. "The Geology of Mankind? A Critique of the Anthropocene Narrative." *The Anthropocene Review* 1 (2014): 62–69. Print.

Mohandass, D. and P. Davidar. "Floristic Structure and Diversity of a Tropical Montane Evergreen Forest (Shola) of the Nilgiri Mountains, Southern India." *Tropical Ecology* 50.2 (2009): 219–229. Print.

Morrison, K. D. "Conceiving Ecology and Stopping the Clock: Narratives of Balance, Loss, and Degradation." *Shifting Ground: People, Animals, and Mobility in India's Environmental History.* Ed. M. Rangarajan and K. Sivaramakrishnan. Delhi: Oxford University Press, 2014. Print.

———. "Discourses of the Remnant: Peoples, Forests, and Sacred Groves of Southern India." *Nature Today: Studies in Ecology and Environment.* Ed. A. Baviskar. Delhi: Oxford University Press, in press. Print.

———. "Provincializing the Anthropocene." *Seminar* 673 (2015): 75–80. Print.

Morrison, Kathleen D. and M. T. Lycett. "Constructing Nature: Socionatural Histories of an Indian Forest." *The Social Lives of Forests: The Past, Present, and Future of Woodland Resurgence.* Ed. M. Rangarajan and K. Sivaramakrishnan. Delhi: Oxford University Press, 2014. University Press Scholarship Online. Web. 3 September 2016.

———. "Forest Products in a Wider World." *Connections and Complexity.* Ed. S. Abraham, T. Raczek, and U. Rizvi. Walnut Creek: Left Coast Press, 2013. 127–142. Print.

Mosse, D. "Colonial and Contemporary Ideologies of Community Management: The Case of Tank Irrigation in South India." *Modern Asian Studies* 33.2 (1999): 303–338. Print.

O'Brien, W. E. "The Nature of Shifting Cultivation: Stories of Harmony, Degradation, and Redemption." *Human Ecology* 30 (2002): 483–502. Print.

Prasad, A. *Against Ecological Romanticism.* New Delhi: Three Essays Collective, 2003. Print.

Rajagoplan, G., R. Sukumar, R. Ramesh, R. K. Pant, and G. Rajagopalan. "Late Quaternary Vegetational and Climatic Changes from Tropical Peats in Southern India." *Current Science* 73.1 (1997): 60–66. Print.

Ramankutty, Narman and J. A. Foley. "Characterizing Patterns of Global Land Use: An Analysis of Global Croplands Data." *Global Biogeochemical Cycles* 12 (1998): 667–685. Print.

Rudel, Thomas K., Oliver T. Coomes, Emilio Moran, et al. "Forest Transitions: Toward a Global Understanding of Land Use Change." *Global Environmental Change* 15 (2005): 23–31. ScienceDirect. Web. 3 September 2016.

Saberwal, V. K. "Science and the Desiccationist Discourse of the 20th Century." *Environment and History* 4.3 (1998): 309–343. Print.

Sharma, S. M., M. Joachimski, H. J. Tobschall, I. B. Singh, C. Sharma, and M. S. Chauhan. "Correlative Evidences of Monsoon Variability, Vegetation Change and Human Inhabitation in Sanai Lake Deposit: Ganga Plain, India." *Current Science* 90 (2006): 973–978. Print.

Shiva, Vandana. *Staying Alive: Women, Ecology, and Development in India.* London: Zed Books, 1989. Print.

Smith, A. *The Wealth of Nations.* 1776. New York: Modern Library, 2000. Print.

Sukumar, R., R. Ramesh, R. K. Pant, and G. Rajagopan. "A Delta 13C Record of Late Quaternary Climate Change from Tropical Peats in Southern India." *Nature* 364.19 (1993): 703–706. Print.

Sutton, D. *Other Landscapes: Colonialism and the Predicament of Authority in Nineteenth-Century South India.* New Delhi: Orient Blackswan, 2011. Print.

Thapar, R. "Perceiving the Forest: Early India." *Studies in History* 17.1 (2001): 1. Print.

Zalasiewicz, J., M. Williams, A. Smith, T. L. Barry, A. L. Coe, P. R. Bown, P. Brenchley, et al. "Are We Now Living in the Anthropocene?" *GSA Today* 18.2 (2008): 4–8. Print.

27

MULTIDIRECTIONAL ECO-MEMORY IN AN ERA OF EXTINCTION

Colonial whaling and indigenous dispossession in Kim Scott's *That Deadman Dance*

Rosanne Kennedy

That great America on the other side of the sphere, Australia, was given to the enlightened world by the whaleman … The whale-ship … the true mother of that now mighty colony … cleared the way for the missionary and the merchant.

—Hermann Melville, *Moby-Dick*

Literary remembrance: the environments of cultural memory

In this chapter, I aim to demonstrate that cultural memory studies can contribute to post-colonial environmental humanities by introducing concepts, approaches, and texts that bring histories of the decline and resilience of human and animal populations into an expanded commemorative frame. Remembering these intertwined histories may, I argue, enable us to think more fruitfully about connections between animal suffering and human suffering, and between extinction and genocide today. Building on Michael Rothberg's (2009) concept of "multidirectional memory," I introduce the concept of "multidirectional eco-memory," which has particular relevance in an era of extinction. He proposes multi-directional memory as an alternative to competitive conceptions of memory in which, for instance, commemorating the Holocaust is seen as obscuring memories of other atrocities such as slavery (3). Memory need not be viewed as competitive in a zero-sum game, he argues; rather, "[w]hen the productive, intercultural dynamic of multidirectional memory is explicitly claimed … it has the potential to create new forms of solidarity and new visions of justice" (5). Rothberg is concerned with how social groups articulate histories of victimization, and the dynamic transfers between public memories in a multicultural, transnational world (2). Multidirectional eco-memory, as I conceive it, would link human and nonhuman animals and their histories of harm, suffering, and vulnerability in an expanded multispecies frame of remembrance.

What I am calling "eco-memory," I should stress, encompasses but differs from the memory of place, which is typically associated with the anthropocentric concept of collective identity. In contrast, I propose eco-memory as grounded in a deep memory of a habitat, conceived as an ecological assemblage in which all elements, human and nonhuman, are mobile, connected, and interactive. Eco-memory, as I elaborate it, is not tied to the usual local or national landscapes of personal or collective memory; rather, it is compatible with an indigenous conception of "country." Eco-memory requires critics to expand outwards to a multispecies horizon that includes the oceans and their creatures, and to examine how events, actions, and processes affect elements in the assemblage. Multidirectional eco-memory places memories of the violence against and dispossession of particular human populations in complex, nuanced relation to memories of the suffering, slaughter, and endangerment of animal populations. It means seeing ecological vulnerability neither exclusively in human animal nor in nonhuman animal terms but as interconnected.

I illustrate the productivity of multidirectional eco-memory through a reading of Kim Scott's novel *That Deadman Dance* (2010). Writing in the wake of *Moby-Dick*, which haunts all later fictions on whaling, Scott takes the largely forgotten history of the whale-ship as a "machine of empire" and an "engine of commerce" (Russell 19) as the vehicle for his historical narrative of cross-cultural and cross-species encounters on the Western Australia maritime frontier. His novel imaginatively remembers, within the settler colonial framework of a narrative of first contact, the contribution of commercial whaling to the "becoming precarious" of both indigenous and whale populations. An act of literary remembrance that creates "new structures of cultural perception" (Goodbody 58) through its representation of human and animal suffering and survival, the novel mediates how we remember the past in the present. The novel, I argue, offers an expansive "multidirectional" alternative to environmental activist memories of the near extinction of the whale, which tend to ignore the ways in which whaling also facilitated the dispossession of indigenous peoples and the destruction of their country.

Although acts of remembrance commemorate past events, they intervene in the present and shape understandings of contemporary issues. As an interdisciplinary field spanning humanities and social sciences, cultural memory studies is concerned with how societies remember their past, and how those memories shape identities, issues, and discourses in the present (Erll and Nunning). Cultural memory studies analyzes the cultural forms and media—literature, film, photography, museums, memorials, and commemorative rituals—through which public memory is produced and shared on local, national, and global scales. A concern in the field is the backward orientation of memory, at a time when the planetary challenges identified under the rubric of the Anthropocene demand attention (Huyssen; Crownshaw et al.). While memory scholars have recently begun to address these issues—for instance, in critical analyses of fiction and film that address climate change, carbon emissions, oil futures, and ruined landscapes—they have not yet engaged extensively with the extinction crisis, for instance through memories of the destruction of species and their habitats (for exceptions, see Rose; Kennedy).

To date, the few scholars forging connections between ecocriticism and cultural memory studies have tended to conceptualize the environment in terms of place understood as land or landscape. For instance, Lawrence Buell values literature as a crucial archive for articulating an "environmental memory" of landscapes such as gardens. As Stephanie LeMenager and Stephanie Foote observe, "without the environmental memory that literary archives provide, we might never know what places *looked or felt like* before their injury, the extent of habitat destruction or the baseline of ecological health" (575). It is precisely the felt dimension of place that Axel Goodbody foregrounds in considering how the insights of cultural memory scholarship can enrich ecocriticism. He argues that literature conveys the affective investments that individuals and local communities have in "real places" and can thereby provide a vital resource for developing a counter-memory of place. Attention to local geographies and places of memory provides a means of challenging the collective memory of dominant social groups, which often coalesces as national memory and excludes the collective memories of small or marginalized communities. *That Deadman Dance* is set in a real physical and affective environment, the remote south coast of Western Australia near Albany, home to the Noongar people. Drawing on archives, archaeological sites and oral storytelling traditions, the novel activates a regional indigenous memory to address a national and transnational audience. Emerging from an indigenous understanding of "country," the novel extends the concept of place offshore, to include islands, the sea, and its creatures.

Remembering whaling in the settler colonial present

That Deadman Dance spans the twenty years from 1826 to 1844, a period during which British settlers arrived in Western Australia, claimed land, and made the region their home. Scott, of mixed Noongar and British heritage, affirms that the novel "is inspired by the history of early contact between Aboriginal people—the Noongar—and Europeans in the area of my hometown of Albany, Western Australia, a place known by some historians as the 'friendly frontier'" (397). In the nineteenth century, Yankee and other international whaling ships regularly hunted sperm whales in King George Sound, with its whale-friendly bay, and King George Town (now Albany) developed to service the industry. In 1840, there were over six hundred whaling ships in the Pacific, and many sailed on to the Southern Ocean (Russell). Whaling contributed to colonial settlement, as whaling ships brought settlers out to Australia and left with whale oil. Settler colonialism led to high rates of indigenous death through introduced diseases, starvation, and violence and irrevocably altered the ways of life of the Noongar. It also depleted the sperm whale population.

Although the story Scott tells ends in the 1840s, whaling in Albany continued into the twentieth century. After World War II, whaling began again in earnest in the 1950s, when Norwegian factory ships were used to kill and process thousands of whales (see Frost). In the 1950s, the International Whaling Commission set a quota on the size and number of whales that could be caught, and in the 1970s, Albany whalers killed around eight hundred whales a year. The town adopted the whale as an icon, with tourists invited to "have a whale of a time" in a "whale of a town." By the 1950s, there were alternatives to whale oil, and whaling could no longer be justified. The industry argued that whaling was sustainable, but in practice prioritized its own profits and the economic livelihood of Albany over the fate of the whales. With the whale population on the brink of extinction, whaling in the 1960s and 1970s was an unfolding ecological catastrophe. Whale biologists extolled the "unique" characteristics of the whale, such as their song, intelligence, enjoyment of life, attachment to their young, and signs of grieving when a calf was killed. Conservation groups such as Project Jonah built on these discoveries and, through education campaigns, encouraged the public to see whales as in some ways "like us" (see Frost).

As the last site of commercial whaling in the Western world, Albany had transnational symbolic value. (The other nations still whaling were Japan and Russia.) In 1977, when Kim Scott was twenty years old, Albany was the site of a protest to end whaling. An Australian anti-whaling campaign enlisted the help of Canadian Greenpeace activists, who went into shark-infested waters in rubber dinghies (Pash; Zelko). Placing themselves as human shields between whales and whale-ships, they created a global media spectacle. Anti-whaling activism in Albany gave birth to Greenpeace in Australia, which is today active in the fight against Japanese whaling. In response to these events, Malcolm Fraser, then prime minister, appointed an independent inquiry to determine whether whaling should be banned in Australia.

The inquiry's attention to the inhumane methods of killing whales signals the emergence of a cross-species imaginary. Peter Singer, then a young philosopher at Monash University in Melbourne, considered as unusually cruel and morally problematic the gruesome procedure known as "flensing"—tearing the whale's blubber away from its body to boil it down and render it into profitable oil—which appeared to inflict extreme suffering. He identified the whale's capacity for suffering as morally relevant: "If a being is capable of suffering, any suffering it might experience as a result of our actions must count in our ethical deliberations irrespective of whether the being is a human or non-human animal" (183). The inquiry's final report records a shift in Australian attitudes toward whales, as observed by the former chairman of Marine Mammals Commission, Dr. Scheffer: "If I understand what men and women are saying today about the whales it is 'Let them be'" (189). (This phrase is echoed in That Deadman Dance, when Menak, an indigenous elder, reflects: "Be the whale.") The Inquiry recommended that Australia outlaw whaling and oppose whaling internationally, thereby shifting whales from the category of animals that could be killed with impunity to the category of accountable killing.

Today, in the Australian settler colonial present (Hinkson), the agency of the white, middle-class environmental movement in ending whaling is remembered as a heroic confrontation to combat and reverse the looming disaster of extinction (see also Pash; Zelko). This public remembrance of the history of whaling—in the museum, in popular nonfiction, on blogs—often fails to acknowledge colonial whaling both as an economic industry in which indigenous people worked and as part of a colonial regime which dispossessed them of their traditional "country." For instance, in Chris Pash's The Last Whale (2008), indigenous people are valued for their spiritual ties to dolphins and whales, but do not otherwise feature. Pash narrates the aftermath of the end of whaling as a story of reconciliation between anti-whaling activists and whalers, without considering how this story might be expanded to facilitate reconciliation between indigenous people and the descendants of British settlers. (The Last Whale and Pash's website have both received significant media attention.) Settler colonial environmental fictions such as Tim Winton's The Shallows have also advanced the memory of the campaign to save the whale as a "white issue," displacing indigenous people from this history (Helff). An exception is Danielle Clode's history of the killer whales of Eden. In this rare case of interspecies collaboration, killer whales worked with humans to hunt baleen whales. While acknowledging that whaling contributed to indigenous dispossession, she documents the significant role indigenous whalers played in the industry and the benefits their special understanding of whales brought to the whaling enterprise. She argues that their practices—for instance, letting the killer whales eat their fill before harvesting the whale—contributed to the collaborative relationship that developed between the killer whales and whalers.

Emerging from an indigenous conception of country, *That Deadman Dance* introduces an expanded multispecies frame of remembrance. Crucial to its vision, I argue, is the novel's juxtaposition of an indigenous imaginary of interspecies kinship with an Anglo-European hierarchical view, in which animals (and some humans) are viewed as lesser beings available for exploitation. The concept of the "creaturely" is productive for identifying the novel's critique of the way in which this Western hierarchy positions whales and certain humans as creatures that may be killed with impunity.

Figurations of memory: whaling as allegory

That Deadman Dance opens with Bobby Wabalanginy, a child on a hill looking out for whales for his British settler patron, Chaine. The opening contrasts a Judeo-Christian cultural memory of human–animal relations as conveyed through the story of "Jonah and the Whale" with a Noongar story. In the former, God punishes Jonah by having a whale swallow him whole. *That Deadman Dance* remediates the Jonah story with a Noongar story that Bobby's uncle Menak, the tribal leader, hands down to him "wrapped around the memory of a fiery, pulsing whale heart" (2). Whereas the Jonah story inspires fear of the whale, the Noongar story tells of a human playfully slipping inside the whale's body through its spout and merging with it:

> Two steps more and you are sliding, slide deep into a dark and breathing cave that resonates with whale song. Beside you beats a blood-filled heart so warm it could be fire. Plunge your hands into that whale heart, lean into it and squeeze and let your voice join the whale's roar. Sing that song your father taught you as the whale dives, down, deep … look through the whale's eyes and you see bubbles slide past you.
>
> (Scott 2–3)

Menak's story from the Dreamtime is an example of what Jan Assmann calls a "communicative memory" that is passed down orally, from one generation to the next, to describe relations of connection and accountability.

The merging and collaboration of human and animal, conveyed in Menak's Dreamtime story of the whale, is a feature of indigenous kinship systems. As Deborah Bird Rose explains, the "Dreaming or totemic way of being in the world is a form of animism which recognizes that 'the world is full of persons, only some of whom are human, and that life is always lived in relationship with others'" (18, quoting Graham Harvey). Bobby and his kin, who come "from ocean and whale," are descendants of and "brothers" with the whale (Scott 33), and there is historical evidence for a whale totem amongst the Noongar. For British settlers, animals and land are exploitable resources rather than kin. Noongar views of "country" also differ from British understandings of "land" and "landscape." As historian Steve Kinnane explains, "The concept of country does not allow for a separation of people, land and waters. In an Indigenous vision of country, economy, spirituality, knowledge and kin are all interrelated" (25). Deborah Bird Rose observes that in an indigenous worldview, "'country' is not just the homeland for humans, but … for all the living things that are there, and care is circulated through country in cross-species relationships of responsibilities and accountability" (86). One of the central themes of the novel is the settlers' greed and their disregard for country and the obligations it entails, which compromise the "friendly" relations between Noongar and the settlers.

The difference in Western and Noongar approaches to "country" and its human and nonhuman inhabitants is conveyed by a juxtaposition of scenes at the heart of the novel. These scenes contrast Menak's attentive regard for a beached whale as it lay dying with the suffering inflicted by whalers (Scott 243). Noticing the firelight from a campfire reflected in the whale's eye, Menak merges with the whale, and feels "himself dissolving there" (245). By contrast Bobby, on board a whale-ship for the first time, witnesses the cruel slaughter of a mother and her calf:

> Harpooned ... the mother was returning to her calf. The silver spear at the bow of the boat stabbed again and again ... The mother whale's tail repeatedly rose and struck the water close to her dead calf ... The boat's lifted oars were a row of spikes, and the man at the bow drove and twisted his steel spear into the whale.
>
> (250)

Identifying with the whale's suffering, "Bobby groaned, thinking he heard a whale groan, too ... The young whale, the mother: each had a flag flying from its spout, and the boat which killed them was already after another pod" (251). Bobby painfully witnesses the effects of the capitalist logic of whaling, in which the whale figures as a commodity harvested for profit rather than as kin to whom certain obligations are owed.

The counter-scene represents Menak's horrified response to the whale slaughter. Standing near the stranded whale, Menak is "deep in the whale story of this place" when he notices that "[f]urther around the beach something was being savaged by sharks and seagulls. A whale carcass, the inner part of a whale, but still fresh." He sees the evidence of "flensing"—the whale has been decapitated and skinned to extract its valuable oil. Shortly after, Menak notices "young [Bobby], rowing from that ship to shore along with the horizon men" (254). His gaze implicates Bobby, who optimistically embraced the new opportunities whaling brought and used his special gift for spotting whales to aid the whalers. As an older man, Bobby realizes that his generous friendship with the "pale men" has been betrayed, and recognizes his own complicity in the demise of the whale population and, by extension, his own people and culture: "Once he was a whale and men from all points of the ocean horizon lured him close and chased and speared and would not let him rest until ... Bobby led them to the ones he loved, and soon he was the only one swimming" (160). The status of the whales in a colonial economy is an allegory for Noongar people: while the whales are hunted, the people die of diseases brought by the settlers, starve, are pushed off their land, and are killed.

Creaturely life and multidirectional eco-memory

Considering the relationship between the killing of humans and the killing of animals—and by extension, the relationship between the extinction of species and human genocide—Deborah Bird Rose foregrounds the issue of accountability. Heidegger, she observes, has provided philosophers with a strong endorsement of the idea that an animal death is a "mere death" and animal life is a "mere life" (Rose 22). Acknowledging that humans are animalized so as to be more easily killed, she argues that the relevant dividing line is not between human and animal per se, but rather between those creatures that can be "killed with impunity" and those whose deaths must be answered for. Rose recounts a powerful episode from the 1940s, recorded by an anthropologist working in central Australia, to illustrate how settlers have killed animals without fear of legal or moral reprisal and how the assumed "right to kill" animals has functioned as a threat to indigenous survival. As she retells it, a white Australian policeman went into an Aboriginal camp and massacred a number of dogs, while the people watched in terror and wailed in grief: "For people who had already been subjected to massacres, the dog shooting was a clear message of the [white] right to kill with impunity. The power and terror show us a darker porosity to the West's human-animal boundary: one in which humans are animalized so as to be killed with impunity" (25). This logic was used, most notoriously, in the Nazi depiction of Jews as "vermin," but indigenous people in Australia have also, at times, been depicted as animals and massacred. Rose argues that challenging the view that it is acceptable to kill animals with impunity is an ethical imperative necessary for countering human genocide and extinction. She advocates that ethical considerations of care and accountability be extended to animals and humans.

A scene from the middle of *That Deadman Dance* powerfully conveys the terrorizing logic of the settlers' assumption of the "right to kill with impunity." When Bobby is on an expedition with Chaine—the British entrepreneur who runs the shore whaling company and who "adopts" Bobby until he is no longer useful—Chaine murders two Aboriginal youths. Significantly, the murder takes place in the bush, figured as a "state of exception" or space outside the law (Agamben). Eric Santner's concept of the creaturely is useful for identifying the biopolitical connotations of this scene. In a Western philosophical and literary imaginary, the human is typically opposed to the "creaturely." For Santner, "creaturely life is just life abandoned to the state of exception/emergency, that paradoxical domain in which law has been suspended in the name of preserving law" (22); as Hal Foster explains, "this is close to Agamben on bare life, yet … Santner imagines this condition from the position of *homo sacer*, from the place of the beast, as it were" ("Human Beasts" 121). In the scene above, the indigenous youths, reduced to "bare life" through starvation and thirst, take on "the cringed posture of the creature" (Santner 35). They are described in nonhuman terms as "two pale vertical objects shimmering on the otherwise featureless plain" (Scott 229) and compared to animals—their heads above the bushes like "seals in the water" and "calling out like wild dingoes … their voices … plaintive and wailing" (231). "Caught between human and nonhuman states, or stranded in the vertiginous space of exile" (Foster, "Decider"), the youths are in a state in which "life takes on its specific biopolitical intensity" (Santner 35). The scene of murder can be read as an allegory of the paradox of settler colonial law (Derrida), which is founded on an illegal act of violence that is "forgotten."

The message conveyed by Chaine's murder is not lost on Bobby, who has witnessed the slaughter of whales. The full critical significance of the scene only becomes clear, however, at the end of the novel, when Bobby is punished for stealing sheep and stores to feed his people. Bobby and his Noongar kin see their "theft" as a quid pro quo, in exchange for the settlers' hunting of whales and kangaroo. Bobby is jailed, an act through which British sovereignty in the new colony is asserted. He is only released when he threatens to expose Chaine's murder and thereby make visible the foundational paradox of sovereign law. In threatening to expose the exclusions and illegality of settler colonial law, the novel draws readers into recognizing the shifting boundary in the category of the human—between *bios* (citizen) and *zoe* ("bare life" or *homo sacer*, to use Agamben's terms)—as it has operated in Western philosophy and cultural imaginaries. Only those whom the colonial regime recognizes as citizens are granted the protections of settler law; those reduced to bare life, such as the youths, can be killed with impunity. After his confrontation with settler law, Bobby is suspended in an in between space—neither citizen nor "creature"—and lives out his days on the margins of the colony.

That Deadman Dance is self-reflexive about the construction of memory, and thus invites us to read it as an intervention into public memory today. The compartmentalization of memory in the public sphere—which separates indigenous rights from animal rights, collective memory from eco-memory—is destabilized by the novel's multispecies imaginary of decline and resilience. As a lonely old man marginalized in the new British colony, Bobby asserts: "my country is here, and belonged to my father, and his father, and his father before him, too" (106). These words resonate in the settler colonial present, in which a land claims settlement has recently been negotiated between Noongar people and the West Australian government, nearly two hundred years after British settlement. Rather than isolate memories of indigenous dispossession, Scott links this colonial history to whaling. When considered in this expanded multispecies frame, settler colonialism's effects on indigenous peoples and their ways of life also productively remembers colonialism's effects on whale populations. This multidirectional memory of the linked fates of the whale and of indigenous people could, I propose, productively re-shape the emergent activist memory culture of the end of whaling and enable it to include promoting justice for indigenous people.

Conclusion

What does this case study—concerned with whaling in a remote corner of the world—deliver, and what is its conceptual significance, for the environmental humanities? To date, scholarship in cultural memory studies has focused almost exclusively on man-made disasters that inflict violence and suffering on human populations, with the Holocaust serving as both paradigm and trope. In *Humanity's Footprint*, a science book written for a popular audience, Walter Dodd describes the current extinction crisis, in which thousands of species will be lost, as an "ecological holocaust." While "holocaust" literally means "destruction or slaughter on a mass scale, especially caused by fire or nuclear war," in common parlance it invokes the mass killing of Jews under the Nazi regime. Dodd's idiom extends the moral imperative associated with the Holocaust, with its mantra of "never again," to the extinction crisis. Although he takes a multidirectional approach, many may reject his yoking together of genocide and extinction, human suffering and animal suffering, and his figuration of humans as "executioners," as simplistic and crude. Thus, his implied analogy may alienate readers from the urgent issues he addresses. While Dodd's rhetoric may lack sensitivity and nuance, the issue he raises—human responses to the mass extinction of species—merits ethical as well as scientific consideration.

The ongoing extinction event compels us to recognize that anthropogenic changes to the planet are forms of "slow violence" (Nixon) that have disastrous effects for humans as well as nonhumans. Rather than heavy-handed analogies between extinction and genocide, narrative and literary imagination may prove more successful in conveying the ways in which human and animal pasts and futures are intertwined. *That Deadman Dance*, I have argued, draws our attention to the slow violence that colonial whaling and settlement inflicted on indigenous and whale populations, but it also conveys the survival and resilience of both. The novel transmits a deep indigenous eco-memory of "country" as a multispecies habitat, thereby bringing memories of human and animal vulnerability and survival into a single frame. By representing "the more-than-human and multispecies world, while at the same time identifying the hierarchical processes that led certain humans to be reduced to 'nature' (or other species)" (DeLoughrey et al. 11), the novel invites readers to connect social justice and environmental justice, human suffering and animal suffering, dispossession and extinction. These insights have been facilitated by a hermeneutics informed by the concept of multidirectional eco-memory. In articulating the mutually imbricated histories of human and animal precarity, multidirectional eco-memory extends cultural memory studies beyond the human and foregrounds the development of a multispecies approach to ethical issues of suffering and harm as vital for the environmental humanities.

References

Agamben, Giorgio. *Homo Sacer: Sovereign Power and Bare Life*. Stanford: Stanford University Press, 1998. Print.

Assmann, Jan. "Communicative and Cultural Memory." *Cultural Memory Studies*. Ed. Astrid Erll and Ansgar Nunning. Berlin and New York: de Gruyter, 2008. 109–118. Print.

Buell, Lawrence. "Environmental Memory and Planetary Survival." University of California, Santa Barbara. November 2007. Presentation. YouTube. Web. 23 April 2015.

Clode, Danielle. *Killers in Eden: The True Story of Killer Whales and Their Remarkable Partnership with the Whalers of Twofold Bay*. Sydney: Allen & Unwin, 2002. Print.

Crownshaw, Richard, Jane Kilby, and Antony Rowland. *The Future of Memory*. New York: Berghahn Books, 2010. Print.

DeLoughrey, Elizabeth, Jill Didur, and Anthony Carrigan. "Introduction." *Global Ecologies and the Environmental Humanities: Postcolonial Approaches*. Ed. Elizabeth DeLoughrey, Jill Didur and Anthony Carrigan. New York and London: Routledge, 2015: 1–32. Print.

Derrida, Jacques. "The Force of Law: The 'Mystical Foundation of Authority.'" Trans. Mary Qaintance. *Cardoza Law Review* 11 (1990): 920–1045. Print.

Dodds, Walter K. *Humanity's Footprint: Momentum, Impact, and Our Global Environment*. New York: Columbia University Press, 2008. Print.

Erll, Astrid and Ansgar Nunning, ed. *Cultural Memory Studies: An International and Interdisciplinary Handbook*. New York: de Gruyter, 2008. Print.

Foster, Hal. "Human Beasts." *Asger Jorn: Restless Rebel*. Munich: Prestel Verlag, Random House, 2014: 110–125. Print.

———. "I Am the Decider." *London Review of Books* 33.6 (17 March 2011): 31–32. Web. 3 March 2016.

Frost, Sydney, and Inquiry into Whales and Whaling (Australia). *The Whaling Question: The Inquiry by Sir Sydney Frost of Australia*. San Francisco: Friends of the Earth, 1979. Print.

Goodbody, Axel. "Sense of Place and Lieu de Memoire: A Cultural Memory Approach to Environmental Texts." *Ecocritical Theory: New European Approaches*. Ed. Axel Goodbody and Kate Rigby. Charlottesville: University of Virginia Press, 2011: 55–67. Print.

Helff, Sissy. "Sea of Transformations: Re-writing Australia in the Light of Whaling." *Local Natures, Global Responsibilities: Ecocritical Perspectives on the New English Literatures*. Ed. Laurenz Volkmann. Amsterdam and New York: Rodopi, 2010: 91–104. Print.

Hinkson, John, Paul James, and Lorenzo Veracini, ed. *Stolen Lands, Broken Cultures: The Settler-Colonial Present*. North Carlton, Victoria: Arena Publications, 2012. Print.

Huyssen, Andreas. *Present Pasts: Urban Palimpsests and the Politics of Memory*. Stanford: Stanford University Press, 2003. Print.

Kennedy, Rosanne. "Humanity's Footprint: Reading *Rings of Saturn* and *Palestinian Walks* in an Anthropocene Era." *Biography* 35.1 (2012): 170–189. Print.

Kinnane, Steven. "Recurring Visions of Australinda." *Country: Visions of Land and People in Western Australia.* Ed. Anna Haebich, Mathew Trinca, and Andrea Gaynor. Perth: Western Australian Museum, 2002: 21–31. Print.

LeMenager, Stephanie and Stephanie Foote. "The Sustainable Humanities." *PMLA* 127.3 (2012): 572–578. Project Muse. Web. 26 January 2016.

Melville, Hermann. *Moby-Dick: or, the Whale.* Foreword by Nathaniel Philbrick. London: Penguin, 2009. Print.

Nixon, Rob. *Slow Violence and the Environmentalism of the Poor.* Cambridge, MA: Harvard University Press, 2011. Print.

Pash, Chris. *The Last Whale.* North Fremantle: Fremantle Press, 2008. Print.

Rose, Deborah Bird. *Wild Dog Dreaming: Love and Extinction.* Charlottesville: University of Virginia Press, 2011. Print.

Rothberg, Michael. *Multidirectional Memory: Remembering the Holocaust in the Age of Decolonization.* Stanford: Stanford University Press, 2009. Print.

Russell, Lynette. *Roving Mariners: Australian Aboriginal Whalers and Sealers in the Southern Oceans, 1790–1870.* Albany: State University of New York Press, 2012. Print.

Santner, Eric. *On Creaturely Life: Rilke, Benjamin, Sebald.* Chicago: University of Chicago Press, 2006. Print.

Scott, Kim. *That Deadman Dance.* Sydney: Picador, 2010. Print.

Singer, Peter. "The Ethics of Whaling." *Whales and Whaling: Report of the Independent Inquiry Conducted by Sir Sydney Frost.* Canberra: Australia Government Publishing Service, 1978. Print.

Winton, Tim. *The Shallows.* Sydney: Allen and Unwin, 1984. Print.

Zelko, Frank. *Make it a Green Peace!: The Rise of Countercultural Environmentalism.* New York: Oxford University Press, 2013. Print.

28

THE CARIBBEAN'S AGONIZING SEASHORES

Tourism resorts, art, and the future of the region's coastlines

Lizabeth Paravisini-Gebert

Those wishing to understand the growing impact of tourism development and climate change on the islands and populations of the Caribbean region could find no better starting point than the tiny island of Petite Martinique, a dependency of the nation of Grenada. A territory of a mere 2.37 square kilometers with a local population of approximately nine hundred, it has been losing one and a half to two meters yearly from portions of its seashore over the last two decades. This territorial loss has resulted from erosion caused by the ceaseless pounding of the Atlantic's waves, which remove the sands from the seashore just as quickly as they deposit them, exposing the soft ash-cinder layers of rock underneath and threatening the island's precarious infrastructure, from its single coastal road to its handful of failing retaining walls (Richards). The once-protective coral reefs have been bleached and are now dead or dying, no longer able to protect the seashore from the ocean's relentless buffeting.

The crisis facing Petite Martinique is a harbinger of things to come for the extended Caribbean region as it faces the compounding effects of climate change—impacts worsened by decades of seashore development as tourism dollars replaced the dwindling profits of the sugar plantation as the source of precarious incomes. The Caribbean is one of the most tourist-dependent regions in the world, its coastal zones threatened by "hotel and resort construction, beach sand mining, marina channel development, waste disposal from yachts and shipping, non-indigenous factory fishing vessels" and now the potentially disastrous effects of climate change (Pulwarty et al. 16). The first line of defense, the ailing coral reefs, had already sustained devastating bleaching events long before the rising temperatures produced by climate change led to coral die-offs that have now reached epidemic levels; thousands of acres of mangrove forests, the second line of defense against pounding waves, have been sacrificed in the name of tourism jobs.

The coastlines of the Caribbean region, as Brian Fagan argues for island chains around the world in *The Attacking Ocean*, find themselves acutely "vulnerable to the ocean and its whims in ways unimaginable even one or two centuries ago" (126) and are facing chronic issues like coastal erosion and persistent flooding "not as an abstract problem for the future, but as a sobering reality" (163). In "Sea Trash, Dark Pools, and the Tragedy of the Commons,"

Patricia Yaeger writes of the crisis that has emerged through the transformation of oceans—and I would add, seashores—into capital, calling for an "oceanic ecocriticism$ [sic]" that can draw on "narratives in a state of emergency, a crisis that demands unnatural histories written by unnaturalists who limn the fleshy entanglements of sea creatures, sea trash, and machines" (529). In the Caribbean, a geography that encompasses both the largest number of small island states and the highest number of maritime borders in the world (Pulwarty), this crisis plays out in the spaces where the sea and the shore meet, where oceanic resources have been marshaled in the name of tourism development, destroying, in the process, the natural features that provided a defense against violence from the sea. Caribbean tourism relies on the sea—as aesthetic background and space for recreation—but imposes measures that contribute to the degradation of its coasts, the "watery realm where exploitation and overconsumption" converge (Yaeger 532). It should not surprise, then, to find that the creation of textual and visual "narratives in a state of emergency" about the plight of the coasts has become a central concern of Caribbean writers and artists. As Elizabeth DeLoughrey argues, "we are on the cusp of an entirely new development in this oceanic imaginary in work that is specifically responding to the threat of sea-level rise, adding a new dimension to how we might theorize our relationship to the largest space on earth, which until recently for most, was imagined as always external, which is to say outside of ordinary terrestrial orbits until it comes to flood our cities and homes."

I focus here on how Caribbean art speaks to the plight of coastal spaces in the region—more concretely in Cuba—through a discussion of *Adrift Patrimony: The Baths (A Tribute to Frédéric Mialhe)*, a 2007 photographic series by Atelier Morales, Cuban architects Juan Luis Morales and Teresa Ayuso. The photographs address the Caribbean seashore as a site where tourism, degraded coastal ecologies, politics, and ideology confront the violent force that is the sea. The artwork engages with the loss of portions of Cuba's architectural patrimony to institutional neglect and coastal degradation in the region's most environmentally resilient island-state—the "ecological crown jewel of the Caribbean" (Whittle 74)—the nation best suited to propose and enact appropriate remediation. In the work of Atelier Morales, imagination, memory, and ideology mediate the artists' representation of Cuba's coastal deterioration. This work, although avowedly environmentally focused, prompts questions about whether its engagement with material conditions responds to an informed environmental aesthetic. Marcia Mueder Eaton has argued that the development of an environmental aesthetics requires "ways of using the delight that human beings take in flights of imagination, connect it to solid cognitive understanding of what makes for sustainable environments, and thus produce the kind of attitudes and preferences that will generate the kind of care we hope for" (180). *Adrift Patrimony: The Baths*, in privileging historical memory and nostalgia for a past of seashore recreation enjoyed amidst iconic coastal architecture, clashes with urgent calls for science-driven environmental remediation measures adopted by the Cuban government—measures that in some cases call for the removal of the very iconic seashore resorts "mourned" in these photographs. As a result, reading Atelier Morales's photographs against the aggressive measures for coastal conservation developed by the Cuban government unveils the complex dichotomies that control environmental action in Cuba and throughout the Caribbean.

During the first decades of the twentieth century, a number of seaside resorts were built throughout Cuba in spaces that had earlier become known for their beauty, healing waters, or provincial charm after being featured in nineteenth-century prints and photographs—early examples of Cuban landscape representation that had fostered the incipient tourism industry on the island. These privately built resorts or *balnearios*, which in many cases

Figure 28.1 Frédéric Mialhe, *Cojimar, Cerca de La Habana* (Cojimar, Near Havana, 1839–1842).

boasted striking art deco architecture, were the foundation of local and international sea-side tourism. Initially privately owned, they fell under state control following the Cuban Revolution (1959) and were subsequently (and in some cases, controversially) neglected or abandoned to the power of the sea, which over the decades that have elapsed since the 1959 Cuban Revolution has brought these neglected structures to ruin. Built originally right on the sandbank, they were extremely vulnerable to the pounding waves as well as contributors to the vulnerability of the shore—both victims of and collaborators with the sea. The series consists of twenty-five lenticular photographs of the ruined resorts interlaced with poignant images of cemetery sculpture to underscore the mourning for these once beautiful spaces. (Lenticular printing produces images with an illusion of depth and the ability to change or move as the image is viewed from different angles.) The series was conceived by Atelier Morales as a dialogue with the iconic engravings of the baths' settings created between 1838 and 1842 by French lithographer Frédéric Mialhe (Figure 28.1). Using the camera as a "technology of memory," these images emerge as "visual monuments to vanishing places" that engage the island's cultural, political, and economic history (Dunaway xviii).

The Baths is the third iteration of a project the artists have titled *Adrift Patrimony*, which included an earlier series called *Bohíos* (2003) that addressed the disappearance of the traditional peasant huts that had dotted the Cuban countryside and had roots in Amerindian culture, and *The Sugar Mills* (2004), which focused on the decline of Cuba's once prosperous sugar industry. All three series open a visual dialogue with nineteenth-century French artists who produced work—primarily prints—of Cuban landscapes. Atelier Morales's working method has been to revisit the spaces illustrated by these earlier artists to capture their present ruined, deplorable state. In the case of *The Baths*, Atelier Morales's interlocutor is Mialhe (1810–1868), the author of dozens of prints collected in two volumes, *La isla de Cuba pintoresca* (1838) and *Viaje pintoresco por la isla de Cuba* (1842).

The baths that interest Atelier Morales had been built in many of the places painted earlier by Mialhe (see Figure 28.1 above, a fragment of Mialhe's *Cojimar* depicting the area near the site of Atelier Morales's *Guanabo Désolé 1*, Figure 28.4). They were small baths with regional charm, built for pleasure and curative purposes primarily for the use of the local population. Most were located on the seashore, especially along the coastline east and west of Havana, which boasted splendid natural beaches. The development of seaside baths in these locations was driven primarily by a plethora of social clubs (the Havana Yacht Club, foremost among them) and was followed after the Revolution by the construction of resort hotels for international travelers. Here I focus particularly on two of these seaside locales: Marianao, the site of the Havana Yacht Club, then about six miles west of the center of Havana (an area now fully integrated into the city), and Guanabo (Cojimar), about three times that distance to the east of the capital. I will specifically highlight the buildings photographed by Atelier Morales: the Havana Yacht Club and the Syndicate of Telephone Worker's Club. The Yacht Club presided over a number of exclusive clubs built in the Miramar and Marianao stretch of coast west of Havana, while the Sindicato Telefónico did the honors for the working and lower middle-class club sprawl along the beaches of Guanabo, which included a number of seaside clubs belonging to various workers' guilds and syndicates, among them those for news reporters, electrical and telephone workers, healthcare labor unions, and bank clerks.

In their work, Atelier Morales prioritize what they call the "provincial allure" of these clubs and resorts over the clear class differences that separated them, finding them to be unified first through their local origins and construction and second by suffering the same neglect under the Cuban Revolution. Nationalized after the Revolution, they were later abandoned and left to be ruined by nature while the government concentrated on the building of what Atelier Morales describe as "massive and impersonal tourist complexes" intended to bring much-needed foreign currency into the island. Hence their representation of these spaces in their work as being "in agony": "Sculptures of angels and figures found in the cemeteries of the world are incorporated in all their pictures, creating a dialogue with space and the lost patrimony. Death has taken over to give lieu to the gross and global tourism industry" (Menocal).

In the description of their work, Atelier Morales focus on their assessment of Mialhe as the "discoverer" of Cuba as a touristic and therapeutic space, and on these early resorts as offering an organic connection between nature and tourist—direct, personal, and potentially healing. Hence their objective of "bringing attention to the deplorable state of such an important architectural patrimony." As Juan Luis Morales writes:

> Given the state of abandon we witnessed during our return visits to Cuba, Teresa and I felt like those sculptures that cry at the feet of tombs in European cemeteries, only that instead of crying over a loved one, these sculptures placed in these old tourist baths were crying for the irremediable death of such an important patrimony. These sculptures represent those travelers responsible and sensitive to local patrimonies and ecosystems whose visits are in counterpoint to the irresponsible, insensitive and consumerist tourist of "groups" and "package tours" of the "all-inclusive" type of globalized tourism. (Morales and Morales)

The work evolved in two parallel voyages, their return to Cuba to photograph the Cuban baths, and various trips to museums and cemeteries in the places that supply tourists to Cuba, such as Spain, France, Italy, Germany, and England, to photograph funerary sculpture. "Our work," Juan Luis Morales concludes, "is a double reflection about a nation's responsibility to safeguard its patrimony and that of tourists in respecting it" (Morales and Morales). In their concept of "patrimony," the artists conflate natural and architectural "landscapes," naturalizing the built environment and equating it with the nation's communal "natural" inheritance.

Through this method, Atelier Morales underscore the dialogic nature of their work and their travels between Europe and Cuba, both in its "conversation" with Miahle's earlier work and in the "doubling" or layering necessary for the creation of lenticular images. Echoing the objective of Mialhe's influential images, which had been to celebrate the natural beauty of Cuba and its "picturesque" quality, the photographs focus on the forlorn quality of the now abandoned infrastructure, simultaneously aestheticizing the ruins and critiquing the loss they represent. The superimposition of the funerary sculpture accentuates the sense of loss.

At the center of Atelier Morales' concerns with Cuba's disappearing patrimony is a deep preoccupation with the integrity of the island nation, founded on their assessment that the island's iconic architectural legacy is central to national identification or *cubanía*. Their work, therefore, by showcasing pre-revolutionary iconic buildings as national heritage, expresses their unease about how to address creatively the problematic legacy of the foreign tourism economy in Cuba, which has led Cuba's revolutionary government to replace earlier iconic buildings with architecturally insignificant hotels and other mass tourism facilities. Moreover, in reading the island's architectural heritage as emblematic of the power of past and present political regimes, they extend this concern with the destruction of the earlier resorts to their home island's relationship with what is in their view an unsympathetic state. This discourse of power also embraces the not-always-benign sea that has shaped their nation's destinies. Atelier Morales, in *The Baths*, address both their anxieties about governmental neglect of architectural patrimony and speak to the Cuban coast as a site of aggressive erosion.

In 2013, Cuba's Center for Coastal Ecosystems Research announced the imminent demolition of thousands of structures built on the island's sandbanks—among them most of the structures photographed by Atelier Morales. The plan calls for the removal of "some 900 coastal structures [that] have been contributing to an average 4 feet (1.2 meters) of annual coastal erosion" on the Varadero peninsula alone, and thousands of similar structures throughout Cuba (Rodríguez). Inspectors and demolition crews are well on their way to razing thousands of houses, restaurants, and improvised docks "in a race to restore much of the coast to something approaching its natural state" (Rodríguez). This decision is the result of scientific projections that rising sea levels will seriously damage or altogether destroy 122 coastal Cuban towns, submerging beaches, tainting freshwater sources, and rendering croplands infertile. The research has projected that seawater will penetrate up to 1.2 miles inland as oceans rise three feet by 2100. As a result, Cuba has moved to undo decades of haphazard coastal development threatening sand dunes and mangrove swamps that protect Cuban shores against rising seas.

What sets Cuba and its coasts apart from other Caribbean islands under comparable environmental threats is its access to substantial expertise and technical resources that the Cuban government makes available to its communities. The product of significant state investments in education and training, environmental policy in Cuba is supported

Figure 28.2 Atelier Morales, *Havana Yacht Club Blessure* (2007).

by an impressive array of governmental and nongovernmental agencies unlike any others in the Caribbean region. As a result, decisions concerning environmental protection are founded on substantial data gathered from many studies that record coastal archeology, economy, history, culture, fisheries, and ecology. These studies form the basis of a plan of action to address the impact of climate change on Cuba's 3,500 miles of coasts. Areas like Havana and Guanabo, sites of significant tourism development and the focus of Atelier Morales' work, have received intensive attention, most recently in the form of a *Plan de Ordenamiento* (Organizing Plan) to monitor development that included local and national government, business and commercial agents, and representatives from civil society.

Figure 28.3 Atelier Morales, *Havana Yacht Club Contusion* (2007).

Cuba's Socialist government also wields a unique advantage no other country in the region claims, since the government controls the island's entire hotel stock, owning at least 51 percent of all tourism facilities and sometimes teaming up with minority foreign partners for management agreements.

Atelier Morales' photographs of the ruins of the Havana Yacht Club and Guanabo evoke a sense of patrimonial and historical loss, especially through their depiction of the architecture's easy integration into the surrounding landscape. But they do not speak to the environmental science behind the Cuban government's abandonment (and planned destruction) of the architectural patrimony they so value. The stone foundations emerging from the deeply blue sea and the rigidity of the sculptures replacing the vibrant vulnerability of flesh force us to confront the poignancy of the decaying infrastructure of an abandoned and superseded way of life (see Figures 28.2 and 28.3). The ambivalent quality of this nostalgia—as Morales asserts—is clearly imprinted in the composition of images like that of the remains of the once powerful Yacht Club, the exclusive institution open only to whites which dared blackball the island's dictator, Fulgencio Batista, on the grounds of questionable racial origins. In *Havana Yatch Club Contusion* (Figure 28.3), the ruins point to loss and romantic regret, while the inclusion of the black youth responds to the former exclusion of blacks from the Club. The image also functions politically as a not-too-subtle critique of the Castro government for allowing this part of Cuba's historical legacy to go to ruin. The critique, Morales argues, is not centered on ideological concerns, but on a preoccupation with erasure and, concomitantly, with rescue. As he explained in an interview with Fabiola Santiago, speaking of the sugar cane series that preceded *The Baths*, whose interlocutor was nineteenth-century printmaker Edouard LaPlante, these images were meant "to rescue the romantic charge that [these earlier artists] had brought to their lithographs in the 19th century with illumination, composition and color," elements of landscape representation that were part of the aesthetic of the sublime. This aesthetic quality lacks, however, the "narrativization of science" so necessary to the development of a modern environmental aesthetic, which Suzi Gablik proposes as a relational model of art practice based on a "participatory paradigm" in which "the world becomes a place of interaction and connection, and things derive their being by mutual dependence" (8). In a world defined as dynamically interconnected, in which art directly engages the environment, Gablik argues, "the old polarity between art and audience disappears" (8).

The photographs of the ruined resorts, with their brooding, poignant portrayal of the collapsed, waterlogged structures, on the other hand, come tantalizingly close to "ruin porn" with their aestheticization of the abandonment and decline of architectural and natural spaces that were once central to Cuba's national iconography (see Leary; Mullins). Their ghost-like appearance recalls figures associated with zombies or revenants from a forgotten past—the ruins also emerge from the sea as unfathomable, unnatural debris. The dialogue with the nineteenth-century images that precede them, however, allows Atelier Morales to transcend the simply maudlin and to transmute Mialhe's romanticized view of seaside leisure into a contemporary "tale of loss and an ode to poetic memory" (Santiago, "The Art"). As Morales and Ayuso described in an interview with Fabiola Santiago: "An entire industry destroyed, a way of life lost and no one thought to at least preserve some of these historical relics and turn them into museums for the generations" ("Artist Pair" 53). Stoler speaks of the trauma behind the treatment afforded to "sites of decomposition that fall outside historical interest and preservation" (13), a sentiment echoed in Atelier Morales' project to bring to the spaces they photograph the honor due to them as ruins of an earlier and mourned (political) past. What perhaps separates Atelier Morales' work from the controversies surrounding "ruin porn" is twofold: both their chosen technique of lenticular photography, which superimposes the funereal sculpture and its political message on the visual field, and a third interlocutor always present although, unlike Mialhe, never explicitly acknowledged in the work. This interlocutor is the earlier occupant of the visualized space, the local and international tourist, whom we can attempt to recover through more traditional analyses of seaside photography. Atelier Morales' concern with social justice—with the exclusion of ordinary Cubans from the privileged spaces the Revolution has allocated to foreign tourists—is conceived here as environmental justice, to the extent that it redefines the environment "to mean not only wild places, but the environment of human bodies, especially in racialized communities, in cities, and through labor" (Ziser and Sze 401).

Ironically, Mialhe's prints were always full of details of those enjoying the seaside. These former deeply racialized "uses" of the landscape are behind Atelier Morales' sense of nostalgia, although perhaps they are more clearly articulated through numerous examples of seaside photography still barely examined in Cuba—or indeed anywhere else in the Caribbean. Poorly catalogued and inconsistently available to researchers, this photographic archive illustrates not only the nature of the patrimony whose loss is regretted so poignantly in Atelier Morales' work but also the richness of its earlier print and photographic representations, as we can ascertain through postcards of the Havana Yacht Club before its slow demolition by the sea, or through numerous photographs of the club in its elegant heyday of racial exclusivity.

It is in this still unexamined photography that we can see clearly the ambiguous, and perhaps problematic, lack of explicit emphasis on race and class in Atelier Morales' project, as well as the absence of an engagement with the ways in which the spaces they photograph contributed to the deterioration of the Cuban coastline. In *Havana: Two Faces of an Antillean Metropolis*, Joe Scarpaci and Roberto Segre show how the segregation that structured Cuba's development of its seashore clubs and resorts reflects the island's intense class and race divisions. We can see it in the relative class discord between the photographs of those enjoying the yacht club and the more modest entertainments and amenities of the clubs of Guanabo, particularly in a series of family photographs from Jaime Leygonier, whose father worked for the telephone company and whose family enjoyed yearly stays at the facilities of the Sindicato Telefónico during the 1950s.

In the 1950s, Leygonier writes, the "home" of the "telephone family" was the Club on the beach of Guanabo:

> A two story building in the shape of a C built on the natural sand, some 100 meters from the sea, a strong structure with more windows and jalousies than walls, a restaurant, changing rooms for men and women, a bar, a clinic, a playground, a gym and small rooms with bunk beds for four people that a family could rent for $3 Cuban pesos a night and was affordable. In summer, or Saturdays and Sundays, the syndicate made available a bus service that departed from Aguila street, although many workers drove their own car.

Michael Ziser and Julie Sze have argued that an effective approach to "tell[ing] the climate story in its historical complexity" can be found "in narrative forms that combine individual biography with environmental history in order to provide concrete examples of environmental damage that can become the basis for redress and reform" (404). Leygonier stresses

Figure 28.4 Atelier Morales, *Guanabo Desolé 1* (Sindicato Telefónico, Guanabo, 2007).

his own condemnation of the neglect of Cuba's architectural patrimony, but represents it through the loss of his family's personal and organized labor connections and through an understanding of that history's relationship to the environment. His personal narrative of dispossession encompasses not only the loss of a building and the vital connection it offered to the enjoyment of the seashore, but also an understanding of how that building was placed "on the natural sand" and designed to be as open as possible to the surrounding elements. From his own middle-class position, he reacts with irony to the Cuban government's stated reasons for neglecting the buildings that previously housed luxury clubs like the Havana Yacht club, purportedly for its commitment to "return the beaches to the people and end the privileges of the bourgeoisie," as Fidel Castro proclaimed in the early 1960s. As symbols of capitalist excesses and privileges, buildings like the Havana Yacht Club exemplified the class and race structures that the Revolution sought to upend. Not so the neglect of the syndicates' beach clubs, which responded, in Leygonier's assessment, to an impulse to destroy a union movement that may have represented workers, but that ideologically had distanced itself from the aims of the revolution.

The Telephone's syndicate opted to preempt dispossession by giving their club voluntarily to the state, which nonetheless could not find a way to reuse or preserve what to the former members was an important institutional and personal space—an important part of Cuba's historical and architectural legacy. The structure of the building remains—perhaps still reparable—"strong against the sea and Castro," Leygonier claims, but windows, stained glass, doors, floor tiles, and bricks have been removed over time, "recycled" by those in need of hard-to-come-by construction materials (see Figure 28.4). In Leygonier's assessment, the building fell prey to the voraciousness of those needing scarce building supplies in Cuba and to the rapaciousness of the sea.

Leygonier, like Atelier Morales, notes the irony of some of these spaces becoming sites for all-inclusive resorts as the Castro government has sought foreign income through tourism, creating a particularly Cuban form of apartheid, where Cubans are kept from tourist beaches in the same way that blacks, mulattoes, and laborers were kept off the beaches of the Havana

Yacht Club and relegated to the beaches of Guanabo. When the spaces have not been reappropriated, Leygonier reminds us, indiscriminate mining for beach sand (a diminishing resource throughout the Caribbean) has destroyed the quality of the natural spaces, with a loss of fifty to sixty yards of sandy shore. Atelier Morales, echoing Leygonier, speak of their work as integrating an ecological dimension that stems from their admiration of Mialhe's attention to nature and its curative potential through healing waters and herbal treatments that are both an Amerindian and an African legacy. However, the ecological dimension of their work is still at odds with local environmental needs as defined by a revolutionary government, armed with scientific assessments, and bent on the erasure of buildings, and their history, from national territory in the name of environmental survival. As environmentally aware photographers and exiled Cubans, Morales and Ayuso have created an art stemming from a space of nostalgia, a space of in betweeness from which they address the Cuban regime's neglect of the island's historical and architectural patrimony while simultaneously capturing the environmental results of the systemic misuse of the land and the early indicators of climate change affecting Cuba. This nostalgia is clearly at odds with state decisions to curb the damage caused by climate change, bringing state goals and artistic project into ideological conflict: both claim an environmental foundation, yet only one acts with the authority of science and political might.

In Cuba, the history of coastal spaces has been reconstructed, archived, and, in the work of Atelier Morales, brought to the fore as an artistic tribute to an endangered patrimony that is both a symbol of loss and a statement of the perils of earlier coastal misuse. For the people of Cuba, the loss of these baths and multiple other structures built on precarious sandbanks represents a loss of architectural heritage, but may ultimately bring protection from the sea and the increased power it has been given by climate change and rising levels. The work of Atelier Morales memorializes the baths' role in an island imaginary as markers of history and of a superseded notion of the nation, ultimately destroyable (and destroyed) by climate change.

The possible fate of islands like Petite Martinique remains more poignantly precarious. Dependent on aid from the US Agency for International Development among other international organizations, Petite Martinique is struggling to put in place a plan to curb coastal erosion and reduce the compounding impacts of climate change. Climate adaptation interventions remain outside of their immediate control and depend on land reclamation, the placement of a 390-foot seawall to halt ongoing erosion, and the construction of a retaining wall on the northern headland to withstand storm surges and strong wave action. With no Atelier Morales to chronicle its endangered beauty, the image that remains is of a little island in battle gear, armored as best it can for its losing battle against an attacking sea.

References

DeLoughrey, Elizabeth. *Allegories of the Anthropocene*. Durham, NC: Duke University Press, forthcoming.

Dunaway, Finis. *Natural Visions: The Power of Images in American Environmental Reform*. Chicago: University of Chicago Press, 2005. Print.

Eaton, Marcia Muelder. "Fact and Fiction in the Aesthetic Appreciation of Nature." *The Aesthetics of Natural Environments*. Ed. Allen Carlson and Arnold Berleant. New York: Broadview Press, 2004. 170–181. Print.

Fagan, Brian. *The Attacking Ocean: The Past, Present and Future of Rising Sea Levels*. New York: Bloomsbury Press, 2013. Print.

Gablik, Suzi. *The Reenchantment of Art*. New York and London: Thames and Hudson, 1991. Print.

Leary, John Patrick. "Detroitism." *Guernica* 15 January 2011. Web. 4 July 2014.

Leygonier, Jaime. "Cuba: Las ruinas del sindicato telefónico, el club de la playa Guanabo." *Hablemos Press* 12 May 2011. Web. 1 October 2015.

Menocal, Nina. "The Sugar Mills/Los Ingenios." Atelier Morales. 2004. Web. 23 August 2013.

Morales, Juan Luis and Teresa Morales. Personal email communcation. July 2014.

Mullins, Paul. "The Politics and Archaeology of 'Ruin Porn'." *Archeology and Material Culture*. Paul Mullins' blog. 19 August 2012. Web. 1 October 2015.

Pulwarty, Roger S., Leonard A. Nurse, and Ulric O. Trotz. "Caribbean Islands in a Changing Climate." *Environment: Science and Policy for Sustainable Development* 52.6 (November/December 2010): 16–27. Web. 1 October 2015.

Richards, Peter. "Saving the Tiny Island of Petite Martinique." *Inter Press Service News Agency* 5 February 2014. Web. 1 October 2015.

Rodríguez, Andrea. "Cuba Girds for Climate Change by Reclaiming Coasts." *Associated Press* 12 June 2013. Web. 12 June 2013.

Santiago, Fabiola. "Artist Pair Portrays Ruins of Cuba's Sugar Industry." Knight Ridder Newspapers. 7 Jan. 2005. Cubanet. Web. 6 September 2016.

——. "The Art of the Americas Is Hot." *Hispanic* 18.4 (2005): 52–53. EBSCOhost. Web. 1 October 2015.

Scarpaci, Joseph L., Roberto Segre, and Mario Coyula. *Havana: Two Faces of Antillean Metropolis*. Durham, NC: University of North Carolina Press, 2002. Print.

Stoler, Anne. *Imperial Debris: On Ruins and Ruination*. Durham, NC: Duke University Press, 2013. Print.

Whittle, Daniel and Orlando Rey Santos. "Protecting Cuba's Environment: Efforts to Design and Implement Effective Environmental Laws and Policies in Cuba." *Cuban Studies* 37 (2006): 73–103. Print.

Yaeger, Patricia. "Editor's Column: Sea Trash, Dark Pools, and the Tragedy of the Commons." *PMLA* 125.3 (2010): 523–545. JSTOR. Web. 1 October 2015.

Ziser, Michael and Julie Sze. "Climate Change, Environmental Aesthetics, and Global Environmental Justice Cultural Studies." *Discourse* 29 (2007): 382–410. Project Muse. Web. 1 October 2015.

29

BEAR DOWN

Resilience and multispecies ethology

Brett Buchanan

On the morning of July 30, 2015, a black bear was discovered dead on a driveway at the north end of Sudbury, a city of 150,000 humans in northeastern Ontario, Canada. When officers from the Greater Sudbury Police, along with an official from the province's Ministry of Natural Resources and Forestry (MNRF), appeared on the scene, they discovered that the bear had died from a bullet wound. It had been killed. According to one report, the bear's death was a "vigilante killing" by a city resident who took matters into his or her own hands (Moodie, "Vigilante"). The "vigilante" label stems from the fact that Sudbury—though located in the heart of Ontario's black bear country—was experiencing an unprecedented rate of bear sightings and "nuisance" behavior within city limits, and a proportion of the general public, though divided on the reasons behind recent bear behavior, blamed their unusually high presence on a lack of management and protection by qualified officials.[1] Even though there were no visible signs or reports of threatening behavior by the bear, it was nevertheless shot and left to die in what the police called an "inhumane" manner.

Characterizations of bear behavior as a nuisance, unpredictable, or threatening are of course nothing new, and certainly not in regions accustomed to dealing with bears. They also suggest a certain one-sidedness in human observation. The summer of 2015, however, is something of an anomaly, as black bear sightings, photos, and videos domi-nated social media, regional news, and even the city council as councillors were forced to address the perceived "crisis" affecting its residents (Moodie, "City Council"). Whatever borders were thought to separate bears and humans—geographical, psychological, histori-cal, developmental—they were showing their true permeability, and city residents were being forced to acknowledge that bears live and forage *within* their communities, just as much as the bears were being forced to come out of their preferred seclusion. Though bears are still popularly seen as "wild" animals, the families and populations living on the fringes of Sudbury are much closer to what Sue Donaldson and Will Kymlicka call "liminal" animals (210), animals that are neither wild nor domesticated but rather part of a growing category of animals that are not without some human contact. Liminal animals, in this sense, are those animals (e.g., racoons, mice, seagulls, chipmunks) that co-exist with humans in shared spaces. And humans are themselves "liminal" in their own hybrid ways, particularly in the sense that Donna Haraway contends "we have never been human" (Haraway, *When Species*; see also Puig). Sudbury residents were thus being forced to recognize the inclusion of bears as a form of liminal animal, and just as impor-tantly, how to address this new development.

As a resident with bears roaming my own backyard, I am interested in these human–bear encounters not only as an instance of place-based thinking, but especially as they highlight the practical complexities of resilient behavior among and between two species that coexist in hesitant and uncertain ways. Both the bears and humans alike were tentatively co-existing in ways that rendered the lives of each uncomfortable, right up to the point of death (so far only the deaths of bears). Further thought and action are needed to preserve the livelihoods of each as mutually meaningful, particularly since these encounters give no appearance of letting up. The environmental humanities provide insightful discursive interventions and critical engagements with environmental issues of *all* kinds, from the local to the global, from the real to the imagined, from the hopeful to the apocalyptic. My focus in this chapter is to briefly address how behavior—and specifically human–animal behavior—can provide a stimulating and ultimately necessary means of engaging with the kinds of multispecies relations that are increasingly strained within our changing environments. As I hope to show, philosophical ethology provides one such way to engage with the resilience of entangled multispecies relations and the imminent threat of their unravelling.

Resilience is often framed by a dynamic juxtaposition, namely "the ability," as Stacy Alaimo puts it, "to snap back and soldier on after adversity" (par. 1). It is the ability to bounce back *and* to continue forward, to come back to a relatively steady state only to find that that state has changed. Resilience thus entails weathering a crisis, but coming out of it changed, altered, and affected. Paul Outka suggests that resilience "is a post-despair environmentalism, which isn't (at all) the same as optimism or thinking it's all going to work out fine—it's finding yourself still alive the day after, with some fight left. ... Rather than trying to get it right, or get it right again, resilience is trying to keep it going here, there, anywhere, everywhere" (par. 4). The scenario recounted above, which is but one example of the kind of resilient behavior found around the world, features the entanglement of one group, humans, that is resilient in maintaining the status quo in the face of perceived adversity, the close proximity of bears, with another group, the bears, who demonstrate resilience in adapting to changing conditions for the sake of survival by foraging farther, deeper, and longer within city limits. Humans' anthropocentric denial and resistance to environmental change, couched in protectionist practices aimed at maintaining human security, confronts the bears' transformative behavioral change as they risk their lives by coming into human settlements for the sake of their survival. Both demonstrate resilience in the face of an environmental threat, albeit they do so differently, and with different outcomes, and all the while emerging transformed from these human–bear encounters.

Thinking about these issues from an ethological point of view—ethology here considered as the study of animal behavior, as advanced by scientists such as Jakob von Uexküll, Konrad Lorenz, and Niko Tinbergen, but also of *ethos*, a practice of habit, conduct, and ultimately ethics (see Burkhardt Jr.; Buchanan, *Onto-Ethologies*; Rose and van Dooren, "Encountering")—provides a unique perspective on human–animal co-existence, especially with respect to forms of behavior that prove to be both more and less resilient with respect to one another. Behavior that is more resilient is capable of adapting to changing conditions, as is increasingly necessary in regions affected by environmental disturbances. There are countless cases of this, from the black bears encroaching more frequently on my city of Sudbury, to Pacific salmon migrating farther north in order to spawn, to marmots coming out of hibernation several weeks earlier than normal. Each represents an adaptation in behavioral habits. In situations where resilience is less strong, and thus where behavioral patterns are unable to adapt, or unable to adapt quickly enough to keep up with significant environmental changes, we witness a counter-response: not only that of weak resilience but also that of succumbing to environmental pressures where the worst-case scenario is species endangerment, potential extinction, and the resulting impact on ecosystem stability. In this scenario, adaptation has been unsuccessful and forms of animal and plant life perish as a result.

My interest here is twofold. On the one hand, I want to suggest that resilience is only possible in a multispecies world where active agents co-learn, co-exist, and ultimately, co-evolve together. This entails understanding that our behaviors—human, animal, nonhuman, more-than-human—are formed in co-constitutive ways, and that we form both explicit *and* implicit, or intentional and unintentional, partnerships with other species. For this reason many human–animal theorists have taken to thinking of community as a network not solely composed of human members but also as necessarily open to multiple species. Writing in 1979, Arne Naess speaks of extending community beyond biological interdependencies to include "mixed communities" wherein "wild animals, domesticated animals, and humans" (238)—for example, wolves, bears, sheep, and human—live together. For Naess it is a matter of opening up human–animal relations to political and ethical questions wherein the mutual benefit of all species involved are being accommodated as best as possible. Similarly, Dominique Lestel has written about the shared meaning, interests, affects, and lives that overlap within human–animal "hybrid communities" ("Hybrid" 63). From the standpoint of his philosophical ethology, living beings cooperate and affect one another as intentional, active agents, and as such, are responsive and responsible one to the other (Lestel et al. 156; Lestel, "Fingers" 68). Lestel's thought shares much in common with Donna Haraway's on this matter—specifically, Haraway's appeal to ethological ethnography in philosophical discourses on "the animal" (*When Species* 21). Haraway argues that co-evolution is an active and not a passive process, and that "co-constitutive companion species and co-evolution are the rule, not the exception" (*Companion* 32). Animal ethologies, particularly in this expanded context of hybrid communities, constitute part of the budding area of multispecies ethnographies that are among the most fertile and exciting developments to come along in environmental humanities, as found in this collection (Chaudhuri; Rose and van Dooren, "Encountering") and beyond (Kirksey; van Dooren and Rose).

What philosophical ethology provides, then, are alternative ways to think through animal, and human–animal, relations (Buchanan et al., "General" 1; Chrulew, "Philosophical"), specifically insofar as they counter the historical and still enduring tendency to address animals as literary tropes or philosophical mirrors. Philosophical ethology begins with *real* animals as opposed to imagined or fictional ones—this forms part of Haraway's critique of philosophers like Derrida and Deleuze, who, in her reading, never address real animals or the robust work of ethologists (*When Species* 19–27)—and addresses such issues as animal subjectivity, friendship, communication, and sexuality, all as meaningful modes of animal comportment. Agency and embodied subjectivity is returned to animals as bearers and givers of meaning, as active participants in the enchantment of the world, and as beings who continually surprise and confound our expectations (Despret, *What Would Animals*; Buchanan et al., "Philosophical"). Resilience is ethologically noteworthy inasmuch as it removes animals and human–animal co-habitation from preconceived limitations such as the reduction of animal intellect to "instinct" or the labeling of domesticated animals as "pets" or economic "resources," and instead allows animals to show themselves as curious, knowledgeable, and social beings. Resilience is a mode of ethological adaptation, and one that demands attentiveness to and recognition of our co-dependency on other animal lives. As Vinciane Despret writes, we are obliged to show animals a degree of politeness and courtesy as a sign of commitment towards our ontological and ethical entanglements wherein resilience comes to be a sign of commitment toward each other's well being (*What Would Animals*; see Buchanan, "Metamorphoses").

When behavioral patterns are unable to sustain themselves, on the other hand, a sharp decline of specific forms of life ensues. In our current time, which many now herald as a new geological period called the "Anthropocene" (Waters et al.)—a label that many others still critique and refuse (see Neimanis et al.; Crist, "Poverty")—species endangerments and extinctions are on the rise. In terms of both national and international status reports, the world's species of animals and plants are increasingly classified as threatened, endangered, or extinct. In Canada, the national Committee On the Status of Endangered Wildlife In Canada (COSEWIC) reports a notable increase in endangered species over its previous national registers, from 500 species at risk in 2006 to 712 in 2015, and the International Union for Conservation of Nature (IUCN)'s Red List of Threatened Species continues to show a striking increase in extinction rates in their global species evaluations.[2] Increasing consensus suggests that we are witnessing extinction rates that surpass normal "background" rates by 10–1000 times; although the vast discrepancy in this spread is due to scientific uncertainties (e.g., extinction of species that have not yet been "discovered"), dependent on geographical region, and other factors, the increase in extinctions is almost unequivocally due to human activities (Rose and van Dooren, "Extinctions"; Heise; Sandler). Concern for world biodiversity is so pronounced that the IUCN has even extended their Red List, as of May 2014, to include threatened ecosystems, where the rough equivalent of species extinction is the "collapse" of an ecosystem (International Union for the Conservation of Nature).

In addition to the efforts to pull species back from the brink of extinction, one can also find engaging stories and narratives constructed around and about these "last animals"—stories that are just as important inasmuch as they tease out the ethical and political complexities involved, and inasmuch as they lend critical meaning to the otherwise discouraging trends. More than anything else, then, it is the ability of these stories to enliven and qualify the lives being lost that is particularly noteworthy, especially in the midst of the obsession with quantifying the escalating rates of environmental catastrophes, of which species extinctions is but one (details in the paragraph above are no exception). The importance of narrative cannot be underestimated in this context. Many theorists in the environmental humanities have taken to documenting stories of extinction so as to, in the words of Thom van Dooren, "hold open simultaneously a range of points of view, interpretations, temporalities, and possibilities" (8). Storytelling invites deeper engagements with endangered species than those typically found in scientific studies because stories feature lively protagonists and antagonists with whom the reader can sympathize and identify. Extinction stories take many shapes, focus on species the world over, and raise many complicated and often competing issues, including love and grief (Rose), ethical accountability (van Dooren), problems in the elegiac narrative (Heise), and many others (see Rose et al.).

In this framework, philosophical ethology can help us think through and engage with the worsening of biodiversity and the loss of species. Instead of focusing on extinct species as symbolic images of life that have disappeared, which tends to reify animals according to the Cartesian model—the familiar lists of extinct animals and plants, either as names or images—an ethological approach accentuates the particular ways that animals fill the world with their idiosyncratic behaviors. It is not just that certain populations or species are disappearing, which is far too narrow a focus, but that what is being lost are meaningful relations with other species, as well as the seemingly endless ways that animals express the possibilities of life differently. Animal behavior—whether a lyrebird's mimicking voice, whale migration, orangutan culture, or a tick's opportunism—demonstrates the full expression of what the world is, in all of its richness, but it equally permits glimpses of what the world is capable of becoming, namely one that is a little less rich and a little less full with every decline in animal numbers. "Animals," Eileen Crist writes, "do not simply grace landscapes and seascapes with their stunning beauty: they electrify the world with—and, indeed, display the world as—species of mind. Nor do animals simply inhabit our houses, fields, and farms as animated bodies; they charge human dwellings with their forms of awareness" ("Ecocide" 56). Clearly this does not apply only to domesticated or liminal animals, for the loss of any form of awareness is an extinguishing of our earth's diversity: "the obliteration of biological wealth in the sense of species, subspecies, populations, ecosystems, and gene pools, the destruction of the diversity of minds (or modalities of aware perceiving, being, and experiencing) is impoverishing the Earth (and ourselves) in ways we do not even begin to comprehend and can barely imagine" (48–49).

Along similar lines, Despret writes of the nostalgia and forgetting that occurs when unique behavior, such as the stunning flight of millions of passenger pigeons, no longer exists: "When a being is no more, the world shrinks at once, and a section of its reality subsides. Every time an existence disappears, it's a small piece of the universe of sensations that becomes blurred" ("Passenger Pigeon" par. 9). Attention to specific animal behavior provides a different and often unique angle on extinctions, and it does so with respect to phenomena that are difficult to classify and address. For example, the plight of Monarch butterflies is well known in North America, and their yearly migration from the eastern United States and Canada to Mexico is drawing attention due to its instability. But the Monarchs present an interesting contrast: whereas the species is "secure" since their numbers exist in the millions, and thus are classified only as "not evaluated" according to the IUCN's Red List parameters for threatened species, it is nonetheless the case that their very being *is* endangered.[3] Because the species cannot be classified as threatened, the IUCN has therefore taken the unprecedented measure of referring to the Monarchs' migration as a "threatened phenomenon" (Wells et al.). It is not the species that is threatened, or at least not yet officially, but the *phenomenon* of the butterflies' migration that is at risk, mainly due to lack of milkweed production and habitat fragmentation along their migratory routes and in their principal wintering site in Mexico. If the populations of Monarchs are unable to become resilient against the changes to their environments, then it is not simply the loss of their beauty that is disturbing but the awe-inspiring, multigenerational migration of such delicate creatures across thousands of kilometers. This unparalleled phenomenon would disappear, and so too would an irreplaceable sense of the world (see Wilcove; Ackerman).

Even in scientific discussions about bringing back the extinct—for example, the current interest in de-extinction—attention to ethological knowledge is for the most part lacking. Remnants and traces of animals that are now long extinct, such as cave bears or woolly mammoths, can still be found in teeth, bones, and even ecosystem types (see Bieder), but the paleolithic behaviors of these animals no longer form part of living ethological knowledge. While public fascination turns to the latest technologies and aspirations to clone and resurrect lost forms of life, others are far more critical of the environmental aspirations for these advances. The proposed de-extinction of the woolly mammoth as a renewed hope to re-wild the barren landscapes of Siberian tundra, for instance, is in practice handicapped by the pragmatic problem of figuring out how their ancient behavioral patterns, learned and passed down over decades and centuries of adaptation, would integrate into contemporary environments, including existing animal forms (Chrulew, "Reversing"). It is one thing to imagine the resurrection of the woolly mammoth; it is another thing entirely to prepare for how these forms of life would bring about renewed forms of behavior, culture, and thought. This is to say nothing of how present-day species within these habitats would adapt to the intrusion of these strange, resurrected creatures. Talk about resilience. The techno-scientific advance of de-extinction is wildly fascinating, to be sure, but it does little more than boil rich and complex animal lives down to their constituent DNA, transferable across time and place. Without considering the ethological dimensions of how animals fill and express the world through their active, entangled, multispecies lives, de-extinction is surely missing the point of what it means to be an animal, let alone why and how resilience is needed within our world today.

Coming back to the present day and my initial example of black bears, it is clear that more attention needs to be paid to human–animal relations in the context of our increasingly unstable environments, and the entanglements of our co-existing behaviors ought to be part of this conversation (see Alagona's *After the Grizzly* for a persuasive and nuanced example of this). Though black bears are far from endangered in northern Ontario, it is precisely the obvious human resistance to adapt our behaviors to those of other species that leads their lives, and our own, to be threatened, marginalized, and possibly eliminated. This was the case in a subsequent human–bear encounter within Sudbury later this same summer. Rather than learning to modify our behaviors so as to co-exist with local black bear populations, Sudbury residents steadfastly refused to adapt to our cohabitants, who were too quickly seen as menaces. Part of this surely has to do with a behavioral reluctance (whether intentional or not) to see the bears as active agents, with different desires, thoughts, and meanings from those we might ascribe to them. Rather than reacting to their presence as one of a "nuisance" or "menace," more appropriate and considered responses are called for, including the simple recognition that they *look* at us too (see Derrida). To recognize this is to admit that our relations with the bears are not one-directional, but reciprocal. On the night of Monday, August 10, 2015, a mother bear and cub were shot dead by police officers who were called to a disturbance in an unoccupied house at the south end of Sudbury. As reported the following morning, police arrived at the scene and discovered that the house had been broken into by the mother and cub, apparently following the scent of fresh baking (the house occupants were not currently at home). With officers from Ontario's MNRF unavailable to assist, the police officers waited outside the house for the two bears to leave, and proceeded to "dispatch" them "in the interest of public safety" (Moodie "South End" par. 5). As reports about the killings unfolded on social media, it quickly became known that there were two other cubs in the vicinity, now orphaned and at large. Over the ensuing days, efforts were put in place to trap the two remaining cubs in order to hopefully transport them to a bear sanctuary with an eye toward eventual relocation once they were deemed mature enough. For the moment, however, they remain without their mother and sibling, and in need, more than ever, of shared resilience in a world that is not yet willing or able to keep up with the changes taking place.

Notes

1 Michael Commito, who recounts the historical relations between humans and black bears in northern Ontario in his doctoral dissertation ("Our Society") as well as in his twitter feed (@mikecommito), emphasizes that the historical management of black bear populations stems in large part from political policies that led to the elimination and re-introduction of the spring bear hunt in Ontario. Though the spring bear hunt is one of the reasons, at least in public opinion, for human–bear encounters during the 2015 summer, the main culprits are a combination of a significant shortage in wild blueberries for the bears to forage on, and residential garbage attracting the bears to Sudbury.
2 At its last meeting in 2015, COSEWIC reported 712 species at risk, which includes 15 extinct, 23 extirpated, 316 endangered, 167 threatened, and 205 of special concern. This is by comparison to 2006 (*Annual Report* 25), when 521 species were found to be at risk, which included 13 extinct, 22 extirpated, 212 endangered, 136 threatened, 151 of special concern, and 41 data deficient species.
3 At the time of writing, the protective status of Monarch butterflies is currently under review by the United States' Fish and Wildlife Service and the Endangered Species Act. See Platt.

References

Ackerman, Diane. *The Rarest of the Rare: Vanishing Animals, Timeless Worlds*. New York: Vintage, 1997. Print.

Alagona, Peter. *After the Grizzly: Endangered Species and the Politics of Place in California*. Berkeley, CA: University of California Press, 2013. Print.

Alaimo, Stacy. "Bring Your Shovel!" *Resilience: A Journal of the Environmental Humanities* 1.1 (2014). Web. 25 Nov. 2015.

Bieder, Robert E. *Bear*. London: Reaktion Books, 2005. Print.

Buchanan, Brett. *Onto-Ethologies: The Animal Environments of Uexküll, Heidgger, Merleau-Ponty, and Deleuze*. New York: State University of New York Press, 2008. Print.

——. "The Metamorphoses of Vinciane Despret." *Angelaki* 20.2 (2015): 17–32. Print.

Buchanan, Brett, Jeffrey Bussolini, and Matthew Chrulew. "General Introduction: Philosophical Ethology." *Angelaki* 19.3 (2014): 1–3. Print.

Buchanan, Brett, Matthew Chrulew, and Jeffrey Bussolini, eds. "Philosophical Ethology II: Vinciane Despret." *Angelaki* 20.2 (2015): 1–185. Print.

Burkhardt Jr., Richard. *Patterns of Behavior: Konrad Lorenz, Niko Tinbergen, and the Founding of Ethology*. Chicago, IL: University of Chicago Press, 2005. Print.

Chaudhury, Una. "Interspecies Diplomacy in Anthropocenic Waters: Performing an Ocean-Oriented Ontology." *The Routledge Companion to the Environmental Humanities*. Ed. Ursula K. Heise, Jon Christensen, and Michelle Niemann. London: Routledge, 2017. 144–52. Print.

Chrulew, Matthew. "Reversing Extinction: Restoration and Ressurection in the Pleistocene Rewilding Projects." *Humanimalia* 2.2 (2011): 4–27. Print.

——. "The Philosophical Ethology of Dominique Lestel." *Angelaki* 19.3 (2014): 17–44. Print.

Commito, Michael. "Our Society Lacks Consistently Defined Attitudes Towards the Bear: The History of Black Bear Management in Ontario, 1912–1987." Ph.D. Dissertation, McMaster University. 2015. Print.

Committee on the Status of Endangered Wildife in Canada. *COSEWIC Annual Report 2006–2007*. 2007. Web. 25 Nov. 2015.

——. "Summary Table of Wildlife Species Assessed by COSEWIC." 2015. Web. 25 Nov. 2015.

Crist, Eileen. "Ecocide and the Extinction of Animal Minds." *Ignoring Nature No More: The Case for Compassionate Conservation*. Ed. Marc Bekoff. Chicago, IL: University of Chicago Press, 2013. 45–61. Print.

——. "On the Poverty of Our Nomenclature." *Environmental Humanities* 3 (2013): 129–147. Print.

Derrida, Jacques. *The Animal That Therefore I Am*. Ed. Marie-Louise Mallet. Trans. David Wills. New York: Fordham University Press, 2008. Print.

Despret, Vinciane. "P is for Passenger Pigeon." *Extinction Studies: Stories of Time, Death and Generations*. Ed. Deborah Bird Rose, Thom van Dooren, and Matthew Chrulew. New York: Columbia University Press, forthcoming. Print.

——. *What Would Animals Say if We Asked the Right Questions?* Trans. Brett Buchanan. Minnepolis, MN: University of Minnesota Press, 2016. Print.

Donaldson, Sue and Will Kymlicka. *Zoopolis: A Political Theory of Animal Rights*. Oxford: Oxford University Press, 2011. Print.

Haraway, Donna. *The Companion Species Manifesto: Dogs, People, and Significant Otherness*. Chicago: Prickly Paradigm Press, 2003. Print.

——. *When Species Meet*. Minneapolis, MN: University of Minnesota Press, 2008. Print.

Heise, Ursula K. "Lost Dogs, Last Birds, and Listed Species: Cultures of Extinction." *Configurations* 18 (2010): 49–72. Print.

International Union for Conservation of Nature. *Red Lists of Ecosystems: A New Global Standard*. 2014. Web. 1 June 2014.

Kirksey, S. Eben. *The Multispecies Salon*. Durham, NC: Duke University Press, 2014. Print.

Lestel, Dominique. "Hybrid Communities." Trans. Brett Buchanan. *Angelaki* 19.3 (2014): 61–73. Print.

——. "Like the Fingers of the Hand: Thinking the Human in the Texture of Animality." Trans. Matthew Chrulew and Jeffrey Bussolini. *French Thinking About Animals*. Ed. Louisa Mackenzie and Stephanie Posthumus. East Lansing, MI: Michigan State University Press, 2015. 61–73. Print.

Lestel, Dominique, Florence Brunois, and Florence Gaunet. "Etho-Ethnology and Ethno-Ethology." Trans. Nora Scott. *Social Science Information* 45.2 (2006): 155–177. Print.

Moodie, Jim. "City Council Confronts Bear Crisis." *The Sudbury Star* 12 August 2015. Web. 25 Nov. 2015.

——. "Two South End Bears Dispatched By Police." *The Sudbury Star* 11 August 2015. Web. 25 Nov. 2015.

——. "Vigilante Killing of Black Bear in Sudbury." *The Sudbury Star* 31 July 2015. Web. 25 Nov. 2015.

Naess, Arne. "Self-Realization in Mixed Communities of Humans, Bears, Sheep, and Wolves." *Inquiry* 22 (1979): 231–241. Print.

Neimanis, Astrida, Cecilia Åsberg, and Johan Hedrén. "Four Problems, Four Directions for Environmental Humanities: Toward Critical Posthumanities for the Anthropocene." *Ethics & the Environment* 20.1 (2015): 67–97. Print.

Outka, Paul. "Environmentalism After Despair." *Resilience: A Journal of the Environmental Humanities* 1.1 (2014). Web. 25 Nov. 2015.

Platt, John R. "Monarch Butterflies Could Gain Endangered Species Protection." *Scientific American* (2015). Web. 30 Dec. 2015.

Puig de la Bellacasa, Maria. "Ethical Doings in Naturecultures." *Ethics, Place and Environment* 13.2 (2010): 151–169. Print.

Rose, Deborah Bird. *Wild Dog Dreaming: Love and Extinction*. Charlottesville: University of Virginia Press, 2011. Print.

Rose, Deborah Bird, and Thom van Dooren. "Encountering a More-Than-Human World: Ethos and the Arts of Witness." *The Routledge Companion to the Environmental Humanities*. Eds. Ursula K. Heise, Jon Christensen, and Michelle Niemann. London: Routledge, 2017. 120–28. Print.

——. "Extinctions." *Encyclopedia of Geography*. Web. 22 Sept. 2010.

Rose, Deborah Bird, Thom van Dooren, and Matthew Chrulew, eds. *Extinction Studies: Stories of Time, Death and Generations*. New York: Columbia University Press, forthcoming. Print.

Sandler, Ronald. *The Ethics of Species*. Oxford: Oxford University Press, 2013. Print.

Van Dooren, Thom. *Flight Ways: Life and Loss at the Edge of Extinction*. New York: Columbia University Press, 2014. Print.

Van Dooren, Thom and Deborah Bird Rose. "Storied-Places in a Multispecies City." *Humanimalia* 3.2 (2012): 1–27. Print.

Waters, Colin N. et al. "The Anthropocene is Functionally and Stratigraphically Distinct From the Holocene." *Science* 351.6269 (2016). Web. 8 Jan. 2016.

Wells, S. M., R. M. Pyle, and N. M. Collins, eds. "Monarch Butterfly: Threatened Phenomenon, California Winter Roosts and Mexican Winter Roosts." *The IUCN Invertebrate Red Data Book*. Gland, Switzerland: International Union for Conservation of Nature and Natural Resources, 1983. 463–470. Print.

Wilcove, David. *No Way Home: The Decline of the World's Great Animal Migrations*. Washington, DC: Island Press, 2008. Print.

Part V

ENVIRONMENTAL ARTS, MEDIA, AND TECHNOLOGIES

30

CONTEMPORARY ENVIRONMENTAL ART

James Nisbet

Since the end of World War II, visual art has undergone a series of major transitions in the way that it addresses issues of the environment. Prior to this paradigm shift, artwork dating back to the Renaissance had contended with the environment principally through the genre of landscape painting. Whether Pieter Bruegel the Elder's paintings of village life and labor, Nicolas Poussin's idealizations of mythical terrains, or John Constable's studies of rain and wind in the craggy English countryside, centuries of landscape painting consistently construed the domain of the visual arts as one of *picturing* environmental conditions. Beginning with the rise of new approaches to artistic production in the 1950s, however, the fundamental role of visual art transformed from that of showing environmental conditions to participating more directly in them. This is to say that while the academic discipline of art history has adopted more expressly ecocritical concerns during the early twenty-first century (Baum; Boetzkes; Brattock and Ater; Brattock and Irmsher; Cheetham; Demos; Lippard; Nisbet; Scott and Swenson), the practice of art itself has a longer history of shaping key questions for the environmental humanities. The following chapter will address these developments over the last half-century through three important artists and artistic partnerships, beginning with the work of Robert Rauschenberg.

The horizontal turn

In 1972, the art historian and critic Leo Steinberg published an essay on Rauschenberg that announced a sea change for reading both the artist's work and that of postwar art more generally. Titled "Other Criteria," Steinberg's text set out to re-visit the terms with which experimental art dating back to the mid-nineteenth century was evaluated and interpreted. Clement Greenberg, the main object of Steinberg's critique, had recently solidified his own position on the formal advancement of what he called "modernist painting" through a series of influential essays in which he claimed that the most important art of the nineteenth and twentieth centuries progressed through the purification of specific media (by which he meant mainly painting and to some extent sculpture) in the visual arts. Greenberg's position and the numerous criticisms it has garnered—including Steinberg's own—are now well established in the art historical literature (Frascina; Guilbaut; Jones; Krauss; Leja). What has not been appreciated, however, is the way in which Steinberg's argument not only cast aside a formalist and medium-specific approach to modern art, but also significantly broaened a more environmentally conscious one.

In addition to simply declaring the need for "other criteria," Steinberg offered his own alternative principles of analysis. Identifying a tendency toward a "horizontal" orientation in recent works by Rauschenberg, Steinberg set out to define what he called "the flatbed picture plane." A term borrowed from the printing industry, the flatbed picture plane is rooted in "hard surfaces such as tabletops, studio floors, charts, bulletin boards—any receptor surface on which objects are scattered … —whether coherently or in confusion" (84). From the fifteenth century through the mid-twentieth, Steinberg notes, Western painting had consistently used the "picture plane"—that dividing surface between the illusion of painterly depiction and the real space surrounding the canvas—to "represen[t] a world, some sort of worldspace which reads on the picture plane in correspondence with the erect human posture" (82). Such is the address, for instance, of landscape painting, which represents an environmental scene illusionistically, as if the spectator could see the depicted landscape through a window.

But beginning around 1950, a different kind of work arose in which painting did not orient itself so much to the verticality of the spectator as to its shared horizontal alignment with her. "The pictures of the last fifteen to twenty years," Steinberg explains, "insist on a radically new orientation, in which the painted surface is no longer the analogue of a visual experience of nature but of operational processes" (84). Such processes in Rauschenberg's art include using a car tire to ink a long, black print on twenty consecutive sheets of paper or asking the artist Willem de Kooning to create a drawing for the stated purpose of Rauschenberg then erasing it. Both *Automobile Tire Print* (1953) and *Erased de Kooning Drawing* (1953) were understood by many critics at the time of their making as deeply negative gestures, a means of simply rejecting the tradition of finely skilled artistic craftsmanship out of hand (Joseph). In the face of this interpretation, Steinberg astutely realized that such works are not merely negations of painterly tradition, but are instead active operations. In each, Rauschenberg set out a task and carried it through. Whether the resulting work of art offers a visible form to its viewer or not—a horizontal line of tread or a seemingly blank page, respectively—depends on the operation. What both works share, Steinberg reiterates, is a different "psychic address of the image" in that this new work's "special mode of imaginative confrontation [is] expressive of the most radical shift in the subject matter of art, the shift from nature to culture" (84).

Steinberg's articulation of this "shift from nature to culture" might appear an odd place to locate a major turn *toward* environmental concerns in experimental art, but that is precisely what it represents. By defining the emergence of what he would also call "a pictorial surface that let the world in again," Steinberg positioned Rauschenberg's art at the precipice of a major transition from painting that addressed itself only to visual inspection and to the fictitious space depicted within the frame of the canvas, to a profoundly new way of understanding art as important because it shares the same existential space as its spectator. In environmental terms, this shift was especially significant in that it acknowledged that the work of art is, at its core, situated within a horizontal world, meaning one that is spatially coterminous with its subject matter and therefore of a single, shared subsistence with all beings. While subsequent works that Rauschenberg produced did not feature any explicitly organic or ecological subjects— his "combine paintings" of the later 1950s were in fact three-dimensional collages of found debris, commercial logos, splattered paint, and household goods—what these works

Figure 30.1 Robert Rauschenberg, *Black Market*, 1961. Canvas 50 in × 59 1/8 in × 4 in. Valise 6 1/2 in × 24 1/4 in × 16 3/8 in. Museum Ludwig, Cologne. © Robert Rauschenberg Foundation/ Licensed by VAGA, New York, NY.

instigated was an integration of artistic production with an ecological notion of materiality. The pillow hanging down from *Canyon* (1959) acknowledges the incessant demands of gravity, just as the quilt used in *Bed* (1955) and the license plate in *Black Market* (1961) insist on the connection of artistic production to everyday patterns of behavior, consumption, and waste (see Figure 30.1). The quilt, after all, was stripped from the artist's own bed. In all, Rauschenberg's work did not envision its own occupation of space apart from the environmental conditions of our shared world. Instead, it employed the "flatbed surface" as a means to reunite the material life of art with the material environment in which it arises.

Of course, the greater significance of Rauschenberg's flatbed picture plane lies in the fact that its mode of addressing the spectator was not unique to his art alone. In "Other Criteria," Steinberg also notes the work of Jean Dubuffet as another artist who introduced a horizontal orientation to his work by incorporating dirt, rocks, cement, and tar into the surfaces of his paintings (84). Beyond Steinberg's own specific examples, we might also take note of contemporaries such as Jackson Pollock, whose paintings were splattered on the floor of an old shed, and Arman, who compiled used items like high heels and alarm clocks under Plexiglas for his series of "Accumulations." Taken as a whole, these practices introduce a kind of work that relies on its means of production and materiality to establish meaning. Earlier in the twentieth century, the advent of abstract painting had done away with the impetus to connect artistic meaning to interpretations based on subject matter and symbolism. With the subsequent horizontal turn in the postwar period, advanced artwork further disengaged from metaphorical interpretation. No longer did subject matter or its reference to a pre-standing image bank govern the impact of an artwork. The generation following World War II presented a more immersive outlook for visual art, albeit one still expressed through the traditional media of painting and sculpture.

The artist as adviser

During the 1960s, the stakes of environmental visual art experienced yet another major change with the emergence of a movement known as "land art." Exemplified by artists such as Walter De Maria, Michael Heizer, Robert Smithson, Nancy Holt, and Richard Long, land art appeared nearly simultaneously in Europe, Asia, and North America as a means of working directly with earth and other organic materials outside in the open air. If the move enacted by the horizontality of Rauschenberg's art had turned away from the importance of visual representation, land art moved even further away from the established history of landscape art by breaking down the apparent isolation of the artistic object from its environment. In constructing "earthworks," artists made large-scale interventions in an existing outdoor site. In the most prominent examples of these earthworks—including De Maria's *Lightning Field* (1977) in New Mexico, Heizer's *Double Negative* (1969) in Nevada, and Smithson's *Spiral Jetty* (1970) in Utah—these artists built the form of their work directly onto the surface of the landscape so that the work of art and its location became inseparable from one another. An earthwork can neither be moved from the place of its making nor detached from the environmental conditions of its site. While exhibitions in 1968 and 1969 that showed photographs of these earthworks and moved heaps of dirt into the gallery space helped to establish land art as one of "the most talked about and publicized" artistic developments of those years, the movement itself was shortlived (Meehan 12). By the early 1970s, almost all of the grand, outdoor earthworks had either been designed or already completed. For practical issues, such as the loss of funding from influential gallerists and private collectors, and more philosophical concerns, such as the ecological impact of creating earthworks, the first wave of land art ran its course within a decade. But rather than focus on the rise and fall of land art, the second case study I would like to examine is the work of Helen Meyer Harrison and Newton Harrison—better known simply as the Harrisons—whose shared artistic project took shape in the very transition between the first wave of land art and the more politically engaged environmental art that crystallized in the 1970s. As such, the Harrisons' work provides a telling transition between the environmental *address* of painting and sculpture of the 1950s and environmental *actions* carried out by artists of the later twentieth century.

The Harrisons' pivot away from land art occurred in the summer of 1971 through their creation of *Shrimp Farm*, also known by the more cumbersome title *Notations of the Ecosystem of the Western Salt Works with the Inclusion of Brine Shrimp*. Consisting of four salinated ponds filled with brine shrimp, this installation was made for the famed "Art and Technology" exhibition at the Los Angeles County Museum of Art (LACMA). It has since become the single work that many identify as the point of transition from land art to more ecologically conscientious practices (Bijvoet; Green; Heartney; McGeevy). Against the tendency of earthworks to expend large outlays of energy in carving up the landscape and introduce industrially produced materials into these environments, works by the Harrisons and many artists after them have claimed a distinction, in Newton Harrison's words, between "[using] earth as material" and "[dealing] with ecology in the full sense of the term" (Adcock 35). The Harrisons' *Shrimp Farm* at LACMA was conceived, in fact, as a direct rejoinder to Smithson's recently completed *Spiral Jetty* in Utah's Great Salt Lake. As the story goes, when Newton Harrison first met Smithson, they talked about the tendency of the water around the *Jetty* to periodically adopt a reddish hue, an effect of carotene produced by algae in the Great Salt Lake. The higher the salt level in the water, the greater the red hue produced by this algae. With this tendency in mind, Harrison proposed to introduce brine shrimp at *Spiral Jetty*, which feed upon the algae, and could be harvested at the site. It was a suggestion that apparently held little interest for Smithson (Levin 136–137).

In response, Newton and Helen Harrison decided that if they could not turn *Spiral Jetty* into a shrimp farm, they would create one in Los Angeles instead. The work they made for "Art and Technology" employed different levels of salt in each of the four ponds outside of LACMA, generating an effect of subtle shifts in color from one to the next. As such, *Shrimp Farm* carries forward the artistic developments of the preceding decades in that the Harrisons had created a "horizontally" oriented work that produced abstract compositions based on the actual operation of the life in each container. In another sense, though, the Harrisons also charted new ground, because *Shrimp Farm* resulted in a genuine harvest, and was in fact the second in their series of six so-called *Survival* projects to do so. In addition to shrimp, other examples in the series include farming ecosystems for fish, trees, potatoes, salad greens, and worms. These pared-down living environments—perhaps none more structurally and visually elegant than the shrimp, salt, and algae at LACMA—announced a conception of art whose aim is to produce ecosystems.

By creating the biological conditions of highly specific and localized ecosystems, the Harrisons' *Survival* works highlight the interrelations among organisms and climate conditions required to sustain life in simplified, enclosed environments. But as their practice would come to demonstrate, the idea of any ecosystem being truly closed was a fallacy that relied precisely on the short-term constraints of the *Survival* pieces, which were created only for the two- or three-month duration of a single exhibition. A significant change in the Harrisons' approach arrived through the formation of a multifaceted work called *The Lagoon Cycle* that occupied the couple from 1972 to 1984. The project began as yet another *Survival* piece—using a crab species native to Sri Lanka and an indoor, simulated estuary—and gradually transformed into a much larger and more ambitious investigation on the scale and integration of ecosystems across the entire Pacific Rim (Figure 30.2). This is to say that the Harrisons' original plan to cultivate the *Scylla serrata*, or mud crab, in a closed laboratory environment had similar goals to previous *Survival* works in that it sought to produce an edible food source in controlled, replicable conditions and to study the behavior of the resulting ecosystem (Burnham). This particular ecosystem, however, would not be limited to a predetermined period of time. Rather than being commissioned for an exhibition, the mud crabs' tanks were set up in the Harrisons' Southern California studio using timed lights,

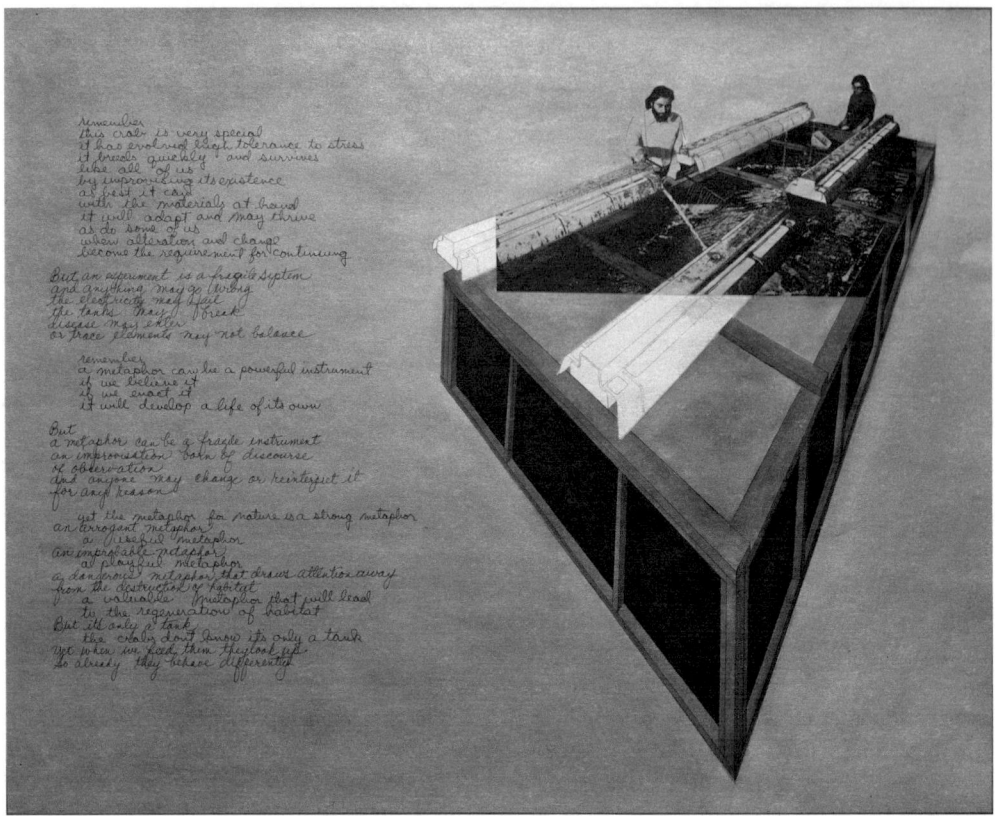

Figure 30.2 Newton Harrison and Helen Mayer Harrison, *The Lagoon Cycle: The Second Lagoon,*
Panel 2, 1974–1984. 8 ft × 7 ft 6 in. Courtesy of the artists.

limited temperature fluctuation, and controlled salinity in the aquatic environment to mimic the conditions of a tropical, estuarial lagoon.

As it turned out, the very duration of this piece proved critical. Late in 1972, during the same time when monsoon season would have arrived in the mud crabs' native Sri Lanka, the Harrisons, drawing on advice from colleagues at the nearby Scripps Institute for Oceanography, attempted to simulate monsoon conditions by decreasing salinity and increasing available food. As a result, they triggered mating behavior among the crabs, the first known instance for this species in an artificial environment. Through this simple discovery that the crabs required periodic changes in the steady state of their ecosystem to survive, the artists were forced to reconsider the type of enclosed, artificial ecosystems they had simulated to date. In response, the Harrisons moved up and out of the laboratory.

Under the new title of *The Lagoon Cycle*, they expanded their experimentation with estuarial environments through seven, increasingly expansive "cycles" and, importantly, introduced new ways of presenting their work that included narrative voices, performance, and printed matter. *The Lagoon Cycle* is exhibited as a series of seven large panels incorporating text, maps, and cut photo-collage. The text on these panels tracks the Harrisons' engagement with estuarial lagoons for mud crabs as told through an oblique dialogue between two characters, a Lagoonmaker and a Witness. The first three panels examine the move from indoor tanks to outdoor environments; the fourth describes transplanting the crabs to a series of ponds cultivated in California's Salton Sea; the fifth and sixth introduce and then reflect upon a bold proposal to regulate salinity, herbicide, and pesticide levels in the Salton Sea by connecting it to the Gulf of California and the Pacific Ocean; and the seventh considers the Pacific Ring of Fire that connects Sri Lanka to the American West Coast. When presenting *The Lagoon Cycle* to live audiences, Newton Harrison performs the role of Lagoonmaker and Helen Meyer Harrison that of Witness. While not free of problems and contradictions in mythologizing the natural world and gendering active behavior as masculine and passive behavior as feminine, *The Lagoon Cycle* took the major step of suggesting that artistic "work" on the environment need not result in an autonomous object. It could instead exist through performance and analysis.

We might look to an extended passage from the opening of *The Second Lagoon* as a key formulation of this approach. The Witness speaks first (in italics) and the Lagoonmaker responds:

> *But*
> *the tank is not a lagoon*
> *nor is it a tidal pond*
> *nor does the mixing of fresh and salt waters*
> *make it an estuary*
> *Filters are not the cleansing of the tides*
> *water from a hose is not a monsoon*
> *lights and heaters are not the sun*
> *and crabs in a tank do not make a life web*
>> But
>> the tank is part of an experiment
>> and the experiment is a metaphor for a lagoon
>> if the metaphor works
>> the experiment will succeed
>> and the crabs will flourish
>> after all
>> this metaphor is only a representation
>> based on observing a crab in a lagoon
>> and listening to stories
> *If*
> *the experiment isolates parts of a real lagoon*
> *and places them in a tank*
> *then the metaphor also refers to alienation*
> *to violation*
> *to breaking the integrity of a real system*

(Harrison and Harrison 44–45)

Through this exchange, we can gather that *The Lagoon Cycle* reframed the Harrisons' practice as both ethically committed to ecological care and increasingly sensitive to the local and even chaotic contingencies of particular ecosystems.

Indeed, beginning with *The Lagoon Cycle*, the Harrisons conceived of their work as both a means of investigating ecological issues and of prescribing solutions. Since the 1980s, this practice has addressed a broad array of topics, ranging from local issues pertaining to Baltimore's pedestrian promenade and the Netherlands' central ring of agricultural land, to more global ones including glacier depletion and international afforestation. Throughout this time, the Harrisons have continued to present their work in galleries and museums—showing large maps and aerial imagery along with descriptive text. These texts would not continue to be dialogic in structure following *The Lagoon Cycle*, but poetics nonetheless remain central to the Harrisons' imperative to communicate their findings. In this respect, the Harrisons' work is indicative of several developments in environmental art of the late twentieth century. The first of these they share with artists such as Patricia Johanson, Agnes Denes, and Mel Chin, who create living environments to remedy and reflect upon ecological imperatives such as waste treatment and flood management. Likewise, the Harrisons have been instrumental in shaping an additional vein of contemporary practice by artist collectives such as the Center for Land Use Interpretation (better known simply as CLUI), spurse, and The Otolith Group, whose varied projects are driven primarily by research and whose exhibitions operate through a combination of analytics and poetics. For these artists and many others, the move initiated by the Harrisons away from land art's shaping of earth-based forms to analyzing current ecological problems has been highly influential. Beginning with their *Survival* series and especially with the lengthy *Lagoon Cycle* that followed directly from it, the Harrisons have modeled a kind of environmental practice that casts the artist as a specialized consultant and treats the work of art as means to mediate a definable issue and course of action for its community of spectators.

Environmental abstraction

For the most part, the moves initiated by the Harrisons in the late twentieth century remain central to environmental art practices in the early twenty-first. As is true of numerous fields, contemporary art has been strongly affected by a push for practical application. No doubt influenced by waves of economic crises, disastrous ecological events, and ever-fracturing political rancor, there has been a decisive move by artists dating from the 1990s onwards to comment directly and incisively on current social and environmental circumstances. The best of this work, including that of the Harrisons, has made substantial contributions; the worst of it merely instrumentalizes the news as a stand-in for aesthetic engagement. This impulse toward practical application in the arts, however, has not taken over the entire arena of environmental art.

While other chapters in the present volume address additional recent trends, such as artwork that grapples with the ecology of big data (Houser) and refuse (Zubiaurre), the third and final development that I will more briefly discuss pertains to work that treats the environment as a sphere best addressed through non-representation. Such "environmental abstraction" suggests that rather than eliciting political policy or acting as the public relations wing of the hard sciences, the arts are better suited to probing the resistance of environmental concepts to partition, definition, and quantifiable analysis. For all of the contemporary art that attempts to capture the scale of our ecological peril—and there is much of it filling the halls of today's international circuit of biennial exhibitions,

Figure 30.3 Teshima Art Museum. Photo: Ken'ichi Suzuki.

from large-scale photographs of eco-waste to documentary exposés on oil extraction—environmental abstraction acknowledges that some concepts are grossly reduced when presented through such formats. Foregoing the sharp edge characteristic of agitprop, this latter category of artwork instead proposes that deeper solutions arise slowly from more ingrained, fluid ways of understanding. Rather than trapping complex ecological issues within a single image or film, this body of work is more meditative in its outlook. One striking example of this environmental abstraction is a permanent work situated on the island of Teshima in Japan's Seto Inland Sea.

The Teshima Art Museum, completed in 2010, is indistinguishable from the singular artwork it displays, named *Matrix* (Figure 30.3). A collaborative project realized by artist Rei Naito and architect Ryue Nishizawa (a co-founder of the firm SANAA), building and art are symbiotic in their unity and inseparability. The structure itself is a single, bare concrete shell shaped like a droplet of water bulging outwards at the moment it strikes a surface. A limited number of spectators are admitted at a time (usually thirty to forty), each of whom is directed to remove footwear and refrain from taking photographs. The interior, on first blush, appears to be a rather desolate, vaulted room with circular openings in the roof at either end, which on many days of the year admit rainfall into the space. There are no objects hung on the walls or freestanding on the floor. But the main attraction is nonetheless on the ground, for the floor of the entire building is equipped with a carefully calibrated hydration system that emits and absorbs water at gradual and measured intervals through pores and tiny vessels. While such an account of the space may cast it as bluntly mechanized and even dull, the reality of the work is anything but.

Settling in to observe a small patch of ground during my time there, I slowed to the decelerated perception this museum encourages. What one is watching is the gradual pooling of water upon the floor, some of it supplied by the apparently glacial seep of an unseen pore, some supplied by tiny marble spheres that irregularly spit and spatter droplets on the ground. In such an arrangement, the name of the game is consolidation. Watching the accumulation of small pools on the floor of the Teshima Art Museum for a prolonged period of time provokes reflection on the tendency of all bodies of water, however big or small, to join with other bodies of water. It is a tendency exhibited quite plainly in *Matrix*: whenever a puddle gains enough critical mass, it begins to move along otherwise imperceptible contours of the floor. If provided enough momentum, this newly mobile stream of water may keep going until it reaches one of the two larger "oceans" located under the antipodal roof openings. But in most instances, a small pool tends to find a larger pool and then stop there. Rather than mountain ranges or deserts keeping these bodies of water in place, each waits only for the measured accumulation of more liquid in order to continue the slow but inevitable drive into progressively larger pools.

Thus, the Teshima Art Museum offers something as ostensibly straightforward as "the water cycle" as subject matter. But unlike the early ecosystems created by the Harrisons, the environment fashioned by the interaction of Nishizawa's architecture and Naito's installation does not exist for the purpose of harvesting a crop or eliciting prescribed behavior. The Teshima Art Museum offers a means of immersing its spectator in an environment through forms that are discernible to the senses but remain unfixed. In this exhibition space, water behaves instead as an object of contemplation as well as an affective element, remaining forever liminal to our attempts at containing it. Water, in other words, can be perceived, collected, and emitted, but ultimately not controlled. Through the complex mechanical system underfoot in the Museum and the island of Teshima's newly established place within a global circuit of art world traffic, the presence of water in this space is clearly not a regression to falsely conceived notions of natural purity or Edenic escapism. Instead, the impact of the work is to stimulate ways to perceive and understand connectivity. It is a seemingly simple undertaking but one crucial to long-term thinking that is too often eclipsed in more politically charged, "topical" projects.

Yet artworks that engage with the environment abstractly need not do so with elemental materials alone. The notion of "environmental abstraction" is a mirroring condition—a state of reflectivity in relation to an object of reflection. Some expanded cinema and psychedelic art of the twentieth century falls within this mode of observation, as does an important tradition of photo-based environmental art that runs from Eadweard Muybridge through Man Ray to Michael Snow. And while the experience of this abstract work might be wholly different than the horizontal painting and sculpture and research-based installation discussed above, all three categories nonetheless underscore the immersive impulse of recent visual art. From Rauschenberg's flatbed operations to the Harrisons' advisory commissions and Nishizawa and Naito's meditative interior, the most important contemporary environmental art of the last half-century actively partakes in the organic and social worlds that it addresses.

The contribution of contemporary art to the environmental humanities is twofold. On the one hand, the operations of contemporary art cannot be separated from those of the environment. As a result of this work breaking loose from the static medium of painting and engaging both local and global scales of ecology, scholars addressing the complexity of environments must necessarily include artistic production within the scope of these environments. On the other hand, contemporary art also provides a critical range of perspectives for tackling a number of pertinent concerns across the humanities, which include such topics as the dynamics of materiality, terrestrial design, and resource consumption. But with these issues in mind, it is crucial to clarify that the contribution of the visual arts is to stage situations for critical thinking rather than to provide conclusions. Instead of supplying the larger field of the environmental humanities with assertions or theories, contemporary art uniquely participates in the socio-ecological worlds that it also frames for analysis and insight.

References

Adcock, Craig. "Conversational Drift: Helen Mayer Harrison and Newton Harrison." *Art Journal* 51.2 (1992): 35–45. Print.

Baum, Kelly, ed. *Nobody's Property: Art, Land, Space, 2000–2010*. New Haven, CT: Yale University Press, 2010. Print.

Bijvoet, Marga. "Helen Mayer Harrison – Newton Harrison: The Ecological Argument." *Art as Inquiry: Toward New Collaborations Between Art, Science, and Technology*. New York: Peter Lang, 1997. 136–148. Print.

Boetzkes, Amanda. *The Ethics of Earth Art*. Minneapolis, MN: Minnesota University Press, 2010. Print.

Brattock, Alan and Renée Ater, eds. *American Art* 28.3 (2014). Special issue on ecocritical American Art History. Print.

Brattock, Alan and Christoph Irmsher, eds. *A Keener Perception: Ecocritical Studies in American Art History*. Tuscaloosa, AL: Alabama University Press, 2009. Print.

Burnham, Jack. *Great Western Salt Works: Essays on the Meaning of Post-Formalist Art*. New York: Braziller, 1974. Print.

Cheetham, Mark A., ed. *Nonsite.org*. March 13, 2013. Special issue on "Ecological Art: What Do We Do Now?" Web. 1 August 2015.

Demos, T. J., ed. *Third Text* 27.1 (2013). Special issue on "Contemporary Art and the Politics of Ecology." Web. 1 August 2015.

Frascina, Frances, ed. *Pollock and After: The Critical Debate*. New York: Harper and Row, 1985. Print.

Green, Charles. "Memory and Ethics: Helen Mayer Harrison and Newton Harrison." *The Third Hand: Collaboration in Art from Conceptualism to Postmodernism*. Minneapolis: Minnesota University Press, 2001. 97–124. Print.

Guilbaut, Serge, ed. *Reconstructing Modernism: Art in New York, Paris, and Montreal, 1945–1964*. Cambridge, MA: MIT Press, 1990. Print.

Harrison, Newton and Helen Meyer Harrison. *The Lagoon Cycle*. Ithaca, NY: Herbert F. Johnson Museum of Art, 1985. Print.

Heartney, Eleanor. "Ecopolitics/Ecopoetry: Helen and Newton Harrison's Environmental Talking Cure." *But is it Art?: The Spirit of Art as Activism*. Ed. Nina Felshin. Seattle, DC: Bay Press, 1995. 114–119. Print.

Jones, Caroline. *Eyesight Alone: Clement Greenberg's Modernism and the Bureaucratization of the Senses*. Chicago, IL: Chicago University Press, 2005. Print.

Joseph, Branden W. *Random Order: Robert Rauschenberg and the Neo-Avant-Garde*. Cambridge, MA: MIT Press, 2003. Print.

Krauss, Rosalind. *The Optical Unconscious*. Cambridge, MA: MIT Press, 1993. Print.

Leja, Michael. *Reframing Abstract Expressionism: Subjectivity and Painting in the 1940s*. New Haven, CT: Yale University Press, 1993. Print.

Levin, Kim. "Reflections on Robert Smithson's 'Spiral Jetty.'" *Arts Magazine* 52.9 (1978): 136–137. Print.

Lippard, Lucy R. *Undermining: A Wild Ride Through Land Use, Politics, and Art in the Changing West.* New York: New Press, 2014. Print.

McGreevy, Linda. "Improvising the Future: The Eco-Aesthetics of Newton and Helen Harrison." *Arts Magazine* 62.3 (1987): 68–71. Print.

Meehan, Thomas. "A Non-Art Article on Art." *Horizon* 13.4 (1971): 4–15. Print.

Nisbet, James. *Ecologies, Environments, and Energy Systems in Art of the 1960s and 1970s.* Cambridge, MA: MIT Press, 2014. Print.

Scott, Emily Eliza and Kirsten J. Swenson, eds. *Critical Landscapes: Art, Space, Politics.* Berkeley, CA: California University Press, 2015. Print.

Steinberg, Leo. "Other Criteria." *Other Criteria: Confrontations with Twentieth-Century Art.* New York: Oxford University Press, 1972. Print.

31

SLOW FOOD, LOW TECH

Environmental narratives of agribusiness and its alternatives

Allison Carruth

Slow Food International formed in 1989 when a countercultural group of Italian food lovers led by Carlo Petrini mobilized in opposition to the global industrial food system, the emblem of which had become a McDonald's franchise that opened on Rome's Spanish Steps in 1986 (Andrews 29). Twenty-five years later, Slow Food International now counts over one hundred thousand members organized into thirteen thousand chapters, or *convivia* (Lindholm and Lie 57). While some *convivia* arguably operate as gourmet eating clubs, the parent organization's mission is to cultivate what Susie O'Brien terms a "vast networking organization" of small farmers, artisanal food producers, peasant communities, and anti-globalization activists (222).

Environmental humanities scholars are likely to be sympathetic with this mission where it fosters systemic alternatives to industrial agriculture and agribusiness—with the attendant infrastructure of seed patents, confined animal feedlots, and energy- and chemical-intensive commodity farming (a.k.a. "monoculture"). It is this chapter's contention, however, that the interplay of ecological, technological, and cultural motifs in slow food discourse—and especially in a genre of nonfiction that has helped to popularize slow food in the United States—merits critical scrutiny.[1] On the one hand, slow food narratives tend to align locally rooted, low-tech foodways with both nature and culture, while advancing a critique of recent agricultural technologies—above all transgenic seeds (or GMOs)—as a priori harmful to local ecosystems and cultures, and as categorically different from earlier agricultural technologies.[2] On the other hand, these entwined motifs mask the interlaced relationships in nearly all modern food systems of local and global, ecology and industry, and biodiversity and biotech.

Although slow food writers no doubt think deeply about the ecological and cultural impacts of agricultural technologies, their nonfiction narratives demarcate sharply between organic and engineered food systems, aligning the former with localism and the later with globalization. However well founded the concerns about recent agricultural technologies are, this ethos would do well to differentiate among individual technologies and the distinct problems they raise. For meanwhile, biotech corporations like Monsanto and DuPont frame their inventions to policymakers, food producers, and consumers as safe and vital implements for the world food supply (see Lipton on the sophistication of the biotech industry's global public relations campaigns). Focusing on the recent nonfiction of Michael Pollan, this chapter aims to imagine a middle way between slow food and agribusiness. I replace the technological determinism that troubles the former as much as it enlivens the latter with a vision of agricultural technologies as uneven in how they interface with the ecological and cultural dimensions of food systems.

Slow food nonfiction: the ecological claims for local foodsheds

Over the last decade, a nonfiction genre has taken shape in American print culture around the slow food movement that includes popular books such as botanist Gary Paul Nabhan's *Coming Home to Eat* (2002), novelist Barbara Kingsolver's *Animal, Vegetable, Miracle* (2007), and Pollan's *The Omnivore's Dilemma: A Natural History of Four Meals* (2006), along with small press primers like the *Urban Farm Handbook* (Cottrell and McNichols 2011), *The Quarter-Acre Farm* (Warren 2011); and the elaborately titled *The Feast Nearby: How I Lost My Job, Buried a Marriage, and Found My Way by Keeping Chickens, Foraging, Preserving, Bartering, and Eating Locally (all on $40 a week)* (Mather 2011).[3] From the blockbuster to the self-published text, these works share in common their occasion: a lament about and commitment to opt out of industrialized agriculture's technologies—from supermarkets and processed foods to growth hormones, pesticides, patented seeds, and monoculture farms. Inspired by a somber account of agribusiness-as-usual, these texts also craft hopeful narratives that illustrate slow food ideals by featuring artisanal food makers, organic farmers, urban gardeners, and grass ranchers who together represent alternative food systems that nourish local ecologies and cultures. Their authors act as proxies for their readers in that they chronicle firsthand experimentation with meeting alimentary needs and desires not as consumers in "the industrial complex" but as active participants in local, ecological food production (Pollan, *Omnivore* 142, 201).

Central to the genre is a thematic connection between knowing one's local food producers and limiting one's complicity with industrialized agriculture. Although most slow food writers recognize that the distance foods travel from "farm to table" is an insufficient indicator of production methods and research extensively the practices of the individual producers they patronize and profile, their narratives still tend to depict locally sourced and locally adapted food ("local food" for short) as more ecologically sustainable, culturally significant, and low-tech than globally distributed food. In *Manifestos on the Future of Food and Seed*, which includes contributions by Petrini and Pollan, organic agriculture advocate Prince Charles defines slow food as "traditional food" that is "also local," going on to reason that "local cuisine is one of the most important ways we identify with the place and region where we live" (Shiva 29). Pollan elaborates on this principle in the same collection, arguing that "farmers stand for the specific" (Shiva 33). The tacit notion underlying such arguments is that local foods are more likely than foods from afar to be in sync with natural ecosystems and cultural traditions, in part because their productions methods are easier to apprehend.

There is a tension in slow food narratives between this affinity for agricultural localism and a countervailing celebration of local foods from *around the world*. In this vein, the *Manifestos* begin by invoking "the farmers, fishers, breeders, nomads from the Peruvian Andes to the Argentine pampas, from the Amazon jungle to the Chiapas mountains, from California vineyards to First Nation reserves, from the shores of the Mediterranean to the seas of northern Europe" (Shiva 11). Likewise reporter Rowan Jacobsen frames his North

American slow food compendium as spanning "the continent, from salmon in Alaska and apples in Washington's Yakima Valley to chocolate in Mexico, coffee in Panama, cheese in Vermont, and wild mushrooms in Quebec" (3). Although O'Brien contends that slow food writers hereby articulate a "sophisticated understanding of the mutual implication of local places and global space in the neoliberal world economy" (223), such rhetoric arguably differentiates too starkly between "good" and "bad" models of this enmeshment.

For his part, Pollan conveys an awareness that many farmers no longer "stand for" ecological and cultural specificities of local place but are instead full participants in global agribusiness. In *The Omnivore's Dilemma*, for example, Pollan's narrative of researching the food systems that underlie four different meals returns repeatedly to an Iowa farmer named George Naylor who cultivates five hundred acres of hybrid corn and soy. In a lengthy interlude, Pollan first follows a steer he has purchased as part of his "educational" experiments (66) from a South Dakota ranch to a Kansas feedlot that feeds cattle the same F-1 hybrid corn Naylor grows, and then spends several days observing slow food hero Joel Salatin, who runs, in his website's words, a "family owned, multi-generational, pasture-based, beyond organic, local-market" Virginia farm called Polyface. After this detour, the narrative returns to Naylor:

> Naylor participates in an infinitely more complex industrial system [than Salatin], involving not only corn (and soybeans), but fossil fuels, petrochemicals, heavy machinery, CAFOs, and an elaborate international system of distribution to move all these elements around: the energy from the Persian Gulf, the corn to the CAFOs, the animals to slaughter, and their meat finally to a Wal-Mart or McDonald's near you. Considered as a whole this system comprises a great machine, transforming inputs of seed and fossil energy into outputs of carbohydrate and protein. And, as with any machine, this one generates streams of waste: the nitrogen and pesticides running off the cornfields; the manure pooling in the feedlot lagoons; the heat and exhaust produced by all the machines within the machine—the tractors and trucks and combines.
>
> (*Omnivore* 130)

Pollan here establishes two opposed models of food production: on the one hand, Salatin's grass-centered livestock ranch that "involves so many variables, and so much local knowledge, that it is difficult to systematize" (202) and, on the other, Naylor's commodity farm that "comprises a great machine" plugged into global agricultural markets. Yet, this dichotomy breaks down under scrutiny. However admirably Salatin draws on ecological science in managing livestock to nourish soil health and maintain grasslands and in finding alternatives to synthetic fertilizers, pesticides, growth hormones, and gas-powered machinery, Polyface still contributes to global climate change. The Food and Agriculture Organization, to this point, estimates that livestock "supply chains" contribute 5 percent of worldwide anthropogenic CO_2 emissions and a full 44 percent of methane emissions (15; see also Gerber). Moreover, although Polyface only sells edible products within what they define as their local foodshed, that region stretches some two hundred and fifty miles in all directions from the farm, which simultaneously maintains an online store of books, DVDs, apparel, and paper products shipped internationally.

My aim is not to suggest that all food systems are equally complicit in the economic structures and environmental consequences of globalization. Rather, my claims are twofold.

First, for some communities, a slow food commitment to the local may come up empty, by which I mean that no matter how well one gets to know the practices and impacts of food producers nearby, there may not be many viable alternatives to the inventory stocked at one's local supermarket (which may well include organic produce and minimally processed foods, if from afar). Second, the dichotomy between "industrial corn" and "pastoral grass" that Pollan constructs through his education in how Naylor and Salatin each farm maps onto a wider binary of "big ag" versus small foodshed in slow food nonfiction that downplays how difficult it is for modern food systems to have no adverse environmental consequences.

Heirloom seeds: the rhetoric of biodiversity versus biotech

The organizing motif of the ecocentric local foodshed in slow food nonfiction finds an echo in the theme of open-pollinated heirloom seeds (i.e., nonpatented, locally adapted seeds that can be replanted) as seeds of ecological and cultural diversity. This motif emerges from the slow food movement's self-image as, in Petrini's words, "waging a battle ... to defend the biodiversity of the planet" that is also "a battle for civilization" (Shiva 13). By civilization, Petrini means not human civilization in the abstract but rather what he terms the "diversities" of food cultures around the world: "Feijoada in Brasil, couscous in the Mediterranean, tamales in Latin America, fufu in Africa, dried reindeer meat in Lapland, pasta in Italy" (12–13). As the movement's founder does here, slow food rhetoric consistently links ecological and cultural diversity—an apposition that Ursula K. Heise has also identified in American environmental fiction of the past three decades (including Kingsolver's 1990 novel *Animal Dreams*). In that literary context, fictional narratives centered around multiethnic characters who embody the "oppositionality" of transnational environmental activism (think: Slow Food International) "remain resolutely local in their opposition to globalization" (387). These narratives employ "biological and cultural diversity as direct metaphors for each other," Heise writes, but this equivalence falters where it disregards cultural "diversities emerging from other than ecological factors" including forms of "technological innovation" (388). Heise persuasively concludes that this alignment of cultural diversity with biodiversity unwittingly simplifies the latter concept, which in the biosciences has been "notoriously broad in its reference to different kinds of variation all the way from genes to species and ecosystems—variations that are not in all cases commensurate with each other" (401).

Such a critique does not entirely unsettle the slow food vision that the conventional food system is harmful to both cultural diversity and biodiversity. It is arguably the case that industrialized agriculture has advanced a few crops to monopoly status, the cascading effects of which have included a diminishment of culinary variety and attendant cultural histories and identities. For many slow food writers, the apotheosis of this monoculture-based system is the scant variety yet growing volume of transgenic crops alongside the GMO monopoly that a single corporation enjoys. To quote Pollan again, "The impact of [GMOs] on seed diversity as well as the overall biodiversity in those areas [where they've been planted] is devastating. A single multinational company, Monsanto, holds the patents for 90 percent of all commercial GM plant traits" (*Food Rules* 84). Pollan here eschews logical transitions between his claim that the cultivation of a small variety of patented transgenic seeds is anathema to biodiversity and his observation that Monsanto owns the intellectual property rights to "90 percent" of those seeds' genetic traits. Ecological monoculture and economic monopoly are in lockstep in this formulation. Pollan offers a provocative symbol for this "monoculture paradigm" (*Food Rules* 90) when he describes a field planted with a single variety of hybrid or transgenic corn as "a most orderly mob" (*Omnivore* 37).

As an alternative to the "monoculture paradigm," slow food writers hold up the perceived diversity and naturalness of heirloom seeds. For Kingsolver, heirloom seeds defy agribusiness in their wildness: "like sunshine, heirloom seeds are of little interest to capitalism if they can't be patented or owned" (47). Rather than commodities, by this logic, heirloom seeds are akin to native plant species threatened with extinction due to introduced species, and the communities that propagate, plant, and preserve them—like Nabhan's Native Seeds organization in Tucson—are akin to conservation biologists. As "collector[s] and preserver[s] of ... endangered traditional seeds," heirloom seed advocates see themselves as protecting domesticated plants that have evolved in specific locales over long periods of time and that the industrialized food system has threatened only very recently with extinction by introducing a small number of engineered commodity crops into the world food supply (Native Seeds/SEARCH). Fellow travelers to slow food *convivia*, heirloom seed banks and greenhouses accordingly tout the large variety of their collections. For instance, Seed Savers Exchange underscores the "over 600 heirloom and open-pollinated varieties" (from Oaxacan Green Dent maize to the Christmas Lima Bean) that sprout at the organization's Heritage Farm in Iowa, while the Baker Creek Heirloom Seed Company highlights its "1,750 varieties of vegetables, flowers and herbs."

These ideas of agricultural biodiversity and endangered seed conservation are paradoxical on two counts. First, in borrowing from conservation biology contested concepts like biodiversity and native species, heirloom seed proponents paper over the fact that domesticated plants are in a sense always invasive. Second, there is a tension within the heirloom seed advocacy of writers like Kingsolver between the local and the global that we have seen more generally at work in slow food nonfiction. On the one hand, heirloom seed varieties are by definition locally adapted, even if some have been cultivated in multiple locales or have become significant in diverse cuisines. On the other hand, the heirloom seed network is a global one in which seed packets travel long distances to be cultivated in greenhouses, gardens, and farms. In this sense, the analogy between heirloom seeds and wild native species arguably does not ecologically hold much water.

Pleasurable politics: the cultural and aesthetic claims for slow food

Of course, the ecological claims about heirloom seeds—and the alternative food systems they emblematize and germinate—are not the sole claims made for them in slow food nonfiction. Following the analogy of biodiversity and cultural diversity, heirlooms also are imagined to be seeds of distinctive cultural histories and pleasurable aesthetic experiences. To wit, Kingsolver: "Heirloom vegetables are irresistible, not just for the poetry in their names but because these titles [like 'Cajun Jewel okra'] stand for real stories. Vegetables acquire histories when they are saved as seeds for many generations" (46). This trope recurs in slow food writing: from a metaphor of rare seeds as rare books in Ruth Ozeki's fictional narrative of a community in conflict over GMOs (*All Over Creation*) to Nabhan's observation that planting edible desert seeds in *Coming Home to Eat* feels like the performance of a religious "canticle" (32) and farmer-writer David Mas Masumoto's reflection that heirlooms like the

Sun Crest peach "communicate their histories through flavors and traditions" (14). Taken together, such metaphors and images work to make the case that the biological building blocks of environmentally conscientious food systems bear not just ecological and somatic advantages but also cultural fruits: they carry place-based histories and stories, nourish creative practices, and afford aesthetic pleasures.

These specific claims made for heirloom seeds are a subset of a more general rhetorical gambit in slow food nonfiction: the yoking of the political resistance to agribusiness with the aesthetic pleasure of local and slow foods (Shiva 12). As Dan Philippon shows, writers like Pollan hereby dissolve "the all-too-common distinction between aesthetics or pleasure on the one hand and politics on the other" (172). Contra monoculture farming that, as a "well-oiled machine of Late Capitalism," generates a monotony of bland and unhealthy calories (Kingsolver 14), writers suggest that to practice slow food is to marry environmental politics with aesthetic pleasure. Or as Kingsolver writes to entice her reader to the movement: "food is the rare moral arena in which the ethical choice is generally the one more likely to make you groan with pleasure" (22). Nabhan captures this happy marriage in the very subtitle for *Coming Home to Eat: The Pleasures and Politics of Local Foods*. Contemporary slow food writers in this linking of pleasure and politics invoke Wendell Berry's 1989 essay "The Pleasures of Eating," which juxtaposes "industrial eaters" with the ethical figure of one who views "eating as an agricultural act" and thus experiences the "extensive pleasure" of knowing "the lives and the world from which food comes" (145–146, 51). Although Berry's body of nonfiction writing exhibits varied ideas about the ecological, cultural, and technological contours of different food systems, "The Pleasure of Eating" establishes a framework for the slow food tenet that conscientious eaters mingle environmental politics with aesthetic pleasures. This idealized eater accordingly finds a foil in Berry's industrial eaters, whom he disparagingly portrays as "*mere* consumers—passive, uncritical, and dependent," all but "strapped to a table with a tube running from the food factory directly into his or her stomach" (146).

The rhetorical move here is to cast industrial agriculture and industrial eaters as not just environmentally and culturally deleterious but also unpleasant and unhappy to boot, while alternative food systems and environmentalist eaters embody a kind of high-minded pleasure. This Epicurean sensibility is particularly evident when writers dwell on the cognitive and sensory pleasures of their slow food-inspired practices. Pollan, for instance, links his own practices variously to "the pleasure of variety," the "pleasures ... deepened by knowing," the "pleasures of eating by the season," the "pleasures of traditional foods enjoyed communally" and the "pleasures mushrooming affords" (*Omnivore* 295, 12, 53, 59, 385). Slow food's ethical stakes gain rhetorical appeal, then, as writers reframe not just environmental activism and everyday habits but also food labor as pleasurable, including the labor of small-scale organic farming, backyard gardening, animal butchering, cooking from scratch and, in Pollan's case, "knowing" itself (or what Charles Lindholm and Siv Lie critically term "educated pleasure").

The slow food test kitchen: Pollan's *Cooked*

The organizing motifs of slow food nonfiction—local foodsheds, agricultural biodiversity, cultural identities, and pleasurable politics—migrate from the farm to the kitchen in Pollan's latest bestseller, *Cooked: A Natural History of Transformation* (2013). Structured around the four elements, *Cooked* unfolds as a partly autobiographical narrative of Pollan's "quest to learn the art of cooking," a kind of nonfiction *bildungsroman* that follows the journalist around the United States as he apprentices with regional chefs, bakers, cheese makers, brewers, picklers, and barbecue pitmasters (*Cooked* 13). At the outset, Pollan promises the reader an epic story of food that will delve into the cultural histories and ecological marvels of smoking, braising, baking, and fermenting. An underlying anxiety about food and agriculture technologies troubles this venture, apparent in Pollan's association of "good" food with preindustrial societies (indeed, with the four elements) and "bad" food with the late modern world in which the average American spends far more hours per week in front of screens than involved in cooking and eating communally (128, 184–189). In response to this crisis (in Pollan's terms), *Cooked* chronicles laborious forms of artisanal, artful cooking that also revive long-standing technologies, like the wood oven and copper pot, but that untether individuals from the recent technologies of agribusiness, on the one hand, and the digital economy, on the other.

Although Pollan, as a seasoned science writer, holds multivalent views on both ecology and technology, in *Cooked*, the reader digests a too-tidy narrative that links environmental and cultural alternatives to agribusiness with ancient implements and artisanal practices. Pollan uses the term "technology" often in the book to lionize timeworn methods of cultivating and preparing food. Building on Claude Lévi-Strauss's thesis in *The Raw and the Cooked*, for example, *Cooked* opens with the claim that cooking was once "the new technology [that] cracked open a treasure trove of calories" and defined human cultures (6). Pollan goes on to group the elemental modes of cooking the book explores under the rubric of an "ancient vernacular 'technology'" that blends the "wonders of nature and culture" (228). As he puts it about cheese making, such a technology is "not so much an invention as a discovery … a form of 'biomimicry'" (348) that contrasts sharply with the "tools of modern food science" (122). *Cooked* thus conveys the messiness of the technology–ecology–culture circuit in slow food rhetoric and, perhaps, in American environmentalism writ large. Pollan is interested in the knowledge and creativity that food and agriculture technologies can facilitate, and yet he jettisons those technologies that have arrived more recently on the scene. At the same time, *Cooked* never really articulates *why* the tools and techniques (not to mention the manual labor) that go into making whole wheat breads, fermented vegetables, and aged cheeses are salutary on ecological and cultural grounds while newer innovations—from white flour to GMOs—will only ever be cogs in the "industrial complex" of agribusiness.

The alternative food futures of bioart

A rejoinder to this open question that *Cooked* leaves readers with can be found in a group of multimedia, food-focused bioartists who work against the grain of both agribusiness and slow food. Environmentally attuned, their projects exemplify "tactical media" as Rita Raley has defined it: avant-garde uses of new media and emergent technologies on the part of artist-activists who disturb the logic of global capitalism through a "practice of designing rather than saving the world" (30). Examples of this practice that fall under the heading of *alimentary bioart* range from Natalie Jeremijenko's "Cross-Species Adventure Club" (a reinvention of the urban supper club that integrates foraged plants and fungi important to local nonhuman diets with molecular gastronomy tools and techniques) to Critical Art Ensemble's "Free Range Grain" (a mobile lab installed in art spaces that allows visitors to test foods from their pantries for GMO traces). Such projects question the slow food imperative to restore heirloom seeds and ancient food cultures, and instead advocate for what the late Beatriz da Costa called "public amateurism" by parodying corporate biotech and food science (373).

Less well known than Jeremijenko and Critical Art Ensemble are bioartists Zack Denfeld and Cat Kramer, a Portland-based duo who co-founded the irreverently named Center for Genomic Gastronomy (CGG) to explore "alternative culinary futures." CGG mixes amateur biology and product design with performance art and speculative writing in projects such as their mobile vending machine for heirloom seeds (the Seed-o-Matic) and schematics for a community-run biotech lab that would produce in-vitro meat as an alternative both to feedlots and to corporate agribusiness. The aesthetic mode of this work is ludic in that CGG facilitates playful participation on the part of audiences to engage serious environmental issues related to industrialized agriculture. Testing out technological responses to such issues that are informed by an attention to global as well as local ecologies and cultures, CGG models what I term "techno-curiosity": a way of conceptualizing alternative food systems that charts a middle way between slow food and agribusiness.

We see this way of thinking not only in the group's built prototypes and multimedia designs but also in the nonfiction writing they do—as especially evident in a little magazine (or "zine") they developed and released in 2013. Titled *Food Phreaking*, the inaugural issue takes the form of thirty-eight vignettes, each of which includes an image, a snippet of text, and an invented hashtag. The zine is structured around four categories: "A: Legal and Open," "B: Illegal and Open," "C: Illegal and Closed," and "D: Legal and Closed." These categories create a structure for exploring a wide range of food technologies, from raw milk vending machines (under "illegal and open") to proprietary GMOs (under "legal and closed"). Although *Food Phreaking* is a print object, its content, form, and mode of circulation all mark it as tactical media in Raley's sense. The pink-and-brown color palette and eclectic exempla are tactics for disturbing the teleological narratives about the past and future of agricultural technologies that slow food and agribusiness each articulate. The four sections of the zine thus gather together a dizzying set of materials that includes instructions for cooking beet peels, advertising for Kraft cheese products, a satire of the Asian fusion restaurant group Momofuku's move to trademark Korean culinary words, and, finally, a reference to CGG's own wry use of the trademarked transgenic GloFish (intended for decorative aquariums) to make sushi. The zine's miniature case studies open onto a multiplicity of alternative food futures that collectively imagine modern agricultural technologies as many-sided rather than predictable in their ecological and cultural consequences, in turn suggesting that contemporary modes of resistance to industrialized agriculture may need to be versed in, rather than averse to, technological innovation.

While not without its own blind spots, this approach to imagining alternative food futures helps to "prototype" high-tech environmentalism: an ethic that reclaims a sense of wonder at the natural world and all we do not know about it by imagining technologies as culturally situated and ecologically enmeshed. So too do projects like *Food Phreaking* tap into the praxis of American natural history dating back to the eighteenth and nineteenth centuries that, as Christoph Irmscher has shown, made natural history a mode of "art as play" and "serious fun" (3). Bioart collectives like CGG stage their contemporary acts of "serious fun" at once to disrupt the economics and ideologies of agribusiness and to incite a new environmental rhetoric that offers a rich field of inquiry for the environmental humanities. In contrast to the investments in manual labor, heirloom seeds and traditional foodways seen in slow food nonfiction, bioart food projects seek to foster ecological resilience, value environmental design as much as ecological conservation, and envision multispecies communities and multimedia art as equally crucial to addressing the twenty-first-century's environmental challenges.

Notes

1 I use capitalization when referring to the organization proper (i.e., Slow Food International), lower case when referring to the wider social movement and its discourses (e.g., slow food nonfiction).
2 Following the use of this concept in the social sciences, I take "foodways" to signify the sociocultural, political, aesthetic, and technological practices of food production and consumption within particular cultural and geographic contexts.
3 I have elsewhere examined this body of nonfiction works under the heading of "the locavore memoir" (Carruth 154–165). See also the work of David Cleveland, David Goodman and E. Melanie DuPuis, Julie Guthman, and Steven Schnell.

References

Andrews, Geoff. *The Slow Food Story: Politics and Pleasure*. Montreal: McGill Queens University Press, 2008. Print.

Berry, Wendell. "The Pleasures of Eating." *What Are People For?* 1989. New York: North Point Press, 1990. 145–152. Print.

Carruth, Allison. *Global Appetites: American Power and the Literature of Food*. New York: Cambridge University Press, 2013. Print.

Cleveland, David A., Allison Carruth, and Daniella Niki Mazaroli. "Operationalizing Local Food: Goals, Actions, and Indicators for Alternative Food Systems." *Agriculture and Human Values*. 2014. Web. 12 October 2014.

Cleveland, David A. et al. "Effect of Localizing Fruit and Vegetable Consumption on Greenhouse Gas Emissions and Nutrition, Santa Barbara County." *Environmental Science and Technology* 45.10 (2011): 4555–4562. Print.

Cottrell, Annette and Joshua McNichols. *Urban Farm Handbook: City Slicker Resources for Growing, Raising, Sourcing, Trading, and Preparing What You Eat*. Seattle, DC: Mountaineers Books, 2011. Print.

da Costa, Beatriz. "Reaching the Limit: When Art Becomes Science." *Tactical Biopolitics: Art, Activism, and Technoscience*. Ed. Beatriz da Costa and Kavita Philip. Cambridge, MA: MIT Press, 2008. 366–385. Print.

Gerber, P. J., et al. *Tackling Climate Change through Livestock: A Global Assessment of Emissions and Mitigation Opportunities*. Rome: Food and Agriculture Organization, 2013. Print.

Goodman, David and Melanie E. DuPuis. *Alternative Food Networks: Knowledge, Practice and Politics*. New York and London: Routledge, 2012. Print.

Guthman, Julie. "Bringing Good Food to Others: Investigating the Subjects of Alternative Food Practice." *Cultural Geographies* 15.4 (2008): 431–447. Print.

——. "Fast Food/Organic Food: Reflexive Tastes and the Making of 'Yuppie Chow'." *Social and Cultural Geography* 4.1 (2003): 45–58. Print.

——. *Weighing In: Obesity, Food Justice, and the Limits of Capitalism*. Berkeley, CA: University of California Press, 2011. Print.

Heise, Ursula K. "Ecocriticism and the Transnational Turn in American Studies." *American Literary History* 20.1 (2008): 381–404. Print.

Irmscher, Christoph. *The Poetics of Natural History: From John Bartram to William James*. New Brunswick, NJ: Rutgers University Press, 1999. Print.

Jacobsen, Rowan. *American Terroir: Savoring the Flavors of Our Woods, Waters, and Fields*. New York and London: Bloomsbury, 2010. Print.

Kingsolver, Barbara. *Animal, Vegetable, Miracle: A Year of Food Life*. New York: HarperCollins, 2007. Print.

Lindholm, Charles and Siv B. Lie. "You Eat What You Are: Cultivated Taste and the Pursuit of Authenticity in the Slow Food Movement." *Culture of the Slow: Social Deceleration in an Accelerated World*. Ed. Nick Osbaldiston. Basingstoke and New York: Palgrave Macmillan, 2013. Print.

Masumoto, David Mas. *Wisdom of the Last Farmer: Harvesting Legacies from the Land*. New York and London: Free Press, 2009. Print.

Mather, Robin. *The Feast Nearby: How I Lost My Job, Buried a Marriage, and Found My Way by Keeping Chickens, Foraging, Preserving, Bartering, and Eating Locally (All on $40 a Week)*. Berkeley, CA: Ten Speed Press, 2011. Print.

Nabhan, Gary Paul. *Coming Home to Eat: The Pleasures and Politics of Local Food*. New York and London: Norton, 2002. Print.

Native Seeds/SEARCH. www.nativeseeds.org. Web. 16 November 2016.

O'Brien, Susie. "Anti-Fascist Gluttons of the World Unite!: The Cultural Politics of Slow Food." *Cultural Autonomy: Frictions and Connections*. Ed. Petra Rethmann, Imre Szeman, and William D. Coleman. Vancouver: University of British Columbia Press, 2010. 219–239. Print.

Ozeki, Ruth L. *All Over Creation*. New York: Penguin Books, 2003. Print.

Philippon, Daniel J. "Sustainability and the Humanities: An Extensive Pleasure." *American Literary History* 24.1 (2012): 163–179. Print.

Pollan, Michael. *Cooked: A Natural History of Transformation*. New York: Penguin, 2013. Print.

——. *Food Rules: An Eater's Manual*. New York: Penguin Books, 2009. Print.

——. *The Omnivore's Dilemma: A Natural History of Four Meals*. New York: Penguin Books, 2006. Print.

Raley, Rita. *Tactical Media*. Ed. Katherine Hayles, Mark Poster, and Samuel Weber. Electronic Mediations. Minneapolis and London: University of Minnesota Press, 2009. Print.

Schnell, Steven M. "Food Miles, Local Eating, and Community Supported Agriculture: Putting Local Food in Its Place." *Agriculture and Human Values* 30 (2013): 615–628. Print.

Shiva, Vandana, ed. *Manifestos on the Future of Food and Seed*. Cambridge, MA: South End Press, 2007. Print.

The Center for Genomic Gastronomy. "About." 2013. Web. October 13 2013.

Warren, Spring. *The Quarter-Acre Farm: How I Kept the Patio, Lost the Lawn, and Fed My Family for a Year*. Berkeley, CA: Seal Press, 2011. Print.

32

MATTRESS STORY

On thing power, waste management rhetoric, and Francisco de Pájaro's trash art

Maite Zubiaurre

"Art is Trash" is a desperate howl against the human condition. I usually paint this condition using trash because this is where it deserves to be. My intention is basically to paint badly and quickly, so that the visual venom is lethal to all the people who feel implicated by it. What I do with trash is not art; it is Art is Trash.

—Francisco de Pájaro, *Art is Trash* 12

The sheer accumulation of waste and the monumentality of open-pit garbage dumps and sanitary landfills has awakened the curiosity of scholars and writers (Boo; Engler; Humes; Nagle; Rathje and Murphy; Rogers; Royte; Urrea), and has become a steady and productive source of inspiration to a number of renowned international artists—Vic Muniz, Chris Jordan, Daniel Canogar, and Ha Schulte among them. This chapter, however, is more interested in a much less explored subject, namely, the artistic representation of refuse at its early stages, when trash is still relatively small and unassuming, still city- and street-bound (George; Thill) and has not yet initiated its journey to faraway transfer stations, recycling centers, and landfills. More specifically, it reflects upon one particular type of bulk refuse, discarded mattresses, and upon Spanish street artist Francisco de Pájaro's artistic manipulation of this particular artifact as found in back alleys and on sidewalks.

It is important to stress that urban solid waste goes relatively unnoticed and more often than not only attracts the attention of urban dwellers, city authorities, and even artists if it gets out of hand; in other words, if it accumulates, overflows, or stays for too long. Trash is invisible, unless it becomes an overpowering nuisance, and it is precisely the concealed nature of garbage before it "degenerates" into threatening and in-your-face monumentality that awakens the curiosity of street creators like Francisco de Pájaro (see de Pájaro, "Francisco de Pájaro"). A Spanish, Barcelona-based artist, de Pájaro soon became disenchanted with the artistic milieu and what he perceived as its entrenched elitism. As a struggling painter of few means and strong leftist convictions, in 2009 de Pájaro decided to trade the canvas for trash bags, cardboard boxes, and tossed furniture pieces and home appliances. Armed with a backpack where he kept his scarce painting utensils—a few brushes and paint tubes, spray cans only used sparingly, and a roll of duct tape—de Pájaro would roam the streets of Barcelona at night, on the lookout for garbage, and then quickly proceed to paint faces and big, expressive eyes on the discarded objects at hand. His signature, "El arte es basura," has since then spread across many metropolises, and occasionally alternates with its English version,

"Art is Trash," in cities like London, Madrid, and New York, and lately, San Francisco, San Diego, and Los Angeles.

Francisco de Pájaro's art, which started small and as an act of defiant desperation, has now spread internationally. De Pájaro's rapidly growing fame, however, has not altered his art in any substantial way. Faithful to his political convictions and strong idealism, the Spanish artist's main canvas is still trash. He wants his art to be ephemeral and to become—or rather remain—garbage. More importantly, his main objective remains to "animate the inanimate," and to have trash intently look at, and speak to, us humans. As Tommy Blaquiere points out, de Pájaro's "use of rubbish—the by-product of our consumerist society—as the canvas for his art makes us look at what we so desperately want to ignore and holds up to us our habits" (6).

The animation of the inanimate is a long-standing convention that popular culture and mainstream Hollywood cinema have exploited successfully in recent years. *Toy Story* (1995) (Ackerman) and its different sequels immediately come to mind, and so does the lesser-known but related film *The Brave Little Toaster* (1987) (Stetz), where a group of "cute" obsolete home appliances led by a spirited toaster barely manage to escape the landfill. Lately, "animated trash" has also awakened the interest of environmental scholars. In her influential volume, *Vibrant Matter* (2010), Jane Bennett calls discarded artifacts "strangely vital things," and quotes W. T. Mitchell's cunning reflection on when things suddenly:

> become the Other, when the sardine can looks back, when the mute idol speaks, when the subject experiences the object as uncanny and feels the need for what Foucault calls "a metaphysics of the object, or, more exactly, a metaphysics of that never objectifiable depth from which objects rise up towards our superficial knowledge."
>
> (156–157)

According to Bennett, it was writers and philosophers such as Mitchell, Thoreau, Spinoza, and Merleau-Ponty who opened her up to the sentient nature of objects and in particular of discarded artifacts. But, foremost, it was her intense experience of what I like to term a "real life trash-collage" that revealed the ultimately "animated" nature of all things lifeless. On a June morning, Bennett stumbled upon "one large men's black plastic work glove; one dense mat of oak pollen; one unblemished dead rat; one white plastic bottle cap; and one smooth stick of wood" (4). "Thing power," as she forcefully puts it, "rose from a pile of trash. Not Flower Power, or Black Power, or Girl Power, but *Thing Power*: the curious ability of inanimate things to animate, to act, to produce effects dramatic and subtle" (6).

"Thing power," however, does not come automatically from the thing itself: it needs intense looking on the part of humans to come to fruition. Bennett saw the power of things and thus endowed them with agency because she chose to look at them intently. As John Berger perceptively puts it, "whenever the intensity of looking reaches a certain degree, one becomes aware of an equally intense energy coming towards on through the appearance of whatever it is one is scrutinizing" (539). These reflections are relevant in order to understand the animation process that takes place when Francisco de Pájaro looks at trash and immediately acknowledges that trash is looking back at him. Conversely, the trash pieces of the Spanish artist—as ephemeral and compelling as the trash heap that transfixed the American environmentalist thinker—shed further light on Bennett's insights. They make us wonder if "any" artifact would have triggered a similar experience, or if the fact that these were discarded objects was crucial to the intensity and wisdom of Bennett's revelation. Is trash perhaps more prone to "animation" than artifacts still deemed useful and valuable? Or, in other words, do unwanted artifacts speak more loudly, do they appeal more strongly to our emotions and to our intellect (in that order), than desirable ones? Is "thing power" really "trash power"? Bennett's epiphanic litterscape, films such as *Toy Story* and *The Brave Little Toaster*, and de Pájaro's street art seem to suggest that obsolete artifacts indeed are more powerfully "animated" than "useful" ones. Suddenly void of desirability and functionality, the discarded object transcends its utilitarian meaning and adopts an identity of its own. Now that nobody wants it or finds use for it, now that it is trash, it becomes an animated object that speaks and remembers (Thill).

De Pájaro's creative genius has no difficulty understanding that the power of trash lies precisely in the triple circumstance of being unwanted, on the verge of death, and memory-laden. Its new eloquent being or "thingness" firmly stands on the gravitas of its past. Things that we toss away have lived, like the dead we bury, and we attribute memories to them. Let's take a closer look, for example, at one particular discarded item that Francisco de Pájaro routinely finds on both sides the Atlantic: a mattress. Cast-off mattresses regularly litter the urban landscape across the globe, probably because it is so cumbersome to get rid of them. The webpage of 1-800-GOT-JUNK, one of the better-known, private full-service junk removal companies in the United States, offers the standard picture of the "what to do with my old mattress" story: "If you've ever tried it, you've likely discovered that mattress disposal is not an easy task. You can't simply throw them in a dumpster or leave them on the curb with your garbage. Perhaps that's why many old mattresses unfortunately become abandoned eyesores in alleyways" (Rubbish Boys Disposal Service). The paragraph ends on a consoling note: "But it doesn't have to be that way," the last sentence reads, for 1-800-GOT-JUNK can help.

As always, private companies find lucrative niches in the shortcomings of municipality and state. New York City, for example, meets the needs of its citizens with further demands. It requires from them that they "seal any mattress or box spring in a plastic bag before placing it out with regular garbage for bulk collection. This rule will help prevent the spread of bed bugs. If you do not dispose of mattresses or box springs properly, they will not be collected, and you may receive a $100 fine" (City of New York). (See Figure 32.7, which shows a mattress in a plastic bag, as stipulated by New York City regulation.)

Recent data tell us that "an estimated 20 million mattresses and foundations [are] hauled away from homes and hotels each year as a result of new bedding purchases," which means that approximately "50,000 mattresses [are] discarded each day in the United States occupying as much as 23 cubic feet of landfill space apiece" (James). Californians alone "buy about 4 million new mattresses and box springs a year" (James) and more than two million of the old mattresses end up in landfills. As James points out, " fewer than 1 in 10 of them is recycled for wood, plastic, fiber batting and springs to be used in other products, such as steel and carpet padding." Although recycling seems to be the sound alternative to used mattresses and spring boxes littering the urban landscape and crowding landfills (particularly since mattresses and spring boxes display a rich conglomerate of reusable materials), again and again research on mattress disposal stresses that the recycling process is cumbersome and riddled with fraud. Ryan Trainer, president of the International Sleep Products Association, notes that very often buyers of used bedding "are unscrupulous renovators, who often just sew a cover over a filthy used mattress, making no effort to properly sterilize the old bed or meet national fire safety standards, and then deceive consumers into thinking they are buying an all-new mattress" (James). Moreover, recycling is a complicated and costly operation, which explains the meager number of successful mattress recycling programs in California and the Unites States at large. In North America, the mattress disassembly process heavily depends on hand labor, thus often "the cost of handling the old mattresses, tearing them down and preparing the recyclable materials for sale usually exceeds the value of the recycled materials, so most programs charge a processing fee just to break even" (James). Furthermore, it is not always easy to find buyers for the components and the prices of the raw materials are subject to the up and downs of the market.

What I have presented above is the condensed version of the life, or rather, decay, death, and even resurrection of a mattress in the Western world, seen through the eyes of manufacturers, recyclers, and policy makers. I have done so deliberately, to highlight how even a "dry" and non-artistic account of the fate of a discarded object has the power to "animate" it, by way of tinting it with the shadow of death. For this is what we have learned thus far: that twenty million mattresses "die" and are discarded yearly in the US, and that certain private enterprises/undertakers (1-800-GOT JUNK being only one of them) will pick up our old mattresses for a fee nationwide; that, if we happen to live in New York, we will have to buy a mattress cover/body bag, or pay a fine if we do not tightly seal the mattress/corpse; that, even so, many "dead" and uncovered mattresses find their way into back alleys, their exposed "bodies" riddled with bedbugs; and that although mattresses have components/organs that can be recycled and brought back to life, the process of repurposing/resurrecting is fraught with difficulties, and subject to unscrupulous "renovators"/dissectors. Finally, even at the dump/cemetery, mattresses/bodies remain invasive and destructive forces, and have a sinister tendency to resurface. As Lifsher recounts, "old mattresses … are nightmares for landfill operators because each piece takes up 23 cubic feet, doesn't decompose, and 'floats to the top' of dumps because of its flexible construction … Steel springs can wreck a $50,000 piece of equipment in a second."

Exposed to increasing political, legal, and social pressure, municipal authorities are putting more effort into cleaning back alleys and side walks of discarded mattresses. In a parallel effort, the sleep industry worldwide is keen on investing in costly mattress recycling programs and facilities. De Pájaro's street art, however, makes it very clear that "mere" cleaning and recycling has its own dangers and blind spots. To keep our urban landscape free from trash will not make it disappear; instead, "displaced" garbage will fester forever in landfills and contaminate our natural surroundings. There is always the risk, on the other hand, that recycling becomes yet another uncritical way of making more stuff out of stuff and of irresponsibly fueling the formidable machine of consumerism. De Pájaro's art forces us to confront the truth of the physical and moral enormity of refuse head on. It teaches us to look garbage straight in the eye and identify it for what it is, namely, the massive and destructive consequence of boundless consumer greed. Like Bennett, de Pájaro is able to see the danger, and the beauty, in trash "vibrant" with life. In fact, Bennett's important question resonates with the Spanish street artist's redundant exercise in animating the already animated: "How would political responses to public problems change if we were to take seriously the vitality of (nonhuman) bodies? … How, for example, would patterns of consumption change if we faced not litter, rubbish, trash or 'the recycling,' but an accumulating pile of lively and potentially dangerous matter?" (Bennett viii).

Among all things discarded, mattresses certainly seem particularly "lively" and "dangerous." They are highly skilled at acquiring anthropomorphic features and speaking to us, since so much of their content and nature is organic already. Humans spend one-third of their lives on a mattress, and their prolonged stay does not go unnoticed. More often than not, cast-off mattresses are covered in human-made stains and grow heavy with bodily fluids and bed bugs thriving on shed skin. Mattresses are intensely private objects. By the time they are deemed trash and expelled into the streets, they have many intimate and profoundly human stories to tell. That is why discarded mattresses appeal to de Pájaro and have an immediate effect also on the audience: even before the artist arrives at the scene and completes the animation process, cast-off mattresses already show signs of life, for they are pregnant with the life (and death) of others. After all and more often than not, mattresses are the silent witnesses to conception, birth, and death, the place where we let go of inhibition. On mattresses, we fart, snore, drool, and sweat, get lost in disjointed dreams, and go about the messy business of fucking, birthing, and dying.

Municipal discourses and policies around mattress disposal use vigorous rhetoric and means to free urban landscapes from impurity. Nothing, however, could be more alien to de Pájaro's sensibility than cleanliness; that is why his street art thrives on discarded mattresses as the epitome of accumulated dirt. In de Pájaro's eyes, the human condition is as filthy as abandoned mattresses, and the history of humankind is equally layered with grime. It is common knowledge that mattresses double in weight after ten years, fattened with our dead skin (on which whole colonies of dust mites thrive), skin oil, blood, semen, and sweat. Thus when de Pájaro paints on mattresses, he is truly painting on human skin and organic matter, and therefore, "animating" the already animated. In his art, the grime of mattresses serves as redundant canvas to the "grime" of the human condition, and foremost, of human history understood as a chain of injustices and exploitative acts. In his prologue to the only book available so far on de Pájaro's street art, Blaquiere volunteers a "quick little breakdown of the Art is Trash characters and symbols," and categorizes them "into three broad camps":

> Firstly, we have "The Indians," who fight for their independence and freedom with their bicycles, bows and arrows, tipis and so on. They fight "The Authorities," their enemies: police, soldiers, bankers, conquistadores, politicians, helicopters, dollar signs, bag of euros, gold and so on. And lastly, there is the "No-Man's land" of the general public: shit, flies, farts [and] cobwebs.
>
> (7)

On the surface of cast-off mattresses, "real" stains (of feces, blood, and semen) stand for the "symbolic" and indelible stains of human misery and colonizing violence (see Figures 32.6, 32.7, and 32.8). But in broader terms, discarded mattresses fattened with human detritus are powerful symbols of life and death, and it is the messiness of Eros and Thanatos publicly frolicking on dirty bedding that fuels de Pájaro's street art. It is important to add that the Spanish artist makes a point of painting fast and, more importantly, of not moving trash and bulk pieces around:

> When I am working in the rubbish, I try not to touch or move the content much. It is the accidental or arbitrary way in which people leave waste that inspires in me a reason to speak through trash. I try to work very fast and without dwelling too much on details. And nor do I take pleasure in technicalities, because I rely on the time factor—the time when the police or the rubbish lorry arrives.
>
> (47)

Figure 32.1 Francisco de Pájaro, "Art is Trash," London, 2012. Courtesy of the artist.

Hence, in his piece shown in Figure 32.1, de Pájaro once again takes advantage of seren-dipity and the already existing wall painting (by another artist), and seamlessly "integrates" it into his "mattress collage." "Integration," in fact, is the Spanish painter's fundamental modus operandi.

Unlike other "mattress artists," such as Wade Guyton and Kelley Walker (Frank), who use sterilized thrift store mattresses as substitutes for canvases and cover up the whole surface with thick layers of bold colors, de Pájaro makes sure that the stains on the mattress become a fundamental and semantically charged component of his drawings. De Pájaro paints the naked model and gaping vagina, but it is the mattress itself that readily provides the vaginal discharge (Figures 32.1).

If "integration" is an important concept in de Pájaro's art, so is "horizontality." De Pájaro is always keen on forcing viewers to lower their eyes. More often than not, de Pájaro needs to kneel down, to squat, to sit, and even to lie down when he is at work, for it is trash on the ground that comes alive and "animated" under his brush strokes, and it is the lives of the disenfranchised and the lower classes and not of the privileged and the upper classes that stir his heart and fuel his talent. De Pájaro works feverishly and in a hurry, before the trash truck arrives and carries his pieces to the landfill. His art is deliberately ephemeral and rushed, but, paradoxically, the scenes he depicts seem timeless. Poverty, one of de Pájaro's

Figure 32.2 Francisco de Pájaro, "Art is Trash," Barcelona, 2011. Courtesy of the artist.

Figure 32.3 Francisco de Pájaro, "Art is Trash," New York, 2014. Courtesy of the artist.

Figure 32.4 Francisco de Pájaro, "Art is Trash," San Diego, CA, 2015. Courtesy of the artist.

main themes, appears as permanently nailed to the floor, even etched into the pavement. The sanitary workers will eventually haul away the mattresses in Figures 32.2 and 32.3, but the arm and the hand of the woman (Figure 32.2) and the arm and legs of the male holding a precarious balance on the edge of the mattress (Figure 32.3) will remain. Trash never really vanishes. It leaves indelible traces, and so does social injustice.

De Pájaro's art has an intense allegorical quality to it. It does not depict poor people or the homeless; it depicts "poverty" and "homelessness." It is not interested in particular trash items that a landfill will soon engulf, but in the ontology of the discarded as a metaphor of the disenfranchised and the unwanted, and as a consequence of a savagely consumerist society. Thus against the background once more of a discarded mattress, the shopping cart in Figure 32.4 is the grotesque counter-image to shopping bliss. Instead of containing "happy" goods and shopping bags, it is overloaded with two terrified animated trash bags anxiously holding on to each other. A naked man carrying a youngster on his back, who in turn carries a trash bag, pulls at the shopping cart, but it has no wheels. Standing on spiky ends, once again it seems permanently nailed to the pavement, one more timeless scene of urban misery under a shady tree.

Trees play an important role in de Pájaro's art, not only because in cities trash often accumulates around trees and plants lining the sidewalks, but also because trees are the only

Figure 32.5 Francisco de Pájaro, "Art is Trash," Barcelona, 2011. Courtesy of the artist.

sign of hope and compassion in de Pájaro's otherwise bleak palette. Trees provide a protective shade against the sun (Figure 32.4), but they also have managed to escape despairing immobility. In de Pájaro's upside down world, shopping carts do not have wheels, and trees have feet (Figure 32.5). I am standing here, the tree seems to say, in front of that parked white car, but I will be gone pretty soon. In the meanwhile, the mattress will remain, since trash never really leaves the city.

I like to call de Pájaro's compassionate environmentalism "oblique": his interest in nature and the environment derives from his main preoccupation with social inequality, and from his conviction that capitalism, mindless consumption, and endless generation of waste is destroying nature and humankind. I also like to call it "historic" environmentalism since de Pájaro's environmental awareness has its roots in a deep-seated nostalgia for pre-conquest America. Somewhat ironically, de Pájaro's ingrained belief that, contrary to our contemporary society, indigenous communities were able to live in harmonious communion with nature, comes from the many Western movies de Pájaro watched during his childhood in Spain. Francoist dictatorship overfed the masses with mainstream Hollywood popular culture, which young de Pájaro was quick to read against the grain: he always took the side of the Indians in cowboy films. The naïve idealism of his childhood years strongly impregnates de Pájaro's art. When asked by an interviewer why so many of his paintings on Manhattan and Brooklyn trash recreate "movie" scenes of cowboys and Indians, de Pájaro volunteers:

> The images of [Native Americans] are born out of Hollywood's lies. As a child, I never believed that the bad guys in the movies were the Indians. At school, teachers also propagated a number of falsehoods about the Spanish Conquest as a wonderful and glorious event of our History. I was born in Extremadura, a beautiful region that was the cradle also of the most sanguinary conquistadors of the American continent. Also, I closely identify with the way of life, the respect, and the philosophy of Native Americans. We only need to look around to see how much harm we are inflicting on Nature with our way of life. My spirit is that of a warrior who paints so that we don't forget that we are children of nature, and that we all have to keep up our internal fight to preserve our natural world. I am aligned with all the native people.
>
> (The Dusty Rebel)

Sure enough, de Pájaro not only paints in defense of Native Americans but on a number of occasions he even dresses up as one while producing his street art work (Figure 32.6).

Figure 32.6 Francisco de Pájaro, "Art is Trash," New York, 2014. Courtesy of the artist.

Figure 32.7 Francisco de Pájaro, "Art is Trash," New York, 2014. Courtesy of the artist.

One could certainly argue that de Pájaro "uncritically" adopts an "indigenous identity" flawed with the stereotypes of the "ecological Indian" in deep communion with Nature. This is closely tied to the fact that, as an artist, de Pájaro wishes to remain firmly anchored to the imaginary world of his childhood, which owes a great deal to Hollywood. As noted above, as a child de Pájaro watched countless Western movies, and always identified very strongly with the disenfranchised, though heavily clichéd, "Indian." Hence eager to remain faithful to the ideals and idols of his childhood, de Pájaro not only dresses up as a "Hollywood Indian" when painting trash on the streets of Brooklyn and Manhattan, but also even appears depicted as such in some of his rare autobiographical portraits, like in the piece in Figure 32.7. On the surface of a discarded mattress fitted into a white plastic cover, as New York municipal regulations demand, De Pájaro portrays himself as a "piel roja" (red-skin) fleeing police and still brandishing the brush with which he painted his signature ("Art is Trash") in hurried green letters.

De Pájaro's street art is both ephemeral (it is trash) and timelessly allegorical. By the same token, it is also simultaneously local and global. De Pájaro's themes are universal—poverty, social inequality, the destruction of nature and the trashing of the planet—but also locally infused. Many of his transient art works on the streets of Barcelona, for example, are created in solidarity with the rural immigrants (de Pájaro himself immigrated into rich Catalonia from impoverished Extremadura) and the undocumented workers from sub-Saharan Africa that live in precarious conditions on the Iberian Peninsula. In the same way, de Pájaro

Figure 32.8 Francisco de Pájaro, "Art is Trash," Los Angeles, 2015. Courtesy of the artist.

paints in solidarity with Native Americans when he encounters trash on the streets of Manhattan and Brooklyn, and even went a step further during his recent trip to California. The piece reproduced in Figure 32.8 once again shows a Native American and an American soldier painted on a mattress, the two opposing worlds of victims and perpetrators separated by a tree. But this time, the gulf between colonizers and the colonized is made apparent also via language. "All of it is mine," the soldier utters in English; "Desea nada y lo tendrás todo" (don't wish for anything and you will have everything), the Native American muses in Spanish, which is the prevalent language of undocumented immigrants, many of them of indigenous origin, in California.

In de Pájaro's street art, new elements (in this case, the sentence in Spanish) that acknowledge local cultures and circumstances go hand in hand with certain "universal" marks or recurrent symbols from the animal world, such as flies (Figures 32.1, 32.4, 32.7, and 32.8) and galloping Indian horses (Figures 32.7 and 32.8). The latter is one of the few positive elements in de Pájaro's art installations. As he explains:

> One night, I angrily headed out into the street with a few markers. I was deter-mined to express myself on anything, and did it on a metal door. I painted a galloping Indian horse; I meant to represent my courage and my escape from everything that surrounds me. Unluckily that same night I was caught by three [undercover] police officers while painting the door. My childish attempts at vandalism ended there.
>
> (20)

In de Pájaro's street installations, galloping Indian horses, a symbol for untamed freedom and unfiltered communion with nature, always stay as far away from flies as possible.

Flies stand for what is dirty, contaminated, and corrupt, such as trash (Figure 32.4), political and royal power (Figure 32.1), and repressive authority (Figure 32.8). Predictably, and to further emphasize two diametrically opposed worlds and moral orders, the mattress painting in Figure 32.8 depicts two flies almost touching the gun held by the soldier, and an Indian horse on the tip of the arrow, covered with a green-striped blanket and freely galloping, almost flying, up into the sky.

With the help of the Spanish street artist thus mattresses in particular and rubbish in general become in-your-face vibrant matter, an indestructible animated mirror that, despite—or, rather because of—its grimy surface, sharply reflects humankind, its misery, and its insatiable greed. More importantly, when de Pájaro "animates" trash and turns it into a living creature that intently looks at us, we not only see ourselves, but also see trash, probably for the first time. "Art is Trash" sheds a glaring light on rubbish as the dirty consequence of relentless consumption, and as "an accumulating pile of lively and potentially dangerous matter," in Bennett's words. What you see is what you get. And it is ugly, terrifying, and only beautiful because it is honest.

References

Ackerman, Alan. *Seeing Things: From Shakespeare to Pixar*. Toronto: University of Toronto Press, 2011. Print.

Bennett, Jane. *Vibrant Matter: A Political Ecology of Things*. Durham, NC: Duke University Press, 2010. Print.

Berger, John. "A Professional Secret." *John Berger: Selected Essays*. Ed. Geoff Dyer. New York: Vintage Books, 2001. 536–540. Print.

Blaquiere, Tommy. "Foreword." *Art is Trash*. Francisco de Pájaro. Barcelona: Promopress, 2015. Print.

Boo, Katherine. *Behind the Beautiful Forevers: Life, Death, and Hope in Mumbai Undercity*. New York: Random House, 2014. Print.

City of New York. "Mattress or Box Spring Disposal." Official Website of the City of New York. 2016. Web. 7 January 2016.

De Pájaro, Francisco. *Art is Trash*. Barcelona: Promopress, 2015. Print.

——. "Francisco de Pájaro – Art is Trash – El Arte es Basura." 2015. Web. 7 January 2016.

Engler, Mira. *Designing America's Waste Landscapes*. Baltimore, MD: Johns Hopkins University Press, 2004. Print.

Frank, Priscilla. "The 10 Hottest Artists at Art Basel Miami Beach." *Huffington Post*. 25 January 2014. Web. 7 January 2016.

George, Rose. "The Blue Girl: Dirt in the City." *Dirt: The Filthy Reality of Everyday Life*. Ed. Kate Forde. London: Profile Books, 2011. 133–174. Print.

Humes, Edward. *Garbology: Our Dirty Love Affair with Trash*. New York: Avery, 2012. Print.

James, Gary. "This Is No Place for a Mattress: Industry Taking Lead to Recycle." *BedTimes: The Business Journal for the Sleep Products Industry*. March 2013. Web. 7 January 2016.

Lifsher, Marc. "California Weighs Mattress Recycling Fee." *Los Angeles Times* 23 March 2013. Web. 7 January 2016.

Nagle, Robin. "The History and Future of Fresh Kills." *Dirt: The Filthy Reality of Everyday Life*. Ed. Kate Forde. London: Profile Books, 2011. 187–205. Print.

Rathje, Wiliam and Cullen Murphy. *Rubbish: The Archeology of Garbage*. Tucson, AZ: University of Arizona Press, 2001. Print.

Rogers, Heather. *Gone Tomorrow: The Hidden Life of Garbage*. New York: New Press, 2005. Print.

Royte, Elizabeth. *Garbage Land: On the Secret Trail of Trash*. New York: Back Bay Books, 2005. Print.

Rubbish Boys Disposal Service Inc. "Mattress Disposal." *1-800-GOT-JUNK?* 2016. Web. 7 January 2016.

Stetz, Margaret, "The Brave Little Toaster from Print to Film: Obsolescent Appliances and Capitalist Allegories." *Opticon1826* 14 (Autumn 2012): 21–26. Web. 7 January 2016.

The Dusty Rebel: "Interview with Art is Trash." *The Dusty Rebel.* 27 June 2014. Web. 7 January 2016

Thill, Brian. *Waste.* New York: Bloomsbury, 2015. Print.

Urrea, Alberto. *By the Lake of Sleeping Children.* Harpswell: Anchor, 1996. Print.

33

TOUCHING THE SENSES

Environments and technologies at the movies

Alexa Weik von Mossner

Human technology enjoys a somewhat ambivalent reputation within the environmental humanities. It has often been framed as a hindrance to harmonious human–nature relationships, an inextricable part of cultural and economic practices that alienate us from nature and put us in a position of power and dominance. However, as Sean Cubitt reminds us, not all technologies are "used as instruments of domination over nature or other humans. Instead … both scientific and entertainment media rely on technology to communicate between human and natural worlds" (4). Technology, in this understanding, functions as a mediator, enabling and often defining our experience of environments near and remote, familiar and strange, actual and virtual. A typical example is the medium of film, which relies on a vast array of technologies to immerse viewers in environments other than those they physically inhabit. Studies such as Nadia Bozak's *The Cinematic Footprint* (2012) have turned our attention to the extreme wastefulness of these technologies, and some scholars have argued that they make film an unsuitable medium for environmental communication regardless of its content. Yet, its wasteful "material ecologies" notwithstanding (Ivakhiv 90), film—and in particular documentary film—has often been the medium of choice for artists and activists who want to raise awareness of ecological problems, precisely because it allows people to engage on the sensual and emotional level with an environment that is not actually present and that they may have never personally experienced.

This chapter explores how the virtual environments of films *come alive* for viewers on the sensory and emotional level through the use of cinematic technology. It pays particular attention to the ways in which our engagement with cinematic storyworlds is both embodied (in a physical body) and embedded (in a physical environment), and, to do this, it draws on ecocritical film analysis on the one hand and on the insights of cognitive science on the other. The first part of the chapter considers the ways in which the technical conventions of film production and exhibition take into account the embodied and embedded nature of human perception and cognition. I argue that processes of *embodied simulation* (Gallese "Bodily Selves") are of particular importance in our engagement with cinematic environments because they mimic actual, physical experience of such environments, thereby cueing powerful emotional responses. The second part of the chapter then uses the example of one specific environmental film—Jeff Orlowski's climate change documentary *Chasing Ice* (2012)—to explore how technology is both enabling and limiting when it comes to transforming an actual environment into the sensory spectacle of a cinematic one. In this context, the results of a qualitative reception study conducted in Germany and Austria shed some light on the relationship between emotional engagement in a film and its potential impact on environmental attitudes and behavior. As Jane Stadler has pointed out, "the most powerful films have an afterlife, an influence that remains with us when we are affected by the sensory impact of films, captivated by story, character, and conflict, and left wondering about the issues they raise" (2). A better understanding of how films achieve this powerful effect on our embodied minds is of vital importance not only for the critical analysis of eco-cinema but also for the environmental humanities more generally.[1]

Environments at the movies: cinematic storyworlds and embodied simulation

Cinematic experience depends on motion in more than just one way. As Gilles Deleuze once put it, "cinema not only puts movement in the image, it also puts movement in the mind" (qtd. in Flaxman 366). Both of those movements are enabled by technology, and both of them critically depend on the specifics of human cognition. The biology of the human brain and perceptive system, and the limitations of its processing capacity, allows us to see actual movement in motion pictures. Katherine Thompson-Jones reminds us that the impression of movement in a projected filmstrip

> is currently understood to rely on two psychological mechanisms: critical flicker fusion, which involves our seeing a rapidly flashing light as a continuous beam, and apparent motion, which involves our seeing motion in a rapidly changing visual display. Thanks to the first mechanism, the movie seems to be continuously illuminated—rather than flashing rapidly in response to the opening and closing of the projector shutter every 1/48th second. Thanks to the second mechanism, there appears on the screen a persistent moving image—rather than a succession of static images.
>
> (116)

What moves, or at least *seems* to move, in that persistent moving image is first and foremost the protagonists, and it is also the protagonists who generate a good deal of movement in the viewer's mind, both literally and metaphorically. As neuroscientist Jeffrey Zacks points out, "[o]ur brains didn't evolve to watch movies: Movies evolved to take advantage of the brains we have" (4), in particular of the fact that "we mimic whether we intend to or not, often without noticing" (5). Complex sets of information, communicated not only through dialogue, but also through body postures and facial expressions, cue viewers to *mimic* or *simulate* a character's embodied experience in their minds.

However, there are other important elements of film that move us. Recent insights in cognitive neuroscience suggest that processes of embodied simulation play a central role not only in our identification with human characters but also in our sense of immersion in the virtual environment of a film. As neuroscientist Vittorio Gallese and film scholar Michele Guerra point out, "[r]ecent studies within cognitive film theory, visual psychology and neuroscience bring out strong evidence of a continuity between perceiving scenes in movies and in the world, as the dynamics of attention, spatial cognition and action are very similar in direct experience and mediated experience" (183). Gallese's influential research on the role of the mirror neuron system in our brains in processes of embodied simulation has far-reaching implications for our understanding of the role of the viewer's body in the engagement with film (see "Mirror Neurons and Art"). Gallese reminds us that "some quarters of cognitive neuroscience are still today strongly influenced, on the one hand, by classic cognitivism [which basically treats the brain as a living computer] and, on the other, by evolutionary psychology" ("Bodily Selves" 2). As an embodiment theorist, Gallese proposes an alternative approach that focuses on what he calls "the lived body" (2). Rather than considering the brain in isolation, such an approach focuses on the "tight interrelated connections" (2) between the brain and the body and also pays attention to the body's situatedness in and interaction with a given environment.

Long before anyone had ever heard of Gallese's theory of embodied simulation, filmmakers and distributors figured out that the illusion of narrative transportation is strongest when viewers' actual environment has very few affordances they must pay attention to. That is why, at least in the context of Western exhibition traditions, everything about viewers' actual environments—the dark room of the theater, the big screen that fills out most of the visual field, the comfortable seats—is geared toward keeping their eyes focused on the screen and the rest of their bodies motionless, making them the perfect sounding boards for the embodied simulation of the storyworld of the film. Significantly, we mentally simulate not only the motions and emotions of human protagonists, but also those of nonhuman life forms. Recent fMRI (functional magnetic resonance imaging) research suggests that we even map the movement of inanimate objects onto the motor systems of our brains, simulating and thus understanding them in relation to our own bodies (Keysers et al. 339).

Only a fraction of these mapping processes are conscious, and yet most film viewers remember moments in which they crouched in their seats, jumped back in fright, or teared up in response to a particularly moving film scene. As Zacks explains, "[y]our eyes and ears are telling you that something exciting is happening in front of you and your brain is preparing to react. Of course, you *know* it's just a movie. But large parts of your brain don't process that distinction" (4). The history of film has been a history of developing technical means that trick our brains into a powerful response to create highly immersive film viewing experiences. Many of these means are "transparent" in the sense that they go unnoticed as long as the viewer's attention is focused on the content of the story. Technical progress has always strongly influenced and sometimes radically changed the medium of film. From D. W. Griffith's infamous *The Birth of a Nation* (1915) onwards, an increasingly standardized set of continuity editing techniques has worked in concert with those technological innovations to create the illusion of spatial and temporal unity and achieve transparency with regard to the cinematic techniques themselves. Most film viewers do not pay attention to camera movements, cuts, or sound design in a mainstream Hollywood production or even most of documentary film unless they are specifically asked to do so.[2]

Despite the fact many cinematic techniques are not consciously registered, they do have a significant impact on viewers' cognitive understanding and emotional response. In the early 2000s, social psychologists Melanie Green and Jeffrey Brock developed the "the transportation imagery model of narrative persuasion" which stated that narratives tend to be more persuasive when they elicit from recipients a state of psychological transportation ("The Role of Transportation"; "In the Mind's Eye"; see also Green). Psychological transportation into a storyworld is understood in this context as an integrative melding of attention, imagery, and emotion, focused on story events (Mazzocco et al.). Green and Brock's research, which relates to both literature and film, indicates that individuals who feel more transported into a storyworld exhibit greater attitude and belief change in response to stories. We must assume that these insights hold as true for environmentalist narratives as they do for other kinds of social communication. Climate change documentaries emerge as a particularly fruitful site for an investigation of these issues because the makers of such documentaries tend to believe strongly in the urgency of their subject matter and put considerable effort into creating narratives that are emotionally powerful and rhetorically convincing. At the same time, the peculiar nature of climate change poses considerable challenges to filmmakers.

One challenge is the vast spatial and temporal scope of climate change. Many of the ecological processes involved cannot be directly perceived by humans because they are either too gradual to be visible to the human eye or take place in areas that are remote and inaccessible. A second and related problem for the makers of climate change documentaries is that they are confronted with the paradox of needing to "document" future developments that they want to help prevent. As Julie Doyle reminds us, climate change campaigning in general necessitates "action to prevent climate change *before its effects [can] be seen*" (280), and that conundrum creates problems for a medium that crucially relies on visual documentation for its argument. Both of these challenges relate to the use of cinematic technology, which can help viewers to experience environments and ecological processes they would not normally have sensory access to, but at the same time also severely limits what can be shown.

The third challenge that has to be addressed is emotional salience. A 2006 survey of American attitudes toward climate change conducted by Anthony Leiserowitz found that experiential rather than analytical processes are crucial for people's climate risk perceptions because they are "holistic, affective and intuitive," encoding "reality in concrete images, metaphors and narratives linked in associative networks" (47–48). A medium that provides viewers with simulated experience thus emerges as an effective choice for climate change campaigning. An intriguing example is Jeff Orlowski's *Chasing Ice*, a documentary film that is built on the assumption that experiencing the dramatic ice loss in the Arctic region in a way that is emotionally salient will lead to increased risk perceptions and thus, ultimately, to action.

Movies of the environment: technology, perception, and engagement in *Chasing Ice*

Chasing Ice chronicles *National Geographic* photographer James Balog's quest to capture the gradual disappearance of the Artic glaciers in such a way that the process becomes visible to the human eye and therefore more easily comprehensible to the human mind. His Extreme Ice Survey (EIS) drives sophisticated photographic technology to the limits of what it can do while at the same time depending on that technology for its success (see Balog). When Balog first went to Greenland to shoot highly aesthetic photographs of translucent ice formations and towering icebergs, he was dissatisfied with the results because the pictures did not communicate the urgency of the situation. In the completed film, these still photographs nevertheless play an important role, because they capture the visual beauty of the Polar region, thus providing viewers with something to admire and, eventually, to mourn. In a qualitative reception study we conducted in Germany and Austria in the winter of 2013–2014, viewers reported feelings of fascination and wonder in relation to these photographic representations of the Artic environment, and they frequently connected these positive feelings to the shock and outrage they felt at the end of the film.[3] This combination

340

of admiration and wonder on the one hand and shock and outrage on the other is exactly the emotional response that the filmmakers aim to provoke. The positive emotions that are cued at the beginning of the film in relation to its beautiful environment serve to intensify the negative emotions that are cued later on through the presentation of visual evidence of its fast and severe degradation. The film's narrative is framed as a quest to get that evidence and to make visible what is invisible to the naked human eye so that viewers can experience and feel it.

Balog's plan is established early on in the film: rather than isolated snapshots in time, or even the "before and after" pictures that have often been used in climate change communication (Doyle), he wants to provide the world with sophisticated time-lapse photography that will compress the gradual change occurring over a period of several years into just a few seconds, presenting it to the human eye as a coherent motion picture. His stated belief is that faced with this visual evidence, humans will understand the magnitude of the anthropogenic impact on the Arctic region and feel motivated to take action toward mitigating climate change. Ironically, what stands in the way of getting that evidence is the Arctic environment itself, which, as one of Balog's team members points out in the film, "is not the nicest environment" for the dozens of time-lapse cameras they install all over the Artic region. In order for the emotional drama of the film to unfold, the harshness of that environment must be communicated to viewers in a way that allows them to experience it on the sensual and affective level.

Thus the documentary must give viewers a sense of sensual immersion, which presumably is why Robert Redford has said of the film that "it deserves to be seen and felt on the big screen" (quoted in "Reviews"). As an audiovisual medium, film can only ever engage two senses directly—sight and hearing—while all other senses must be engaged on the imaginary level.[4] A film about the climate of the Arctic must communicate that this climate is very harsh and very cold for the story to function. The sensual perception of temperature can be communicated visually if the viewer has enough personal experience with snow and ice to know that they feel cold. Sound design can aid in that process of imagined perception, and in Chasing Ice we often hear the sound of strong winds in combination with shots of objects that are almost blown away by the wind gusts (on theories of the soundtrack in relation to perceptual processing and the mind, see Branigan). Most important, however, in evoking the feel of the Arctic, are the film's human protagonists. Unlike nature documentaries that depict the Arctic deliberately as devoid of human life—an omission that tends to include the filmmakers themselves—Chasing Ice foregrounds the struggle of Balog's team as it tries to set up the time-lapse cameras in a way that will allow them to keep functioning for months in the icy landscape of the Arctic. Seeing them cope with the elements creates "movement" in the viewer's mind, both literally in the sense that processes of embodied simulation help them to imagine what that environment feels like on the sensory level and metaphorically in the sense that they are cued to feel with and for Balog and his team, hoping more or less anxiously that they will be able to succeed.

The plotline of Chasing Ice is a familiar one, well anchored in the genre of melodrama (as is that of Davis Guggenheim's An Inconvenient Truth, with which it shares a number of rhetorical and narrative features). Balog emerges as the melodramatic hero who risks both his financial security and his health in order to document a disappearing environment and perhaps save it from annihilation. This "obsession," as his wife calls it in the film, drives him to the limits of what he can do. He is shown to take upon himself great suffering, both physically and emotionally, when it turns out that the time-lapse cameras cannot function in these climatic conditions and that the work of many months has been in vain. In climactic moments of disappointment and near-despair his face is captured in close-up, allowing viewers to simulate his feelings through processes of emotional contagion (Plantinga). Importantly, the film makes clear that Balog does not suffer for personal gain but for the greater good. This well-tried narrative strategy eventually culminates in several spectacular successes as the Balog team is not only able to get the cameras to work but also is serendipitously present when a chunk of ice the size of Manhattan breaks off the Jakobshavn Glacier in Greenland and captures the event on film.

The resulting footage, just like the time-lapse photography of the retreating glaciers that Balog presents in the final minutes of the film, is nothing less than breathtaking. Viewers see the mighty glaciers change shape, crumble and race backwards within seconds. Several close ups of audience members watching the footage during Balog's slide show cue viewers to feel shock along with them as enormous icebergs shoot up in the air before thundering down into the water. Psychologists Ed Tan and Nico Frijda have argued that our emotional responses to imagery portraying "an environment in which one feels tiny and insignificant" and other visually overwhelming scenarios may be independent from our investment in character and narrative. Such imagery triggers both fascination and anxiety in an experience that, in other contexts, we would call the sublime (62). Embodied simulation is nevertheless crucial in such moments because we tend to map even the movement and touch of nonhuman entities onto the motor systems of our brains in order to understand them. And in case viewers are not impressed by these moving images alone, the film also provides them with a deeply melancholic music score that cues feelings of loss and mourning (see Schiavio on embodied processes and perception of music).

Balog thus achieves at the end of the narrative what he had set out to do, and given the success of the film and the popularity of his related lecture show, one could speculate that his photography and its documentation in a nonfiction film has indeed helped raise awareness. Social anthropologist Kay Milton has observed that an aesthetic appreciation of natural beauty can play an important role in motivating people to act on behalf of an environment, exactly because perception and emotion are crucial parts of our meaning-making faculties. "As we engage with our environment," she writes, "we perceive meanings in it ... These meanings become known to us through the emotions they induce, which we then experience as feelings" (100). In this view, cuing strong emotions in a climate change documentary makes a lot of sense, and yet the results of our reception study were somewhat more ambiguous. In part, this may be related to the filmmakers' choice to put Balog at the center of the film, a choice that had to do with the gestation of the project.[5] While this worked well for many viewers, who felt both curiosity and compassion in relation to Balog's quest, others complained that the personal troubles of the photographer had no bearing on the issue of climate change and should therefore have been left out of the picture.

A larger problem, however, is inherent in the climate change thematic itself. Our hypothesis was that people would mostly name the spectacular footage at the end of the film as what had touched them the most. It was indeed a response we received frequently. Some people reported a combination of fascination and sadness in response to the time-lapse photography and the calving event, others a deep sense of shock. Interviewees also noted that they had been aware of the ice loss in the polar regions but added that they had not realized that it was so dramatic before seeing the film. Virtually all respondents stated that they would recommend the film to others, many of them because they felt that people needed to see the time-lapse images and understand how bad the situation is. What the film does seem to have achieved, then, at least with the groups that we interviewed, was the raising of awareness and the evocation of emotions. Some of what we heard is echoed in the testimonials the filmmakers themselves collected ("Chasing Ice Testimonials"). Many of our interviewees shared a mixture of sadness, mourning, and anger at capitalism, society, mankind, or no one in particular. In some cases, this led to an expressed desire to help bring about social change through personal engagement. Others, on the contrary, expressed a sense of helplessness and the belief that there was nothing they could do to stop or change the ice loss in the Arctic or climate change more generally. Some even stated that they would probably forget the film before long because they were too busy getting on with their normal lives, thereby echoing the results that Kari Marie Norgaard presents in *Living in Denial* (2011), her study of public attitudes toward climate change in Norway and the United States. Awareness alone, Norgaard found out, is not necessarily sufficient for social change. Confronted with the over-whelming evidence for the changing climate on the one hand, and with the well-engrained behavior patterns of our normal lives on the other, we tend to block the topic from our minds in order to avoid painful feelings of fear, guilt, and helplessness. Rather than leading to behavioral change, awareness might thus lead to apathy or outright denial if the received information remains disconnected from political, social, and private life. This is a problem that cannot be solved by any one film, no matter how powerful or well intended it might be.

However, what we perceive, feel, and understand when we watch climate change docu-mentaries or other environmentally oriented forms of communication still has a bearing on our private and social lives. As the climatologist Mike Hulme has pointed out, climate change is an idea as much as it is a physical phenomenon, and over the past decades it has "moved from being predominantly a physical phenomenon to being simultaneously a social phenomenon" (xxv). The social dimension of the phenomenon is affected by the way in which narratives about the physical phenomenon are framed, and the impact of such nar-ratives is to a large degree dependent on how convincing and emotionally engaging they are. A film like *Chasing Ice*, which uses cinematic technology to present the melting of the Arctic at a time scale people can see and understand, might not be able to change viewers' attitudes instantly, and yet it most likely contributes to a slower, longer cultural change in accepting, facing, and responding to global climate change. This mediating function makes environmental film a relevant research subject for the environmental humanities which, like other activist forms of scholarship, are ultimately invested in bringing about cultural change.

Acknowledgments

Research for this chapter was supported by a Fellowship for Advanced Researchers from the Swiss National Science Foundation and by a grant from the Research Council of the University of Klagenfurt.

Notes

1 In Paula Willoquet-Maricondi's relatively narrow definition of ecocinema, the genre involves "consciousness-raising and activist intentions, as well as responsibility to heighten awareness about contemporary issues and practices affecting planetary health" (45). I myself prefer a looser definition of the term that also includes the mainstream Hollywood productions that Willoquet-Maricondi excludes from her definition.

2 Experimental film, on the other hand, often foregrounds the filmic means, thereby disrupting the illusion of transparency and calling attention to the film as film. Some ecocritical film scholars, among them Scott McDonald and Paula Willoquet-Maricondi, have argued that such films are more conducive than mainstream productions to provoking critical reflection and behavior change.
3 I conducted the reception study in the winter of 2013–2014 together with the media studies scholar Brigitte Hipfl. Our subjects consisted of several groups of students at the University of Klagenfurt in Austria and the visitors of a screening of the film as part of the Rachel Carson Center for Environment and Society's Green Visions film series in the Gasteig Cultural Center in Munich, Germany. Subjects received questionnaires directly after the viewing of the film, and the Klagenfurt students also talked about their viewing experience in moderated group discussions.
4 There have been several attempts to add scents to the film-viewing experience, among them Smell-O-Vision and Odorama, but they have remained mostly unsuccessful. In recent years, 4DX technology has added selected scents as well as motion, air gusts, and sprays of water to the viewer's seat, claiming that this makes the experience more immersive.
5 Director Orlowski was an undergraduate student at Stanford University at the time he met Balog, and his initial idea was to make a biopic about the famous photographer. Only later, after he had become a member of the EIS-team, did he and Balog decide that his film would foreground the issue of climate change.

References

Balog, James. "About EIS." Extreme Ice Survey, Earth Vision Institute. Boulder, CO. 2014. Web. 9 December 2015.

Bozak, Nadia. *The Cinematic Footprint: Lights, Camera, Natural Resources.* New Brunswick, NJ: Rutgers University Press, 2012. Print.

Branigan, Edward. "Soundtrack in Mind." *Projections: The Journal for Movies and Mind* 4.1 (2010): 41–67. Print.

Chasing Ice. Dir. Jeff Orlowski. Submarine Deluxe, 2012. DVD.

"Chasing Ice Testimonials." Exposure Labs. YouTube. 20 May 2013. Web. 9 December 2015.

Cubitt, Sean. *EcoMedia.* New York: Rodopi, 2005. Print.

Doyle, Julie. "Seeing the Climate? The Problematic Status of Visual Evidence in Climate Change Campaigning." *Ecosee: Image, Rhetoric, Nature.* Ed. Sidney Dobrin and Sean Morey. Albany: SUNY Press, 2009. 279–298. Print.

Flaxman, Gregory, ed. "The Brain Is the Screen: An Interview with Gilles Deleuze." *The Brain Is the Screen: Deleuze and the Philosophy of Cinema.* Minneapolis, MN: Minnesota University Press, 2000. 365–374. Print.

Gallese, Vittorio. "Bodily Selves in Relation: Embodied Simulation as Second-person Perspective on Intersubjectivity." *Philosophical Transactions of the Royal Society* (April 2014): 1–10. Print.

——. "Mirror Neurons and Art." *Aesthetics and the Embodied Mind: Beyond Art Theory and the Cartesian Mind-Body Dichotomy.* Ed. Alfonsina Scarinzi. Dordrecht: Springer Science+Business Media, 2015. 441–449. Print.

Gallese, Vittorio and Michele Guerra. "Embodying Movies: Embodied Simulation and Film Studies." *Cinema: Journal of Philosophy and the Moving Image* 3 (2012): 183–210. Print.

Green, Melanie C. "Transportation into Narrative Worlds: The Role of Prior Knowledge and Perceived Realism." *Discourse Processes* 38 (2004): 247–266. Print.

Green, Melanie C. and Timothy C. Brock. "In the Mind's Eye: Imagery and Transportation into Narrative Worlds." *Narrative Impact: Social and Cognitive Foundations.* Ed. Melanie C. Jeffrey, J. Strange, and Timothy Brock. Mahwah, NJ: Lawrence Erlbaum, 2002. 315–341. Print.

——. "The Role of Transportation in the Persuasiveness of Public Narratives." *Journal of Personality and Social Psychology* 79 (2000): 701–721. Print.

Hulme, Mike. *Why We Disagree About Climate Change: Understanding Controversy, Inaction and Opportunity.* Cambridge: Cambridge University Press, 2009. Print.

Ivakhiv, Adrian. *Ecologies of the Moving Image: Cinema, Affect, Nature.* Waterloo: Wilfrid Laurier University Press, 2013. Print.

Keysers, Christian, Bruno Wicker, Valeria Gazzola, Jean-Luc Anton, Leonardo Fogassi, and Vittorio Gallese. "A Touching Sight: SII/PV Activation during the Observation and Experience of Touch." *Neuron* 42 (2004): 335–346. Print.

Leiserowitz, Anthony P. "Climate Change Risk Perception and Policy Preferences: The Role of Affect, Imagery and Values." *Climatic Change* 77 (2006): 45–72. Print.

Mazzocco, Phil J., Melanie C. Green, Joe A. Sasota, and Norman Jones. (2010). "This Story is Not for Everyone: Transportability and Narrative Persuasion." *Social Psychological and Personality Science* 1 (2010): 361–368. Sage Journals. Web. 1 August 2015.

Milton, Kay. *Loving Nature: Towards an Ecology of Emotion.* London and New York: Routledge, 2002. Print.

Norgaard, Kari Marie. *Living in Denial: Climate Change, Emotions, and Everyday Life.* Cambridge, MA: MIT Press, 2011. Print.

Plantinga, Carl. "The Scene of Empathy and the Human Face on Film." *Passionate Views: Film, Cognition, and Emotion.* Ed. Carl Plantinga and Greg M. Smith. Baltimore, MD: Johns Hopkins University Press, 1999. 239–255. Print.

"Reviews." *Chasing Ice.* Boulder, CO: Chasing Ice, LLC, 2015. Web. 9 December 2015.

Schiavio, Andrea. "Action, Enaction, Inter(en)action." *Empirical Musicology Review* 9.3–4 (2014): 254–262. Print.

Stadler, Jane. *Pulling Focus: Intersubjective Experience, Narrative Film, and Ethics.* New York: Continuum, 2008. Print.

Tan, Ed S. and Nico Frijda. "Sentiment in Film Viewing." *Passionate Views: Film, Cognition, and Emotion.* Ed. Carl Plantinga and Greg M. Smith. Baltimore, MD: Johns Hopkins University Press, 1999. 48–64. Print.

Thompson-Jones, Katherine J. "Sensing Motion in Movies." *Psychocinematics: Exploring Cognition at the Movies.* Ed. Arthur P. Shimamura. Oxford: Oxford University Press, 2013. 115–131. Print.

Willoquet-Maricondi, Paula. "Shifting Paradigms: From Environmentalist Films to Ecocinema." *Framing the World: Explorations in Ecocriticism and Film.* Ed. Paula Willoquet-Maricondi. Charlottesville: University of Virginia Press, 2010. 43–61. Print.

Zacks, Jeffrey M. *Flicker: Your Brain on Movies.* Oxford: Oxford University Press, 2015. Print.

34

CLIMATE, DESIGN, AND THE STATUS OF THE HUMAN

Obstacles and opportunities
for architectural scholarship in the
environmental humanities

Daniel A. Barber

Scholarship in architecture has a complex relationship with the environmental humanities. This has to do with three important ways that architectural history and theory have edged close to concerns now seen as central to the environmental humanities, but have not effectively engaged them: the capacity for humanistic scholarship to inflect scientific and professional practices; the focus on the entanglement of human and nonhuman systems and flows, and subsequent re-conceptualizations of the human; and the importance of imagery—literary, artistic, scientific, economic, and so on—as a site for communication about environmental histories and possible futures. In this chapter, I will first summarize some of the obstacles that this apparent closeness presents, and then describe a historical episode to indicate possible ways forward for an environmental humanities analytic within the field.

Obstacles

Architectural historical and theoretical research has long had the opportunity to cross divides between critique and action—scholarly work in architecture, taking place in professional schools and published, at times, in journals read by practitioners, has often been framed as a means to operate on the profession itself. Given that those professions concerned with designing the built world are also, in part, engaged in managing flows of materials and resources, this apparent closeness can be amenable to environmental discussions. However, while the architectural profession appears to be a fertile field for deploying environmentalist ideas, there are obstacles, rooted in disciplinary traditions, which complicate that prospect.

The framework of environmental humanities offers new opportunities for understanding relationships between architecture and life. This emergent perspective helps to clarify that architecture is not only about buildings as physical manifestations in the environment, but also about the generation of concepts and ideas—concepts that may be critical or progressive, future looking or conservative, but which elucidate the possibilities implicit in the design and production of the built landscape. This notion of architecture as a realm of ideas has been central to architectural scholarship for at least the past two decades (Colomina); my invocation of it here is not in itself innovative. However, the environmental humanities suggest that a critical analysis of design can focus not only on tangible objects, but also on the resonance of architectural concepts across a diverse array of concerns related to environmental change.

This resistance to buildings—or, better, a willingness to move beyond them in architectural-environmental scholarship—emerges from a number of insights. Foremost, economic and industrial pressures have, in general, pushed architects to ignore rather than integrate environmental externalities. As the Marxist architectural historian Manfredo Tafuri demonstrated in 1968, architects, and the scholarly discipline of architectural history, tend to be in the service of capitalist development even when they claim to be resistant or critical. Tafuri's polemic that such a condition made architecture irrelevant to the class struggle can be transposed, *in toto*, to struggles for the rights of nature and environmental justice. Given an economic system that exploits the environment to accumulate capital, the built objects that are produced through that system also exploit the environment. At the same time, such Marxian absolutes must be tempered with the potential for envisioning new futures, in which innovations in the built environment can play a substantive role. Inherited conceptions of architecture and its practices need to be reconsidered in order to reconceptualize relationships between the built world, humans, and natural processes.

Insofar as architects specify materials and work with the construction industry to produce buildings, there are manifest opportunities to minimize environmental damage. However, the momentum, to put it mildly, goes the other way. An ossified supply chain, a profit-driven and highly competitive construction and consulting industry, the complexities of corporate and contractor agreements, and industry or state regulations make most attempts toward sustainable building one of the worst kinds of green washing. Industry innovations such as wood products certified by the Sustainable Forestry Initiative, or, more generally, attention to environmental benchmark systems such as LEED or BREEAM, complicate but do not disrupt these dominant patterns.

Even more pernicious are the regulatory imperatives, softwares, and management systems that erode the possibility of creating models for new, and possibly more environmentally engaged, built worlds. Structural conditions render individual attempts at energy efficiency or sustainable materials generally ineffective. The well-established regulatory regime of ASHRAE (American Society of Heating, Refrigeration, and Air-Conditioning Engineers), which stipulates specific terms for the conditioning of air in buildings through the use of mechanical systems, is the most extreme. Since the late 1950s, when ASHRAE was formed as an organization with international ambitions to support the growth of the HVAC (Heating, Ventilation, and Air Conditioning) industry, the group has effectively promoted standards to the point that they are global and nearly universal. Compliance with ASHRAE is law in many countries, states, and municipalities. No matter how a building is designed, in other words, adherence to ASHRAE standards is generally stipulated, along with the fossil-fuel powered heating and cooling plants thereby required. Furthermore, ASHRAE standards have, over the decades, significantly increased expectations for the experience of interior climates, which has led to a need for increased capacity of mechanical conditioning systems. Even though such systems have become more efficient, and architects have developed

347

innovative design components that intensify this efficiency, the rise in expectations has far outpaced technological or design innovation. Because of this, a so-called energy-efficient building today is often using more throughput of fossil fuel energy than a building from a few decades ago (Oldfield et al.). Though these regulations have been the subject of some criticism, their global application in new construction has not been questioned in any substantive way.

Thus design of the built environment remains a site of intensely structured practice, one with little prospect for immediate transformation relative to environmental costs. If, on the one hand, as I have just summarized, profit motive and regulatory systems make real environmental adjustments difficult, on the other hand scholarly discussions and methodological innovations in the field have largely, until quite recently, been focused on new formal expressions rather than on creative approaches to exploring intersections between the design of the built environment and environmental or social change (Hays, *Architecture Theory*).

While these regulatory regimes and structural conditions limit the possibilities of environmental engagement in built work, debates internal to the discipline are perhaps more challenging for both scholars and practitioners. Following to some extent Tafuri's 1968 intervention—particularly his insistence that design only excelled when it recognized its "pure uselessness" relative to social struggles (iii)—the discursive parameters of architectural history and theory have for the past few decades focused on formal innovation. Though many architects have attempted to approach building projects with attention to potential environmental efficiencies, such potentials have, generally speaking, been secondary to innovations in formal arrangements and the affective conditions such new aesthetic possibilities are purported to allow (Cohen and Naginski). In recent years, the historical discussion of architecture has become increasingly diverse in geographic, technological, and theoretical terms, though the reification of architecture and of architects has, too often, persisted (Heynen and Vandevyvere). Debates in schools, journals, and to a great extent in the popular imagination have largely focused on these formal possibilities, even more so since digital tools have provided architects with a dizzying range of softwares and strategies. Nonformal attributes of a building, such as the systems and services that characterize a building's material and energy metabolism, have until very recently been relegated to engineering consultants rather than being envisioned as a site for creativity in the design process, despite many discursive interventions to the contrary (Banham). Environmental concerns have been lumped together with behavioral and, at times, economic constraints, scientist or technocratic determinants that limit the expressive capacity of the designer (Hays, *Oppositions Reader* xii; Frampton).

Despite these obstacles, many firms, schools, and critics have argued for a more robust environmentalist disposition in architectural scholarship and practice. Discussions of ecological urbanism (Mostafavi; Hagan), considerations of the posthuman (Ingraham; Grosz), and engagements with actor-network theory (Latour and Yaneva) have begun to productively complicate scholarly parameters and some theoretically informed practices. Digital tools allow for a multitude of means to break through the apparent formal-environmental divide (Benjamin). Symptomatically, however, most of these interventions are framed as resistance to the status quo. One of the more robust proposals in a large volume on *Ecological Urbanism* edited by Moshen Mostafavi, the Dean of the Harvard University Graduate School of Design, makes this clear: Mostafavi cautions that "we need to view the planet and its resources as an opportunity for speculative design innovations rather than as a form of technical legitimation for promoting conventional solutions" (17). Environmental engage-ment in architecture still requires theoretical and pragmatic frameworks that can subvert or disrupt conventional practices. Despite the closeness of architectural discourse to theory and speculation on the one hand, and the built environment on the other, a focus on environmental impacts remains elusive.

Opportunities

The environmental humanities offers a tantalizing framework through which architectural scholarship can look beyond these entangled professional and discursive norms and focus, instead, on the relationship of architectural ideas to social and environmental change. Such scholarship can examine how concepts developed in architectural discourse have engaged, reinforced, or subverted ideas about possible transformations to the environment and new ways of life. It is this interrogation of "life" that I find most promising in the inter-section between architectural history and the environmental humanities, as both a concept and a pragmatic intervention to assess how the conditions for human existence have been modeled and framed. Two central aspects of environmental humanities discourse—the entanglement of human and nonhuman systems and an interest in environmental imaginaries—provide openings for an architectural scholarship at once engaged in the history of the field and in the complex pressures of the present. New narratives of architec-ture's history, as much as technological innovations, can offer an opportunity to creatively reframe design practices.

An important clarification, relative to the discussion above, is that architectural practices and ideas have, in fact, long been engaged with the complexities of human and nonhuman relations and with outlining possible futures. Examples abound including Cedric Price's cybernetic-informed Potteries Thinkbelt project of 1966 proposed an early and creative brownfield redevelopment scheme (Aureli); the global planning projects of Constantin Doxiadis, while portending neoliberal management scenarios, nonetheless put forward a global ecological imaginary (Pyla); and a number of architects active during the oil crises of the 1970s saw the end of fossil fuels as an opportunity for design interventions (Borasi and Zardini). Indeed, many of the celebrated speculative practices of the post-World War II period have been engaged in or responded to emergent knowledge about the entanglement of biotic and human systems and warnings of environmental decline. Increased interest in environmental theories and concepts has revealed these previously unknown or suppressed histories, but many other examples could be listed. Historical scholarship is demonstrating the extent to which architects have already rethought the parameters for human agency in environmental change.

One promising direction for this emergent realm of interdisciplinary scholarship is the analysis of the history of architecture as, in part, a history of the environmental imaginary—an archive of drawings, photographs, and diagrams about experiments that have sought to redefine the relationship between social and natural systems. First principle of this ana-lytic framework: recognize that social and natural systems are inextricably entangled, to the extent that distinctions between them become meaningless (Morton). Second prin-ciple: make these entanglements more legible. In historical terms, a number of architectural experiments since World War II—one of which I discuss below—have in this sense elabo-rated on the cybernetic modernisms of systems theory, ecologies, and networks.

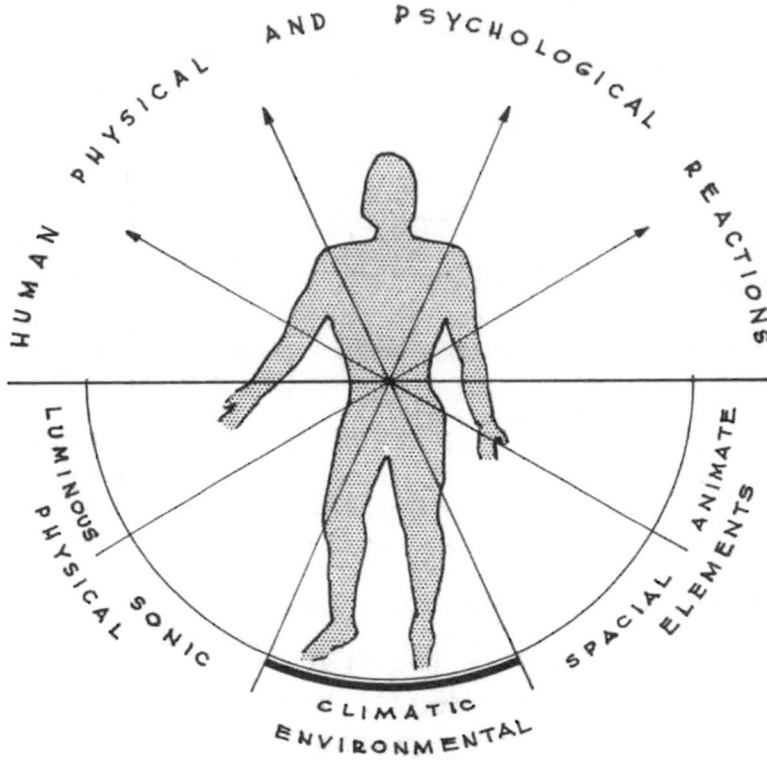

Figure 34.1 Victor Olgyay. "Man as the Central Measure in Architecture" from *Design with Climate*, 1963.

At stake, then, are design experiments that reconceive human and natural systems and human and nonhuman life. From the analytic perspective of the environmental humanities, the discursive resonance of architecture since World War II is less about the capacity to maximize resource or material efficiency, and more about encouraging awareness of the intermingling of social and biotic systems, articulating new means for living, and disseminating methods, ideas, and strategies on these terms (Barber, "World Solar Energy Project"; Gonzalez de Canales). Such concerns are often masked; again, that there are significant discursive challenges to analyzing architectural environmentalisms, in that the appearance of environmental interests has often been identified, by architects and critics, as a collapse into imperatives for technocratic efficiency rather than an opportunity for thinking about emergent conditions for life.

Case in point: the climate design methods developed in architectural discussions of the 1940s and 1950s. As part of the global expansion of architectural modernism in the postwar period, such methods borrowed scientific knowledge and applied technologies to reconsider building practices. In the period just before the regulatory imperatives of ASHRAE were widely used, architects around the world began to look at how to best correlate a building to its regional climatic conditions (Jackson; Chang and King). Exemplary here is the research of Hungarian émigré architects Victor and Aladar Olgyay, who helped to develop the Princeton Architectural Laboratory in the 1950s. At the Lab—itself a largely unknown site for the intensification of architectural research—the twin brothers developed a complex design method. Their project was aggressively instrumental, attempting to articulate new ways of building that sought to improve on received modernist tenets; my analysis of it here elides their largely frustrated attempt to improve a building's performance and focuses instead on conceptions of human and nonhuman life that were implicit in their proposals.

The Olgyays offer a fertile field for such an inquiry. In the plans, diagrams, and methods they developed, the figure of the human is central: literally, as it appears in the middle of images illustrating a climatic architectural approach; and conceptually, as the subject of the new possibilities that these architectures allow (Figure 34.1). The human is drawn amidst a range of factors—moral, historical, thermal, sonic, and spatial. The role of architecture, their diagrams proposed, was to filter these factors and align them with human needs through new design strategies. "Man," as the Olgyays wrote, "with his intimate physical and emotional needs, remains the module—the central measure—in all approaches. The success of every design must be measured by its total effect on the *human* environment" (*Solar Control* 5, emphasis in original). Such diagrams accompanied their architectural drawings and methodological proposals, serving both to illustrate the method and to foreground the goal of producing a universal comfort zone that they saw as most amenable to human habitation. The Olgyays here participated in the wide-ranging postwar interest in humanism, which was especially potent in architecture; their research is one of the more elaborate attempts, in the architecture of the period, to re-think the human in relationship to the environment.

The Olgyays' climate design method was elaborate (Figure 34.2). In brief summary, its basic premise was to collect climatic data, evaluate them, integrate them into new diagrammatic representations, and then use these diagrams as parameters for formal and material choices in the design process. The design proceeded in several phases. The first was to engage climatologists to gather data on the building site in question. The second phase involved evaluation, and led to the development of new kinds of representational tools such as a "bioclimatic charts" and a "timetable of climatic needs," which clarified the problem and the means they proposed in order to solve it (*Solar Control* 81ff). The third phase focused on calculation, with extensive data gathering and processing techniques relative to building orientation, wind, and the use of different materials. The fourth phase then explored the findings of the various calculations from phase three, compared them with each other, and evaluated the differences through a range of diagramming techniques. For example, the "timetable of climatic needs" was mapped on to orientation studies to determine the best building shape. Data-driven analysis yielded formal interpretations, sometimes suggesting balanced or harmonious patterns, sometimes reflecting specific shapes appropriate to the design process. Finally, the fifth and sixth phases led the architect closer to familiar territory, bringing all of these factors to bear on the specific design needs of a given project.

The point of the method was the design of a comfort zone—a careful correlation between building and climate so that interior conditions would be optimized for human habitation (*Climate Data* 4). To conceptualize this zone, the Olgyays relied on the work of the physiologist Douglas H. K. Lee. In the "Schematic Bioclimatic Index" it becomes clear that designing a building according to these parameters was, in fact, quite a lofty goal. Here again the human is figured in the center—relaxing on a modern chaise longue, smoking a pipe, reading the newspaper and completely at ease (Figure 34.3). Without irritation, without having to experience any of the possible climatic conditions that surround and threaten

351

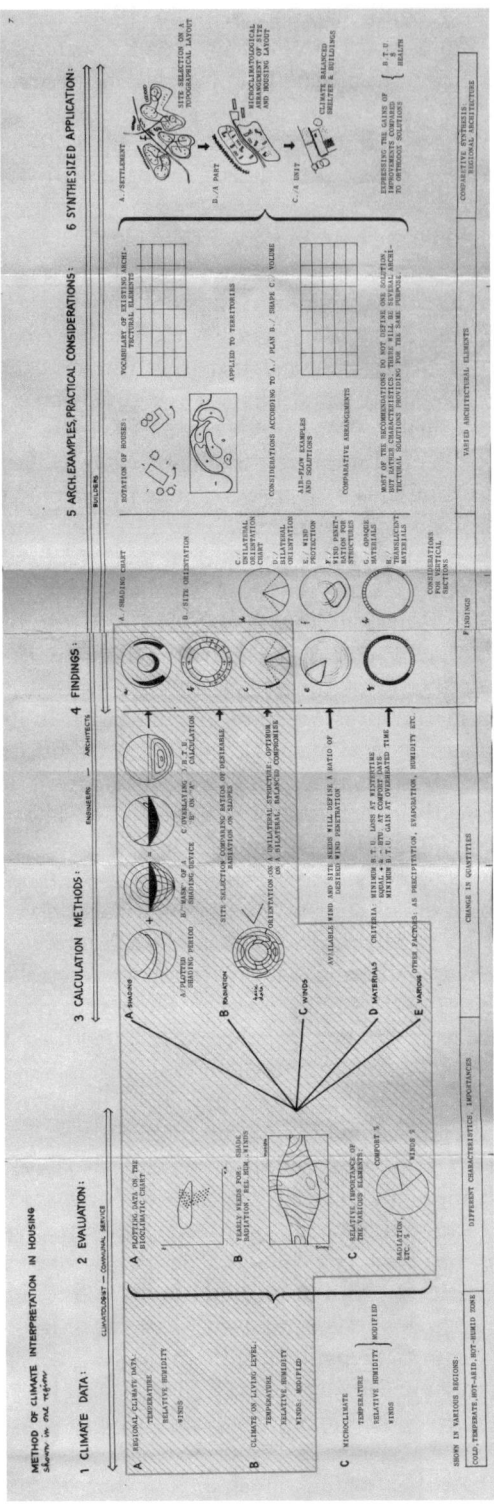

Figure 34.2 Aladar and Victor Olgyay, "Method of Climatic Interpretation in Housing" from *Solar Control and Shading Devices*, 1957.

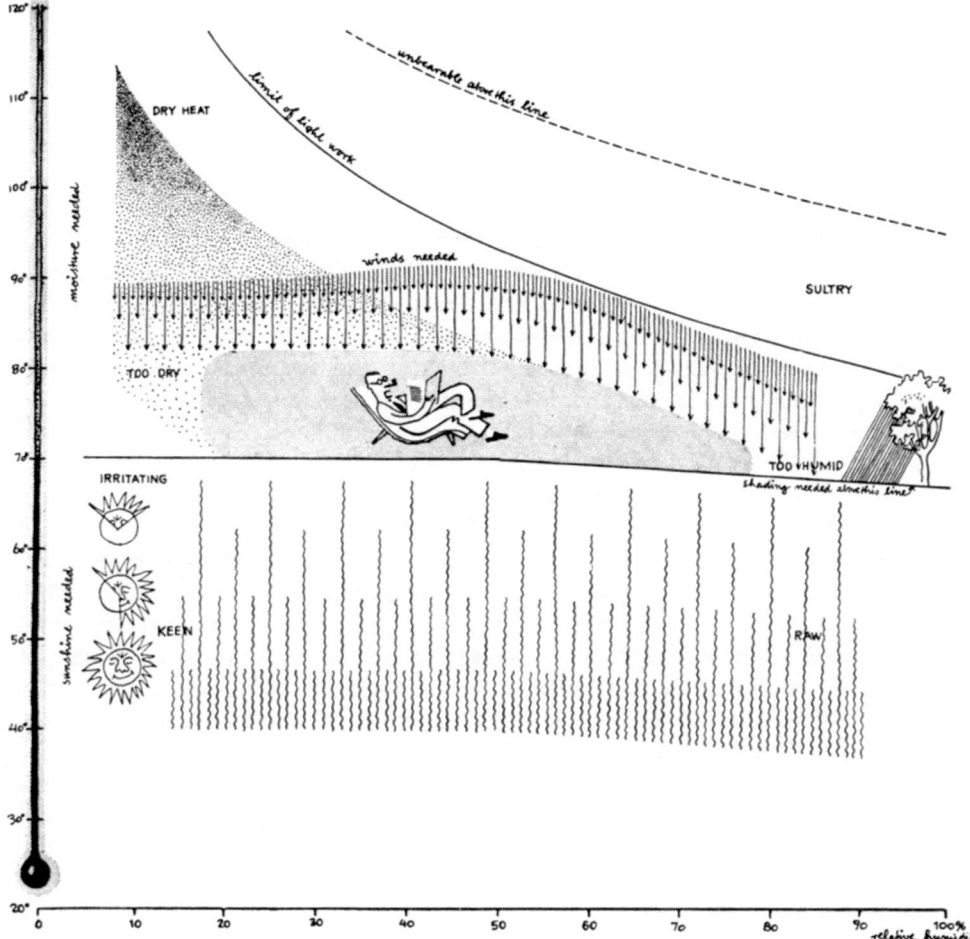

Figure 34.3 Aladar and Victor Olgyay, "Schematic Drawing of the Comfort Zone" from *Solar Control and Shading Devices*, 1957.

him, the human is imaged and imagined as a stable, protected figure. He—it is clearly a male figure—is solid in the experience of well-designed space, a consistent subject across changes in the elements that the past, present, or future can bring.

What can this image of stasis tell us about the status of the human in postwar architectural discussions? First of all, it becomes clear that these diagrams—expressive interpretations of technical knowledge—were aspirational, drawn in order to identify and celebrate the new possible conditions for the human that a climate-focused architecture could bring about. On the one hand, this static human figure was the correlate of a static climatic interior, which had long been the dream of modern architectural interventions. In 1930 the Swiss-French architect Le Corbusier famously predicted a time when "every building, around the globe, will be 18 degrees [celsius—about 65 Fahrenheit]" (66). On the other hand, the Olgyays' method is indicative of how difficult it was to achieve climatic consistency solely through design—that is, without mechanical conditioning systems, which, in the 1950s, were just beginning to be widely available and affordable. In these quasi-scientific diagrams, the image as scientific explication and the image as subjective desire were conflated, in the hope that a well-presented idea about how to build would lead to a well-developed notion of how to live.

The Olgyays' diagrams thus sit between the poles of affect and instrumentality. As methodological diagrams they activate a disciplinary agenda and make an appeal for architecture as a site of a possible re-alignment of relationships: between the interior of a building and its site; between the inhabitants of a building and the weather outside; and between climatic analysis and the forms, materials, and orientations that can lead to a balanced climatic condition. As an example of an emerging architectural–environmental imaginary, their images appealed to both the technological disposition and the aesthetic intentions of mid-century designers, encouraging them to realize the potential of architecture as a shelter, the provision of comfort that can, in turn, improve the lives of the humans who inhabit it. It is a utopian premise: an argument, in diagrammatic form, for containing the human and isolating the species from the unpredictability of the natural world.

Human life is, in the Olgyays' work, conceptualized as a normative state, though the biological and physiological assumptions on which the Olgyays relied were themselves the subject of much debate in the immediate postwar period. The physiologist Douglas H. K. Lee developed his research in the context of improving life conditions in so-called developing economies. In his work and elsewhere, the optimization of the designed interior went hand in hand with the integration of new subjects and economies into an increasingly managed global social order. The reification of the human as a stable subject was crucial to attendant conceptualizations of global flows—of money, ideas, wind currents, biota, and disease (Mitchell). Normativity, as Georges Canguilhem was arguing in this same period, has no concern for the future; irritation or disease, instead, is "the source of speculative attention which life attaches to life by means of man" (101). The Olgyays did not participate in these debates so much as they reinforced conclusions from them in an architectural context, establishing a constructed milieu in which this seemingly stable human subject could, it was proposed, most effectively operate.

The Olgyays' research had a decidedly mixed reception in postwar architectural culture: their method was taught in numerous schools and they were embraced by many prominent architects, but their influence was limited by concern that the technological determination of any aspect of building design frustrated architectural expression. They also operated, at Princeton, in the context of an emergent architectural postmodernism in which new forms would be derived from historical references rather than from innovations in the understanding of complex building functions. The Olgyays' focus on design research did, however, help to reposition architecture in relationship to other fields of knowledge, especially those sciences that were also engaged in climatic and environmental research. The physicist John von Neumann's Meteorology Project, exploring computational methods of weather prediction at the Princeton Institute for Advanced Study, just down the street from the Olgyays' own lab, was beginning to take into account the complexity of the climate system, and the difficulty in representing or modeling it (Nebeker 135ff; Edwards 113–115). Most of the Olgyays' research was funded by outside organizations such as the Rockefeller Foundation, the Ford Foundation, the National Academy of Sciences, and industrial organizations, establishing a model of the architectural scientist that has persisted in the academy to the present.

Related images of the static human subject were also developed in other fields, including the science of ecology. Ecological analyses were interested in tracing "man's" capacity to disturb an existing set of interconnected energetic pathways. Chains of energy as they flowed through the ecosystem were in part the result of human influence; at the same time the impacts of these interventions by "man as a manipulator" also affected humans in their animal state, according to their biological needs (Van Dyne 3). Ecology, as it was schematically figured, relied on a dual positioning of "man," now partially de-centered, and a new image of the species as both manipulated and manipulator. By the early 1960s, new research also began to elaborate alternatives to the premise that all ecosystems inherently worked toward a state of balance (Barbour). Instead of a progression to a climax state, such models conceived of a world in constant flux. The "world model" imaged by the "Group for the Study of the Predicament of Mankind," working with the Systems Dynamics Group at MIT in the 1960s, for example, presented a system of variable inputs and outputs, subject to bottlenecks and stoppages, with complex and ultimately unpredictable consequences (Meadows). This "limits to growth" model would come under much scrutiny, largely for its assumptions about the technological capacities of emerging economies (Taylor and Buttel). The figuration of the human both as a subject with agency in economic and environmental conditions and as an object of flows and conditions beyond its control, persisted. One partial consequence of the integration of "man" into the flows of the global ecological system was the simultaneous disappearance of the human figure as a central focus in the production of methodological images in architectural science, for better or worse.

The Olgyays' work is an important trace of the status of the human in architectural practices focused on the environment. Though the foundation of their research—its intent to use architectural design, rather than mechanical systems, to temper interior climates—was soon eroded by the increasing affordability of mechanical HVAC systems, their invocation of a stable human subject and of the comfort zone persisted as a premise in technological and aesthetic models of a building's performative capacities. Indeed, aspects of their method have since been integrated into software platforms that simulate a building's potential energy efficiency. In the Olgyays' work, the premise that the human is static and stable was accompanied by the assumption that the environment was available for manipulation through technology. Architecture continues to address the environment as a set of external factors that must be managed so that human life can persist unchanged (de Graaf; Barber, "Visualizing").

At the same time, the Olygyays' research helps open up architecture to analysis on humanistic rather than purely technical and instrumental terms (Hulme). Rather than recovering the Olgyays' research as an instigation to design interventions in the present, the stakes here are immaterial and cultural. Re-scripting architectural histories and theories through the concepts and methods of the environmental humanities opens up new perspectives on the resonance of architectural ideas. In other words, though the Olgyays' concepts may have been flawed, they were nonetheless engaged with a wide range of efforts to think differently about relationships between human and biotic systems. Their history suggests a narrative of the development of architectural ideas that includes the engagement of scientific knowledge, considerations of human and environmental well-being, and many other factors not yet integrated into familiar histories. The obstacles with which I began this chapter, including especially the primary emphasis on novel form, can begin to be dissolved as new architectural historical narratives, attendant to the entanglement of the field with other ideas and practices focused on the environment, emerge and accumulate. Today there is again an urgent need for architects to become image-makers, articulating new ways of situating the human in relation to the environment.

References

Aureli, Pier Vittorio. "Labor and Architecture: Revisiting Cedric Price's Potteries Thinkbelt." *Log* 23 (2011): 97–118. Print.

Banham, Reyner. *Architecture of the Well-Tempered Environment*. Chicago: University of Chicago Press, 1969. Print.

Barber, Daniel. "The World Solar Energy Project, ca. 1954." *Grey Room* 51 (2013): 64–93. Print.

——. "Visualizing Renewable Resources." *Architecture and Energy: Performance and Style*. Ed. William Braham and Dan Willis. New York: Routledge, 2013. 256–279. Print.

Barbour, M. G. "Ecological Fragmentation in the Fifties." *Uncommon Ground: Rethinking the Human Place in Nature*. Ed. William Cronon. New York: Norton, 1995. 233–256. Print.

Benjamin, David. "Experiments with Architectural Activities." *A+U: Architecture and Urbanism* 12.531 (2014): 6. Print.

Borasi, Giovanna and Mirko Zardini. *Sorry, Out of Gas: Architecture's Response to the 1973 Oil Crisis*. Montreal: Canadian Center for Architecture, 2008. Print.

Canguilhem, Georges. *The Normal and the Pathological*. 1966. New York: Zone Books, 1989. Print.

Chang, Jiat-Hwee and Anthony D. King. "Towards a Geneaology of Tropical Architecture: Historical Fragments of Power-Knowledge, Built Environment and Climate in the British Colonial Territories." *Singapore Journal of Tropical Geography* 32 (2011): 282–300. Print.

Cohen, Preston Scott and Erika Naginski, eds. *The Return of Nature: Sustaining Architecture in the Face of Sustainability*. New York: Routledge, 2014. Print.

Colomina, Beatriz. *Privacy and Publicity: Modern Architecture as Mass Media*. Cambridge, MA: MIT Press, 1996. Print.

de Graaf, Rainier. *Roadmap 2050: A Practical Guide to a Prosperous, Low-Carbon Future*. 2009. Web. 1 August 2015.

Edwards, Paul. *A Vast Machine: Computer Models, Climate Data, and the Politics of Global Warming*. Cambridge, MA: MIT Press, 2010.

Frampton, Kenneth. "On Reading Heidegger." *Oppositions* 4 (1974): 1–4. Print.

Gonzalez de Canales, Francisco. *Experiments with Life Itself: Radical Domestic Architectures between 1937 and 1959*. Barcelona: Actar, 2013. Print.

Grosz, Elizabeth. *Architecture from the Outside: Essays on Virtual and Real Space*. Cambridge, MA: MIT Press, 2001. Print.

Hagan, Susannah. *Ecological Urbanism: The Nature of the City*. New York: Routledge, 2015. Print.

Hays, Michael. "Introduction." *Architecture Theory Since 1968*. Cambridge, MA: MIT Press, 1998. Print.

——. "Introduction." *The Oppositions Reader*. Ed. Michael Hays. New York: Princeton Architectural Press, 1998. ix–xv. Print.

Heynen, Hilde and Han Vandevyvere. "Sustainable Development, Architecture, and Modernism: Aspects of an Ongoing Controversy." *Arts* 3 (2014): 350–366. Print.

Hulme, Mike. "Climate." *Environmental Humanities* 6 (2015): 175–178. Web. 1 August 2015.

Ingraham, Catherine. *Architecture, Animal, Human: The Asymmetrical Condition*. New York: Routledge, 2006. Print.

Jackson, Iain. *The Architecture of Edwin Maxwell Fry and Jane Drew: Twentieth Century Architecture, Pioneer Modernism, and the Tropics*. Burlington, VT: Ashgate, 2014. Print.

Latour, Bruno and Albena Yaneva. "Give Me a Gun and I Will Make All Buildings Move: An ANT's View of Architecture." *Explorations in Architecture: Teaching, Design, Research*. Ed. R. Geiser. Basel: Birkhauser, 2008. Print.

Le Corbusier. *Precisions on the Present State of Architecture and City Planning*. 1930. Cambridge, MA: MIT Press, 1986. Print.

Lee, Douglas H. K. *Physiological Objectives in Hot Weather Housing: An Introduction to the Principles of Hot Weather Housing Design*. Washington, DC: US Housing and Home Finance Agency, 1953. Print.

Meadows, Donatella et al. *The Limits to Growth: A Report for the Club of Rome's Project on the Predicament of Mankind.* New York: Universal, 1974. Print.

Mitchell, Timothy. "Fixing the Economy." *Cultural Studies* 12.1 (1998): 82–101. Print.

Morton, Timothy. "Architecture without Nature." *Tarp: Architecture Manual* 3 (2012): 20–25. Print

Mostafavi, Moshen. *Ecological Urbanism.* Baden, Switzerland: Lars Müller Publishers, 2010. Print.

Nebeker, Frederik. *Calculating the Weather: Meteorology in the 20th Century.* New York: Academic Press, 1995. Print.

Oldfield, Philip et al. "Five Energy Generations of Tall Buildings: An Historical Analysis of Energy Consumption in High-rise Buildings." *The Journal of Architecture* 14 (2009): 591–613. Print.

Olgyay, Aladar and Victor Olgyay. *Application of Climate Data to House Design.* Washington, DC: US Housing and Home Finance Agency, 1954. Print.

——. *Design with Climate: A Bioclimatic Approach to Architectural Regionalism.* Princeton, NJ: Princeton University Press, 1963. Print.

——. *Solar Control and Shading Devices.* New York: Reinhold, 1957. Print.

Pyla, Panayiota. "Planetary Home and Garden: Ekistics and Environmental Development Politics." *Grey Room* 36 (2009): 6–35. Print.

Tafuri, Manfredo. *Architecture and Utopia: Design and Capitalist Develoment.* 1968. Cambridge, MA: MIT Press, 1976. Print.

Taylor, Peter and Frederick Buttel. "How Do We Know We Have Global Environmental Problems? Science and the Globalization of Environmental Discourse." *Geoforum* 23.3 (1992): 405–416. Print.

Van Dyne, George M. *Ecosystems, Systems Ecology, and Systems Ecologists.* Oak Ridge, TN: Oak Ridge National Laboratory, 1966. Print.

35

CLIMATE VISUALIZATIONS
Making data experiential

Heather Houser

In environmental discourse, information visualizations can relay data and create critical flashpoints for establishing public opinion, corporate agendas, and policy positions.[1] Arctic ice melt and rebound, the rate of ocean acidification, correlations between temperature and carbon dioxide (CO_2) concentrations: these phenomena rarely move out of their localities into a discursive space unless they are quantified and visualized. These visualizations have become staples of environmental discourse, especially on climate change. However, in Johanna Drucker's assessment, "we seem ready and eager to suspend critical judgment in a rush to visualization" (126). There is no doubt that this rush is on. Visualization consulting firms and in-house visualizers abound, and their information graphics now make up a significant portion of Internet users' media consumption.

It is certainly easy to let quantitative graphics such as charts and colormaps stream past our eyes as we read a *Washington Post* report on deforestation or an article in *Nature Climate Change* on desertification. Yet environmental visualizations, especially those addressing climate change, cry out for humanistic interpretation because they are not realist translations of natural phenomena. Their representational features bear a great burden of signification, especially as the objects roam from their typical origins in specialized journals, to blogs and policy documents, and even into skeptics' arguments. The interpretive tools the humanities have honed are vital to getting beyond the perceived self-evidence, the transparency, of visualizations in climate discourse. These tools help us understand how quantitative objects that seem stripped of "semantic value" accrue meaning (Drucker 2). If environmental humanists typically ask, "How do cultural works help people *feel* the effects of climate change?," I ask here, "What does climate change feel like as data, and not only as the lived phenomena to which data point?"

With this motivating question, I investigate climate visualizations for the affective qualities of their representational conventions, the tension between realism and invention, and the interplay between abstract data and individuals' understanding. I argue that the aesthetic strategies of scientific and commercial visualizations make data experiential. They activate epistemological procedures, that is, thought processes that highlight how form produces knowledge, and thus foreground what obstructs and what aids understanding of climate data. A humanistic reading of visualizations underscores that epistemology is deeply aesthetic in environmental culture. The formal features of visualizations and the procedures those features set in motion shape knowledge production. Features such as sophistication, cartographic perspective, color, materialization, and genre allusions affect whether climate disturbance becomes palpable or remains in the realm of abstraction. They shape whether audiences apprehend climate change as masterable, indomitable, or a complex amalgam of the two. Whether climate visualizations emerge from scientific institutions or for-profit consultancies, they build worlds whose value inheres not so much in their proximity to quantifiable realities as in the emotional realities visualized data produce, realities that do not always align easily with actionable knowledge of climate change.

Models: between data and figuration

Humanities scholars tend to separate quantification and interpretation into discrete realms of inquiry (see Poovey for a historical account of this separation). Even if one accepts that data are "simple descriptors of phenomenal particulars" (Poovey 4), they are not impervious to figuration and interpretation. In fact, data and the visualizations that register data are rhetorical objects couched in meaning-laden imagery and language. Birgit Schneider's scholarship on environmental visualization reminds us of this in an echo of Drucker: despite the complexity of climate models, their visualizations are largely "looked *through*," rather than *at*, because "they are thought to be either self-evident or mere instruments for mediating ('illustrating') scientific results" (191–192). To treat them as transparent places their meaning solely within scientific discourse. "Suspend[ing] critical judgment" stymies climate communication because it impedes an understanding of how visualizations carry into the public sphere and inform environmental positions (Drucker 126).

Climate change is a particularly apt site for placing quantification and interpretation under the same analysis because models, climatology's primary research tool, muddle the lines separating these domains. What is more, their importance is incontestable because they are public stages on which data and figuration meet. Now, to say models stray into representation and interpretation does not mean they are wild guesses or culturally constructed fantasies. Nor am I espousing a version of the Strong Program within science studies, which holds that "[science] is primarily the congruence of a hypothesis or theory with social interests of members of a scientific community that determines its acceptance by that community" (Longino 11). Climate models are symbiotic with data but also have an inventive dimension. They propose relations—for example, between aerosols and temperature—given a set of factors that may include year, season, geography, and topography. The metric for assessing models is not truth versus falsehood, but whether they are efficacious or fail within specific scientific discourses and institutions. To advance research agendas, climate models use resources of invention such as scenarios and projections that are neither constructed myths nor reproduced realities.

Climate models first developed out of regional weather forecasting when computer processing power ballooned in the 1960s, and they are currently heuristics for understanding the whole Earth's climate system (see Edwards; Dessler and Parson; Weart). They simulate this multivariable system using core principles of physics, chemistry, biology, and the geosciences. Today's most discussed models account for changing anthropogenic forcings, those events or long-term patterns that alter the energy in the climate system. Forcings affect climate sensitivity, which is the change in global mean temperature in response to the doubling of CO_2 concentrations since the industrial revolution. General circulation models (GCMs) are the most important models of this type and are our focus here. Models are technological marvels and morasses: marvels for the staggering data, diverse disciplines, and stochastic mechanisms for which they account; morasses because they are contested reference points in climate policy debates.

Models cross over from number into figuration, from quantification into invention and interpretation, because they are built on scenarios and require tweaks that make them non-realist. I use "realism" as it is used in science studies to refer to "an *objective match* between, on the one hand, statements, beliefs, descriptions or models and, on the other hand, a fixed reality" (Herrnstein Smith 75). For sociologist Andrew Pickering, the question for classical realists is, "does scientific knowledge mirror, correspond to, represent truly, how the world really is?" (181). For a climate model to be realist in the sense of "matching" or "mirroring," it would have to capture accurately the functioning of all systems processes, from cloud formation and ice albedo effect, to jet stream and circulation of aerosols, at all places and points in time. Climate models swerve from realism not only because supercomputers are not up to that task but also, and more importantly, because GCMs incorporate parameters, "a kind of proxy—a stand-in for something that cannot be modeled directly but can still be estimated or at least guessed" (Edwards 338). As hybrids of scientific principles, observational data, and parameters, the worlds that models build are at least partially invented and yet remain generative for climate research and discourse.

Invention enters models through their projective function; they rely on scenarios to simulate futures. The Intergovernmental Panel on Climate Change (IPCC), which synthesizes thousands of climate change studies every six years, defines a climate projection as "the simulated response of the climate system to a scenario of future emission or concentration of greenhouse gases and aerosols, generally derived using climate models. Climate projections are distinguished from climate predictions by their dependence on the emission/concentration/ radiative forcing scenario used, which is in turn based on assumptions concerning, for example, future socioeconomic and technological developments that may or may not be realised" (Planton AIII-6). "Simulation," "scenario," "assumptions": this diction indicates that models straddle the line between invention and realism. Because they are not strictly predictive, models "are not assigned probabilities or other indicators of expectation" (Diffenbaugh and Field 486).[2] Given this, the goal is not mimesis but simulating sensitivity to design better research trajectories.

In sum, models are heuristics; they make knowledge move rather than fixing airtight facts about present and future states of the climate system. This heuristic function speaks to the fact that uncertainty is a certainty in scientific pursuits. Even if we acknowledge this, however, models, with all their uncertainties and use of scenarios, enter charged sociocultural and political discursive spaces. When models are visualized, their epistemologies take aesthetic forms that influence their sometimes contested authority and legitimacy. For these reasons, model visualizations, to which I turn now, are ripe sites for employing the instruments of humanist analysis to address how representational conventions manage information about climate disturbance to shape knowledge and the affects that attach to it.

Scientific visualizations: between mastery and humility

The complexity of such conventions—and their interaction—is on display in a visualization by the US National Oceanic and Atmospheric Administration's (NOAA) Geophysical Fluid Dynamics Laboratory (GFDL), located at Princeton University. GFDL is one of the world's premier climate research and modeling institutions. Aspiring to "understand global and regional climate change," the lab simulates temperature variation over land and sea through computational modeling, and its outputs have a prominent public face in the Internet age (Geophysical Fluid Dynamics Laboratory). On the organization's site, any web user can access the often captivating and perplexing model visualizations. These artifacts fulfill several functions: they deliver quantitative information about climate forcings; they register epistemological desiderata of the modeling enterprise, such as mastery; and their aesthetic features ignite emotions and thought procedures that make those desiderata difficult to attain.

"Surface Air Temperature Anomalies" (2005, hereafter "Anomalies") appears on GFDL's website as a still image, projecting out to 2100, and as a twenty-one-second animation; these visuals lack accompanying explanatory text. Climate models are highly specialized devices built on a substrate of algorithms, code, and physical concepts and laws, but here the viewer's interpretation rests on the visualization alone. The work by scientist Keith Dixon and visualizer Remik Ziemlinski estimates temperature variations out to the year 2100 based on a mean established between 1971 and 2000 (Figure 35.1). The visualization communicates temperature variation through an ensemble of components: a captivating colored map whose palette morphs over the runtime, a static legend linking temperature and color, a line graph charting the upswing in global average temperatures from 1971 to 2100, and another dynamic line graph indicating the latitudinal zones in which those variations occur.

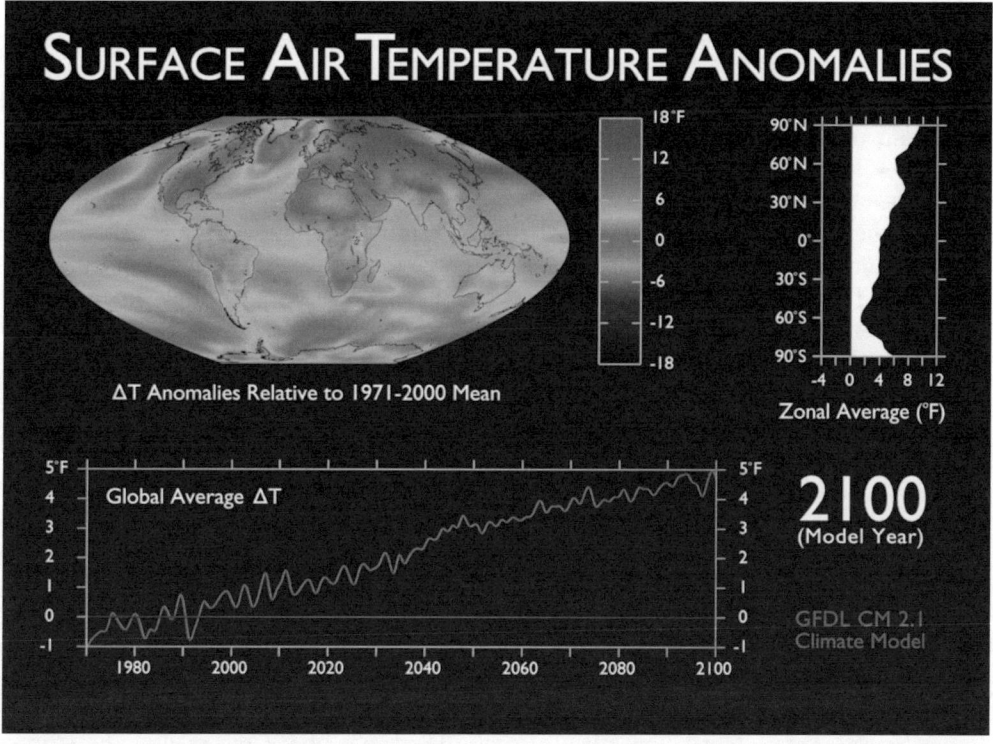

Figure 35.1 "Surface Air Temperature Anomalies." NOAA, GFDL. Scientist, Keith Dixon. Visualizer, Remik Ziemlinski (October 2005).

The elements position the viewer in one unit of measure—degrees Fahrenheit—but across the varied geographical scales of the globe and the region and the temporal scales of year, decade, and even second if we include the time slider on the media player.

Unlike a bald chart, line graph, or standalone map, "Anomalies" is sophisticated in its use of animation and multimodality. The visualization requires many more lines of code on top of the "more than a million lines of program code" that twenty-first-century climate models include (Edwards 146). Attending to its sophistication reminds us that Dixon and Ziemlinski's choice to animate the data across multiple components was just that, a choice and not a given. Climate scientist Mike Hulme explains that these elements bespeaking sophistication contribute to "the performative demonstration of [a model's] epistemic authority" (39). They convey the investments of money, infrastructure, and expertise that modeling requires and that keep the practice in the hands of wealthy nations such as the United States, Japan, and Russia. Sophistication thus suggests the authority of the data, research methods, and results that, for Hulme, is crucial to climate modeling's public viability. He emphasizes that "[c]limate models need to be 'seen' to be performing credibly and reliably. … [They] therefore need to inhabit public venues, displaying to all their epistemic claims of offering credible climate predictions" (33). Arguably, anyone who googles their way to the animation can sense sophistication even if they cannot evaluate the other metrics of credibility mentioned above. The visualization's sophistication also helps us forget that models cannot mirror the physical world and instead rely upon scenarios and parameters. This aesthetic feature thus gives the visualization a patina of objectivity, an epistemic value that, in Lorraine Daston and Peter Galison's account, presumes to "filter out the noise that undermines certainty … [and] aspire[s] to knowledge that bears no trace of the knower" (17).

"Anomalies" sets off epistemological procedures by which its components cohere and are meant to produce knowledge of and concern about temperature projections. Yet the work's aesthetic features draw attention to these same procedures and the very difficulty of comprehending climate data. That is, an impression of sophistication readily yields to an overwhelmed feeling, what I term "infowhelm," as you dwell with the visualization. Because of its multimodality, it is difficult to hold all the visualization's components in view simultaneously. The animation runs quickly, and the media players are not fine-tuned enough to allow easy pausing. It is therefore a challenge to toggle between map, line graphs, and legend. And this is just the first of many steps: a viewer must also check the time stamp in the lower corner while accounting for seasonal fluctuations displayed in the left-hand line graph, correlate the colors to the legend, and compare zonal averages. The piece thus figures infowhelm—data's power to overwhelm rather than enlighten—even as it attempts to wrangle proliferating climate data. Its interactivity imparts a feeling of impotence that matches the feeling the "raw" climate datascape produces. The viewer is not just a vessel into which information is poured; rather, features like multimodality make the data experiential. The aesthetic produces relays between epistemology and affect that mold how climate change potentially comes to matter for individuals.

In addition to sophistication and multimodality, color and aerial perspective activate feelings as they render the data. To illustrate this, let us pause at 2015. The average temperature in all zones is at or above the mean: at the equator, there is only a slight temperature anomaly; toward the North Pole (90°N), it is almost 4°F above the mean. The strip of gray girdling Earth's midline in the colored map signifies little change, and splashes of yellow-orange over Iceland indicate a 4–6°F change. Gray blanketing the oceans may mean that no observational inputs are available or that air temperatures are stable. In addition to toggling between many data elements, the viewer also compares the present back to 1971 and projects into a future—2100—likely outside their lifespan. The planet in the twenty-second century will be on average 5°F warmer, with the northern latitudes entering the "red zone" of a 10°F anomaly. Travel back to 2050, a year some viewers will live to see, and the dominant palette of the map morphs: solid orange and even red in the Arctic; orange over northern Africa, the Middle East, central Europe, and parts of the United States and Australia; and yellow covering much of the rest of the globe.

The specific color patches are key to how viewers experience the data. As the animation traverses the model period in one-year increments, it ranges from cerulean blue to orange-red. The symbolism of these colors has been naturalized. Blue for cold and red for heat are transparent because they seem intuitive and realistic, yet they deliver powerful feelings, as color schemes like the US Department of Homeland Security's now-defunct terror alert system attest (for the "hue-heat hypothesis," see Greene and Bell 949; on alternative color schemes for climate visualizations, see Samsel et al.). Red is not only hot in temperature but also in temperament, signaling danger and provoking agitation and reaction. By contrast, blue is pacifying. Color is, then, not just a practical device for differentiating temperatures, but an affective trigger.

Like the color scheme, the work's aerial perspective feels natural. Where color naturalizes climate phenomena while alarming and alerting viewers, aerial perspective inspires a sense of command even as it reveals human limitations. As scholars of cartography attest, the use of the aerial view in the West has a history that its ubiquity occludes (see Cosgrove). The aerial hails audiences as those privileged to know it all. If "[c]limate simulations are based on the assumption that nature can be quantified" (Lahsen 899), the aerial is based on the assumption that Earth can be laid out for consumption. The planet becomes apprehensible at a glance. But just as uncertainty, error, and incompleteness accompany quantification in modeling, imprecision travels with the aerial. Parts of the globe necessarily distend in a two-dimensional view from above even as they become visible. Shapes distort at the lateral edges, and the heart of the Arctic and Antarctica, regions most endangered by climate change, are not in view. The God's-eye view veers away from mastery of the global space even as it strives for it.

Such unavoidable disjunctions surface when we pause to examine the emotional and epistemological effects of the aesthetic choices that underlie "Anomalies." Strategies of sophistication, multimodality, color, and aerial perspective make data experiential, compelling the viewer to toggle between mastery and humility, control and failure. Especially in combination, the aesthetic strategies activate thought processes without which the components do not cohere, and these processes shape viewers' affects with regard to the information. These affectively charged epistemological procedures give data an emotional reality distinct from classical scientific realism. On the one hand, the work's multimodality, color scheme, and aerial view suggest mastery over the climate datascape. The visualization depicts the latest climatological research the model aggregates, conveying trends and projections in temperature variation but also, as Edwards instructs, developing "new ways of thinking globally" (xx). On the other hand, the very representational features that produce this knowledge emphasize our failure to totalize and display climate data and environmental change more broadly. The information situation, like that of the imperiled planet, cannot be mastered. Visualizations thus unexpectedly record the elusiveness of mastery, and this can lead to a productive inferiority and humility that counters anthropocentrism and environmental domination (see Hastrup 6). Or perhaps a productive perplexity in Latour's sense of a way in which "the collective makes itself attentive and sensitive to" a world not circumscribed by normalized facts (246). Even so, perplexity is not terribly comforting, especially in the face of an existential and physical threat like climate change. As the animation runs into the mid-twenty-first century and yellow, orange, and red splatter most of the globe, few refuges emerge. If the modeling enterprise is always rife with contradictions and tensions because models necessarily swerve from realism, visualizations are rife with contradictions and tensions because their aesthetic features produce emotions that may redirect knowledge away from engagement. When a sense of mastery and control gives way to humility, defeat may ensue. The experience of the overwhelming here does not lead to a triumphant recompense as in the tradition of the sublime. Humanity's dazzling technological acumen does not elevate the human in the face of the geophysical.

Carbon consulting: making data matter

High-powered research centers like GFDL are not the only ones designing visualizations of climate crisis. In fact, such visualizations have become a profitable enterprise for private sector environmental consultancies specializing in infographics. Clients for visualization services range from universities and municipalities, to hydrocarbon behemoth BP, to upstart and established environmental groups like 350.org and WWF. The success of private climate visualization firms suggests that the objects research labs like GFDL produce do not meet all market and policy needs because they aim at scientific communication rather than branding, public relations, or institutional sustainability plans.

"Making the invisible *visible*" is the key task of firms like UK-based Carbon Visuals ("Home"; original emphasis). Whereas the GFDL visualization makes data experiential through the representational conventions that hold mastery and humility in tension, Carbon Visuals' pieces make data experiential by giving data virtual materiality and employing cinematic genre conventions. These commercial products deemphasize knowledge of scientific processes in favor of overt emotional appeals. In imagining what it might be like to be surrounded by solid, rather than gaseous, CO_2, Carbon Visuals paradoxically creates a world in which the human endangerment that clients hope to mitigate has already come to pass.

Carbon Visuals "creat[es] scientifically accurate volumetric images that help audiences make sense of data. We call this 'concrete visualisation'—an approach that provides quantitative insight physically" ("What We Do"). In other words, the firm aims to produce insight about climate change by making data weighty.[3] Partnering with Environmental Defense Fund (EDF), Carbon Visuals created a prize-winning suite of images and animations that depict New York City's 2010 emissions (Figure 35.2). To "make sense of data," they convert statistics about CO_2 emissions per second into a volume of the gas. The data itself does not come under scrutiny in these commercial visualizations. Carbon Visuals cites an online encyclopedia from corporation Air Liquide when providing information on CO_2 mass, but climate science's complexities lie outside the visualizers' purview. Their energies go explicitly into inventing a world to harbor the data.

In the animation, spheres representing—or "concretizing"—carbon bounce through New York's streets to ambient car noise and murmuring voices (Figure 35.2). A charcoal-colored title screen that evokes carbon relates data on the city's emissions: "In 2010 New York City

364

Figure 35.2 Carbon Visuals, "New York City's Carbon Emissions," Flickr.com album (2012). Creative Commons license 2.0: https://creativecommons.org/licenses/by/2.0/

added over 54 million metric tons of carbon dioxide to the atmosphere[.] That's nearly 2 tons every second" (Carbon Visuals "New York's"). At every second of the animation, a translucent blue sphere materializes in a schematized New York City thoroughfare, sometimes popping out of a taxi's exhaust pipe. The statistic ceases to be an airy abstraction and instead disrupts the urban landscape. Carbon Visuals fills in this landscape with buildings resembling clay models. The strategy of using these generic icons rather than a photographic view has divergent effects: for non-New Yorkers, the lack of specificity allows them to transpose the data onto their own streets. Alternatively, it could de-concretize, render abstract, the very data Carbon Visuals wants to make palpable. Though the animation eventually gives an aerial view of these buildings, it begins at street level. This perspective contrasts with the bird's-eye view common to climate model visualizations and corresponds to the piece's local purview.

To make environmental change matter, Carbon Visuals figures data *as* matter. However, though this strategy gives carbon statistics volume, it does not explain why the numbers should concern viewers. For that information, a user must read the city's report on emissions, which is not available on the YouTube or Flickr sites that host the visualizations, and which includes only one sentence on the impacts of rising CO_2 (see City of New York). Instead of contextualizing the data, Carbon Visuals gives data physicality within the domain of the digital. This is more apparent when the street-level animation stops, and the next frames show hourly, daily, and annual emissions. When annual totals for 2010 reach 54,349,650 metric tons, the spheres representing CO_2 have consumed lower Manhattan. After only one day, the Empire State Building is nearly submerged in carbon molecules (Figure 35.3).

In this sequence of frames, Carbon Visuals trades on conventions of horror and disaster films. Undergoing mutant metastasis, carbon spheres suffocate the city's corridors. Though the visualization does not detail the health effects of rising CO_2, viewers sense that their own bodies are at risk from this accumulation. The orbs come to resemble a cluster of insects

Figure 35.3 Carbon Visuals, Screenshot of "New York City's greenhouse gas emissions as one-tonne spheres," Flickr.com album (2012). Creative Commons license 2.0: https://creativecommons.org/licenses/by/2.0/

or an alien invasion once they engulf skyscrapers and thus evoke action blockbusters like Roland Emmerich's *Independence Day* (1996) and *The Day after Tomorrow* (2004), in which the destruction of national monuments symbolizes planetary endangerment. The difference from these films, however, is that this imagined New York is already depopulated. EDF advocates staunching the flow of carbon dioxide into the atmosphere and the text supplementing the visualization emphasizes New York's successes thus far, but the image presents a world without us that suggests mass extinction is a fait accompli.

As the visualization materializes data through conventions of horror and the thriller, it figures data as matter that first clogs our streets, then overpowers our bodies and crowds us out. Through this strategy, the visualization indicates that the data of climate change engulfs us. The information the object is designed to manage still overcomes viewers in the end. While relaying quantitative information about climate crisis, the visualization ultimately privileges affective ways of knowing, suggesting that the feelings the piece produces are tantamount to—or even more important than—grasping the science of how the effluvia of contemporary life could suffocate and depopulate a metropolis. If GFDL's "Anomalies" was overwhelming because of its complex multimodal composition, cartographic perspectives, and color scheme that envision a future from which there is no escape, Carbon Visuals' animation overwhelms by showing that we humans have already succumbed to our own emissions. Employing strikingly different representational strategies and emerging from institutions with varying intentions and investments in the particulars of climate research, both pieces indicate that the information that visualization wrangles exceeds our mastery. They do so through the very aesthetic features designed to manage unruly data.

Multimodality and materiality, views from above and from the street, naturalized color and conventionalized horror: these representational features reveal the affectivity of climate data. What is more, they make it clear just why we need a *"humanistic approach"* to visualizations that quantitative disciplines produce, an approach "rooted in the recognition of the *interpretative* nature of knowledge" (Drucker 128). Artifacts such as climate visualizations are not self-evident translations of scientific data, nor are they "mere" cultural constructions. Their aesthetics make data experiential, which generates a knowledge that reveals as much about the emotional trajectories of form as it does about the climate phenomena under examination. This affective knowledge is one for which the humanities, with their attention to how representation makes data move, are a necessary voice in dialogue with quantification.

Notes

1 For this chapter, I define visualizations as graphical depictions of abstract data aimed at knowledge production and understanding.
2 That said, the language of prediction frequently appears in modeling scholarship. On the heuristic versus predictive functions of models, see Hulme.
3 "Insight" is a term in visualization practice referring to "the act or outcome of grasping the inward or hidden nature of things or of perceiving in an intuitive manner" (Bertini and Lalanne 13).

References

Bertini, Enrico and Denis Lalanne. "Surveying the Complementary Role of Automatic Data Analysis and Visualization in Knowledge Discovery." *Proceedings of VAKCD'09, Paris, 28 June 2009*. New York: ACM, 2009. 12–20. Print.

Carbon Visuals. "Home." *Carbonvisuals.com*. 2015. Web. 3 July 2015.

———. "New York's Carbon Emissions – in Real Time." *Carbonvisuals.com*. 2012. Web. 3 July 2015.

———. "What We Do." *Carbonvisuals.com*. 2015. Web. 3 July 2015.

City of New York. *Inventory of New York City Greenhouse Gas Emissions, September 2011*. New York: Mayor's Office of Long-Term Planning and Sustainability, 2011. Print.

Cosgrove, Denis. *Apollo's Eye: A Cartographic Genealogy of the Earth in the Western Imagination*. Baltimore: Johns Hopkins University Press, 2001. Print.

Daston, Lorraine and Peter Galison. *Objectivity*. 2nd ed. New York: Zone, 2010. Print.

Dessler, Andrew E. and Edward A. Parson. *The Science and Politics of Global Climate Change: A Guide to the Debate*. New York: Cambridge University Press, 2010. Print.

Diffenbaugh, Noah S. and Christopher B. Field. "Changes in Ecologically Critical Terrestrial Climate Conditions." *Science* 341 (2013): 486–492. Print.

Drucker, Johanna. *Graphesis: Visual Forms of Knowledge Production*. Cambridge, MA: Harvard University Press, 2014. Print.

Edwards, Paul N. *A Vast Machine: Computer Models, Climate Data, and the Politics of Global Warming*. Cambridge, MA: MIT Press, 2010. Print.

Geophysical Fluid Dynamics Laboratory. "Welcome." National Oceanic and Atmospheric Administration. n.d. Web. 1 August 2015.

Greene, Thomas C. and Paul A. Bell. "Additional Considerations Considering the Effects of 'Warm' and 'Cool' Wall Colours on Energy Conservation." *Ergonomics* (1980): 949–954. Print.

Hastrup, Kirsten. "Anticipating Nature: The Productive Uncertainty of Climate Models." *The Social Life of Climate Change Models: Anticipating Nature*. Eds. Kirsten Hastrup and Martin Skrydstrup. New York: Routledge, 2013. 1–29. Print.

Herrnstein Smith, Barbara. *Scandalous Knowledge: Science, Truth, and the Human*. Durham, NC: Duke University Press, 2005. Print.

Hulme, Mike. "How Climate Models Gain and Exercise Authority." *The Social Life of Climate Change Models: Anticipating Nature*. Ed. Kirsten Hastrup and Martin Skrydstrup. New York: Routledge, 2013. 30–44. Print.

Lahsen, Myanna. "Seductive Simulations?: Uncertainty Distribution around Climate Models." *Social Studies of Science* 35 (2005): 895–922. Print.

Latour, Bruno. *Politics of Nature: How to Bring the Sciences into Democracy*. Trans. Catherine Porter. Cambridge, MA: Harvard University Press, 2004. Print.

Longino, Helen E. *The Fate of Knowledge*. Princeton: Princeton University Press, 2002. Print.

Pickering, Andrew. *The Mangle of Practice: Time, Agency, and Science*. Chicago: University of Chicago Press, 1995. Print.

Planton, Serge, ed. "Annex III: Glossary." *Climate Change 2013: The Physical Science Basis. Contribution of Working Group I to the Fifth Assessment Report of the Intergovernmental Panel on Climate Change*. Ed. T. F. Stocker et al. New York: Cambridge University Press, 2013. Print.

Poovey, Mary. *A History of the Modern Fact: Problems of Knowledge in the Sciences of Wealth and Society*. Chicago, IL: University of Chicago Press, 1998. Print.

Samsel, Francesca et al. "Colormaps That Improve Perception of High-Resolution Ocean Data." *Proceedings of the 33rd Annual ACM Conference Extended Abstracts on Human Factors in Computing Systems, Seoul, 18–23 Apr. 2015*. New York: ACM, 2015. 703–710. Print.

Schneider, Birgit. "Image Politics: Picturing Uncertainty. The Role of Images in Climatology and Climate Policy." *Climate Change and Policy: The Calculability of Climate Change and the Challenge of Uncertainty*. Ed. Gabriele Gramelsberger and Johann Feichter. Berlin: Springer, 2011. 191–209. Print.

Weart, Spencer R. "The Development of General Circulation Models of Climate." *Studies in History and Philosophy of Science Part B* 41.3 (2010): 208–217. Print.

36

DIGITAL?
ENVIRONMENTAL:
HUMANITIES

Stéfan Sinclair and Stephanie Posthumus

We have deliberately chosen a title for this chapter that is at once understandable and defamiliarizing. If the words are relatively recognizable both individually and in combinations, the punctuation may seem a bit more arbitrary. Readers with some knowledge of computer programming may recognize this syntax as a ternary operator, a very compact way of formulating a conditional expression. Taken literally, at least by a computer, the title says something like "if *digital* is true then consider *environmental*, otherwise consider *humanities*." Except perhaps to the most creative and poetic minds, this statement may seem nonsensical (what programmers might call syntactically valid and semantically garbage), but we keep it here as it has the merit of drawing attention to the atomistic parts and prompts us to reflect on the various logical relationships that might link them. To really understand (or invalidate) the formulation we presumably need a solid understanding of what is the digital, what is environmental, and what are the humanities, terms that are of course complex and contextual from the start, and only grow in complexity as they are combined.

We can go beyond the individual word unit to consider compound forms like "digital humanities" and "environmental humanities," two academic fields where introspective defining of boundaries and identities is characteristic of the scholarly literature. Indeed, this Routledge volume can be considered as exemplary of the effort to define the environmental humanities and its constitutive parts, an ongoing negotiation about such things as agents (human, animal, and nature), scale (local, national, and global), time (past, present, and future), priorities (survival, justice, and conservation), and forms of communication (film, literature, code, and math). The interest in approaches that combine multiple humanities disciplines is more recent for the environmental humanities than for the digital humanities (see Figure 36.1), though of course fields such as environmental history and ecocriticism have a history reaching back to the 1980s and the 1990s, respectively.

Humanities Computing, the term used for Digital Humanities prior to about 2000, has a much longer history as a metadiscipline, reaching back to the middle of the last century (see Hockey for a brief history); we see it register in the Google Books corpus a bit later with the establishment of the journal *Computers and the Humanities* in 1966. It may seem a bit curious to have conferences and publications that combine history, literature, music, and many other disciplines, but for many years Humanities Computing represented a kind of refuge for

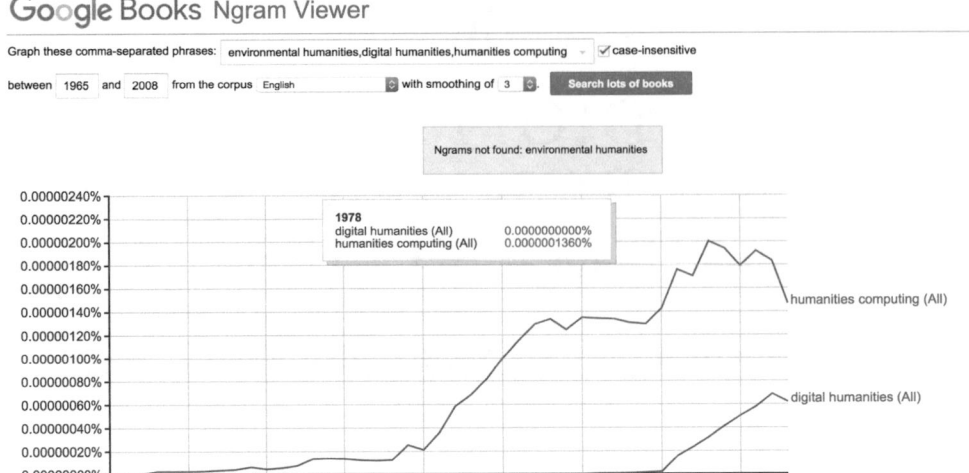

Figure 36.1 A Google Ngram Viewer graph showing occurrences of "humanities computing," "digital humanities" and (no) "environmental humanities" in the Google Books English corpus up to 2008 (the graph can be reproduced and modified at bit.ly/1NXIUKA). We are indebted to Jon Christensen for showing a variant of this graph at the Digital Environmental Humanities Workshop at McGill University in September 2013.

scholars who felt largely marginalized from the traditional approaches of their disciplines. Moreover, Digital Humanities continues to provide a shared space for exploring methods and techniques that can apply to the messy and interpretive enterprise of studying cultural texts. In other words, Digital Humanities has partly consolidated around practical and technical questions of how to study objects of interest rather than what objects to study and what can be said about them specifically. This may contrast with the environmental humanities, where the subject matter is the nexus: the environment as a qualifier and constraint on the much larger set of concerns in the humanities.

The Digital Humanities arguably lack thematic cohesion—the humanities are simply too expansive to be useful as an intellectual rallying point. The Environmental Humanities are thematically more coherent but perhaps lack methodological cohesion—how does one *do* environmental humanities? Perhaps more than anything else, this chapter is a proposal that the Digital Environmental Humanities potentially represent a sweet spot between conceptual and methodological concerns (Figure 36.2).

We recognize that a diagram can simplify and reduce real complexities and challenges (in fact, that may be the principal vocation of diagrams), so in what follows we will provide more details on some of the developments we see happening at the confluence of the digital, the environmental, and the humanities. We are structuring these observations into four main branches (with multiple offshoots), though of course there are many other ways of structuring these phenomena. We are interested in part in tracing continuities between historical context, or what might be called pre-DEH, and more recent activities.

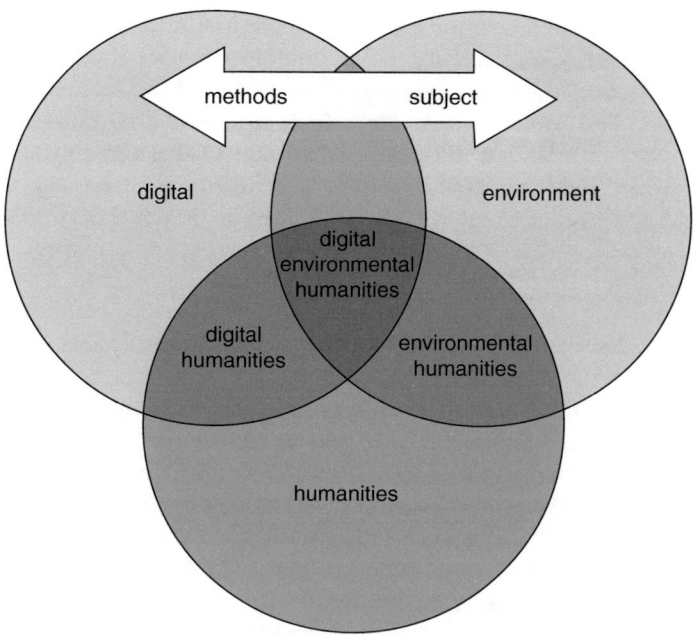

Figure 36.2 A schematic representation of the Digital Environmental Humanities (DEH).

Critical perspectives on technologically mediated experiences of nature

We find critical reflections on the impacts of technologies on human existence nearly throughout recorded cultural history. For instance, in Plato's *Phaedrus*, Socrates relates the story of how Thamus gently reprimanded Theuth about his enthusiasm for the new technology of writing, which would not lead to more knowledge as Theuth the inventor believed, but rather to the mere semblance of knowledge, and indeed to more forget-fulness (274c–275e). Modern echoes of this debate can be found in Nicholas Carr's "Is Google Making Us Stupid?" (2008) and Chad Wellmon's response (2012). Writing, even in nondigital forms, is certainly significant for representing real and imagined natures, transporting the reader to experience Walden Pond or Hogwart's Forbidden Forest. Similar phenomena are at play in other media (though each also has its particularities), such as photography, film, animation, and video games; see for example Alexa Weik von Mossner's contribution about film in this collection, Ursula K. Heise's "Plasmatic Nature" on anime (2014), Alexander Wilson's *The Culture of Nature* on documentaries (1991), and Melissa Bianchia's work on digital game ecologies (2014).

A common thread throughout these works, and the many more of which they are only a sample, is the ways in which technologies, both analog and digital, have led to seeing and experiencing nature differently than would be otherwise possible. This ranges from enhanc-ing a real experience (glasses, telescopes, microscopes), to capturing and representing nature (painting, film), to simulating imaginary natures (animation, video games). DEH needs to be constantly aware of the mediating effects and potential of technologies on perceptions of and access to nature, foregrounding the technologies not for their own sake, but so as to notice their effects on how we understand ourselves and our surroundings. Finn Arne Jørgensen's article "The Armchair Traveler's Guide to Digital Environmental Humanities" (2014) is exemplary of this, drawing parallels between Wolfgang Schivelbusch's work on how railroads shaped perceptions of landscape and how new media can do the same for an armchair nature traveller, combining texts, images, maps, GPS data, video, sound, and social media.

Environmental impacts of digital technologies

This second branch of DEH traces the multiple material ways in which the use of digital technologies affects the environment. This branch has a long history if we consider more generally the ways in which technologies have transformed the state of the planet. Paul Crutzen has popularized the term "Anthropocene" to describe the extent to which human activity has affected Earth systems since the invention of the steam engine and the advent of the industrial age. Humans have of course been modifying their environment since long before the 1800s, but the intensity of this modification has increased exponentially since the 1950s, in what has been called the Great Acceleration.

It is no surprise that digital technologies have played an important role in increasing energy demands in the West. According to a recent study, the digital economy uses a tenth of the world's electricity (Walsh). The problem with the information communications technologies (ICT) ecosystem is that it is becoming more and more extensive. Where a typical middle-class North American family may have had one desktop computer in the 1990s, they now have several tablets, smartphones, and laptops—not to mention the ever-increasing demand for cellphones around the world. Even if a company like Apple claims that its data centers are run one hundred percent on renewable energy sources, the company continues to produce an increasing number of gadgets for consumption, each year releasing new versions of the iPhone, the iPad, or new objects like the Apple Watch.

Critiquing the environmental impacts of digital technologies such as the ecological footprint of cloud computing, servers, and e-waste is an important aspect of DEH. A recent example of this work is Allison Carruth's "The Digital Cloud and the Micropolitics of Energy" (2014). This DEH branch also examines and exposes social justice issues that arise in relation to digital technologies. Beyond the obvious examples of labor conditions in electronics factories in developing countries, there is the problem of phenomena such as crowdsourcing which appeal to a large community of internet users to accomplish repetitive or tedious tasks—for example, Mechanical Turk. (The digital humanities have used crowdsourcing in big data projects where transcription, correction, tagging, and classification of digitized content is an essential step before analysis can begin; see Carletti et al. for example.) While crowdsourcing may help social justice organizations raise funds for new projects, it is plagued by the same problems as other forms of outsourced work that reduce the market value of professional services.

The environmental humanities have much to contribute to the analysis of social and environmental justice issues related to the creation, use, and disposal of digital technologies. While this is a relatively recent development, it is one to which digital humanists are also committed. In her plenary talk "Digital Humanities in the Anthropocene" (2014),

Bethany Nowviskie underscored the need to "attend to the environmental and human costs of DH—from our complicity with device manufacturers and social media manipulators, to the carbon footprint and price tag of conferences like this—and ask ourselves seriously what we might change, or grow to be." It is these kinds of self-critical questions about our use of digital technologies, within the environmental humanities as much as within the digital humanities, that characterize this developing branch of DEH.

Digital technologies and activating publics

Whereas the first two branches we described focus on critical perspectives on technologies, this branch is more interested in experimenting with digital technologies to reach broader publics and make arguments that are unconventional in scholarly terms, though of course critical perspectives and technological experimentation are not mutually exclusive. Characteristic of this branch is a motivation to raise public awareness about environmental issues and in some cases to mobilize action.

For instance, Maya Lin's artistic web project "What is Missing?" experiments with digital media to represent and explore species extinction. Lin's work affords the user the opportunity to navigate interactively between local and global contexts, between past, present, and speculative future, between personal and curated documentation, and between different media. "What is Missing?" provides information to be explored but also invites user contributions ("Add a Memory") and solicits action ("What You Can Do"). The site clearly has an activist intent for its use of technology: "by creating innovative artworks that utilize sound, media, and science, people will connect with both the species and places that have disappeared or will most likely disappear if we do not act to protect them" ("About the Project"). From coverage in CNN and *The New York Times* to the *Huffington Post* and *Fast Company*, the project, sponsored by Bloomberg, has succeeded in garnering media attention. It has also inspired other projects, such as a group of students working in a class with co-author Sinclair on the project Losing52.org, with an attempt to reach local university-aged students with additional infographics and videos about species extinction.

Another example in this vein is a project led by Jon Christensen called *City Nature* which compiles and visualizes green spaces in about three dozen cities in the United States; the availability of data in different regions of the world is of course an important factor in determining what can be done. The project has a research agenda, but it also seeks to inform activism and policy, as Christensen says: "The hope [...] is that urban planners and activists can use *City Nature*'s data to eventually pinpoint the neighborhoods that have the greatest 'nature need'—and take action" (qtd in Lalasz 2013). In some ways, DEH is uniquely positioned to leverage digital technologies to give more visibility and prominence to the humanities: just as the 4Humanities group pursues advocacy of the humanities by harnessing the digital humanities, this branch of DEH leverages digital technologies to reach broader publics.

Digital scholarly tools, environmental content

This final branch of DEH represents the closest form of collaboration between the digital humanities and the environmental humanities as academic disciplines. It involves applying and creating tools from the digital humanities to analyze the environmental humanities, on the one hand, and developing new networks, portals, and curated interactive objects to disseminate research in the environmental humanities, on the other. In terms of this second type of use of digital scholarly tools, environmental history has been a leader. The Rachel Carson Center in Munich has been developing the *Environment & Society Portal* that includes many examples of digital scholarly tools (timelines, curated exhibits, etc.) to bring research in environmental history to a wider public. The Portal's collectively edited blog, *Ant Spider Bee*, expressly aims "to engage academics and practitioners in exploration, discussion, and reflection about digital practices, methodologies, and applications in environmental humanities work." The *Network in Canadian History and Environment* (NiCHE) is another example from environmental history of how digital technologies (blogs, podcasts, etc.) can be successfully used to disseminate environmental content within a scholarly community.

Outside of the field of environmental history, other environmental humanities scholars have been collaborating with digital humanists to look at the possibilities of adopting and adapting digital tools. For example, literary scholar Ursula Heise and DH specialist Elijah Meeks analyze textual data in the IUCN Red List of Threatened Species to determine the kinds of narrative templates used to frame species extinction (Heise, *Imagining Extinction* ch. 2). The co-authors of the present chapter have also harnessed the possibilities of (not so) big data, using text mining, topic modeling and Gephi visualizations, to analyze themes in the environmental humanities (Posthumus et al.). In addition to digital text analysis, this fourth branch of DEH uses mapping tools to critique and explore human perceptions of place and space. Geographical Information Systems (GIS) are being integrated into the environmental humanities to create more complex mappings of cities, to track public awareness of environmental issues, and to critique dominant discourses about the nonhuman world. Examples include the work of the Digital Environmental Humanities research group at Trinity College, Dublin, Ireland, led by Charles Travis, the Participatory Geoweb project led by Renée Sieber at McGill University, and the smartphone applications "(Mis) Guide to Alpine Plants," created by Jill Didur and her research assistant, Ian Arawajo, at Concordia University. What these different examples illustrate is that digital humanities tools can be used at varying scales from individual texts to larger data sets.

This brings us to an important point about DEH. A commonly held misconception about the digital humanities (DH) is that they advocate solely for quantitative methods of analysis and large data sets. Moreover, they are at times stereotyped as the "big, bad (scientific-methods or capitalist-driven or administration-loved) wolf" looking to take over the humanities (for an example of the genre, see Stephen Marche's "Literature is not Data: Against Digital Humanities"). In reality, the digital humanities involve many different approaches that are rooted in traditional humanities scholarship, including such practices as interpretation, critique, exploration, and communication. A DH corpus may be textual, visual, or auditory, and it can include one object of study or many. What sets DH scholarship apart from traditional scholarship in philosophy, literature, and history, for example, is not its desire to take over the humanities, as if such a thing were even possible, but the importance of collaboration. While single-authored work can still be found in DH, most DH projects typically involve at least two scholars working alongside programmers, technicians, designers, students, librarians, and nonacademics. In our own DEH endeavors, we have been indebted to the contributions of undergraduate students who have developed essential digital skills while working on our research projects. DEH then becomes a site for bridging the gap between academics and nonacademics, students and scholars.

Four branches of the same tree?

While our own work in this field is primarily representative of the fourth branch, there are lessons we have learned that are relevant for describing DEH as a whole.

First, the importance of play, exploration, and interaction. DEH offers an opportunity for humanities scholars to take a more experimental stance with respect to their object of study, as they engage with methods they may not have been aware of before. This was the case for our project that used *Neatline* to map the geospatial information and the main character's movements in a contemporary French novel (Michel Houellebecq's *La possibilité d'une île*). This project was an exercise in exploration and play. Working closely with a research assistant, we transformed the novel into an interactive object that students could use to complement their reading of the text. This spatial visualization of the book's narrative also led to some interesting discoveries about the novel that a more traditional reading had not suggested. In short, playing with and exploring the digital tools became part of the hermeneutic process (for a more in-depth explanation of this work, see Posthumus and Sinclair).

DH opens up the possibility for EH scholars to approach their object of study differently, to schedule time for play without needing to learn to program in Rails or Javascript (although this can be fun, too). Our *Neatline* exhibit was born from an initial curiosity about the tool and its possible applications to literary texts (it had previously been used mostly for environmental history projects), but had no pre-conceived research agenda. Working with a research assistant who was also an artist, Amy Goh created a space for creativity and interaction that led to an innovative use of icons, something that *Neatline* designers themselves had not yet explored in full. This process is one of creation and discovery, but also of frustration and asking, "why can't the tool do this or that?"—a well-known aspect of DH.

A second lesson learned in working on DEH projects is the importance of dissemination, curation, and critique. The publish-or-perish model has a stranglehold on many (untenured) humanities scholars. DH has succeeded in moving the humanities toward alternative modes of scholarship by advocating for peer evaluations that include blog posts, tweets, and digital tool development (though more remains to be done; see, for example, the MLA's "Guidelines for Evaluating Work in Digital Humanities and Digital Media"). Given the massive amounts of information available online about environmental issues, DEH has an important role to play in curating and critiquing. Again, DH provides a useful model with sites like *DHNow* and publications like *Debates in DH* that favor a publish first, review after model of scholarly dissemination (see also the PressForward.org project). While also using tweets, blogs, and wikis to disseminate EH research, DEH can help filter and organize the most relevant information for different kinds of publics (see, for example, our website dig-eh.org).

Conclusion
for (Digital; Environmental; Humanities)

Academic disciplines progress both incrementally and cyclically. This section's title could theoretically be parsed by a computer as something like a loop where the initial condition is *digital*, the loop would continue while *environmental* is true, and the value of *humanities*

375

would increment at each cycle. As an expression of the Digital Environmental Humanities, this arguably makes more sense than the chapter's title, but the point again is to draw attention to the constitutive parts and the various ways in which they might logically relate.

We have tried to valorize the Digital Environmental Humanities as a generative confluence of the digital humanities and the environmental humanities, a nexus where shared methodologies are called into service for common thematic concerns (we first wrote about this interaction in a 2003 article "Technology and Ecology: *Dialogue de sourds?*"). We have structured an overview of some relevant activities around more critical perspectives (mediating technologies and impacts of technology) and more applied approaches (activating publics and digital scholarly tools), though of course many activities are multivalent and other structures are possible.

We will conclude by considering some of the DEH characteristics of this article (introspection or navel-gazing, depending on perspective). In many ways this is a very conventional publication: primarily text-based, destined mostly for a specialized academic audience (and not open access—not all that public), and with relatively little opportunity for *in situ* discussion and feedback. It will be frozen into a static state by the time it is published, and hardly lends itself to interactivity. It decidedly does not draw attention to its own medium of expression as a technology, except perhaps because of the disconnect with what is described. As a humanities publication it is noteworthy, though not rare, as a co-authored piece (it is one of only four in the current collection), and the simultaneous authoring capabilities of Google Docs undoubtedly had an effect on the process and product. On the other hand, a pair of scholars is perhaps not as radically collaborative as one might hope; we cannot help but daydream about how this chapter might have looked with the collaboration of other scholars interested in the intersections of our fields. Nor have we tracked the energy consumption of our two laptops during the writing of this chapter or the network and server infrastructure needed to support the online work. Ultimately, we are resigned to the fact that this chapter is *about* DEH, it is *not* DEH (though we have drawn attention to some projects we think are more exemplary). In a last-ditch attempt to salvage the DEH-ness, we will post a digital version of this chapter that will be more conducive to exploration, commentary, outreach, and the integration of other tools (dig-eh.org/deh-routledge/). You can join the conversation there or tweet at #whatifDEH.

References

4Humanities. "Mission." *4Humanities: Advocating for the Humanities*. Web. 1 September 2015.

Bianchi, Melissa. "Rhetoric and Recapture: Theorising Digital Game Ecologies through EA's *The Sims* Series." *Green Letters* 18 (2014): 209–220. Print.

Carletti, Laura, Derek McAuley, Dominic Price, Gabriella Giannachi, and Steve Benford. "Digital Humanities and Crowdsourcing: An Exploration." *Annual Conference of Museums and the Web*. Portland, Oregon. August 2013. Presentation.

Carr, Nicholas. "Is Google Making Us Stupid?" *The Atlantic*. July/August 2008. Web. 1 September 2015.

Carruth, Allison. "The Digital Cloud and the Micropolitics of Energy." *Public Culture* 26.2 (2014). Web. 1 September 2015.

Christensen, Jon et al. *City Nature*. Stanford University Libraries. 2013. Web. 1 August 2015.

Heise, Ursula K. "Plasmatic Nature: Environmentalism and Animated Film." *Public Culture* 26.2 (2014). Web. 1 September 2015.

———. *Imagining Extinction: The Cultural Meanings of Endangered Species*. Chicago, Illinois: University of Chicago Press, forthcoming 2016. Print.

Hockey, Susan. "The History of Humanities Computing." *A Companion to Digital Humanities*. Ed. Susan Schreibman, Ray Siemens, and John Unsworth. Oxford: Blackwell, 2004. Web. 1 September 2015.

Jørgensen, Finn Arne, "The Armchair Traveler's Guide to Digital Environmental Humanities." *Environmental Humanities* 4 (2014). Web. 1 September 2015.

Lalasz, Bob. "Jon Christensen: Why Nature's on the Margins in U.S. Cities—and That Could Be a Good Thing." *The Nature Conservancy*. 18 April 2013. Web. 1 September 2015.

Lin, Maya. *What is Missing?* Web. 1 August 2015.

Marche, Stephen. "Literature is Not Data: Against Digital Humanities." *Los Angeles Review of Books*. 28 October 2012. Web. 1 September 2015.

MLA Executive Council. "Guidelines for Evaluating Work in Digital Humanities and Digital Media." Modern Language Association. 2012. Web. 1 September 2015.

Nowviskie, Bethany. "Digital Humanities in the Anthropocene." Digital Humanities Conference. Lausanne, Switzerland. 10 July 2014. Presentation.

Plato. *Phaedrus*. Perseus Digital Library. Web. 1 September 2015.

Posthumus, Stephanie and Stéfan Sinclair, "Reading Environment(s): Digital Humanities and Ecocriticism." *Green Letters: Studies in Ecocriticism* 18.3 (2014): 254–273. Print.

Posthumus, Stephanie, Stéfan Sinclair, and Veronica Poplawski. "Digital Environmental Humanities: Strong Networks, Innovative Tools, Interactive Objects." *Resilience: A Journal of the Environmental Humanities*, forthcoming.

Sinclair, Stéfan and Stephanie Posthumus. "Technology and Ecology: *Dialogue de Sourds?*" *Culture and State: Landscape and Ecology* 1 (2003): 168–178. Print.

Walsh, Bryan. "The Surprisingly Large Energy Footprint of the Digital Economy." *Time* 14 August 2013. Web. 1 September 2015.

Wellmon, Chad. "Why Google Isn't Making Us Stupid … or Smart." *The Hedgehog Review* 14.1 (2012). Web. 1 September 2015.

Wilson, Alexander. *The Culture of Nature: North American Landscape from Disney to the Exxon Valdez*. Toronto: Between the Lines, 1991. Print.

37

FROM *THE XENOTEXT*

Christian Bök

Editorial note: "Phage φX174" and "The Dire Seed" are excerpts from *The Xenotext*, an ongoing project by the poet Christian Bök. For the last fifteen years, Bök has been working with geneticists to engineer an extremophile bacterium (*Deinococcus radiodurans*) so that its DNA might both encode a poem and prompt the bacterium to produce a benign protein, whose amino acids also encode a poem in response. We include these excerpts here as examples of a project that fuses literature and science with implications for how we might understand humanity, art, and inhuman forms of life.

PHAGE ΦX174

1.

The Virus was first detected on Earth in 1935, when the pathologist Nicolas Bulgakov (brother of the Soviet writer Mikhail Bulgakov) was working at the Pasteur Institute, conducting a census of organic samples collected from the sewers of Paris. The Virus was isolated and cultured by Bulgakov, who learned that the newly found life-form could infect the enteric microbe *E. coli*. The Virus proved to be uncannily geometric in its structure when later seen under an electron microscope, since the Virus resembled an icosahedron – a perfect, faceted capsule of protein, containing an annular plasmid of genetic material. The Virus could inject this package of alien genes into a host cell so as to commandeer the metabolic machinery of the infected organism, replicating the Virus more than one thousand-fold, until the Virus overwhelmed its host – finally killing it.

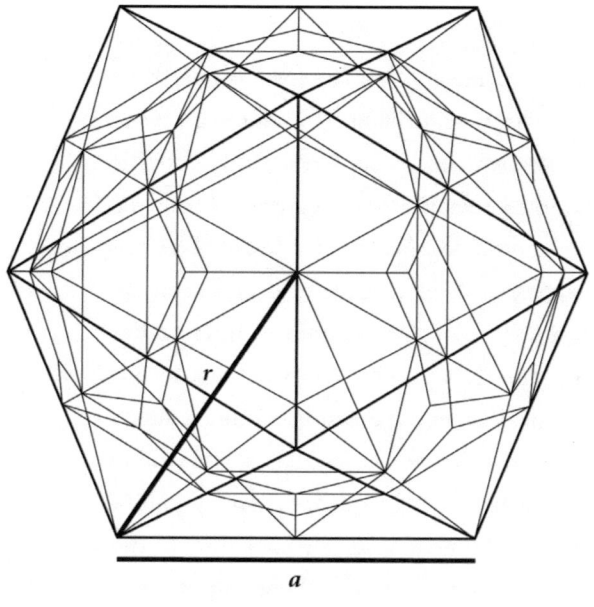

$$\phi = \frac{4r^2}{a^2\sqrt{5}} = \frac{1+\sqrt{5}}{2} \approx 1.61803\ldots$$

THE VIRAL ICOSAHEDRON

2.

The Virus went on to become the first life-form on Earth to have its genome completely sequenced by humans in 1977, when Frederick Sanger (winner of the Nobel Prize for this achievement) determined that the Virus contained only 11 genes, encoded by a mere 5386 bases – far fewer than expected. The Virus appeared to be capable of compressing its entire genome into a short, but dense, chain of information by using overlapping, interleaved sets of instructions to encode more than one protein simultaneously within a single series of nucleotides. At the time, no biochemist could readily explain the origin for such efficient sequences of genes, all of which remained conserved, unchanged, for eons despite the selective pressures of mutation. A single change at any site within such a genome could compromise many functions of the life-form – yet it thrived in the face of these threats.

Frederick Sanger et al. 'Nucleotide Sequence of Bacteriophage φX174 DNA.' *Nature* 265 (24 Feb 1977): 687–695.

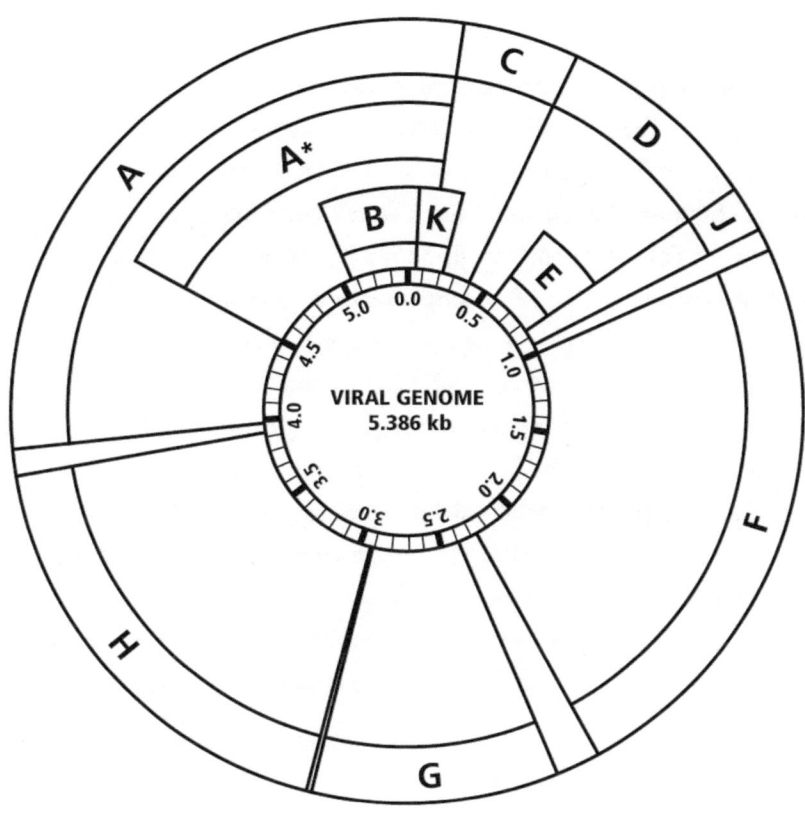

GENE A	3.981 – 0.136	FABRICATION OF VIRAL STRAND
GENE A*	4.497 – 0.136	SHUTDOWN OF HOST CELL DNA
GENE B	5.075 – 0.051	MORPHOGENESIS OF CAPSID (A)
GENE C	0.133 – 0.393	MATURATION OF VIRAL GENOME
GENE D	0.390 – 0.848	MORPHOGENESIS OF CAPSID (B)
GENE E	0.568 – 0.843	LYSIS OF HOST CELL MEMBRANE
GENE F	1.001 – 2.284	CAPSID (MAJOR COAT PROTEIN)
GENE G	2.395 – 2.922	CAPSID (MAJOR SPIKE PROTEIN)
GENE H	2.931 – 3.917	CAPSID (MINOR SPIKE PROTEIN)
GENE J	0.848 – 0.964	CAPSID (CONDENSING PROTEIN)
GENE K	0.051 – 0.221	ENHANCEMENT OF VIRAL YIELD

3.

The Virus seemed so perfectly intricate despite the tiny size of its genome that, for some scientists, only intelligent engineering could account for the complexity of such a life-form; consequently, Hiromitsu Yokoo and Tairo Oshima (working at laboratories in Tokyo) set out to demonstrate the artificiality of this genome in 1979, focussing attention upon a peculiar sequence of 121 codons, shared across Gene A and Gene B – the instructions for two entirely distinct proteins. The scientists mapped these codons onto a grid of pixels, 11 × 11 units in size, doing so in the hope of detecting a bit-mapped, geometric glyph that might signify a message sent from outer space by extraterrestrials. The researchers used six different, but plausible, techniques for decipherment of this genetic segment – but no intelligible, mathematical imagery was disclosed in any of the resulting gridworks.

Hiromitsu Yokoo and Tairo Oshima. 'Is Bacteriophage φX174 DNA a Message from an Extraterrestrial Intelligence?' *Icarus* 38 (1979): 148–153.

GAA TGG AAC AAC TCA CTA AAA ACC AAG CTG TCG

CTA CTT CCC AAG AAG CTG TTC AGA ATC AGA ATG

AGC CGC AAC TTC GGG ATG AAA ATG CTC ACA ATG

ACA AAT CTG TCC ACG GAG TGC TTA ATC CAA CTT

ACC AAG CTG GGT TAC GAC GCG ACG CCG TTC AAC

CAG ATA TTG AAG CAG AAC GCA AAA AGA GAG ATG

AGA TTG AGG CTG GGA AAA GTT ACT GTA GCC GAC

GTT TTG GCG GCG CAA CCT GTG ACG ACA AAT CTG

CTC AAA TTT ATG CGC GCT TCG ATA AAA ATG ATT

GGC GTA TCC AAC CTG CAG AGT TTT ATC GCT TCC

ATG ACG CAG AAG TTA ACA CTT TCG GAT ATT TCT

GRID OF CODONS (11 × 11)

GRID 1

FIRST NUCLEOTIDES IN EACH CODON

PURINE BASES

(A, G) = BLACK

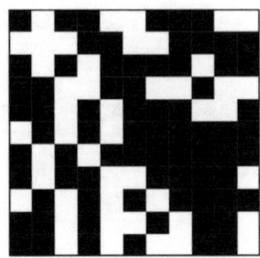

GRID 2

SECOND NUCLEOTIDES IN EACH CODON

PURINE BASES

(A, G) = BLACK

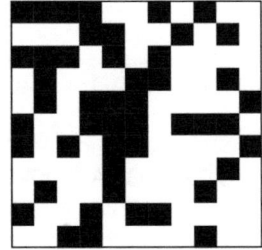

GRID 3

THIRD NUCLEOTIDES IN EACH CODON

PURINE BASES

(A, G) = BLACK

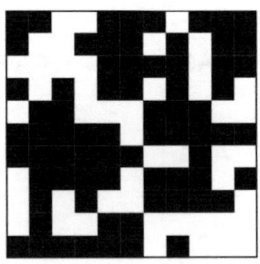

GRID 4

FIRST NUCLEOTIDES IN EACH CODON

DYADIC BASES

(C, G) = BLACK

GRID 5

SECOND NUCLEOTIDES IN EACH CODON

DYADIC BASES

(C, G) = BLACK

GRID 6

THIRD NUCLEOTIDES IN EACH CODON

DYADIC BASES

(C, G) = BLACK

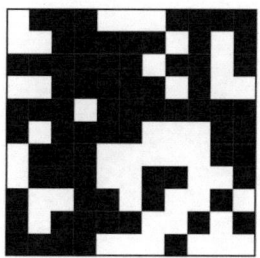

4.

The Virus did not appear to contain a glyph from outer space; however, the scientist Hiroshi Nakamura revisited this research again in 1982, and he suggested that, within the 11 × 11 grid, all the codons containing a triplet of identical oligomers (AAA, CCC, GGG, and TTT) could be connected to create a figure that appeared to correspond to the positions of stars in the vicinity of the constellation Boötes (as seen from the planet Earth), leading him to speculate that the Virus could in fact contain a "star chart" – a secret message indicating, to its readership, the stellar address of its author. He also studied analogous sequences of DNA found in SV40 (a virus capable of causing cancer in apes), and likewise he claimed to have detected evidence for the existence of a map depicting a constellation located in the vicinity of the star Epsilon Eridani (10 light years away).

Hiroshi Nakamura. 'SV40 DNA – A Message from ε Eri?' *Acta Astronautica* 13.9 (1986): 573–578.

GAA	TGG	AAC	AAC	TCA	CTA	AAA	ACC	AAG	CTG	TCG
CTA	CTT	CCC	AAG	AAG	CTG	TTC	AGA	ATC	AGA	ATG
AGC	CGC	AAC	TTC	GGG	ATG	AAA	ATG	CTC	ACA	ATG
ACA	AAT	CTG	TCC	ACG	GAG	TGC	TTA	ATC	CAA	CTT
ACC	AAG	CTG	GGT	TAC	GAC	GCG	ACG	CCG	TTC	AAC
CAG	ATA	TTG	AAG	CAG	AAC	GCA	AAA	AGA	GAG	ATG
AGA	TTG	AGG	CTG	GGA	AAA	GTT	ACT	GTA	GCC	GAC
GTT	TTG	GCG	GCG	CAA	CCT	GTG	ACG	ACA	AAT	CTG
CTC	AAA	TTT	ATG	CGC	GCT	TCG	ATA	AAA	ATG	ATT
GGC	GTA	TCC	AAC	CTG	CAG	AGT	TTT	ATC	GCT	TCC
ATG	ACG	CAG	AAG	TTA	ACA	CTT	TCG	GAT	ATT	TCT

THE STAR CHART IN THE VIRUS

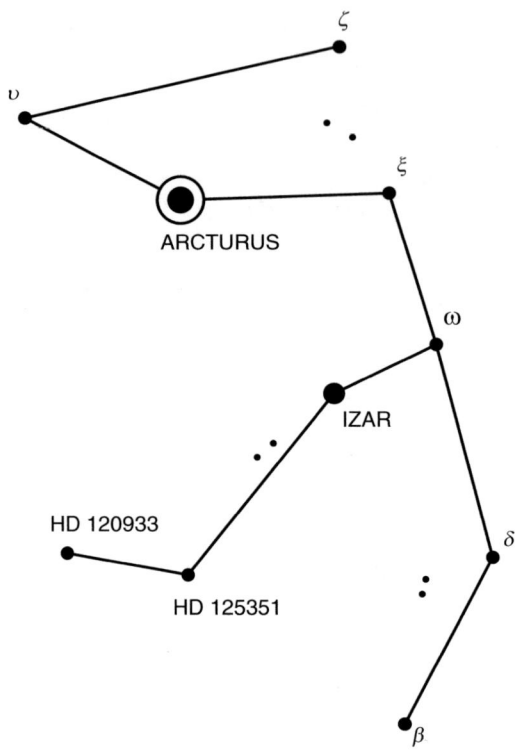

THE ASTERISM OF BOÖTES

5.

The Virus eventually became the first organism in the history of the Earth to be fabricated from scratch by computer in 2003, when Craig Venter used automated chemistry to combine the component molecules of this life-form into a viable genome – and by doing so, he demonstrated that humans now had the skills to design synthetic organisms unseen anywhere in Nature. The Virus, born of this digital process, inspired research later used by Venter in 2010 to create the first synthetic bacterium, nicknamed "Synthia" (a *Mycoplasma* with an artificially manufactured genome). He "watermarked" this newly built germ by grafting into it a gene that enciphered a quote from A *Portrait of the Artist as a Young Man:* "To live, to err, to fall, to triumph, to re-create life out of life." He had thus preserved a sample of modern poetry in the first cells of an embryonic ecosystem.

Craig Venter, et al. "Generating a Synthetic Genome by Whole Genome Assembly: φX174 Bacteriophage from Synthetic Oligonucleotides." *Proceedings of the National Academy of Sciences* 100.26 (23 Dec 2003): 15440–15445.

6.

The Virus continues to persist in the imagination of scientists (who hope to exploit its parasitism in the fight against "superbugs" resistant to antibiotics) – but despite the fact that this life-form has occasionally fomented a kind of apophenia, resulting in a frantic pursuit for an encoded message embedded within its genome, humans have so far failed to read the signature left there by some recondite, but invisible, sentience. We have instead become the alien force whose rubaiyat we seek, since we can now transmit our own messages across stellar distances or epochal intervals by storing them durably within cells. We can harness such life, exploiting its power, not only to recopy itself with few errors, but to modify itself for the better, all in the face of random change. We are now on the verge of writing truly alive books, perusable by other minds around other stars.

The Dire Seed

1.

Deinococcus radiodurans ('the dire seed, immune to radiation') is a vermilion, spherical microbe capable of surviving in the most inhospitable environments. Such a bacterium can live in exsiccated barrens and irradiated deserts, surviving bombardment by gamma rays, even when exposed to the nothingness beyond the ionosphere. Each germ (3.5 microns in diameter) can encapsulate up to ten copies of its genome rather than just one – and with this level of redundancy, the dire seed can repair itself, suturing together fractured sequences of its DNA in less than 24 hours, without mutating or expiring. No one knows the natural habitat for this microorganism (although biologists have found the germ in elephant dung from Bangladesh and in gneissic soil from Antarctica). The germ has undoubtedly colonized the entire planet – and yet we know nothing of its origin.

2.

Deinococcus radiodurans was first detected on Earth by accident in 1956, during the Cold War, when Arthur W. Anderson was conducting research for a laboratory at Oregon State University in Corvallis. Anderson was trying to see if extreme dosages of ionizing radiation from cobalt-60 could be used to sterilize tins of meat for durable storage (since the United States Armed Forces needed a means to preserve food for prolonged refuge in atomic bunkers or for perennial travel in manned rockets) – but despite irradiating cans of beef with 1.75×10^6 rads of exposure (enough to obliterate any existing pathogen), Anderson discovered that one sample still rotted after the experiment, apparently due to some undestroyed contaminant. Anderson isolated the micoorganism responsible for this decomposition, thereby sparking military interest in the study of extremophiles.

Arthur W. Anderson et al. "Studies on a Radio-Resistant Micrococcus: Isolation, Morphology, Cultural Characteristics, and Resistance to Gamma Radiation." *Food Technology* 10 (1956): 575–578.

3.

Deinococcus radiodurans went on to be recruited for tests by NASA in 2002, when a team of astrophysicists (led by the neurosurgeon, Roya Saffary) painted a monolayer of the germ onto a polycarbonate lens, which the team then installed, as a payload, aboard an MK-12 Black Brant Rocket, launched from the White Sands Missile Range in New Mexico. The rocket flew to an altitude of 304 kilometres, where it remained at apogee for 395 seconds, exposing the lifeform directly to the abyss of outer space so that the dire seed could absorb a high dose of ultraviolet irradiation from the Sun. The rocket then fell back to Earth, parachuting to a site, where the lens could be retrieved for analysis. The team found that, under exospheric conditions, the quantity of microbes declined by four orders of magnitude (from 10^8 to 10^4) – yet the colony remained viable. It did not perish.

Roya Saffary et al. "Microbial Survival of Space Vacuum and Extreme Ultra-violet Irradiation: Strain Isolation and Analysis During a Rocket Flight." *FEMS Microbiology Letters* 215 (2002): 163–168.

4.

Deinococcus radiodurans thus demonstrated that it could survive exorbitant hardships never before experienced by any biome on Earth – and for this reason, Anatoly K. Pavlov speculated, in 2006, that the germ could only have gained its remarkable immunities by evolving in a more exotic milieu, possibly offworld in outer space, on a planet like Mars, where the dire seed could undergo periods of cyclical dormancy, adapting to extremes of both heat and cold in a parched barren, where levels of radiation far exceeded any sources on Earth (including even the ancient, nuclear reactors, buried in the uranic oxides at Oklo in Gabon). Ejecta from the planetary collision of a meteorite could have transported an archaebacterium from Earth to Mars, forcing this life-form to adapt to an alien world, before being returned to Earth by yet another, meteoric transfer.

Anatoly K. Pavlov et al. "Was Earth Ever Infected by Martian Biota? Clues from Radioresistant Bacteria." *Astrobiology* 6.6 (2006): 911–918.

5.

Deinococcus radiodurans did go on to provide evidence for such otherworldliness in 2007, when the scientist Ulrike Pogoda de la Vega put the germ to the test by blend-ing the dire seed with granules of Goldenrod B-6090 (an iron oxide, produced by mixing red hematite with yellow goethite) – whereupon he sealed this powder in a vacuum chamber, filled with a gas, similar in composition to the at-mosphere of Mars (95.5% carbon dioxide, 2.5% nitrogen, 1.5% argon, with traces of oxygen, all at a rarefied pressure of 0.7 kilopascals). The chamber was subjected to diurnal changes in both ultraviolet irradiation and hypothermic temperature (typically occurring in summer at the latitude of 60° on Mars). If coated with a dusting of iron oxide, the wildest strains of the germ could endure this anoxic milieu for the weeklong duration of the test, without difficulty.

Ulrike Pogoda de la Vega et al. "Simulation of the Environmental Climate Conditions on Martian Surface and its Effect on *Deinococcus radiodurans*." *Advances in Space Research* 40 (2007): 1672–1677.

6.

Deinococcus radiodurans thus demonstrated that it could likely eke out its existence in the baleful climate of an alien world – and in 2011, the exobiologist Ximena Abrevaya did much to confirm this supposition, when her team of specialists cultured the germ on a resin plate, which was then encased in a vessel, designed to simulate the environment on the hostile surface of the icebound moonland, Europa. The team subjected the sample to a very high vacuum (equal to 10^{-8} kilopascals), all the while irradiating the germ for three hours, using a synchrotron to emit polychromatic, ultraviolet rays at frequencies over 2.5×10^{15} hertz (with a maximal, radiant density of 135 joules per square metre). The germ resisted the hellishness of this onslaught with a survivability of 1%, implying that a remnant of the dire seed could still flourish after such punishment.

Ximena Abrevaya et al. "Comparative Survival Analysis of *Deinococcus radiodurans* and the Haloarchaea [...] Exposed to Vacuum Ultraviolet Irradiation." *Astrobiology* 10–11 (Dec 2011): 1034–1040.

7.

Deinococcus radiodurans was first exploited for archival purposes by humanity in 2003, when Pak Chung Wong (working at a lab for the United States Department of Energy) enciphered the lyrics to *It's a Small World (After All)*, implanting this text as a plasmid of DNA inside the genome of the dire seed, thereby showing that such a message could persist, undamaged and unaltered, through myriad cycles of mitosis, all the while secured for eventual decoding. Wong noted that, in a world of fragile media with limited space for storage, messages embedded in DNA could allow us to preserve our cultural heritage against planetary disasters (including thermonuclear warfare and astrophysical barrage), thus ensuring that future humans could access lost data from the past, simply by extracting such information from storehouses locked inside immortal microbes.

Pak Chung Wong et al. "Organic Data Memory Using the DNA Approach." *Communications of the ACM* 46.1 (Jan 2003): 95–98.

8.

Deinococcus radiodurans has thus prevailed for many eons against the lethality of the universe itself. Such an extremophile has, so far, defied all threats to its existence – (scorch it; freeze it; wither it – and still, the dire seed can bloom). The invincibility of such a microorganism almost brings to mind the barbarian from Cimmeria – the mythical superman who can stride, unscathed, through the slaughterhouse of History, as if immune to magic and jihad. Let us engineer this primitive bacterium so that it might become the symbiote for a poem – an epic song, whose words might subsist, like a harmless parasite, inside the genome of the dire seed, thereby causing the germ to generate, in response, a protein whose molecular structure enciphers yet another text. Let us build not only a durable archive for storing such lyrics, but also an operant machine for writing their echoes.

Part VI

THE STATE OF THE ENVIRONMENTAL HUMANITIES

38

THE BODY AND ENVIRONMENTAL HISTORY IN THE ANTHROPOCENE

Linda Nash

Over the last decade and a half, the humanities have witnessed a material turn, one aspect of which is the emergence of the environment as a key area of study among humanists and a growing group of culturally focused social scientists. The reasons for this are clear enough: the magnitude of the contemporary environmental crisis has challenged any exclusive focus on culture and language. Equally important, the modernist framing of the "environment" as an object of the natural sciences has led to a research program based on the assumption that responses to this crisis will emanate primarily from the natural and policy sciences. The claim for environmental humanities is thus an argument that studies of culture are highly relevant and should not be marginalized within ongoing studies of global change (Palsson et al.). Sometimes it has the feel of a beggar knocking at the door for a few scraps of research funding.

But if the environmental humanities face marginalization, the problem lies not only—and perhaps not even primarily—with the claims of natural scientists and the assumptions of policymakers. Humanists have been equally complicit in the divvying up of the intellectual landscape. In approaching society, culture, and language as entities that exist apart from an environmental and material context and so can be studied independently, we have written the environment out of tens of thousands of narratives. We have assumed repeatedly that the social can be separated from the natural, and that our object of study is the former (Latour). This assumption has had untold costs.

The term "environmental humanities" is thus a necessary oxymoron that stems from a Western intellectual tradition built upon the separation of nature and culture, body and mind, subject and object. It should not be a surprise that such a term has emerged only when the contradictions of that tradition threaten to overwhelm us. Yet even while the necessity of uniting studies of culture with studies of the material world has gained broad acceptance, it is far from clear how we should go about it. Humanistic scholars find themselves struggling to find meaningful ways of incorporating what they once took to be the nonhuman(istic) world—environments, materials, and animals—into human stories. As the environments of the world change ever more rapidly in ways both large and small, the stakes could not be

403

greater. This chapter is an attempt to think through some of the possibilities of the environmental humanities via the specific paths that environmental history has taken and might yet take in the future, drawing particularly on the insights of anthropologist Tim Ingold as well as the potential of practice theory to guide future work in the field.

Models of environmental history: systems and neo-Darwinism

Environmental history was a product of the modern environmental movement. Emerging in the 1970s, it shared with other left-leaning scholarship a critique of Western technoscientific development. But from the outset the field was founded upon two substantially different approaches. The first based itself on the postwar concept of systems and was strongly influenced by sociological functionalism and ecosystem ecology. Taking the social and natural worlds as relatively self-contained but overlapping and interactive systems, the first generation of environmental historians tracked how the social and ecological worlds influenced one another. Changes in the social system were interpreted through existing social scientific and historical paradigms, while changes in the natural system were traced through ecology, geology, and soil science. The interaction between these systems was understood as dialectical (Worster, "Transformations of the Earth"; Cronon 13). Although individual historians approached the task somewhat differently, the key early studies of Richard White (*Land Use, Environment, and Social Change*), William Cronon (*Changes in the Land*), and Donald Worster (*Dust Bowl*) all followed this model.

The other approach was neo-Darwinian at its core, and focused on the relative adaptability of peoples, plants, and animals to new environmental conditions. The key texts were those by Alfred Crosby (*The Columbian Exchange*) and William McNeill (*Plagues and Peoples*). For both of these authors, the influence of natural selection on the disease history of Europeans was crucial in explaining their success in the New World. In these stories, the exposure of Europeans to many more diseases over generations had granted them an immunological advantage over indigenous populations in America. In a subsequent book, *Ecological Imperialism*, Crosby extended the argument to other European species, notably domestic animals and weeds. Many of these species were well adapted to North American conditions and thus able to spread rapidly. Ultimately they displaced much of the continent's native flora and fauna and helped enable European conquest.

Most contemporary environmental history still traces back to one of these approaches. Though the systems approach has been far more flexible and influential, it was seriously challenged by the cultural turn, which called into question both the straightforward reliance on scientific texts as well as any simplistic demarcation between natural and social worlds. In response, environmental historians turned their focus to the "hybridity" of natural and social systems, while disavowing distinct boundaries between them (White, "Environmental History"; Nash). Yet, methodologically, most of this work still chronicled distinct systems even while calling them into question. Moreover, as Paul Sutter has observed, the repeated return to hybridity as a conclusion is not all that satisfying, offering neither analytical nor moral clarity, or even a powerful explanation for historical change. Studies of hybridity seem to have reached a dead end.

Recently there has been renewed interest among historians, including environmental historians, in neo-Darwinian approaches (Russell; J. McNeill; Worster, "The Living Earth").

Partly this reflects the dissatisfaction with cultural studies of environment as well as environmental historians' rediscovery of disease as a topic. But it also reflects the spillover effects of the burgeoning field of neuroscience. Amidst this elevation of the biological brain as an object of study has come a reassertion of neo-Darwinian theories of cultural evolution. Although those committed to neo-Darwinian views of culture have their differences, what unites them is the belief that natural selection accounts for not only the spread of genes but also culture. Culture is understood as that which is *not* genetic—material that is transmitted by learning and mimesis rather than through biological replication. Nonetheless, these cultural "traits" (e.g., having a small family, making a particular kind of basket) are transmitted through societies and are subject to the forces of natural selection. Genetic and cultural transmission may be inherently different, but they follow the same rules and can be understood through similar methods. Biology and culture "co-evolve" (Boyd and Richerson; Mesoudi et al.).

One example is the recent interest in "neurohistory," which brings attention to the role of nonintentional processes in human societies, drawing on both history and neurobiology. Its most prominent exponent is the historian Daniel Smail, who argued that the denser settlements that arose alongside the Neolithic revolution played a crucial role in setting the human brain on a new evolutionary path. In his words, "a long-dormant neurophysiological capacity to recognize and acknowledge dominance hierarchies in human societies was turned back on" (*On Deep History* 166). Whatever we might think of them, Smail argues, dominance hierarchies served a larger purpose. Because they provided a successful means of organizing dense populations, they survived as a key element of human culture.[1] Smail's account is neo-Darwinian in that he assumes that human culture—at least those aspects that persist—is functionally adaptive. In this view, religion, gossip, dominance hierarchies, or even the prevalence of rape can be explained as having offered some survival advantage in an earlier period.

Although neurohistorians claim an affinity with environmental history, their ability to incorporate environments into their stories is ultimately limited by their methods. Much like the disease histories of Crosby and W. McNeill, contemporary neurohistory aspires to overcome the dichotomy between nature and culture by placing humans—specifically human brains and human culture—inside of nature and representing them as subject to the same forces of nature (i.e., natural selection) as other species. Material environments are given a crucial if undetailed role at specific points so as to explain a specific cultural or biological development as adaptive. But most of the time, they play no role at all. In fact, Smail argues that in the case of neurohistory, the relevant environment may lie within the body itself. As he puts it, "[g]iven the fact that the 'environment' in environmental history does not have to consist of nature, it takes just a small leap of the imagination to treat our own nervous system as an ecological niche in which the patterns of human culture have emerged and evolved" ("Neuroscience" 897).

The collapsing of "environment" into "our own nervous system" is a telling move; human cultural development proceeds with little reference to the material world. Thus, even though Smail critiques the harder versions of evolutionary psychology for grounding all key human behaviors in adaptations to the Pleistocene environments of the African savannah, Smail does something quite similar when he points to the Neolithic revolution as creating an environment that set the brain on a new evolutionary course. After that, we hear very little about the diverse environments of the globe and their effects on varied and complex human histories. Gracing the cover of Smail's book is a skull, apparently floating in space over an ancient map of the Middle East; the image is striking and also appropriate because the brain of Smail's neurohistory is a disembedded and disembodied organ.

Not only do neo-Darwinian models treat human culture in simplistic terms, they, somewhat ironically, reinforce an impassable divide between humans and other animals that many humanists themselves have come to trouble. Humans are understood as a unique species because of their brains and their ability to pass on cultural knowledge rather than simply genetics. Consider the classic example of the bee. The bee's genome is held to determine not only its size, shape, color, and food preferences, but also the shape and structure of its hive. In short, the bee's genotype determines its phenotype. The hive, while not part of the bee itself, might be thought of as an "extended phenotype." (The phrase is Richard Dawkins's.) In contrast, when humans construct a house they are presumed to think, then build. The house, unlike the hive, is perceived as a product of culture, not genetics. The problems with this perspective are numerous. For one thing, it requires that there be a moment in time in which the ancestors of humans began to think first and then build. Put another way, it implies that there was an originary moment in which our ancestors stepped out of evolution and into history and culture (Ingold, *Perception* 340). But are the actions of modern humans and animals categorically different, or just distantly related?

Take, for instance, food choice. We explain the varied food choices of humans as largely cultural, overlaid atop a biological substrate of taste perception and stimulation, while we explain the choices of animals as genetic. Humans, of course, consume a vast array of foods and those food habits have changed radically over time; we understand them to be a product of cultural practice, politics, and economics. But of course they are also environmental and mimetic. When British colonists arrived in North America, they found that many of their preferred foods could not be grown, and thus they adopted many Native foods, oftentimes reluctantly. Over time Native North American foods—such as cornbread—became incorporated and celebrated within Euro-American subcultures. But we view the food choices of animals differently—as a product of their genomes. Thus wolves introduced from Canada into Yellowstone National Park were selected based on their pre-existing prey choices: wolves who hunted elk were chosen because they were viewed as genetically predisposed to hunt elk and thus most likely to succeed in the Yellowstone environment. Yet over time, one pack introduced into the park—finding itself in a particularly difficult territory with few elk available in winter—learned to successfully hunt bison (Smith and Ferguson 74–76). Is this a case of a gene for bison hunting turned back on, or something else?

Looking ahead

If systems approaches have foundered on their own contradictions and neo-Darwinism erases the role of the very environments that we are trying to capture, what, then, are our alternatives for writing histories in the Anthropocene?

Within contemporary biological science there are other, potentially more fruitful avenues of connection. Developmental biology and niche construction theory, to name two of the most important, both emphasize the development of the organism in the world, and allow for the foregrounding of epigenetic processes and environments. With the dawning realization that organismic development "relies less on the existence of genes ... than on the regulated expression of those genes, which ultimately depends on a host of environmental factors," many biologists have challenged the dominance of neo-Darwinian explanations in their own field, pointing to the need for a fundamentally new evolutionary synthesis (Stotz).

Humans—and many animals—grow up in environments furnished by the work of previous generations, and as they grow they come to carry certain forms within their bodies—specific skills, sensibilities, and dispositions—that emerge in response to those constructed environments. As Ingold argues, what is passed on from generation to generation is not merely a genome, but also a segment of the world (a territory, for lack of a better word), and an example of how to live in that world (*Perception*). There is no practical way to separate these in the life of an individual.

The neo-Darwinian genotype—even when augmented by the "culture-type"—is, in this reading, a poor model for understanding human—and, for that matter, animal—history because it fails to account for the fact that human life develops in a context that is absolutely crucial to what will emerge. To abstract the genotype from the world is always an intellectual exercise, not a description of reality; and in this exercise the environment is rendered relatively static, limited to activating extant and random genes or memes rather than having an active role in their creation. Potentiality is prefigured, somehow scripted in advance, rather than developed over the course of a life or generations. Without stable environments, the next generation will look and act very differently, even though its genotype remains unchanged. Similarly, when culture is cast in neo-Darwinian terms, patterns of behavior already exist in people's heads with their meanings attached, independent of the world in which they come to be expressed (Ingold, "The Trouble with Evolutionary Biology"). Similarities across generations may be as much, if not more, a product of stable environments than of stable genomes. To return to the bison-eating wolves, rather than ascribing this to the wolves' genotype, we might simply view it as learning and development in a new environmental context.

As environmental historians have demonstrated again and again, the ability of humans to "make their own history" depends, in ways large and small, upon the material worlds that they have encountered. Within biology, the theory of niche construction comes closest to this idea: "If genomic expression and silencing depend on epigenetic processes, and if epigenesis is environmentally sensitive, then our anthropogenic environments and the experiences they may individually or collectively induce have to be taken into account" (Ramirez-Goicoehea 69). From this perspective, humans, when they build houses or freeways or the Internet, shape not only their own history, but also their own evolution, to the extent that we understand evolution in developmental rather than strictly genetic terms. And so, for that matter, do bees and wolves.[2]

Writing in the 1960s, French anthropologist André Leroi-Gourhan speculated that an extra-terrestrial observer, unbiased by the categories of Western philosophers and historians, "would separate the eighteenth-century human from the human of the tenth century as we separate the lion from the tiger or the wolf from the dog" (247). Leroi-Gourhan's insight was that the division that we draw between human beings and their material and cultural environment is arbitrary, a peculiarly Western way of viewing the world, in which the skin marks the definitive boundary of the human animal. If we were to view humans and their lifeworlds as an organic totality instead, we would have to acknowledge that that totality has changed radically over the course of human history despite the fact that human bodies have changed very little. The extra-terrestrial observer "would therefore perceive different species at different periods, whereas we, who insist on cutting human beings out from their sociotechnical context, are inclined to perceive the same species, *Homo sapiens*, throughout" (Leroi-Gourhan 247; see also Ingold, "Tools").

In other words, despite claims to the contrary, the turn to evolutionary biology reinforces the divide between nature and culture, setting humans outside of and apart from their environment and from the larger animal world. Similarly, the problem with neurohistory lies in its very conception—that there might be a history of something called the "brain" that can be uncovered in the laboratory apart from both the body and the wider world.

In contrast, environmental history and the environmental humanities should challenge us to consider the essential role of changing historical environments not merely in constraining and enabling social organization, but also in making us who we are both biologically and socially, undergirding our intellect and culture, and shaping our development over the scale of both individual lives and generations. And yet despite their crucial and ongoing role, environments do not determine outcomes because they are never separable from the beings that have already shaped, and continue to shape, them, intentionally—in order to help secure their own future—and unintentionally, often in quite radical ways. To the extent that these re-shaped environments can be handed down to the next generation, they allow groups to survive and prosper by making effective a whole suite of learned skills and tools.[3] As Ingold suggests, this allows for a quite different understanding of "evolution" as a process that "can occur without reference to genetic change through cumulative transformations wrought through the actions of the organisms themselves on the conditions of the development under which they and their successors grow to maturity" ("Prospect" 12).

To track *this* kind of evolution is the work of environmental history.

In the wake of massive environmental change, existing skill sets ("knowledge") may be rendered largely, or even completely, useless and will necessarily give way to new forms of knowledge; new environments require culture to be rebuilt, often radically, using the materials at hand. In other words, what we mean by "knowledge" or "culture" is not something that humans carry around in their heads; these concepts are fundamentally human–environmental all the way down. As cognitive anthropologist Edward Hutchins has observed, "humans create their cognitive powers by creating the environments in which they exercise those powers" (169).

Part of what has made the ongoing disregard of environmental context possible is the unprecedented ability of certain modern human groups (including academic neuroscientists and evolutionary biologists!) to hold their environments relatively constant and to assume that their science is primarily a product of their own thinking rather than their access to an environment filled with an unprecedented array of tools, materials, and networks. The regulation and homogenization of space that is the outcome of modern technoscience and modern forms of empire has enabled the proliferation of technoscientific—as well as humanist—theories that disregard the role of the environment.

Part of my argument here is that while the claims for neurohistory and evolutionary biology have been much too grand, the claims for environmental history have not been grand enough. If Smail's version of neurohistory is ultimately unsatisfying, it is not because there is nothing to appreciate and respond to in modern neuroscience. As William Reddy points out, much of contemporary cognitive neuroscience reveals an increasingly complex picture of the brain that, in his words, "resoundingly confirms" the claims historians, anthropologists, and others have made about the powerful impact of cultural factors on perception and practice (21). Rather the problem lies in a recurring return to neo-Darwinist models of explanation. Human thought and action require much greater environmental contextualization, while our understandings of "the environment" require human (and animal) contextualization. Rather than looking to a simplistic biological theory to ground history in the Anthropocene, humanists need to forge new ways of thinking about both nature and humans as at once social and biological.

While niche construction theory offers a more sophisticated and richer biological theory, within the humanities we might look to practice theory—a theoretical stance associated most prominently with French sociologist Pierre Bourdieu but developed in various ways by authors such as Judith Butler, Theodore Schatzki, and Bruno Latour. Crucially, practice theory conceptualizes the human as something like a body–mind complex and locates the origins of the social world not within the skull but in "practices"—that is, routinized behaviors that are made up of several elements: bodily activity, thoughts, background knowledge, know-how, emotional states, and things. Any "practice"—for example, cooking, writing, hunting, taking care of oneself—depends upon these multiple elements. Unlike most culturalist social theories, practice theory treats the body as more than an instrument or the site of inscription: it is the carrier of practices (Reckwitz). Practice theory thus overcomes the division between mind and body: both are co-equal and integrated. Social practices preexist individuals and are learned, some voluntarily, some not. Individuals, in turn, are carriers of practices, ensuring their transmission and reproduction.

For environmental humanists, the key attraction of practice theory is that it "insists on drawing our eyes back down to earth" (Reddy 8). It foregrounds the crucial role not only of humans but also of environments and the objects through which practices are carried out. Specific environments afford opportunities for certain practices while foreclosing others (Gibson). The presence of wild game afforded the opportunity for hunting; the presence of swiftly flowing rivers afforded the opportunity to build power-driven looms; the presence of the lac bug (along with other things) afforded the possibility of mass-produced recordings. Environmental change will inevitably bring changes in practice, while practices themselves ramify in the environments that humans occupy.

Toward an environmental history of culture

As an example, we might consider the history of the senses. Thus far, much of sensory history replicates the familiar division between body and environment, with historians emphasizing the preeminent role of culture in shaping sensory perception. While this has yielded important insights, it has also participated in an objectification of the senses. The senses themselves are treated as mere instruments of playback, sending signals to the brain for "processing." Although any given mind may perceive things differently in accord with "culture," the senses themselves (and by extension, the body) are assumed to be somehow independent (Ingold, *Being Alive* 136–140). It is the mind–body split all over again. Sensory capacities are established biologically before birth, and culturally thereafter. The history of the senses then becomes a history of culture. The body itself is a constant—and thus disappears. The environment plays little or no meaningful role.

But to view the senses as static information processors is itself a historical development. In her history of modern sound, historian Emily Thompson points to the role of electronic measuring technologies and building techniques in shaping our perception of not only sound but also hearing. Hearing itself is now technically measured through one's ability to distinguish certain electronic signals, which raises the question, what did it mean to hear in an earlier period? Similarly, the rise of the modern chemical flavoring industry and the quantitative measurement of taste were outgrowths of a particular instrument, the gas chromatograph, which allows for the detection of particular chemical substances. But is the sense of taste reducible to an internal gas chromatograph?

We are well aware that the senses may be trained; we acknowledge that a professional musician or ornithologist has a more developed sense of hearing, an enologist or chef a finer sense of taste and smell. But are these differences *merely* cultural? Or do they reside in the organization of the body itself? Modern neuroscience helpfully suggests the latter: that practice changes not only the organization of the body (changes in muscle mass, joint flexibility, the ability to repeat complex movements accurately) but the brain as well. However, much of our scholarship relies on older notions: that cultural differences are superimposed upon biological givens. Yet scientific work increasingly supports what much anthropological evidence and philosophical speculation had already suggested: sensory capacities are developed, and then decay, over the course of a life. That is, sensory capacity is incorporated as we grow. It changes not only in response to adult guidance and cultural norms, but also in response to the environment, to aging, and to biologic trauma. So, for instance, after the onset of "deafness," the so-called "auditory cortex" of the brain may learn to process sign language. The "loss" of a sense in this case is better understood as a reorganization of the organism.

Sensory development depends upon the changing capabilities of the body, what the environment affords, and how culture guides us to engage with that environment. Paul Carter gives us the example of the Australian bush call, "Coeee"—an aboriginal call that white immigrants adopted in place of their traditional "Hallo!" The reasons, it seems, were environmental—it was possible to produce a larger volume of sound with the word "Coeee" than with "Hallo" and thus the new sound had distinct advantages in the vast open spaces of Australia. The persistence of "Coeee," however, had less to do with its usefulness than with its incorporation into white Australians' sense of national identity. This history of "Coeee" suggests the imbrication of environment and social practice. Steven Feld, who studied the musical expression of the Kaluli people of New Guinea in the 1970s, found that in order to understand Kaluli music he needed to become an expert on the local bird population and on the birdsong from which their music derived. But while this was common knowledge among local people, Feld himself struggled to recognize and differentiate birdsong in the same way (for examples, see Mark Smith; my interpretation is indebted to Ingold, *Being Alive* 136–140). The rich avifauna of the island was an affordance; Kaluli musical skills emerged in a particular environmental context. Body and environment are coupled systems; intelligence and knowledge emerge out of their interaction.

One of the key insights of modern neuroscience and epigenetics is that environmental change is not merely that. It is also human change, through and through, down to the most intimate levels. The emergent field of neuroanthropology has brought attention to the inherent entanglement of biology and culture in the development of the nervous system and bodily praxis. Diverse cultural practices shape us in quite biological ways. But even that does not go far enough because environments and materials enable and constrain cultural practice. In any cultural-historical moment, there is a set of feedback loops operating that cannot be specified in advance of careful historical and ethnographic work. If we recognize this complexity, we also have to acknowledge that genetic mapping and functional magnetic resonance imaging will not yield much in the way of understanding; we need thick, contextual description. As Greg Downey and Daniel Lende put it, what we know as the "human" might be better understood as a "human-environment feedback system":

Although every animals' nervous systems is open to the world, the human nervous system is especially adept at projecting mental constructs onto the world, transforming the environment into a sociocognitive niche that scaffolds and extends the brain's abilities. This niche is constructed through social relationships; accumulating material culture and technology; and the physical environments, ritual patterns, and symbolic constructs that shape behavior and ideas, create divisions, and pattern lives. Thus, our brains become encultured through reciprocal processes of externalization and internalization, where we use the material world to think and act, even as that world shapes our cognitive capacities, sensory systems, and response patterns.

(28)

As we take advantage of the affordances available to us, we change the environment and the affordances available to us thereafter. The environment—or perhaps landscape is the better term—thus becomes a crucial record and reservoir of past development and knowledge that we cannot take for granted. So our children are born into a world shaped by computers and electronic technologies that affords them the ability to develop their visual capacities in new ways. At the same time, other ways of being-in-the-world are no longer available to them. For instance, the potential to develop one's sense of touch, dexterity, or musical voice—capacities essential to human beings the world over until at least the late nineteenth century—is now, in modern societies, available only as a luxury to those with ample leisure time, for arts and crafts such as pottery, woodworking, and vocal performance. From this perspective, the history of both human culture and the human body is, must be, environmental through and through.

Notes

1 In focusing on the Neolithic environments of the Middle East as the "environment of evolutionary adaptation," Smail differs from classic accounts of sociobiology and evolutionary psychology which emphasize the Pleistocene environments of the African savannah.
2 There is a connection here with the work of scholars who study contemporary neuroscience and artificial life and who argue that cognition is best understood as a distributed system. As Katherine Hayles, drawing on Hutchins, has observed, "[m]odern human beings are capable of more sophisticated cognition than cavemen not because moderns are smarter … but because they have constructed smarter environments in which to work" (289).
3 This is reminiscent of what pioneering women's historians labeled the work of social reproduction, though with a strong environmental component.

References

Boyd, Robert and Peter J. Richerson. *Culture and the Evolutionary Process*. Chicago, IL: University of Chicago Press, 1985. Print.
Cronon, William. *Changes in the Land: Indians, Colonists, and the Ecology of New England*. New York: Hill and Wang, 1983. Print.
Crosby, Alfred W. *The Columbian Exchange: Biological and Cultural Consequences of 1492*. Westport, CT: Greenwood, 1972. Print.

——. *Ecological Imperialism: The Biological Expansion of Europe, 900–1900*. Cambridge: Cambridge University Press, 1986. Print.

Downey, Greg and Daniel H. Lende. *The Encultured Brain: An Introduction to Neuroanthropology*. Cambridge, MA: MIT Press, 2012. Print.

Gibson, James J. *The Ecological Approach to Visual Perception*. Boston: Houghton Mifflin, 1979. Print.

Hayles, N. Katherine. *How We Became Posthuman: Virtual Bodies in Cybernetics, Literature, and Informatics*. Chicago, IL: University of Chicago Press, 1999. Print.

Hutchins, Edwin. *Cognition in the Wild*. Cambridge, MA: MIT Press, 1995. Print.

Ingold, Tim. *Being Alive: Essays on Movement, Knowledge and Description*. New York: Routledge, 2011. Print.

——. *The Perception of the Environment: Essays on Livelihood, Dwelling and Skill*. New York: Routledge, 2000. Print.

——. "Prospect." *Biosocial Becomings: Integrating Social and Biological Anthropology*. Ed. Tim Ingold and Gisli Palsson. New York: Cambridge University Press, 2013. 1–21. Print.

——. "'Tools for the Hand, Language for the Face': An Appreciation of Leroi-Gourhan's Gesture and Speech." *Studies in History and Philosophy of Science Part C: Studies in History and Philosophy of Biological and Biomedical Sciences* 30.4 (1999): 411–453. Print.

——. "The Trouble with Evolutionary Biology." *Anthropology Today* 23.2 (2007): 13–17. Web. 1 September 2015.

Latour, Bruno. *We Have Never Been Modern*. Trans. Catherine Porter. Cambridge, MA: Harvard University Press, 1993. Print.

Leroi-Gourhan, André. *Gesture and Speech*. Cambridge, MA: MIT Press, 1993. Print.

McNeill, John. *Mosquito Empires: Ecology and War in the Greater Caribbean, 1620–1914*. New York: Cambridge University Press, 2010. Print.

McNeill, William H. *Plagues and Peoples*. Garden City, NY: Anchor Press, 1976. Print.

Mesoudi, Alex, Andrew Whiten, and Kevin N. Laland. "Towards a Unified Science of Cultural Evolution." *Behavioral and Brain Sciences* 29.4 (2006): 329–347. Web. 1 September 2015.

Nash, Linda. "Furthering the Environmental Turn." *Journal of American History* 100.1 (2013): 131–135. Print.

Palsson, Gisli et al. "Reconceptualizing the 'Anthropos' in the Anthropocene: Integrating the Social Sciences and Humanities in Global Environmental Change Research." *Environmental Science & Policy* 28 (2013): 3–13. Print.

Ramirez-Goicoechea, Eugenia. "Life-in-the-Making: Epigenesis, Biocultural Environments and Human Becomings." *Biosocial Becomings: Integrating Social and Biological Anthropology*. Ed. Tim Ingold and Gisli Palsson. New York: Cambridge University Press, 2013. 59–83. Print.

Reckwitz, Andreas. "Toward a Theory of Social Practices: A Development in Culturalist Theorizing." *European Journal of Social Theory* 5.2 (2002): 243–263. Print.

Reddy, William M. "Saying Something New: Practice Theory and Cognitive Neuroscience." *Arcadia: International Journal for Literary Studies* 44 (2009): 8–23. Web. 6 September 2016.

Russell, Edmund. *Evolutionary History: Uniting History and Biology to Understand Life on Earth*. New York: Cambridge University Press, 2011. Print.

Smail, Daniel Lord. "Neuroscience and the Dialectics of History." *Análise Social* 47.205 (2012): 894–909. Web: 18 January 2016.

——. *On Deep History and the Brain*. Berkeley, CA: University of California Press, 2007. Print.

Smith, Douglas and Gary Ferguson. *Decade of the Wolf: Returning the Wild to Yellowstone*. Rev. ed. Guilford, CT: Globe Pequot Press, 2012. Print.

Smith, Mark Michael. *Sensory History*. Oxford: Berg, 2007. Print.

Stotz, Karola. "Developmental Niche Construction: An Integrative Causal Factor in Evolution." Oregon State University, 2007. Presentation.

Sutter, Paul S. "The World with Us: The State of American Environmental History." *Journal of American History* 100.1 (2013): 94–119. Print.

Thompson, Emily Ann. *The Soundscape of Modernity: Architectural Acoustics and the Culture of Listening in America, 1900–1933*. Cambridge, MA: MIT Press, 2002. Print.

White, Richard. "Environmental History: Watching a Historical Field Mature." *Pacific Historical Review* 70.1 (2001): 103–111. Print.

——. *Land Use, Environment, and Social Change: The Shaping of Island County, Washington*. Seattle: University of Washington Press, 1992. Print.

Worster, Donald. *Dust Bowl: The Southern Plains in the 1930s*. New York: Oxford University Press, 1979. Print.

——. "The Living Earth: History, Darwinian Evolution, and the Grasslands." *A Companion to American Environmental History*. Ed. Douglas Cazaux Sackman. Hoboken, NJ: Wiley-Blackwell, 2010. 51–68. Print.

——. "Transformations of the Earth: Toward an Agroecological Perspective in History." *The Journal of American History* 76.4 (1990): 1087–1106. Print.

39

MATERIAL ECOCRITICISM AND THE PETRO-TEXT

Heather I. Sullivan

Our current geological epoch, provisionally called the Anthropocene, has come into being with the increasingly intensive extraction and utilization of fossil fuels by human beings. One might say that the Anthropocene is the era of humanity writ large across the planet (Crutzen; Crutzen and Steffen; Zalasiewicz et al., "We Are Now Living"; Zalasiewicz et al., "New World"). While it seems that we humans (or industrialized humans) are the power bringing about the changes, it would be more accurate to say that much of industrialized human *power* in this new geological era is enabled by the access to and use of concentrated forms of *energy* driving our technologies, global transportation systems, and modern agricultural practices. Timothy Morton states that however problematic the term "Anthropocene" is, it nevertheless de-emphasizes human exceptionalism so that "the Anthropocene is the first truly anti-anthropocentric concept" ("How I Learned" 6). There are numerous non-human factors influencing our era, including other life forms and all kinds of active matter such as carbon dioxide, various types of pollution and waste, shifting weather patterns bringing drought and floods, and the warming climate, among others. Some of the most significant agents of change in these flows of energy and matter are, to reiterate, fossil fuels; indeed, high-tech industrialized (human) culture since the eighteenth century is now being described as an "oil culture" or a "petroculture" (Barrett and Worden; LeMenager). This chapter addresses the impact of fossil fuels on cultural productions with the help of *material ecocriticism*, the branch of ecocriticism that analyzes how human and nonhuman agencies exchange energy, matter, and information.

To study what Serenella Iovino and Serpil Oppermann call "storied matter" in terms of "power" acknowledges that life follows its energy sources, and human bodies and cultures are no exception ("Theorizing" 451). We are, in many ways, embodiments of the fuel that carries us along our many roads, whether it be food or gas. Fuel, by its very nature, is energetic and should not be thought of as a static object to be consumed. In petrocultures, a petroleum-derived road system carries food itself fueled and protected by petroleum-based fertilizers, herbicides, and pesticides, and transported in vehicles using petroleum or other fossil fuels contained in petro-based plastics. As this chapter will show, petro-foods and petro-paths produce petro-texts. This does not mean, however, that human agency has been erased or flattened out into a distributed form determined solely by the fuel of choice; instead we must think of tandem, joint agencies coalescing into hybrid human–petroleum actions. Our petro-texts emerge from this strange collaboration.

Material ecocriticism combines a study of stories and materiality together. It explores how human discursive practices and "vibrant matter"—in Jane Bennett's term—or "storied matter" together shape our worlds of words and matter. In *Material Ecocriticism*, Iovino and Oppermann describe the field's primary assertions as follows:

> the world's material phenomena are knots in a vast network of agencies, which can be 'read' and interpreted as forming narratives, stories. Developing in bodily forms and in discursive formulations, and arising in coevolutionary landscapes of natures and signs, the stories of matter are everywhere: in the air we breathe, the food we eat, in the things and beings of this world, within and beyond the human realm. All matter, in other words, is a 'storied matter.'
>
> (1–2)

One aspect of material ecocriticism's contribution to ecological scholarship is thus the combined study of human and nonhuman agents, whether the "post-human," animals, plants, living things generally, or fictional beings, too, as well as matter's active, agentic processes at all scales including the cellular, quantum, and even cosmic levels. As Dana Phillips and I have noted, material ecocriticism "insists that human beings are 'actors' operating within material processes that include multitudes of other 'actors,' the majority of which are not human or, for that matter, conscious" (446).

We live in a vibrant world of "actants"—to use Bruno Latour's term—that directly influence their surroundings and create changes. Hence this is not a world in which *passive* matter is merely the *object* of study beheld by *active human subjects*. Karen Barad, quantum physicist, philosopher, and new materialist, states in *Meeting the Universe Half Way: Quantum Physics and the Entanglement of Matter and Meaning*: "Matter is neither fixed and given nor the mere end result of different processes. Matter is produced and productive, generated and generative. Matter is agentive, not a fixed essence or property of things" (137). Bennett, author of *Vibrant Matter*, asserts that agency is distributed and all actions are in fact brought about by "a swarm of vitalities" (32), including bacteria on and in our bodies and electrons coursing (sometimes on their own unexpected paths) through electrical wires energizing the electronics without which we seemingly cannot function culturally.

Material ecocriticism therefore analyzes matter, stories, and power together as co-emergent processes—and with potentially empowering possibilities. An optimistic appraisal of such materialities is offered by Stacy Alaimo and Susan Hekman's introduction to *Material Feminisms*:

> Moreover, thinking through the co-constitutive materiality of human corporeality and nonhuman natures offers possibilities for transforming environmentalism itself. Rather than centering environmental politics on a wilderness model, which severs human from nature and undergirds anti-environmentalist formulations that pit, say, spotted owls against loggers, beginning with the co-extensive materiality of humans and nonhumans offers multiple possibilities for forging new environmental paths.
>
> (9)

Alaimo also writes in *Bodily Natures* that "the material self cannot be disentangled from networks that are simultaneously economic, political, cultural, scientific, and substantial" (20).

She thus speaks of "trans-corporeality," or thinking materiality across bodies both human and more-than-human, so that we gain "a recognition not just that everything is interconnected but that humans are the very stuff of the material, emergent world" (20). Stories—no matter how metaphorical, allegorical, or science fiction(al) they seem—are a part of this emergent, material, and meaningful world.

Additionally, not just human culture but also matter has a history and takes different forms through time. As Dipesh Chakrabarty notes for the environmental humanities generally, we must, in the era of the Anthropocene, study human and natural histories as conjoined and co-emerging. Matter can also have an informational capacity such as cell walls that allow specific substances to enter and exclude others, as Wendy Wheeler, Timo Maran, and other scholars in biosemiotics describe. Iovino and Oppermann's concept of "storied matter" can also suggest that "stories matter" as a discursive power, which need not be mimetic to be effective. "Power" here refers to the ability to have cultural impact as well as to the literal power of energy sources and fuels. It will not come as a surprise to note that "power," in broad terms, affects meanings and produces particular interpretations, nor to note that emerging meanings and analyses most often perpetuate and replicate the very power system that enables them. Thus material ecocriticism's efforts to accommodate the implications of all kinds of "power," especially the energy sources fueling cultural productions, highlight the fact that these are, in the Anthropocene, primarily fossil fuels. The editors of *Oil Culture*, Ross Barrett and Daniel Worden, state that:

> oil culture encompasses the fundamental semiotic processes by which oil is imbued with value within petrocapitalism, the promotional discourses that circulate through the material networks of the oil economy, the symbolic forms that rearrange daily experiences around oil-bound ways of life, and the many creative expressions of ambivalence about, and resistance to, oil that have greeted the expansion of oil capitalism.
>
> (xxvi)

This chapter utilizes this notion of oil as text in considering how recent novels by Margaret Atwood, Cormac McCarthy, and Paolo Bacigalupi function as petro-texts explicitly analyzing our petroculture in terms of the ecological impacts of "peak oil" and "post-oil" cultures.

How does living in a petroculture shape "natureculture" and our stories? Much of humanity's recent population and urban expansion occurs not alongside, but rather because of the capitalist consumption of oil and other fossil fuels. The costs of this "turbo-capitalism," as Rob Nixon describes it in *Slow Violence and the Environmentalism of the Poor*, are often considered by those with *power* (political power, but also the literal access to rich sources of energy, and, of course, capital) to be merely future problems that we will solve with future technology (8). However, the costs are already dramatically spread about the globe in oil spills, pollution, and violence over resource acquisition. Overwhelmed by the shining economic prosperity that fossil fuels bring (to the industry), many in the West and the United States see this petroleum sheen as a cultural enrichment rather than a short-term burst on borrowed time, as Nixon states. In my home state of Texas, the agentic capacity of oil provides massive wealth, but it also continues to wreak ecological harm such as the disastrous impact on beaches and ocean life of the Deepwater Horizon oil spill, which is still producing tar balls, polluted water, and sickened animals ("Deep Water," "Latest NOAA Study," and "Five Years"). Impacts on air, water, soil, and geology (fracking earthquakes are common) across the state are significant (Weingarten et al.; Hildenbrand et al.). More broadly in the world, particularly in the less economically powerful nations in the global South, various forms of "petro-despotism" and "oil hegemony" rule with devastating ecological impact and unspeakable lack of environmental justice for local inhabitants (Nixon 68). Stories of energy are literal tales of nonhuman *agents*, often with surprising twists about who or what has the most impact.

In the face of petro-power's agency and its impact on our naturecultures, Nixon asks why there are not more "Big Oil novels" like the King Coal texts of the nineteenth and early twentieth century. Cultural productions, after all, whether they are extractive practices, agriculture, urban development, economic structures, or the arts and literature, are all shaped and diverted into specific forms by the very energy driving their production and consumption, which in this era are fossil fuels. Thus we might today describe almost all literature from industrialized countries as well as countries with extractive industries as "petro-texts" in line with material ecocriticism's efforts to study human and material agents together as interconnected and interacting energies creating the texts of the world. To use Timothy Morton's term, the "mesh" of the world—both ecological and textual—now has a petroleum sheen.

Despite Nixon's concern about the lack of a great oil novel, there are, in fact, quite a number of excellent recent texts directly addressing oil as cultural agent. Primarily science fiction, they are examples of "cli fi" or climate (change) fiction (Trexler; Canavan and Robinson). The coining of the term "cli fi" is usually attributed to Dan Bloom, and Stephanie LeMenager's contribution to this volume explores the sociological phenomenon of cli fi's constitution as a genre. Three examples in particular serve as representative here, each one exemplifying a more general category, and each exploring oil as if from the future aftermath of a "post-oil" culture. The first category, which I term "post-apocalyptic dark pastoral," could be applied to a broad range of texts, including Margaret Atwood's *MaddAddam* trilogy (*Oryx and Crake* 2003, *The Year of the Flood* 2009, and *MaddAddam* 2013), Octavia Butler's *Parable of the Sower* (1993), Mindy McGinnis' *Not a Drop to Drink* (2013), Jean Hegland's *Into the Forest* (1996), and Simone Pond's *The City Center* (2013), all of which depict the devastation of current industrial culture and a return to the land to start anew in the wasteland or radically altered landscapes. In Atwood's trilogy, the vast majority of humanity is wiped out by a humanly created disease, allowing a "new nature" to emerge and flourish, hence "dark" pastoral (see Sullivan, "Dirty Traffic"). This nature is composed of genetically modified animals (rakunks are raccoon/skunk mixes, wolvogs are wolf/dogs, and pigoons are human/pig), and posthumans (the ecologically friendly posthuman Crakers) who live alongside a few remaining traditional humans who garden, tend bees, develop new stories not based on fossil fuel consumption, and seek alternative fuel sources like solar energy. "Three more functioning solar units have been installed. An existing one has gone out of commission. ... Shackleton and Crozier have experimented with making charcoal" (*MaddAddam* 377).

Atwood has two storylines in the trilogy, though: first this post-apocalyptic dark pastoral in which we follow the few survivors joining the posthumans and pigoons to form new ecocultures and, second, the even darker story of what leads to the total destruction of petrocultures in the first place, including climate change, pollution, and techno-capitalism. In this storyline, Jimmy experiences climatic disasters as the normalized backdrop to his mundane childhood. Jimmy functions as the standard for our culture today: mostly oblivious and complicit. Only in the aftermath does he see the consequences. Atwood calmly paints disturbing climate change and devastation as the barely noticed frame for Jimmy's worries about his lunchmeats:

Still, as time went on and the coastal aquifers turned salty and the northern permafrost melted and the vast tundra bubbled with methane, and the drought in the midcontinental plains regions went on and on, and the Asian steppes turned to sand dunes, and meat became harder to come by, some people had their doubts [about whether the human-pig mixed "pigoons" were not the source of all the bacon and ham in the cafeteria]. Within OrganInc Farms itself it was noticeable how often back bacon and ham sandwiches and pork pies turned up on the staff menu.

(Oryx and Crake 24)

What leads to the demise of petrocultures and "turbo-capitalism" in Atwood's novels is as much human activities as human ignorance, passivity, and focus on trivialities instead of massive planet-wide distress. In *Oryx and Crake*, the first novel in the trilogy, Jimmy's journey for supplies back to "Paradice," the scientific laboratory where both the posthuman Crakers and the disease that eliminated most of humanity were created by his friend Crake, enacts the journey of impending awareness and the move into a new life. This is a path out of the darkness and into a newly re-greened world; it occurs, however, only after the genocide that destroys virtually all of humanity.

The second category of petro-texts is the post-apocalyptic dead-end tale: declensionist novels like Cormac McCarthy's *The Road* that imagine that we have not just passed the infamous tipping point of climate change (or pollution or altered landscapes) but rather exploded—literally and mysteriously—out of any possible human living conditions altogether. Unlike other cli fi texts, this novel fits more closely within the older category of the Cold War novels portraying apocalyptic anticipations of nuclear winter and radioactive wastelands with no hope for some kind of dark pastoral return to the land. Examples include Philip K. Dick's *Do Androids Dream of Electric Sheep* (1968), Ward Moore's *Greener than You Think* (1947), in which suburban lawns conquer the entire earth thanks to a very special fertilizer leaving nothing in its wake but swaying grass as a "green" wasteland, and the short story by Alfred Bester, "Adam and No Eve" (1941). One might also include Thomas M. Disch's *The Genocides* (1965) in this category since the alien invasion that plants its massive, nutrient-sucking trees as a crop across the entire earth eliminates all other living species and, when they disrupt the agricultural system, eradicates human beings as well in a classic example of "reverse colonization." In McCarthy's world of *The Road*, our petrocultures have imploded and all that remains in terms of "fuel" for both our bodies, transportation, and technology are small quantities of gas or oil, random foods like left-over canned goods that are going fast, a few remaining rotting apples, or, as in the vast array of zombie or vampire tales, the flesh and blood of our fellow *Homo sapiens*. Like J. G. Ballard in *The Drought* (1964), McCarthy eradicates any possibility of new dark pastorals or alternative paths toward a new life. *The Road* explores instead the material realities of an almost fuel-less society; even sunlight barely makes it through the gloomy haze to the earth's surface. On foot and with the hope of heading south where it might be warmer, the only goal of the "man" and his son, the "boy," is to get somewhat closer to the sun's energy in the southern United States. The story is as much about their struggles as it is an enactment of *the road* itself as the major remaining feature of the Anthropocene. In McCarthy's *The Road*, there are no more states, but "the roads are still there" (43). The traces of petroculture still exist, though not much else.

The man and boy arrive finally at the coast, where the man dies and the boy joins the one remaining nuclear family apparently left in the novel, one that at least claims not to eat human beings: "You dont eat people. No. We dont eat people" (284). One can only hope that they can continue to avoid the roaming hordes of cannibals and, also, to overcome their own hunger. There is some hope at the very end of the novel, however, a hope of *symbolic*, if not concrete, fuel. The father's final dying words to his son are that he must leave him and go on to "carry the fire." The boy begins:

> I want to be with you.
> You cant.
> Please.
> You cant. You have to carry the fire.
> I dont know how to.
> Yes you do.
> Is it real? The fire?
> Yes it is.
> Where is it? I dont know where it is.
> Yes you do. It's inside you. It was always there. I can see it.
>
> (279)

Fossil fuels, food, other animals, the states, and his father are gone, but he still has "the fire," the symbolic or spiritual energy that remains when all else (except for roads and cannibals) has disappeared. *The Road* takes the petro-text and turbo-capitalist model all the way to a radical end of consumption. In the end, all forms of life, except humanity, inexplicably, and all energy sources, except his symbolic "fire," are gone. The challenge for the new family at the end of the novel is to keep this "fire" alive while avoiding becoming fodder—fuel—for other human beings. In the end, it's still a matter of fuel. Human beings concretely become our own fuel and there is one choice: either the "fire" within or the flesh that feeds others.

The third type of post-oil narrative is somewhat more optimistic than the declensionist end-of-the-road scenarios where only human bodies remain as fuel and the question of why animals die but humans survive is never answered, or the dark pastorals springing "anew" out of the remains of mass genocide and environmental collapse. The final forward-looking category of the "post-oil urban tale" includes works, such as Paolo Bacigalupi's *The Windup Girl*, that describe second- or third-generation post-energy crisis societies with a greatly reduced but recovering world population driven by the never-ending quest for old and new forms of energy or else imagined futures based on new energies. Similar texts include other works from Bacigalupi such as his collected short stories, *Pumpsix* (2008) and his novel *Shipbreaker* (2010), as well as Tobias S. Buckell's novel *Artic Rising* (2012), Andreas Eschbach's *Ausgebrannt* (2007), Saci Lloyd's *The Carbon Diaries* (2008), Kim Stanley Robinson's *2312* (2012), and Antti Tuomainen's *The Healer* (2011; translated from Finnish by Lola Rogers in 2013).

Bacigalupi's *Windup Girl* regales us with creative energy options, whether new or revised, such as energy stored in kink-springs, genetically altered elephant-like "megadonts" so huge as to be able to provide energy to entire factories as they amble along all day in circles like the mules that drove forges of yore, the good-old fashioned use of wind in sailing ships, and, in terms of food, the quest to rediscover lost varieties of fruits and vegetables. Methane is mostly illegal and oil is the rarest of remnants, available only to the military and the richest moguls like the powerful "dung lord," who rules by gathering garbage, waste, offal, and corpses, but there are other possibilities for the bustling urban Bangkok. It's not a utopia; indeed, Bacigalupi's novels and short stories are more accurately described as dystopian horror tales. In terms of energy, however, they envision a world beyond oil, with relatively large urban populations, that does not require either the annihilation of virtually the entire human population or the civilizational dead-end of nothing but human "fuel" remaining.

These three types of tales, the post-apocalyptic dark pastoral, the post-apocalyptic end-of-the-road tale, and the post-oil urban tale, all play with the "end" of fossil fuels and map out possible—or impossible—futures. Atwood's, McCarthy's, and Bacigalupi's stories engage with ecological questions of climate change and fuel crises, but they also re-wire our own ecological discourses and possible materialities with the exploration of innovative alternatives. One thing such stories can do is help our petro-cultures to overcome the blindness to our "enabling conditions," in Val Plumwood's terms (17). Dominant ideological and economic systems tend to overwrite the significance of ecology for our own survival, and often reject different energy trajectories as being impractical or idealistic options irrelevant to "economic realities." These petro-texts imagine the end of petro-capitalism and thereby reveal to various degrees the material ecocritical mesh of stories and materialities.

The strategies of the three kinds of tales vary quite dramatically. The "post-apocalyptic dark pastoral" works assume a massive population reduction and a return to land-based living; in many ways this is the most utopian and troubling strategy since it begins with vast genocide that makes possible new, or actually old, agricultural foundations. Its advantages, however, include documenting the path *toward* the crisis that Atwood in particular so clearly describes. Post-apocalyptic dark pastoral works show how choices we are making right now could lead to greater problems in the future. These warning tales suggest we might act to avoid the crisis and their utopian new beginnings map out some possible options. The post-apocalyptic dead-end tale, on the other hand, warns only that our current path will lead to total destruction. Yet McCarthy's play with themes of energy, the "fire" within, and human bodies as food highlight the very problem that we face currently: our energy choices literally shape our future paths. Finally, Bacigalupi's much more optimistic post-oil urban tale, like the other texts in this category, provide the most grounded assessment of future possibilities without the extreme drama of the apocalyptic-type declensionist narratives. Alternative forms of energy are lined up with alternative forms of human and posthuman and nonhuman beings. Coping with crisis, long-term resilience, and urban ecology are the stuff of these tales, which range from optimistic futures (Robinson) to pessimistic devastation. Across this spectrum, however, post-oil tales include practical strategies like eating locally and riding bikes (Eschbach; Lloyd). These tales make visible the mostly hidden facts of how our current cultures are shaped by petroleum and fossil fuels generally. We have agency, but it is dripping with oil and acting along with the smoggy particulates.

Studying petro-texts and petroleum-based agency reveals significant challenges in our current petro-cultures. Besides the obvious pollution, expanding anthropogenic havoc on the planet, climate change, and the ethical questions that we must face, there is also the fact that much of our impact is inadvertent and unintentional in kind. We certainly do not mean to be the driving force behind what is being termed the sixth extinction event on Earth due to pollution, development, and loss of habitation, for example (Kolbert). Our "agency" in terms of local and global alterations is thus as much accidental as purposeful. What is worse, then: to be the (purposeful) agents of destruction or to be the unwitting destroyers fumbling along in inevitable and endless quests for more energy? Additionally, these processes of change that we let loose are most often feedback loops in nonlinear, non-equilibrium, and nondeterministic systems that take on new and completely unexpected directions and energy beyond our control (see Prigogine; Botkin; Schneider and Sagan; Coole and Frost; Sullivan "Affinity Studies").

Moreover, the greater our technology and knowledge, the more we come to see and understand the significant number of factors that drive and influence *us* beyond our conscious control, including the obvious examples of hormones and coffee as well as the vast array of bacteria that live with, in, and on us, allowing digestion and making up the vast majority of the DNA in our bodies. Our own agency co-emerges with all kinds of energies. Of course, knowing of our co-agents and seeking to work with them offers possible new strategies for action. Barad notes that understanding agency necessitates

> a robust account of the materialization of *all bodies*—"human" and "nonhuman"— including the agential contributions of all material forces (both "social" and "natural"). This will require an understanding of the nature of the relationship between discursive practices and material phenomena; an accounting of "nonhuman" as well as "human" forms of agency; and an understanding of the precise causal nature of productive practices that take account of the fullness of matter's implication in its ongoing historicity.
>
> (66)

What does this interactive notion of agency mean for our techno-petrocultures and the petro-texts of today: is it the human beings seeking and utilizing oil and other fossil fuels, or the siren call of these energy sources that drive us, literally, both to them and to everywhere else we go? These fuels power most of the prevailing industries and dominant world economies today, which is to say that they shape—literally—their cultures and narratives. We seek and use with ever more impressive technological capacities the concentrated energy of fossil fuels, no matter how deep we must go, or how cold, hot, or dangerous their location, and most definitely no matter what the impact and long-term consequences on local, national, and global scales. Or, one might reverse this statement and assert that it appears *they use us*, human beings, to expose and engage or expend them at an ever faster pace, like a cosmic drive toward entropy that reduces high levels of energy deliriously in a compelling enactment of the Second Law of Thermodynamics, which explains the universal flows decomposing "gradients," the areas of difference between high energy and low energy, toward entropy (Prigogine; Schneider and Sagan). Do we in the post-industrial revolution era of the Anthropocene currently have much agency over (high-energy) fossil fuels, or do they now "drive us," or is it a combination of the two? Allan Stoekl writes in the "Foreword" to *Oil Culture*: "We are the slaves of our energy slaves [petroleum fuels], in a surprising (to us) revision of the Hegelian master-slave dialectic. We work to further oil's consumption, imagining ever more wasteful 'uses' for it—from lavish oil-heated homes to gas-guzzling Hummers" (xii–xiii). Furthermore, oil is "an *agent* as well: it calls to us, and we respond to its entreaties, its interpellations" (xii, my emphasis).

Studying petro-texts through material ecocriticism embarks us on a different road, one making overt our current petro-cultural practices. Our human agency to determine the course does indeed have world-shaping impacts, but always in conjunction with the energy sources that fuel it. The road we take from here will depend both on the fuel we utilize and the meanings to which we attribute our energy partnerships. We can—hopefully—choose our partner(s) and take responsibility for them; we might even leave the depths of fossil-fuel extraction and steer our path toward the sun, for example, whose agency continues to fuel most of the biosphere. And there are already many solar texts, of course, but that is another story.

References

Alaimo, Stacy. *Bodily Natures: Science, Environment, and the Material Self*. Bloomington, IN: Indiana University Press, 2010. Print.

Alaimo, Stacy and Susan Hekman, eds. *Material Feminisms*. Bloomington, IN: Indiana University Press, 2008. Print.

Atwood, Margaret. *Oryx and Crake*. New York: Doubleday, 2003. Print.

——. *MaddAddam*. New York: Knopf Doubleday, 2013. Print.

Bacigalupi, Paolo. *The Windup Girl*. San Franciso, CA: Night Shade Books, 2010. Print.

Ballard, J. G. *The Drought*. 1964. New York: Liveright, 2012. Print.

Barad, Karen. *Meeting the Universe Halfway: Quantum Physics and the Entanglement of Matter and Meaning*. Durham, NC: Duke University Press, 2007. Print.

Barrett, Ross and Daniel Worden, eds. *Oil Culture*. Minneapolis, MN: University of Minnesota Press, 2014. Print.

Bennett, Jane. *Vibrant Matter: A Political Ecology of Things*. Durham, NC: Duke University Press, 2010. Print.

Botkin, Daniel. *Discordant Harmonies: A New Ecology for the Twenty-First Century*. New York: Oxford University Press, 1990. Print.

Canavan, Gerry and Kim Stanley Robinson, eds. *Green Planets: Ecology and Science Fiction*. Middletown, CT: Wesleyan University Press, 2014. Print.

Chakrabarty, Dipesh. "The Climate of History: Four Theses." *Critical Inquiry* 35 (2009): 197–222. Print.

Coole, Diana and Samantha Frost, eds. *New Materialisms: Ontology, Agency, and Politics*. Durham, NC: Duke University Press, 2010. Print.

Crutzen, Paul. "Geology of Mankind." *Nature* 415 (2002): 23. Web. 1 July 2015.

Crutzen, Paul Josef and Will Steffen. "How Long Have We Been in the Anthropocene Era?" *Climatic Change* 61 (2003) 3: 251–257. Print.

"Deep Water: The Gulf Oil Disaster and the Future of Offshore Drilling." The National Commission on the BP Deepwater Horizon Oil Spill and Offshore Drilling. Government Publishing Organization, Jan. 2011. Web. 11 July 2015.

"Five Years and Counting: Gulf Wildlife in the Aftermath of the Deepwater Horizon Disaster." National Wildlife Federation. 2014. Web. 8 August 2015.

Hildenbrand, Zacariah Louis et. al. "A Comprehensive Analysis of Groundwater Quality in the Barnett Shale Region." *Environmental Science and Technology* 17 June 2015. 1–27. Web. 10 July 2015.

Iovino, Serenella and Serpil Oppermann, eds. *Material Ecocriticism*. Bloomington, IN: Indiana University Press, 2014. Print.

——. "Theorizing Material Ecocriticism: A Diptych." *Interdisciplinary Studies in Literature and Environment* 19 (2012): 448–475. Print.

Kolbert, Elizabeth. *The Sixth Extinction: An Unnatural History*. New York: Henry Holt, 2014. Print.

"Latest NOAA Study Ties Deepwater Horizon Oil Spill to Spike in Gulf Dolphin Deaths | Response. restoration.noaa.gov." *Office of Response and Restoration*. US Department of Commerce: National Oceanic and Atmospheric Administration; National Ocean Service, 13 Aug. 2015. Web. 14 Aug. 2015.

Latour, Bruno. *Politics of Nature: How to Bring the Sciences into Democracy*. Trans. Catherine Porter. Cambridge, MA: Harvard University Press, 2004. Print.

LeMenager, Stephanie, *Living Oil: Petroleum Culture in the American Century*. Oxford: Oxford University Press, 2014. Print.

McCarthy, Cormac. *The Road*. New York: Vintage, 2006. Print.

Maran, Timo. "Semiotization of Matter: A Hybrid Zone between Biosemiotics and Material Ecocriticism." *Material Ecocriticism*. Ed. Serenella Iovino and Serpil Oppermann. Bloomington, IN: Indiana University Press, 2014. 141–154. Print.

Morton, Timothy. *The Ecological Thought*. Cambridge, MA: Harvard University Press, 2010. Print.

——. "How I Learned to Stop Worrying and Love the Term *Anthropocene*." *Cambridge Journal of Postcolonial Literary Inquiry* 1 (2014): 257–264. Web. 1 August 2015.

Nixon, Rob. *Slow Violence and the Environmentalism of the Poor.* Cambridge, MA: Harvard University Press, 2011. Print.

Phillips, Dana and Heather I. Sullivan. "Material Ecocriticism: Dirt, Waste, Bodies, Food, and Other Matter." *Interdisciplinary Studies in Literature and Environment* 19 (2012): 445–447. Print.

Plumwood, Val. *Environmental Culture: The Ecological Crisis of Reason.* London: Routledge, 2001. Print.

Prigogine, Ilya. *From Being to Becoming.* San Francisco, CA: W. H. Freeman & Co, 1980. Print.

——. *The End of Certainty: Time, Chaos, and the New Laws of Nature.* New York: Free Press, 1997. Print.

Schneider, Eric D. and Dorion Sagan. *Into the Cool: Energy Flow, Thermodynamics, and Life.* Chicago, IL: University of Chicago Press, 2005. Print.

Stoekl, Allan, "Foreword." *Oil Culture.* Ed. Ross Barrett and Daniel Worden. Minneapolis, MN: University of Minnesota Press, 2014. xi–xiv. Print.

Sullivan, Heather I. "Affinity Studies and Open Systems: A Nonequilibrium, Ecocritical Reading of Goethe's *Faust*." *Ecocritical Theory: New European Approaches.* Ed. Axel Goodbody and Kate Rigby. Charlottesville: University of Virginia Press, 2011. 243–255. Print.

——. "Dirty Traffic and the Dark Pastoral in the Anthropocene: Narrating Refugees, Deforestation, Radiation, and Melting Ice," *Literatur für Leser* 14.2 (2014): 83–97. Print.

Trexler, Adam. *Anthropocene Fictions: The Novel in a Time of Climate Change.* Charlottesville, VA: University of Virginia Press, 2015. Print.

Weingarten, M., S. Ge, J. W. Godt, B. A. Bekins, and J. L. Rubinstein. "High-rate Injection is Associated With the Increase in U.S. Mid-continent Seismicity." *Sciencemag.org* 348.6241 (June 2015); 1336–1440. Web. 1 August 2015.

Wheeler, Wendy. "Natural Play, Natural Metaphor, and Natural Stories: Biosemiotic Realism." *Material Ecocriticism.* Ed. Serenella Iovino and Serpil Oppermann. Bloomington, IN: Indiana University Press, 2014. 67–79. Print.

Zalasiewicz, Jan, Mark Williams, Will Steffen, and Paul Crutzen, "The New World of the Anthropocene," *Environmental Science & Technology* 44.7 (2010): 2228–2231. Web. 1 August 2015.

Zalasiewicz, Jan et al. "Are We Now Living in the Anthropocene?" *Geological Society of America* 18.2 (2008): 4–8. Web. 1 August 2015.

40

FOSSIL FREEDOMS

The politics of emancipation and the end of oil

Hannes Bergthaller

Thinking past petroleum

The refrain of Joni Mitchell's "Big Yellow Taxi" could serve as a motto for much environmentalist thinking: "Don't it always seem to go/That you don't know what you've got/ Till it's gone." But the lines seem particularly apposite with regard to the new interest that humanist scholars have begun to take in what has variously been referred to as "oil culture" (Barrett and Worden), "petroculture" (Szeman 148), or "petromodernity" (LeMenager 71). Oil, these scholars argue, is much more than a culturally inert "energy carrier." Its distinct physical properties have shaped not only the material infrastructure of industrial society in the twentieth century, but also affected what one might call its social and psychological aggregate state. Petroleum has almost twice the energy density of coal. By comparison with the latter, extracting and transporting it requires very little physical labor, putting a correspondingly higher premium on technical expertise and entrepreneurial ingenuity. Not only does burning it produce many fewer particulates but also motorized transport allows for the increased spatial disaggregation of the spheres of production, consumption, and reproduction. This enables many people to live in cleaner, healthier environments—and makes it much easier for them to ignore the local ecological costs of oil extraction and processing. Thanks to advanced organic chemistry, petroleum can be transmuted into a prodigious array of products, from motor fuels and lubricants to plastic wrap and film stock, nylon stockings and cosmetics, pharmaceuticals, pesticides and fertilizers—products which, together, made possible consumer society as we know it. If we ask what it *felt* like to be "modern" in the twentieth century, our answers will invariably lead us back to petroleum. It fueled not only cars, ships and airplanes, but, along with them, visions of the good life, individual and collective aspirations, and a cultural style which, against a backdrop of steadily growing material affluence, valorized risk-taking and the transgression of established boundaries (Buell; Musiol; Sloterdijk).

Many of the distinctive features of petromodernity as they emerge from such accounts are, of course, quite familiar—they are roughly congruent with what earlier scholarship had described under such headings as "the society of the spectacle" (Debord; Watts), "time-space compression" (Harvey), "acceleration" (Virilio; Rosa), "deterritorialization" (Appadurai), or, in the broadest sense, globalization. What *is* novel is the attempt to systematically correlate such phenomena with the energetic base on which modern world society rests—an attempt that derives much of its urgency from the fact that, since the turn of the century, this energetic base has come to seem increasingly troubled and precarious. Predictions that "peak oil" was imminent or had already passed, the Iraq war (ostensibly fought to wrest weapons of mass destruction from a vicious dictator, but widely regarded as an attempt to shore up US control over the world's dwindling supply of hydrocarbons), the Deep Water Horizon oil spill, but above all the growing concerns over climate change—all made it abundantly clear that the industrialized world's continuing reliance on petroleum as a primary source of energy comes at a steep price. In the United States, political pundits frequently spoke of the nation's dependency on oil in terms of an "addiction," a phrasing adopted by none other than President George W. Bush in his 2006 State of the Union address.

To be sure, the metaphor of addiction is a deeply problematic one. It invites a view of the dependency on oil as a sort of character flaw, something that could be shaken off if only we could muster the requisite moral stamina. It also may feed into the above-mentioned belief that oil is inessential to advanced industrialized society, or suggest that the benefits it has yielded are spurious. Nevertheless, the metaphor is instructive insofar as it draws attention to those aspects of the problem with which the new scholarly approaches to oil culture are principally concerned. Addiction is not just a physiological, but also a psychological condition, and it is notoriously difficult to determine where one ends and the other begins. One is never addicted just to the substance itself, but always also the emotions it affords and to the states of being it allows one to experience. Effective treatment requires an approach which focuses not only on the addict's body, but also looks at the ways in which he conceives of himself and his relationships to others. A similar argument can be made about oil dependency: if we wish to get over it, we cannot do without expertise from the natural and social sciences; but surely, we also need to come to grips with how petroleum has come to structure "a whole way of life," to quote Raymond Williams's well-known definition of culture (3). Understanding the ways in which energy systems condition the life-worlds people inhabit, their sense of agency, horizon of expectations, and modes of communication, will therefore have to rank among the principal tasks of the environmental humanities. It requires a multidisciplinary approach that views such systems in light of their historical genesis and examines the symbolic forms—the narratives, tropes, and images—that have co-evolved with them.

What the metaphor of addiction also highlights is the extent to which contemporary debates about oil are charged with profound anxieties over a loss of autonomy. On a geo-political level, dependency on oil has tempted democratic nations into unsavory alliances with petro-dictatorships, thus undermining their ability to conduct foreign policy in accord with the values they publicly uphold. Oil itself, some political scientists have sought to demonstrate, has "antidemocratic properties" (Ross 325). On the individual level, too, petroleum seems to function as a powerful generator of hypocrisy and bad faith, making us do all sorts of things we would rather not be doing, such as generating large quantities of CO_2 every time we attend an international conference. On the national stage in the United States, "energy independence" has become a popular rallying cry across the entire political spectrum, and is called upon to justify policies as different as government subsidies for renewable energy and stepped-up efforts to exploit remaining domestic hydrocarbon reserves. The immense appeal of "energy independence" as a political slogan rests on the understanding, deeply ingrained in public discourse since the oil crises of the 1970s (Huber 114–127), that access to cheap gasoline is an essential component of the "American way of life," and that threats to the oil supply also constitute a threat to the liberty and self-reliance of American citizens.

But what is freedom, one must ask, if it depends on something that is not free? And what is self-reliance really, if it relies on something that is not self? While the political mainstream largely agrees on the need to secure American liberty by subjecting the nation to the equivalent of a twelve-step program—carefully extricating it from its dependence on foreign oil imports, increasing energy efficiency, and gradually shifting toward other sources of energy—there are also those who look forward to the prospect of going cold turkey. Around the middle of the last decade, "peak oil" became a catalyst for survivalist fantasies wherein it was imagined as the apocalyptic event that would usher in a new birth of freedom. When the oil runs out, authors such as Howard Kunstler, Richard Heinberg, or Derrick Jensen predicted, the entire elaborate artifice of modern society would collapse like a house of cards. While this would spell catastrophe for many, it would also purge the world of the evils of corporate power and big government, and allow for the return to a way of life in which individuals and small-scale communities could finally regain control over their own fate. In "Big Yellow Taxi," Mitchell had presented the building of a "parking lot" as a replay of the biblical Fall; in the minds of the "peak oil" crowd, the imminent demise of petroleum-based infrastructure heralded a return to the Garden.

Petromodernity as socio-metabolic regime

Petroleum, as it figures in these debates, assumes something of the ambivalent status of Derrida's *pharmakon* (95–104): flickering between remedy and poison, feeding and frustrating the desire for individual or national self-determination, it is both that which makes liberty possible, and, at the same time, poisons it at the root. If oil can serve as a potent emblem of modernity's original sin, it is because our dependency on it flagrantly contradicts a metaphysics of free will that is an integral component of Western traditions of thought. The Enlightenment had envisioned history in terms of a progressive emancipation of the human spirit from material constraints. Here is how August Baldwin Longstreet expressed this view in his *Patriotic Effusions* of 1819, ventriloquizing the "Goddess of Freedom": "Not with stone or with wood ... /Shall the mansion *of Freedom* henceforth be supplied,/ ... For these ... /Will moulder quite fast, and in time pass away./'Tis my mandate, said she ... /In the hearts of *Columbians* the ground work be laid" (16–17). The edifice of the American nation, the home of the free, is imagined as a gift from providence ("heaven's bounty," 17) which owes its superior durability precisely to the fact that its foundation is immaterial.

Given that Longstreet went on to become a vocal defender of slavery during the antebellum years, one must point out that much of the "ground work" of his mansion was in fact laid by the physical labor of black bodies. In light of the foregoing discussion, however, another layer of misrecognition comes into view, one that puts into question not only the racism which so constrained Longstreet's understanding of liberty, but also the very notion of liberty as an *unconditioned* good, as a property of the human spirit which transcends material conditions and drives historical progress. In his seminal essay on the implications of climate change for the historians' craft, Dipesh Chakrabarty has articulated this challenge with epigrammatic concision: "The mansion of modern freedoms," he writes, "stands on an ever-expanding base of fossil-fuel use" (208). But how are we to spell out the implications of this metaphor? Does the term "mansion" refer to our abstract conceptions of freedom, to the institutions that we think of as safeguarding freedom, to the habits we have acquired in inhabiting them, to the material infrastructure of consumer capitalism—or to all of the above? What does it mean for fossil fuels to form the "base" of this mansion? Exactly how deep does this foundation lie? And, most importantly, does the metaphor imply that with the end of fossil fuels, the modern understanding of freedom must go the way of the dinosaurs and become itself a fossil, as it were?

Any attempt at answering these questions must begin by widening the aperture so as to view petromodernity within the larger context of a history of human energy use. This history can be understood in terms of a succession of different "socio-metabolic regimes" (Fischer-Kowalski and Rotmans). Like biological organisms, social structures require a steady flow of energy in order to reproduce themselves. They absorb structured energy so as to maintain their internal complexity and avoid thermodynamic equilibrium, and release it back into their surroundings in an unstructured, unusable form (i.e., as waste). The availability of energy places fundamental constraints on the kinds of social order that are possible. The energy sources a society relies on, and the technologies it has at its disposal in order to convert them into usable energy, condition "patterns and levels of resource use ... demographic and settlement patterns, patterns of use of human time and labor ... institutional characteristics, and communication patterns" (Krausmann et al. 639). This also explains the many structural convergences between societies that had little or no cultural exchange, for example the parallel development of writing, monarchy, bound labor, priesthood, or astronomy in agrarian civilizations across the world (Sieferle 9). On this basis, one can distinguish three large phases of human history: the hunter-gatherer regime, the solar-agrarian regime, which prevailed from the Neolithic age until roughly 1800, and the still-dominant fossil energy regime.

For my purposes here, the shift from the agrarian to the fossil energy regime is most immediately relevant. It is difficult to overstate the scope and force of this transition. Under the agrarian regime, whatever economic growth occurred was directly tied to increases in population and arable land, because about 95 percent of the energy used was derived from biomass, that is, from cereals, vegetables, wood, and livestock. The energetic needs of society were met almost exclusively by way of photosynthesis and the conversion of biomass into muscle power or thermal energy. While technological innovation during this period led to modest increases in agricultural productivity, these were always quickly offset by subsequent population growth, leading to a corresponding decrease in available energy per capita. Agrarian societies always operate at the limits of scarcity. Wealth accumulation under such conditions is a zero-sum game in which the only way to increase one's own share is to subjugate or expropriate other people. During the roughly ten thousand years that the agrarian regime prevailed, living conditions for the 90 percent of the human population who were engaged in primary production (mostly farmers) changed very little—certainly much less than they have over the roughly two centuries which have passed since it began to be superseded by its successor formation.

The harnessing of fossil fuels allowed modern societies to emancipate themselves from the narrow energetic constraints imposed by the availability of land. Already in 1815, the energy produced from coal in the United Kingdom was equivalent to a doubling of its geographical surface area and allowed it to control, through mercantile and military expansion, a labor force much larger than its national population. The development of technologies that could convert thermal into mechanical energy—the steam engine and, later on, the internal combustion engine—revolutionized production and transportation, and also freed up vast amounts of time which could now be allocated to tasks other than agriculture; for example, to reading and writing. It thus powered new forms of geographical and social mobility. Energy from fossil fuels also thoroughly transformed what remained of the earlier agrarian regime: the Haber-Bosch process of artificial nitrogen fixing, new, often petroleum-based pesticides and insecticides, and gasoline-powered machinery lead to an unprecedented increase in agricultural productivity even as they drastically decreased the need for human and animal labor, which in turn fueled a growth of the human population like nothing the world has ever seen.

What this makes clear, first of all, is that petromodernity must be understood as a distinct phase within the fossil fuel regime. And it was this larger transition from agriculture to fossil fuels that provided the material basis for much of what we mean when we speak of modern history as characterized by processes of emancipation. Throughout the entire agrarian period, land had been the primary means for the acquisition of wealth and power. Accordingly, land ownership was seen as a necessary precondition for political participation. This view also shaped early liberal theories of government, such as that of John Locke, and prevailed even after the official abolition of feudalism: the republican form of government envisioned by the American "Founding Fathers" (as well as by Augustus Baldwin Longstreet) was one in which "liberal gentlemen"—an aristocracy of independently wealthy property owners—would hold the reins of power (Wood 198).

An economy based on fossil fuels opened up many new avenues for the accumulation of wealth and thus rendered obsolete the linkage between land ownership and the ability to hold political office. It also allowed for the rapid growth of cities, and consequently for the emergence of modern mass politics. The wage-labor mechanism fractured the old patron–client networks which had structured social relationships during the agrarian regime. The distribution of the new prosperity was extremely uneven, but, as progress turned more and more from an abstract concept into a matter of everyday experience, even the poorest members of society could expect to reap some of its benefits. The distinctive geography of early coal-based infrastructure, which concentrated flows of energy at a few strategic sites where abundant manual labor was needed (e.g., coal mines, steel mills, railroads, and transshipment points), enabled workers to effectively organize and enforce political demands (T. Mitchell 26–27). The social and ecological consequences of industrialization during this phase were often catastrophic, but they were also locally circumscribed and highly visible, spurring the development of reform movements that in many instances successfully improved environmental conditions and the lives of working-class people. To a significant extent, the mechanisms of social securitization that developed in most industrialized nations between the late nineteenth and the mid-twentieth century were an outcome of this process.

Only against this background do the distinctive features of petromodernity come into view, and they have as much to do with the spread of electricity as with the rise of petroleum—that is to say, they are linked not only to a new energy source, but also with new technologies for energy conversion which did not so much displace coal as radically remold its presence in the geography of everyday life. Environmental historians have described the internal combustion engine and electricity as core elements of a "second industrial revolution" (Kander et al. 251). As already mentioned at the outset, these new technologies allowed industrialization to clean itself up and to spatially disperse its infrastructure. One effect of this was the gradual disempowerment of organized labor (T. Mitchell 173); another, an increasing tendency to "outsource" adverse ecological effects, a process Alf Hornborg refers to as "environmental load displacement" (49–54). It also transformed how individuals experienced their relationship to society. The ease with which energy could now be distributed facilitated the emergence of an infrastructure tailored to smaller, more flexible social units—it produced the "spatiality of single-family homes, cars, yards, and highways" which, as Matthew Huber has pointed out, began to be deliberately promoted as an essential feature of the "American Way of Life" during the New Deal era (17). In these suburban settings, freedom and independence became fully identified with individualized control over one's private circumstances. Gasoline-powered machinery and new electric appliances turned the single-family home into a self-reliant enterprise whose relative prosperity could be experienced as a pure product of hard work and individual life choices. This, in turn, facilitated the increasing privatization of social space, and led people to conceive of their relationship to the state in more and more adversarial terms (Huber 163–164).

However, from the outset, this development has been accompanied by a countervailing tendency Sieferle describes as one of "de-autarkization" (195, my translation). Petromodern infrastructure has made private households ever more dependent on large-scale, collectively organized systems of energy provision which are highly abstract, anonymous, and opaque, and whose stability depends on factors that are entirely withdrawn from individual control. Under the agrarian regime, farmers could directly dispose over most of their means of subsistence, and even during the earlier phases of the fossil fuel regime, most households retained some of this ability: coal-fired stoves, for example, require that large amounts of coal are permanently kept in stock, and can switch to wood if necessary. Under petromodern conditions, by contrast, the consistent trend has been to liquidate local capacities for subsistence and to substitute them with mobile, geographically dispersed flows of energy, material, and information. Nowadays, most households in advanced industrialized society own few tools that would remain of use if access to these flows were suddenly cut off. From this perspective, life under petromodernity presents an oddly ambivalent, contradictory aspect. At the same time that "freedom" figures more centrally in the self-descriptions of contemporary society than perhaps in any earlier social formation, people have become entangled in an increasingly dense network of complex dependencies which exert an overwhelming systemic pressure to conform to social expectations and drastically curtail their ability to live autonomously. To quote Claus Offe's pithy summary of this process: "The more options we make available for ourselves, the less optional becomes the institutional (and particularly the technological) framework with whose help we do so" (104, my translation).

Thinking past liberation

All of this would seem to strengthen the position of those who contend that the emancipatory gains of the fossil fuel regime have been illusory. Hornborg, for example, argues that the apparent decoupling of economic growth from land use disguises an underlying process whereby the rich nations have been appropriating space and time from the impoverished peoples at the periphery of the world system. Fossil fuels "provided a minority of the world's population with an unprecedented source of [thermodynamic and political] *power*" (12, original italics); yet the idea that the living standards thus achieved in the industrialized world represent the future of humanity—rather than the luxuriating tip of a global pyramid of exploitation—was never more than a pleasant conceit. Climate change not only forces us to recognize ecological limits but, by the same token, also makes evident that the equitable distribution of resources is not a problem that, if only given enough time, will solve itself. This is why Naomi Klein can propose that climate change should be treated as an opportunity to combine the forces of social justice around the globe in an effort to dismantle the neoliberal capitalist world order and complete the "unfinished business of liberation" (458).

However, as Elizabeth Kolbert has pointed out in a critical review, Klein's vision of a better world is predicated on the optimistic assumption that energy will remain as cheap and plentiful as it is now even after the use of fossil fuels has been phased out (Kolbert). The changes she advocates in her book—investing in better public infrastructure, such as mass transit and social housing, and extending social safety-nets—sound rather like those which social democracies in Europe have been implementing for many decades. They did so not in order to abandon capitalism, but in order to distribute its fruits more widely and thus shore up popular support. The example of countries such as Denmark or Germany also argues strongly against the hopeful idea that the transition toward decentralized, more small-scale forms of energy provision such as wind and solar will lead to a redistribution of political and economic power. Perhaps most importantly, it must be remembered that these countries still have an unsustainably large ecological footprint, and that there is little indication that their citizenry would accept a significant reduction of its high living standards in order to limit CO_2 emissions.

It is comforting to attribute our difficulties with decarbonization to the machinations of a neoliberal cabal. But such an argument has to downplay just how deeply fossil fuels, and the economic dynamism they made possible, are woven into the fabric of modern industrial society. From the nineteenth century onwards, political systems of all stripes staked their legitimacy on the ability to deliver increasing material prosperity to their citizens. In the emancipatory master narrative of modernity, the advancement of individual liberty and the promise of economic betterment were always closely, albeit conflictually, intertwined. The symbiotic relationship between liberal democracy and consumer capitalism as it prevails today is a product of this development. And as Ingolfur Blühdorn has argued, there are good reasons to believe that at least in their current form, liberal democracies are ill-suited to the implementation of radical reforms such as might be necessary in order to attain real ecological sustainability: they are structurally disposed to "prioritize the interests of today" and "always center on the enhancement of rights and (material) living conditions." The pluralization of values and social norms characteristic of liberal democracy, itself driven by forms of social and physical mobility that, in their turn, were enabled by fossil fuels, makes the subordination of collective life to "categorical ecological imperatives" an unlikely prospect (Blühdorn).

All of this leads to the unpleasant conclusion that in crucial respects, the understanding of freedom which co-evolved with the fossil fuel regime has now become an obstacle to its overcoming. If the challenge for engineers is to develop technologies of energy provision which allow for a decoupling of economic growth from carbon emissions, environmental humanists face a parallel task: to disarticulate our notions of liberty from a life-style centered on the liberal dissipation of energy. To this end, we will have to highlight the inherent contradictions of petromodernity and point out how the individualized forms of freedom cherished by those who live in the "global North" are in fact conditioned by a panoply of barely understood dependencies and coercions which, however, are becoming increasingly difficult to ignore. Finally, we need to remind ourselves that just as democracy did not begin with fossil fuels, there is no reason to believe that it will end with them. To give a biblical twist to Chakrabarty's metaphor: freedom is a mansion with many rooms. With proper precautions, some of them will remain habitable even after fossil fuels are gone.

References

Appadurai, Arjun. *Modernity at Large: Cultural Dimensions of Globalization*. Minneapolis, MN: Minnesota University Press, 1997. Print.

Barrett, Ross and Daniel Worden, eds. *Oil Culture*. Minneapolis, MN: Minnesota University Press, 2014. Print.

Blühdorn, Ingolfur. "The Sustainability of Democracy: On Limits to Growth, the Post-Democratic Turn and Reactionary Democrats." *Eurozine*, 11 Jul. 2011. Web. 24 Aug. 2015.

Buell, Frederick. "A Short History of Oil Cultures: Or, the Marriage of Catastrophe and Exuberance." *Journal of American Studies* 46 (2012): 273–293. Print.

Chakrabarty, Dipesh. "The Climate of History: Four Theses." *Critical Inquiry* 35 (2009): 197–222. Print.

Debord, Guy. *The Society of the Spectacle*. Trans. Donald Nicholson-Smith. New York: Zone Books, 1995. Print.

Derrida, Jacques. *Dissemination*. Trans. Barbara Johnson. London: Athlone, 1981. Print.

Fischer-Kowalski, Marina and Jan Rotmans. "Conceptualizing, Observing, and Influencing Social-Ecological Transitions." *Ecology and Society* 14.2 (2009). Web. 24 Aug. 2015.

Harvey, David. *The Condition of Postmodernity: An Enquiry into the Origins of Cultural Change*. Cambridge, MA: Blackwell, 1990. Print.

Heinberg, Richard. *The Party's Over: Oil, War, and the Fate of Industrial Societies*. Gabriola Island, BC: New Society, 2003. Print.

Hornborg, Alf. *Gobal Ecology and Unequal Exchange: Fetishism in a Zero-sum World*. London: Routledge, 2013. Print.

Huber, Matthew T. *Lifeblood: Oil, Freedom, and the Forces of Capital*. Minneapolis, MN: Minnesota University Press, 2013. Print.

Jensen, Derrick. *Endgame*. 2 vols. New York: Seven Stories Press, 2006. Print.

Kander, Astrid, Paolo Malanima, and Paul Warde. *Power to the People. Energy in Europe over the Last Five Centuries*. Princeton, NJ: Princeton University Press, 2013. Print.

Klein, Naomi. *This Changes Everything: Capitalism vs. The Climate*. New York: Simon and Schuster, 2014. Print.

Kolbert, Elizabeth. "Can Climate Change Cure Capitalism?" *New York Review of Books*, 4 Dec. 2014. Web. 24 Aug. 2015.

Krausmann, Fridolin, Marina Fischer-Kowalski, Heinz Schandl, and Nina Eisenmenger. "The Global Sociometabolic Transition: Past and Present Metabolic Profiles and Their Future Trajectories." *Journal of Industrial Ecology* 12 (2008): 637–656. Print.

Kunstler, James Howard. *The Long Emergency: Surviving the Converging Catastrophes of the Twenty-First Century*. New York: Grove, 2005. Print.

LeMenager, Stephanie. *Living Oil: Petroleum Culture in the American Century*. Oxford, UK: Oxford University Press, 2014. Print.

Longstreet, Augustus Baldwin. *Patriotic Effusions*. New York: Lockwood, 1819. Print.

Mitchell, Joni. "Big Yellow Taxi." *Ladies of the Canyon*. Reprise Records, 1970. LP.

Mitchell, Timothy. *Carbon Democracy*. London: Verso, 2011. Print.

Musiol, Hanna. "'Liquid Modernity': Sundown in Pawhuska, Oklahoma." *Oil Culture*. Ed. Ross Barrett and Daniel Worden. Minneapolis, MN: Minnesota University Press, 2014. 129–144. Print.

Offe, Claus. "Die Utopie der Null-Option: Modernität und Modernisierung als politische Gütekriterien." *Die Moderne – Kontinuität und Zäsuren*. Ed. Johannes Berger. Göttingen. Germany: O. Schwartz, 1986. 97–117. Print.

Rosa, Hartmut. *Social Acceleration: A New Theory of Modernity*. Trans. Jonathan Trejo-Mathys. New York: Columbia University Press, 2013. Print.

Ross, Michael L. "Does Oil Hinder Democracy?" *World Politics* 53 (2001): 325–361. Print.

Sieferle, Rolf Peter. *Rückblick auf die Natur. Eine Geschichte des Menschen und seiner Umwelt*. München, Germany: Luchterhand, 1997. Print.

Sloterdijk, Peter. "How Big is 'Big'?" *Collegium International*, Feb. 2010. Web. 26 May 2015.

Szeman, Imre. "How to Know About Oil: Energy Epistemologies and Political Futures." *Journal of Canadian Studies* 47.3 (2013): 145–168. Print.

Virilio, Paul. *Speed and Politics: An Essay on Dromology*. Trans. Marc Polizzotti. New York: Semiotext(e), 1977. Print.

Watts, Michael. "Oil Frontiers: The Niger Delta and the Gulf of Mexico." *Oil Culture*. Ed. Ross Barrett and Daniel Worden. Minneapolis, MN: Minnesota University Press, 2014. 189–210. Print.

Williams, Raymond. "Culture Is Ordinary." *Raymond Williams on Culture and Society: Essential Writings*. Ed. Jim McGuigan. London: Sage, 2013. 1–18. Print.

Wood, Gordon S. *The Radicalism of the American Revolution*. New York: Vintage, 1993. Print.

41

SCALING THE PLANETARY HUMANITIES

Environmental globalization and the Arctic

Sverker Sörlin

The Arctic is changing. So is knowledge about the Arctic. So are the Arctic humanities, a term that has just emerged. These changes are coupled. Knowledge of the Arctic is increasingly based on scientific observation and the collection of data on environmental change. The Arctic has also forcefully entered the realm of humanities because of the geopolitical upheavals that the region has undergone in the last couple of decades. But even in these the environment holds a central position. The particular strand of globalization that we may call "environmental" indeed has the Arctic as one of its most spectacular showcases.

In this the Arctic humanities make up a microcosm of the environmental humanities at large. As they expand, they are embracing both environmental Big Data and historical and cultural interpretations that rely on these data. Declarations of sympathy toward more biologically and scientifically informed humanities, downplaying *Berührungsangst* vis-à-vis risks of determinism, have also become more frequent in recent years (Adeney Thomas; Hamilton et al.), which is not to deny that there has also been an articulated critique of Dipesh Chakrabarty's plea for a more "natural" history in his influential essay in *Critical Inquiry* (2009), and of similar arguments voiced in the so-called Anthropocene debates. Other evidence of a similar openness to more and new kinds of data is the growing global "deep time" IHOPE project (Integrated History and Future of People on Earth, www.ihope.org), which since 2004 has forged new alliances of archaeologists, geographers, and historians with earth system scientists. In this wave of sympathy for an emerging science-based Anthropocene *Weltanschauung* we also find Bruno Latour, praising the quantification of the world in his 2013 Gifford Lectures as the metaphorical spinning of a "cocoon" of insight around humanity that may build the responsibility we need to avoid transgressing life-upholding planetary boundaries: a sociology turned Gaia-optimism.

At the other end of the spectrum are those, less providentially minded, who connect the growing use of numbers with many problematic antecedents. The argument starts with Theodore Porter's *Trust in Numbers* (1995), drawing a line from its analysis of nineteenth-century social and infrastructural projects all the way to the rise of modern (including environmental) expertise, and the use of data and numbers as ways of universalizing environmental problems as well as their solutions through various processes of what has increasingly been termed "depoliticization" (Swyngedouw): the privileging of (neoliberal) economics, and de-privileging of more complex value formations that are closer to citizens and reach beyond the monetary (Monfreda; Ernstson and Sörlin).

These discussions are also heard across the Arctic. They may prompt us to reflect on what it is that we do when we transform the humanities we knew—sometimes esoteric, often particularistic, rarely interested in policy, advice, environmental challenges, and skeptical of metrics and numbers—into something we find more *of* the world and *for* the world. It is both a tempting and a respectable undertaking, but it is also one that entails the politics of numbers and the perceptions and ethics of scaling and of distant connections in time and in space.

On the emerging Arctic humanities

In the first decades of the twenty-first century, the Arctic is being firmly drawn into new integrative writing about the environment, encompassing culture, science, technology, history, and future through the lens of the politics of the present. There are now humanities for the Arctic way beyond the long-standing traditions of Inuit studies, Eskimology, Arctic anthropology, and other earlier specializations.

For a long time, however, this scholarship was a marginal phenomenon in relation to the massive efforts in what was always called Arctic "science," a concept that followed on the heels of the earlier "exploration" (Levere; Bravo and Sörlin). Despite a certain presence in some academic departments and Arctic museums—related to aboriginal cultures, such as the collections built in Copenhagen through Denmark's colonial relationship to Greenland (Thisted)—the humanities were really very small by comparison. The massive research effort during the International Geophysical Year 1957–1958, which privileged the poles and counted as the third of four "polar years" thus far, mobilized some fifty thousand scholars and scientists from more than sixty countries, virtually none of whom were humanists. The Fourth International Polar Year 2007–2009, on the contrary, counted about one-third of its scholars from the humanities and social sciences (Krupnik et al., "Polar Societies"). Keeping in mind that the total population of the region north of the Arctic Circle is no larger than four million people, less than that even of small polar states such as Denmark and Norway, the Arctic draws considerable scholarly interest, surpassing anything we have seen in the past. Much of that interest, including from the humanities, is related to the drastic changes in the environmental politics of the region.[1]

How could this be explained? An article that reflected critically on the rapid rise of the Anthropocene as a concept and a frame of understanding observed that, in the light of the growth of an environmentally activist "planetary science," "there is no planetary humanities" (Pálsson et al. 11). At the same time, the article was meant as a plea for such a planetary humanities to be articulated, in order to achieve a richer and fairer understanding of the Anthropocene. To understand the discourses of the planetary, and to navigate an increasingly planetary era of integrated fates of nations, cultures, economies, and environments, there seemed to be an urgent need for humanities that could address these issues and decipher their meaning and underlying patterns of power, ideologies, and directionalities. Natural science and conventional predictive social sciences alone would not suffice (Andersson and Rindzeviciute).

At about the same time newer tendencies in the humanities emerged that took precisely this planetary view, proclaiming concepts such as the "environmental humanities" (Rose et al.; Sörlin, "Environmental Humanities") or the return of the "long term narrative" (Guldi and Armitage), all accompanied by an appropriately critical discussion (e.g., Cohen and Mandler). A new kind of humanities seems already to be emerging, to which many contribute, and that relies on the classical virtues of the humanities fields. At the same time, the new humanities insist on a new relevance and a sense of urgency that is crucially, albeit far from solely, about the planetary.

In a similar way, but perhaps more acutely, change happens to the humanities in and of the Arctic. In fact, just as there were no planetary humanities, there used to be no Arctic humanities (Sörlin, "The Emerging Arctic Humanities" and "Cryo-history"). There did not seem to be many common Arctic issues outside national concerns, nor was there any common agenda, at least not one that the humanities deemed fit to deal with (Jørgensen and Sörlin). Planetary change, tremendous challenges, looming threats, and fabulous opportunities have been proclaimed for the Arctic ever since the short post-Cold War period of peaceful region-building in the North (Keskitalo) came to an abrupt end in the first decade of the new century.

Whatever lasting truth there may be to these grandiose statements, there has been enough *Realpolitik* to this rhetoric to make a response meaningful. Just as we have seen the emergence of the environmental humanities as a reflection of the *planetary* crisis, we should see the intense and careful humanities scholarship currently emerging on the Arctic as *also* the legitimate outcome of an *Arctic* crisis. What may seem a new bonanza of minerals and fossil fuel resources on the Arctic rim has in reality been a period of utmost stress and profound shake-up of cultures and societies facing global change, with the added challenge of what has been called an "Arctic amplification." The expression refers to the speed and intensity of climate change and the dramatic rise of temperatures in the region (Bekryaev et al.; Pithan and Mauritsen), but might be extended to a wider range of phenomena. Arctic change is literally on speed, by most indicators, from tourism, geopolitics, and suicide rates to the penetration of modern technologies and, indeed, the presence of global change science, taking the measure of all things. Thus, we begin to see reinterpretations of the standard narrative that renegotiate the relationship of the Arctic local with the global and show the world's northernmost settlements, like those of the *inughuit* of Qaanaaq in Northeast Greenland, as at the same time fully embedded in the web-connected world and increasingly vulnerable to global warming (Hastrup, "Anticipation on Thin Ice" and *Thule*), continuing but also profoundly transforming a long-standing discourse of Arctic "security" (Doel et al.).

The emerging Arctic humanities also hold out the promise of becoming the dusk of the long day of Arctic exceptionalism, the tradition of seeing this part of the world largely as a reserve for those who studied nature, mostly in splendid isolation. Those scientists did admirable work, and it is their ice cores, climate data, sediment measures, and readings of the buoys that now get a second lease on life as they are taken up and integrated into human story lines and social explanations. Likewise, humanities from other regions of the world become relevant, and vice versa. The Arctic of the new humanities faces outward, to the tropics, the oceans, the deserts, the plains, and the cities. Like "every man" in John Donne's poem, the Arctic "is part of the main."

The cryo-historical moment

One important strand of the rising humanities interest in the Arctic is predicated on climate and climate change. Conventional global change discourse, drawing its language from natural science, presents climate as the "driver" of Arctic change. Although one may doubt the soundness of this idea—don't we live in the Anthropocene, where humans are the drivers?—one would expect climate change to be a cross-cutting perspective, linking Arctic studies in all fields, including the humanities. If cold, ice, and snow were always perceived determinants of history, an old trope of the North with imperialist and racist implications that suggested the superiority of sturdy, conquering "northern" peoples, if not "races" (e.g., Stefansson; Paglia), we now also have a discourse on the retreat of cold, ice, and snow.

It has been suggested that warm and cold periods form a historical pattern. There was even a widespread opinion in the nineteenth and early twentieth centuries that Arctic ice may have its gaps and holes (Wright). One perspective portrayed a green, forested central Greenland, fueled by observations of floating timber in the waters near the big island (Örtenblad). As late as 1926, one of the most reputable Russian geologists working in Siberia, Vladimir A. Obruchev, used such a myth of an earthly resort beyond the ice in his science fiction novel *Sannikov Land,* thus justifying Soviet territorial claims for Arctic islands (Frank). The trope of an ice-free Arctic became fashionable when a period of de facto Arctic warming began to be noticed in the 1920s and 1930s. It was still not regarded as an outcome of human forcing, but the strangely behaving sea ice nonetheless resulted in a growing interest in several countries. Empirical findings mixed with research methods and terminological issues (Maurstad; Transehe; Polyak), and after World War II it became a strategic issue in the Cold War to know how reliable the sea ice was (Sörlin, "Cryo-history"; Roberts).

A similar timeline can be drawn for the investigation of terrestrial ice. The very concept *cryosphere* was an invention of the Polish scientist Antoni Dobrowolski in his massive work *The Natural History of Ice* (1923, in Polish; Barry et al.), at about the same time as studies of the mass balance of glaciers started in northern Scandinavia, Spitsbergen, and Greenland (Ahlmann, "Glaciers in Jotunheimen" and "Le régime de glaciers"), professionalizing earlier alpine interest in mountain ice. The word itself derives from the Ancient Greek word 'κρύος' (*cryos* meaning "cold," "frost," or "ice"). It is the term which collectively describes the portions of the Earth's surface where water is in solid form, including sea ice, lake ice, river ice, snow cover, glaciers, ice caps and ice sheets, and frozen ground, including permafrost. By the late 1930s the Swedish scientist Hans Ahlmann even launched a theory based on glaciological data to suggest a long-term trend of "polar warming," albeit with no anthropogenic origins (Sörlin, "The Anxieties of a Science Diplomat").

The concept "cryosphere" was largely forgotten until the 1970s, when it was picked up in earnest as satellites provided a much better understanding of the changes in global ice cover and Global Circulation Models captured the waning albedo effect of shrinking ice to reflect heat and thus hinder global warming. Acknowledging the cryosphere required that it could be *seen* as a sphere, and part of the planet (Cosgrove; Wormbs). The concept is not enough: there needs to be this moment of crystallization when accumulated knowledge turns into a shift in perception. This shift in turn helped accelerate Cold War attempts to use geophysical sciences and environmental and military technologies to survey and control the earth, what we might call *the production of the planetary* (Masco, "Bad Weather" and "The Age of Fallout"; Turchetti and Roberts; Höhler).

The appreciation of cryospheric change gave the Arctic a special position in the narrative of the planetary. The literature has since expanded almost beyond comprehension, and it has fused with the new orthodoxy of climate change based on human climate forcing. What used to be the unavoidable coming and going of ice and snow, the capriciously variable play of climate with culture in some of the most remote stretches of human presence, has become a global predicament. In the twenty-first century, glaciers are melting on all continents. The snows of Kilimanjaro will soon be gone, the Peruvian ice sheets are melting, the icebergs of Puget Sound are calving, and the glistening crystal ornaments of the Himalayas that water the dry lands of the north Indian plains are disappearing. Humanities scholars have started to follow their flow, roaring and dripping (Cruikshank; Carey; Christensen et al.).

What seemed to be unique to the Arctic has become part of the human condition. Being used to looking at the Arctic from the outside in, as the microcosm at the end of the world, where change arrived last—the *Ultima* Thule, as the ancients baptized it—the situation is turning almost into its opposite. The rest of the world is undergoing the same kind of Arctic amplification, but later. The world, I would argue, has reached a "cryo-historical" moment, manifested in the retreat of one of its core elements, and directing our attention to the historical powers of human forcing in the Anthropocene.

At the center of this world

At the same time it is a moment when conventional narratives and temporalities are being questioned or even turned on their heads. Ice and snow have been seen as obstacles to the usual grand cultural narrative of an expanding agriculture and a sedentary civilization, according to the theories of domestication that started to flourish in the interwar period (Childe). But, as we are now keen to note, ice and snow have for thousands of years also been the life conditions for Arctic peoples, who live with these elements and whose cultures have formed around them without fitting the standard narrative. The "improvement of climate," an old colonial dream, has now reached a stage that questions human wisdom and threatens cultures more than it sustains them. The old trope speaks to the smallness of the human genius when faced with the powers of Nature. In Mary Shelley's *Frankenstein* (1818), the experimental scientist on a polar expedition meets his own creation, "the Creature," for a final showdown on the sea ice. The Arctic becomes the place where Enlightenment Man has to face his fate, near the end of this world that, it finally turns out, he cannot subdue. At its bicentennial, Shelley's work seems to have turned into a foundational document of the environmental humanities, which identifies the Arctic as the space where science and technology have to navigate differently if they are to be of use to us.

The fate of ice as a sign of the fate of our societies invites new readings and interpretations that can be provided by the environmental humanities. Ice plays a role in narratives of climate change (Bravo). It has come to symbolize the ephemeral and fugitive nature of human existence. In practice, modern humans are all but fugitive: we have a heavy, sometimes fatal impact. At the same time, we can ask where authoritative knowledge about this elusive material rests. Glaciers, scholars have argued, can both speak and listen, and they can and should be gendered, work that humanities scholars have embarked upon (Carey et al.; Sörlin, "Anxieties of a Science Diplomat"). Local epistemic communities with traditional knowledge in many continents live alongside increasingly present ice and glacier experts (Carey). The waning ice links the excesses of modern consumer society and industrialism to disastrous impacts on the innocent original populations, whose vulnerability, despite a long history of successful adaptation, is aggravated when the ice melts (Krupnik et al., *SIKU*; Hastrup, "The Icy Breath"). It also reflects a complex relationship between Arctic residents and modern mainstream science, which provides the best available knowledge of ice but still does not seem to be able to capture the full magnitude of the changes. Mother Nature does, as she certainly must, hold out surprises, two centuries after the ill-fated Frankenstein. Evidence of what has been called "the politics of uneven time" (Jordheim)—that is, when impacts caused in one part of the world in one period are passed on to somewhere else at some other time—pose questions of responsibility in new ways.

Part of the change is an emotional and political mobilization in many regions of the world but perhaps most clearly in the Arctic, where most people are directly affected. A range of scholarly, artistic, and community projects over the last decades have been mapping this mobilization. In one, called SIKU, an acronym but also an Inuit word for sea ice, scholars worked with local residents and did fieldwork together. A volume resulting from the collaboration, *The Meaning of Ice* (Gearhard et al.), demonstrates that ice is a deeply sensory experience that cannot be treated separately from your life world, your personal or local temporalities, or from the world at large.

Environmental geopolitics: scaling and telecoupling

To see this point we need a different optic. As Kirsten Hastrup has emphasized in her work on Qaanaaq in far northwestern Greenland, these so-called "remote" places are in essence not much different in many of their current parameters from global hubs such as Los Angeles, Shanghai, or Lagos (*Thule*). Globalization and electronic communication, but perhaps even more so the acknowledgment of shared environmental framing conditions, for example so-called "planetary boundaries" (Rockström et al.; Steffen et al.), have brought about what Ursula Heise has called a "sense of planet." The realities of earth system science and climate change are as well known in Qaanaaq as anywhere. Things happen in Arctic communities in distant, yet close, relationships with things happening elsewhere. Synchronicities are built into the ontology of the object, the ice.

To use this new optic requires a sensibility for the multiple dimensions that reach far beyond the old tripartite division of historical scaling into the local, national, and global (White). In her book, *Sense of Place and Sense of Planet*, Heise juxtaposes place and planet. They require each other and through the scaling they are connected in multiple ways. They might, as Eric Paglia has suggested with a concept drawn from earth systems science, be called *telecoupling* relationships. They couple the local place where impacts are felt and registered and identities are re-shaped, with the planetary level where the elements of local knowledge accumulate into an earth systems science understanding, and where impacts accumulate also. But where identities have yet to grow, if indeed there can be a planetary identity—citizens not only of the world, but also of the planet (Henderson and Ikeda).

We seem to be talking here about an emerging form of tragedy that might best be understood by comparison with the metaphysical framings that societies had in the past; for example, Renaissance Europe and its general awareness of being in the Last Days of the World. This frame of mind multiplied into the personal and the temporal; one could not be separated from the other. As Hastrup observes, the experience of time itself changes

(*Thule*). The *inughuit* concept of future used to be one of trust and comfort; now it is laden with the growing uncertainties of the present, causing what we might think of as a cultural sadness. Even though we can conceive of political and technological processes that may reverse the impacts in the next generations, these are processes not controlled by the indigenous populations in the Arctic, but by others, elsewhere.

This is precisely why the Arctic is becoming such a significant domain for the environmental humanities. Past humanities interactions were about studying Arctic cultures as distant and different. The current interest is about forming a more comprehensive, multisited understanding of shared cultural experiences and politics of environmental change. This is also why the concept of scaling is drawing attention. Innumerable local actions around the world are scaling upwards to the planetary atmosphere where they form a dangerous, unified force. So far this force has not been downscaled properly, through politics or technology, and therefore it hits disproportionately hard in certain places.

While geopolitical action through an individual state has always had this property of changing places far away, through regulation, trade, extraction, or brute force, the environmental geopolitics of today have far more distributed agency, a kind of crowd-sourcing of global change. It is also increasingly mediated across a range of planetary scales and its manifestations are material and, indeed, environmental. Landscapes, seascapes, icescapes, cloudscapes—ephemera that can be radically altered—have elastic appearances. Arctic winters change. Habits of life will change, as will cultures, experiences, temporalities, artistic expressions—things that humanists are used to thinking about. But the point here is rather different: it concerns the deepened insights into the connectedness of these core humanities with wider geophysical phenomena that scale across the world and enlarge the domains of humanities scholarship.

Note

1 Seminal recent works include but are not limited to: McCannon; Hastrup and Skrydstrup; Martin-Nielsen; Farish; Sörlin, *Science, Geopolitics and Culture in the Polar Region*; Jørgensen and Sörlin; Christensen et al.; Powell and Dodds; Doel et al.; Evengard et al.; Hastrup, *Thule*; Körber et al.

References

Adeney Thomas, Julia. "History and Biology in the Anthropocene: Problems of Scale, Problems of Value." *American Historical Review* 119.5 (2014): 1587–1607. Print.

Ahlmann, Hans. "Glaciers in Jotunheimen and their Physiography." *Geografiska Annaler* 4 (1922): 1–57. Print.

——. "Le régime des glaciers." *Revue de géographie alpine* 29.4 (1941): 537–566. Print.

Andersson, Jenny and Egle Rindzeviciute, eds. *The Struggle for the Long-Term in Transnational Science and Politics: Forging the Future*. New York: Routledge, 2015. Print.

Barry, Roger G., Jacek Jania, and Krzysztof Birkenmajer. "A.B. Dobrowolski – the First Cryospheric Scientist – and the Subsequent Development of Cryospheric Science." *History of Geo- and Space Sciences* 2 (2011): 75–79. Print.

Bekryaev, R.V. et al. "Role of Polar Amplification in Long-Term Surface Air Temperature Variations and Modern Arctic Warming." *Journal of Climatology* 23 (2010): 3888–3906. American Meterological Society Journals Online. Web. 15 December 2015.

Bravo, Michael. "Preface: Legacies of Polar Science." *Legacies and Change in Polar Sciences: Historical, Legal and Political Reflections on The International Polar Year*. Ed. Jessica M. Shadian and Monica Tennberg. Aldershot: Ashgate, 2009. xiii–xvi. Print.

Bravo, Michael and Sverker Sörlin, eds. *Narrating the Arctic: A Cultural History of Nordic Scientific Practices*. Canton, MA: Science History Publications, 2002. Print.

Carey, M. *In the Shadow of Melting Glaciers: Climate Change and Andean Societies*. Oxford: Oxford University Press, 2010. Print.

Carey, M., M. Jackson, A. Antonello, and J. Rushing. "Glaciers, Gender, and Science: A Feminist Glaciology Framework for Global Environmental Change Research." *Progress in Human Geography* 40.1 (2016): 1–24.

Chakrabarty, Dipesh. "The Climate of History: Four Theses." *Critical Inquiry* 35 (2009): 197–222. JSTOR. Web. 1 December 2015.

Childe, G. *Man Makes Himself*, Oxford: Oxford University Press, 1936. Print.

Christensen, M., Nilsson, A.E., and Wormbs, N., eds. *Media and Arctic Climate Change: When the Ice Breaks*. New York: Palgrave Macmillan, 2013. Print.

Cohen, Deborah and Peter Mandler. "*The History Manifesto*: A Critique." *American Historical Review* 120 (2015): 530–542. Web. 15 December 2015.

Cosgrove, D. *Apollo's Eye: A Cartographic Genealogy of the Earth in the Western Imagination*. Baltimore: Johns Hopkins University Press, 2001. Print.

Cruikshank, J. *Do Glaciers Listen?: Local Knowledge, Colonial Encounters, and Social Imagination*. Vancouver: University of British Columbia Press, 2005. Print.

Dobrowolski, Antoni Boleslaw. *Historia naturalna lodu*. [The Natural History of Ice]. Warsaw: Kasa Pomocy im. Dr. J. Mianowskiego, 1923. Print.

Doel, R.E. et al. "Strategic Arctic Science: National Interests in Building Natural Knowledge–Interwar Era through the Cold War." *Journal of Historical Geography* 44 (2014): 60–80. Print.

Ernstson, Henrik and Sverker Sörlin. "Ecosystem Services as Technology of Globalization: On Articulating Values in Urban Nature." *Ecological Economics* 86 (2013): 273–284. Print.

Evengard, B., J. Nyman Larsen, and Ø. Paasche, eds. *The New Arctic*. New York: Springer, 2015. Print.

Farish, M. "The Lab and the Land: Overcoming the Arctic in Cold War Alaska." *Isis* 104 (2013): 1–29. Print.

Frank, S.K. "Arctic Science and Fiction: A Novel by a Soviet Geologist." *Journal of Northern Studies* 1 (2010): 67–86. Print.

Gearhard, S.F. et al., eds. *The Meaning of Ice: People and Ice in Three Arctic Communities*. Hanover, NH: International Polar Institute Press, 2013. Print.

Guldi, J. and Armitage, D. *The History Manifesto*. Cambridge: Cambridge University Press, 2014. Print.

Hamilton, Clive, Christophe Bonneuil, and François Gemenne, eds. *The Anthropocene and the Global Environmental Crisis: Rethinking Modernity in a New Epoch*. London: Routledge, 2015. Print.

Hastrup, Kirsten. "Anticipation on Thin Ice: Diagrammatic Reasoning." *The Social Life of Climate Change Models: Anticipating Nature*. Ed. Kirsten Hastrup and M. Skrydstrup. New York: Routledge, 2013. 77–99. Print.

——. "The Icy Breath: Modalities of Climate Knowledge in the Arctic." *Current Anthropology* 53.2 (2012): 226–244. Print.

——. *Thule: Paa tidens rand*. Copenhagen: Lindhardt og Ringhof, 2015. Print.

Hastrup, Kirsten and M. Skrydstrup, eds. *The Social Life of Climate Change Models: Anticipating Nature*. New York: Routledge, 2013. Print.

Heise, Ursula K. *Sense of Place and Sense of Planet: The Environmental Imagination of the Global*. New York: Oxford University Press, 2008. Print.

Henderson, Hazel and Daisaku Ikeda. *Planetary Citizenship: Your Values, Beliefs, and Actions Can Shape a Sustainable World*. Santa Monica, CA: Middleway Press, 2004. Print.

Höhler, S. *Spaceship Earth in the Environmental Age, 1960–1990*. London: Pickering & Chatto, 2015. Print.

Jordheim, H. "Introduction: Multiple Times and the Work of Synchronization." *History and Theory* 53.4 (2014): 498–518. Print.

Jørgensen, Dolly and Sverker Sörlin, eds. *Northscapes: History, Technology, and the Making of Northern Environments*. Vancouver: University of British Columbia Press, 2013. Print.

Keskitalo, E.C.H. *Negotiating the Arctic: The Construction of an International Region*. New York: Routledge, 2002. Print.

Körber, L.-A., S. MacKenzie, and Anna Westerstahl Stenport, eds. *Arctic Environmental Modernities: Politics and Representation from the Age of Polar Exploration to the Era of the Anthropocene*. New York: Palgrave Macmillan, 2016. Print.

Krupnik, I. et al., ed. *SIKU: Knowing Our Ice*. New York: Springer, 2010. Print.

Krupnik, I. et al. "Polar Societies and Social Processes." *Understanding Earth's Polar Challenges. International Polar Year 2007–2008*. Ed. I. Krupnik et al. Edmonton: CCI Press & ICSU/WMO, 2011. 311–334. Print.

Levere, T.H. *Science and the Canadian Arctic: A Century of Exploration, 1818–1918*. Cambridge: Cambridge University Press, 1993. Print.

Martin-Nielsen J. *Eismitte in the Scientific Imagination: Knowledge and Politics at the Center of Greenland*. New York: Palgrave, 2013. Print.

Masco, Joe. "Bad Weather: On Planetary Crisis." *Social Studies of Science* 40.1 (2010): 1–31. Sage Journals. Web. 15 December 2015.

——. "The Age of Fallout." *History of the Present* 5.2 (2015): 137–168. Print.

Maurstad A. *Atlas of Sea Ice*. Oslo: Cammermeyers Boghandel, 1935. Print.

McCannon, J. *A History of the Arctic: Nature, Exploration and Exploitation*. London: Reaktion Books, 2012. Print.

Monfreda, C. "Setting the Stage for New Global Knowledge: Science, Economics, and Indigenous Knowledge." *Conservation and Society* 8 (2010): 288–297. Print.

Örtenblad, T. *Om Sydgrönlands drifved: Bidrag till Kungl*. Stockholm: Vetenskapsakademiens Handlingar, 1881. Print.

Paglia, E. "The Northward Course of the Anthropocene: Transformation, Temporality and Telecoupling in a Time of Environmental Crisis." Dissertation. Stockholm: KTH Royal Institute of Technology, 2016.

Pálsson, G., S. Sörlin, B. Szerzynski, et al. "Reconceptualizing the 'Anthropos' in the Anthropocene: Integrating the Social Sciences and Humanities in Global Environmental Change Research." *Environmental Science and Policy* 28 (2013): 4–13. Print.

Pithan, F. and Mauritsen, T. "Arctic Amplification Dominated by Temperature Feedbacks in Contemporary Climate Models." *Nature Geoscience* 7 (2014): 181–184. Print.

Polyak, L. et al. "History of Sea Ice in the Arctic." *Quaternary Science Reviews* 29 (2010): 1757–1778. Print.

Powell, Richard C. and Klaus Dodds, eds. *Polar Geopolitics?: Knowledges, Resources and Legal Regimes*. Cheltenham: Edward Elgar, 2014. Print.

Roberts, P. "Scientists and Sea Ice under Surveillance." *The Surveillance Imperative: Geosciences During the Cold War*. Ed. S. Turchetti and P. Roberts. New York: Palgrave, 2014. 125–145. Print.

Rockström, J. et al. "Planetary Boundaries: Exploring the Safe Operating Space for Humanity." *Nature* 461 (2009): 472–475. Print.

Rose, Deborah Bird, Thom van Dooren, Matthew Chrulew, et al. "Thinking through the Environment, Unsettling the Humanities." *Environmental Humanities* 1 (2012): 1–5. Web. 1 December 2015.

Sörlin, Sverker. "The Anxieties of a Science Diplomat: Field Co-production of Climate Knowledge and the Rise and Fall of Hans Ahlmann's 'Polar Warming.'" *Osiris 26: Revisiting Klima*. Ed. J. R. Fleming and V. Jankovich. Chicago: University of Chicago Press, 2011. 66–88. Print.

———. "Cryo-history: Ice and the Emerging Arctic Humanities." *The New Arctic*. Eds. B. Evengard, J. Nyman Larsen, and Ø. Paasche. New York: Springer, 2015. 327–339. Print.

———. "The Emerging Arctic Humanities: A Forward-looking Post-script." *Journal of Northern Studies* 9.1 (2015): 93–98. Print.

———. "Environmental Humanities. Why Should Biologists Interested in the Environment Take the Humanities Seriously?" *BioScience* 69 (2012): 788–789. Web. 15 December 2015.

———, ed. *Science, Geopolitics and Culture in the Polar Region: Norden beyond Borders*. Farnham: Ashgate, 2013. Print.

Stefansson, Vilhjamur. *The Friendly Arctic: The Story of Five Years in Polar Regions*. New York: Macmillan, 1921. Print.

Steffen, W. et al. "Planetary Boundaries: Guiding Human Development on a Changing Planet." *Science* 347.6223 (15 January 2015): 736–746. Web. 15 December 2015.

Swyngedouw, Erik. "Depoliticized Environments: The End of Nature, Climate Change and the Post-Political Condition." *Royal Institute of Philosophy Supplement* 69 (2011): 253–274. Web. 18 January 2016.

Thisted, K., ed. *Grønlandsforskning: Historie og perspektiver*. [Research on Greenland: History and Perspectives]. Copenhagen: Det Grønlandske Selskab, 2005. Print.

Transehe, N.A. "The Ice Cover of the Arctic Sea, with a Genetic Classification of Sea Ice." *Problems of Polar Research*. New York: American Geographical Society, 1928. 91–123. Print.

Turchetti, S. and Roberts P., eds. *The Surveillance Imperative: Geosciences During the Cold War*. New York: Palgrave, 2014. Print.

White, R. "The Nationalization of Nature." *American Historical Review* 86.3 (1999): 976–986. Print.

Wormbs, Nina. "Eyes on the Ice: Satellite Remote Sensing and the Narratives of Visualized Data." *Media and Arctic Climate Change: When the Ice Breaks*. Eds. M. Christensen, A.E. Nilsson and Nina Wormbs. New York: Palgrave Macmillan, 2013. 43–65. Print.

Wright, J.K. "The Open Polar Sea." *Geographical Review* 43 (1953): 338–365. Print.

42

SOME "F" WORDS FOR THE ENVIRONMENTAL HUMANITIES

Feralities, feminisms, futurities

Catriona Sandilands

Feralities

In Toronto, the city where I live, there is an extraordinary place called, variously, Tommy Thompson Park (TTP), the Leslie Street Spit, the Outer East Harbour Headland, or simply "the Spit." This piece of land, stretching five kilometers into Lake Ontario, was (and continues to be) created out of the detritus of Toronto's development. Starting in the 1950s, when it was intended to create a breakwater to support increased shipping on the Great Lakes, the Spit has received tens of thousands of tons of waste. Assembled from everything from building teardowns to subway construction to shipping channel dredgeate, the Spit is a rubbly archive of the city's history. As Watt-Meyer shows, visitors can, with a bit of digging, locate particular urban remains at specific points on the Spit and know that they are walking on the grave of, for example, the Toronto Board and Trade Building (demolished 1958). Moreover, as Schopf and Foster demonstrate, the Spit tells a larger story about urban development and environmental justice. Deposits from 1960s slum clearances contain large numbers of personal artifacts, indicating that "full houses with belongings still inside were demolished, compacted, and then dumped" (1092). Subsequent deposits from the 1980s are "much more uniform and organised" (1095): by this period, "there was considerable planning for the afterlife of the rubble" (1103), a rationalized folding of waste, as it were, into the aesthetic and political matrices of capitalism.

Despite the appearance of orderly management, many of these wastes are decidedly toxic. In 1987, a soil survey on the Spit conducted for the Ontario Ministry of Environment indicated that "mercury, lead and PCB—compounds classified by the ministry as high priority chemicals … with potential human health concerns—had exceeded the guidelines" (Chan 69, 75). Many of the dredgeates that are the foundation of the Spit's beaches and headlands are loaded with hazardous chemicals, including lead and cadmium, as well as construction waste. From 1974 to 1983, over six million cubic meters of sand and silt were placed at the Spit to create four peninsulas, and in 1985 Cell One, the first of three planned "confined disposal facilities" (CDFs), had been completely filled with hazardous dredgeates. But the

Spit is no ordinary dump, and this CDF is no ordinary toxic sludge pile. After the site was filled and capped, beginning in 2003, Cell One was intentionally remodeled as a half-terrestrial, half-aquatic ecosystem, mimicking a nearby bay to approximate lost "natural" Lake Ontario habitats (Chan 77). According to the Toronto and Region Conservation Authority (TRCA):

> Cell One is already providing functional habitat for a wide variety of fish and wildlife species. Juvenile fish species, as well as adults have been recorded in the completed Cell One wetland. The newly created tern island was home to almost 300 nesting pairs of Common Terns in 2005 and a created Bank Swallow habitat hosted a colony just days after completion … Midland painted turtles and Northern map turtles have been observed in and around the cell. Muskrats and mink have also been documented.

At the same time as these 7.7 hectares of wetland are "a valuable ecological resource that offers new opportunities for … environmental enhancement" (Chan 77, 85), it is important to remember what lies beneath. Indeed, another 9.3 hectare CDF site that contains toxic contaminants is currently slotted for "restoration," meaning the creation of "a wetland complex [that will] … further enhance the ecological value of the site as a biological centre of organization within Toronto" (TRCA).

Part of the Spit officially became TTP in 1995; the other part is still an active dumpsite, which means that Park visitors cannot enter the Spit on weekdays during the day. But that does not stop some of the Spit's passionate advocates from calling the site an "urban wilderness." More specifically, despite the fact that it is subject to intensive and ongoing regulation and management, TTP is widely understood as a *feral* landscape. Beginning in the late 1970s, pioneering plant and animal species began to take advantage of the new habitat. As Foster and Fraser note, at that time "thousands of ring-billed gulls recognized this tree-free area as nesting habitat and took up residence in the spring and summer months" (213). By the late 1980s, a cottonwood forest had sprung up over much of the peninsula. A 1990 survey conducted by the "Friends of the Spit" indicated the presence of over 400 indigenous and exotic vascular plant species. The TRCA currently counts "50 species of butterflies, 42 species of moths and 17 species of dragonflies"; several species of herpetofauna have immigrated to the park (turtles are targeted in habitat construction); and there is a sufficient fish population to support a recreational fishery. Perhaps most prominently, given its location as a southern landing in the northern Great Lakes, the Spit has come to provide an important refuge for migratory birds and was, in 2000, declared a Globally Important Bird Area by BirdLife International because of both its importance to migratory songbirds and its support of several species of colonial waterbirds, most famously double-crested cormorants, *Phalocrocorax auritus*.

Although it is clear that spontaneous ecological succession is only part of the complex life of the Spit, it is also clear that the site involves a mélange of agencies that visibly (and audibly and olfactorily) trouble any neat bifurcation of wildness from domesticity. At what point does the Spit stop being a dump and start being a refuge, especially when elements such as lead and mercury refuse to respect the line between the two? At what point does the abundant garlic mustard (*Alliaria petiolata*) change from a vegetal agent of habitat creation to a pernicious invasive species? At what point do cormorants cease to be an environmental success story, in which the Spit brought them back from virtual extirpation from the Great

Lakes, and become a massive, habitat-destroying nuisance because of their colonial exuberance (nearly twelve thousand nesting pairs in 2013)? That the landscape is thus feral—in the sense of a place *in between*, or *both/and* nature and culture, wilderness and domesticity—is not all that unusual: one might argue that ferality is a constant companion to human settlement (Swyngedouw). Rather, what is interesting about the Spit is that it appears to be a public *celebration* of ferality in a context in which most popular ecological discourse prefers distinction: wilderness versus city, indigenous versus exotic, natural versus artificial.

There is reason to be suspicious of an *uncritical* celebration of the feral, and particularly of the idea that "feral" means a sort of rewilded return to a state of pre-anthropogenic grace. As Chan puts it, "the more unkempt the headland appears, the more authentic and legitimate it becomes" (7). Here, popular discourse about the Spit as having "gone wild" is problematic as it erases both the ongoing, large-scale human alteration of its physical and biological landscape and the complex social, political, and economic histories that created the Spit and that continue to play a role in the unfolding of life there. Considering the Spit as a "man-made wilderness" also has the unfortunate effect of providing an alibi for the unfettered continuation of capitalist urban development: if a toxic sludge pile, left alone for long enough, can become a celebratory zone of multispecies fecundity, then why worry about the species and spaces that development will irrevocably alter or destroy along the way?

In calling it "feral," then, I aim instead for an understanding of the Spit as always giving human control the slip even as it is constantly subject to (re)incorporation into capitalist regimes of accumulation, regulation, aesthetics, and politics. Cormorants, for example, refuse to do what we want them to do, which is flourish *successfully* (i.e., in ways that confirm the possibility of environmental success within a capitalist metabolism) but not quite so *excessively* (i.e., in ways that remind us of the fact that we are not in control). Cormorant agency defies the desired limits we have placed around it: the fact that they are transforming the Spit according to *their* practices of development does not sit well with the injunction that they conform to *our* urban aesthetic and ecological demands. In this light, the cormorants' ferality does not confirm the possibility of wilderness within capitalism; instead, it challenges the ways anthropogenic desires are naturalized as the only game in town. Perhaps especially because cormorants are what Foster and Fraser call a "landscape transforming" colonial species (in the same register as humans), the fact that cormorants refuse to be obedient citizens on the Spit highlights the ways in which "good" multispecies citizenship is often understood in anthropocentric terms, and "bad" citizenship invites managerial intervention in order to reconstitute the semblance of an affirmative order so that "the wild" can, ironically, inhabit its proper place *within* capitalist spatial formations.

On the Spit, citizenship is a contested process and ferality does not look like a nicely balanced wilderness; rather, the feral is an agent of politicization in that it demands a conversation about who and what belong where (Garside). Neither, however, is ferality purely a matter of disruption and slippage: it is constitutive and dynamic in a way that reminds us of our co-implicated indebtedness in particular places and times. As Anne Milne writes, "at some level, the feral … is the energy that … throws place open to its ineffable vicissitudes. And ineffability is what ultimately characterizes place: … that ephemeral, darting spirit that makes one place like no other on earth even as, just in the nick of time, it pulls the rug out before the taxonomizers glue it down and try to name that place and tell us definitively what it is and what belongs there" (331). Arguing that the Spit is a feral landscape, then, gestures toward a view of a world that is both demonstrably a multiagential co-production (involving multiple feralities in concert: cormorants, coyotes, lead, dog-strangling vine, carp, tent caterpillars, surfacing remains of 1960s kitchens) and continually re/subject to the biopolitical interventions of capitalism.

Feminisms

As Dana Luciano and Mel Y. Chen describe in their essay "Has the Queer Ever Been Human?," the recent "nonhuman turn" in the humanities has both extended and problematized older antihumanist currents, including feminist, antiracist, queer, disability, and decolonizing movements. Where feminism has, for example, long since called into question the falsely universalizing humanity of the "Man" around which liberal-humanist conceptions of rights, individuality, autonomy, and rationality circulate, recent moves to understand the human materially and ecologically as a *species*—rather than as the attainment of something beyond mere species-life—has opened up a new realm of understanding of power, agency, and politics for intersectional anti-oppressive work. Luciano and Chen build on the productive tension in histories of queer theory concerning whether the "queer" has ever been (or has ever desired to be) human—that is, the simultaneity within queer politics of a critique of the ways in which LGBTQI people(s) have been dehumanized and a countervailing skepticism with "the politics of rehabilitation and inclusion to which liberal-humanist values lead" (188). They consider that queer, feminist, and other concerns with injustice and oppression must accompany any explorations of our post-, non- and in-humanity. At the same time, however, recent movement into *anthro-decentric* thought opens up new understandings of the ways in which power relations no longer primarily operate (and never only operated) at the levels of discipline, interpellation, and/or subjection. Following, for example, Jasbir Puar's observation that "societies of control tweak and modulate bodies as matter" (63)—and riffing, of course, on Donna Haraway's much earlier observations about the informatics of domination and global corporeal webs of biopolitical connectivity—Luciano and Chen write that "analyses of neoliberalism show how fantasies of possessive individualism and sovereign agency have worn thin in a new labor economy ... [and] critical discussions of the commercialization of 'life itself' illuminate the breaking down of beliefs in species individuality" (191). In other words, even as anti-oppressive movements must now attend, ever more closely, to the ways in which "control societies" exercise power at the inhuman level of bodies, flows, affects, and assemblages, this attention "does not mean wholly abandoning the ethical investments and methodological frameworks that drove ostensibly 'human-centered' fields of inquiry based in identity and social location" (191).

Some recent work under the broad (and polyvocal) heading of "material" feminism has tended to downplay these earlier ethical and political concerns as part of a turn to an ontologically driven inquiry focused on the generativity, interactivity, and porosity of matter itself, without the (necessary) intervention of such power relations as gender, sexuality, class, race, and ability. For example, Elizabeth Grosz creates an almost complete caricature of a feminism obsessed with equality, identity, and autonomy in order to make a "new" case to direct feminist thought to "the problematic of sexual difference, the most fundamental concern of feminist thought at its most general, in the context of both animal becomings and the becomings microscopic and imperceptible that regulate matter itself" (86; for an extended critique, see Sandilands, "Feminism and Biopolitics").

More accurately, however, the notion that we relate as *organisms* to the biotic, chemical, and physical liveliness of the world in ways that transcend our conscious appreciation, and especially that "*all* life is now precarious life" (Luciano and Chen 193), is one of the most fundamental lessons that ecology has taught the humanities at least since the publication of *Silent Spring*. At the same time, the fact that these ecological relations must be understood in the context of colonialism, race, gender, sex, class, and ability is a crucial part of what the intersectional feminist humanities have long since offered back to ecology, especially by way of ecofeminism and other forms of environmental justice. Perhaps responding to oversimplistic rejections of feminism's concern with "social construction," new materialist Claire Colebrook specifically announces her affinities with ecofeminism, which she sees not as a "minor offshoot" of feminist theory but as part of the structure of its genealogy. Where liberal feminism began by questioning the exclusion of women from the rights of humanity in general, ecofeminism "takes feminism from a mode of human-human combat (women fighting for their rights, for the sake of all humanity) to a war on the man of reason; for it is man whose drive to mastery for the sake of his own self-maintenance has resulted in an unwitting suicide" (Section I). Thus, in her evaluation:

> Feminism's recent turn to life (in environmentalism and "new materialisms") should not appear as an addition or supplement but as the unfolding of the women's movement's proper potentiality. Indeed, this is just how eco-feminism has presented itself. It makes no sense to strive to transform our relation to the environment without transforming our own mode of being. Feminist criticisms of man would not be add-ons to environmentalism but would be crucial to any reconfiguration of ecological thinking.
>
> (Section I)

As Sarah Ahmed points out in a powerful essay critiquing the gathering of new materialist feminisms around the (Oedipal) rejection of earlier feminisms, it is not just ecofeminists that have continued to remember the mattering of matter. Not only have second-wave "feminists … produced very different kinds of critique of the role of biology, not all of which depend upon the rejection of the biological as a sphere of life" (28), but also "the commitment to rich description of biological processes was generated by women's activism: the women's health movement, for instance, involved new understandings of women's bodies, which required engaging with the biological sciences often critically, but also very closely" (30). Although it is not reasonable to argue that feminism was ever as opposed to materialist understanding as some (such as Grosz) claim it to have been, it is crucial to remember that, for anti-oppressive politics, matter matters *critically*, and bodies and flows are always already implicated in (if obviously never reducible to) power relations that can be named, following Judith Butler's important work, as *specific* constellations of precariousness and precarity. "In this light," note Luciano and Chen, "the palpable resistance by many critical race, feminist, and queer thinkers to … the nonhuman turn is not the effect of some recalcitrant or retrograde attachment to the human. Rather, it illuminates a concern over the critically and politically limiting effects of much recent critical insistence on the 'positive,' of calls to turn away from 'critique' as such" (194).

Here, Puar is especially aware of the tension between, on the one hand, a focus on contemporary political conditions and, on the other, the simultaneous need to move toward a new politics in order to pursue a more expansive ecofeminist agenda. Such a tension signals what "might be thought of as a dialogue between theories that deploy the subject as a primary analytic frame, and those that highlight the forces that make subject formation tenuous, if not impossible or even undesirable" (49). In her article "I Would Rather Be A Cyborg Than a Goddess"—a critical homage to Haraway—she writes: "on the one hand I have been a staunch advocate of what is now commonly known as an *intersectional* approach: analyses that foreground the mutually co-constitutive forces of race, class, sex, gender, and nation ... At the same time, encountering ... fatigue with the now-predictable yet still necessary demands for subject recognition, I argue ... that intersectionality as an intellectual rubric and a tool for political intervention must be ... complicated and reconceptualised ... by a notion of assemblage" (49). Puar's concern is thus to bring into conversation material feminist concerns with flows, fluxes, processes, inter- and intra-actions with more traditionally intersectional concerns about subjects and the ways they are created and positioned in multiple webs of power. As she writes, "there are different conceptual problems posed by each; intersectionality attempts to comprehend political institutions and their attendant forms of social normativity and disciplinary administration, while assemblages, in an effort to reintroduce politics into the political, asks what is prior to and beyond what gets established" (63). In sum, then, I would argue that without a strong understanding of *biopolitics* as well as becoming, *inequality* as well as intra-activity, and *precarity* as well as porousness, the material turn will fail to live up to its promise. In particular, there needs to be a more sustained feminist conversation, as Puar and others have initiated, about how capitalism involves matter's vitality in specific ways that are both familiar and excessive to earlier feminist accounts of power, bodies, subjectivation, activation, and domination.

Futurities

What then, one might ask, are double-crested cormorants and queer-materialist feminists doing together in this chapter? Are there ways in which cormorant politics illuminate materialist feminism, and vice versa? Are cormorants "good to think with" for the environmental humanities? Certainly, I am arguing that scholars in environmental humanities might want to think and act generally with a *feral feminism* in mind, an approach that both insists on the attempt to give human-centered thinking, identity, and power configurations the slip *and* remembers or embodies the ways in which these anthropogenic configurations cannot really ever be escaped on this earth. However, I am also arguing that this approach can help us develop biopolitical and intersectional insight about the multispecies relations of the Leslie Street Spit: how might this understanding lead to potentially life-affirming alternatives to the death-dealing solutions generally associated with cormorant management? In these questions, I hope to draw attention to the potential of feminist environmental humanities to probe larger questions of multispecies *futurities*, in which the complex agencies of nonhuman others are understood as part of the *same* roster of biopolitical relations as those that entangle more conventionally understood political subjects.

Not surprisingly, a feral feminist analysis reveals quickly that much of the "problem" of cormorant excess lies in their *reproductivity*. According to Taylor, Andrews, and Fraser, "by 2007, a total of 7,240 nesting pairs were [sic] recorded, making the Spit home to the largest cormorant colony on the lower Great Lakes" (381). That number of nesting, fish-eating, colonial waterbirds spells huge deposits of acidic, tree-killing guano in addition to fatal physical damage to trees caused by the removal of branches and leaves for nest material: just at the moment the Spit was beginning to look like a wilderness, the cormorants came along and developed it. Not surprisingly, in other jurisdictions the cormorant problem has been addressed by interfering with this mode of reproduction, as well as by shooting adult birds under the names of "culling" and "hunting." Reproduction is disrupted by harassing cormorants with flare guns so that they nest elsewhere, and especially by egg oiling, meaning coating their unhatched eggs with mineral oil so that the embryo is starved of oxygen and fails to hatch. This method is considered the most "humane" way of controlling ground nesting birds and, if done at the right time, may be doubly effective because the parents will still tend to the coated egg, preventing them from laying another until it is too late in the season for a new chick to survive (Shonk et al.). Egg oiling was experimentally tested on the Spit and deemed effective (Taylor and Fraser): parent birds sat on their dead eggs long enough to preclude re-nesting.

Clearly, then, cormorants are companions in the queer and feminist questioning of a *capitalist reproductive futurity* in which some lives (human and nonhuman) are biopolitically encouraged, and others, decidedly, are not. Having transgressed the unwritten rule that colonial waterbirds are supposed to be good ecological citizens in their new habitats, the birds moved easily into the political realm of excess, in which the health of the ecological body requires clinical intervention. Indeed, the use of the term "humane" to discuss embryonic cormorant death by asphyxiation shows that cormorants are not considered meaningful lives so much as bodies to be manipulated in various ways according to specific environmental ideals; they are, in other words, "bare life" in Agamben's sense and thus always already killable (Sandilands, "Dog Stranglers"). Cormorants have an extensive history of negative cultural association with rapaciousness and overabundance (see Wires): they have long since been eminently killable in the context of discourses tying their bodies to negative social conditions. But what is especially clear in the present biopolitical moment is that cormorant lives and deaths show the porousness of the boundary between human and nonhuman matter. The birds are objects of an invasive reproductive practice that is not at all dissimilar to the involuntary sterilization of racialized women and people with disabilities (hysterectomies for young women with intellectual disabilities are still popularly championed as "humane"; see Browne). Many people will find the equation of cormorants and people with disabilities objectionable. But the fact is that the reproductive control strategies are identical: the direct manipulation of humans and cormorants as *matter* is as much a part of the equation as the manipulation of people's *desires* for a certain kind of family, body, parent, Spit, and futurity.

As Heather Latimer and others have demonstrated, capitalist reproductive futurism is not about making the world safe for children or biodiversity. It is, instead, organized around the "fantasy that we may somehow return to our own innocence or childhood, to a time that never-quite-was, through constant attempts to protect our future world and our future children" (147). The figure of The Child, here, becomes a regulatory telos, in which the phantasmic future Child must be protected from the prolific complications of a corporeally and politically diverse (multiracial, multispecies) present. The effect of this heteronormative telos is both that queerness comes to be figured as the "'unfuture' or limit of the system, as death itself" (148) *and* that the reproductivities of others can come to be posed as a threat to the future rather than as part of its flourishing. On top of the directly racist and able-ist dimensions of this dynamic, the fantasy of return inherent in reproductive futurism bears a strong resemblance to that animating the idea of *wilderness*: if we can keep safe a phantasmic

449

wild, there will be room for The Child to play. But not all lives are allied to this future. In particular, feral and other complicated bodies that tangibly remind us of the non-innocence of the present, the constant dynamics of regulation and resurgence, the lack of a clear ecological or corporeal telos, are figured as abject and dispensable.

Unlike many abjected bodies and subjects, cormorants on the Spit had advocates, some of whom clearly recognized the extreme bad faith of killing birds to support the idea of a wilderness that (obviously) never existed: as Taylor, Andrews, and Fraser note, the Spit is a specifically *urban* nature and, "with the number of annual [human] site users in the hundreds of thousands, [it] cannot be free from human presence, and consequently, human manipulation" (389). Although they do not choose the term, they underscore that the *ferality* of the site needs to be taken into account and perhaps even encouraged as part of the complex unfolding future of the Spit: from extensive public consultation about different options, the TRCA eventually chose a management strategy of cormorant/human "para-habitation," in which particular zones are set aside for cormorants to practice their desires at some remove from the equivalent zones for people, but in which no direct bio-manipulation of cormorant (or human) bodies would take place. Although spatial regulation is still a form of biopolitical intervention, as they rightly argue, this strategy "allows the cormorant colony to flourish" as something other than a "static object of conservation" (389): cormorant lives and agencies are acknowledged in this choice. At the same time, cormorant landscape transformations are very much *visible* to the park visitor, part of a larger educational strategy emphasizing ecological change which, "while sometimes aesthetically undesirable, is an inevitable outcome, and examples of [other] species modifying their landscape can be used in experiential learning" (389). Together, these factors indicate a very promising direction: an affirmation of lives—perhaps even of the *need* for lives—that do not conform neatly to capitalist reproductive futurities.

But not all cormorants—and not all bodies more generally—are so lucky. The fact that the Spit has been a deeply *politicized* site from the beginning has helped to make the question of cormorant/human biopolitics visible and the list of management options publicly accountable. The environmental humanities could make a significant contribution by publicly inserting a feral feminist perspective into similar situations, where biopolitical controls manipulate reproductive and other relations in conditions of much less scrutiny (e.g., many other wildlife management and medical decisions). Especially, perhaps, as we think in and through the idea of the Anthropocene, the intervention of feral feminism into other sites and at other scales may prove useful as we attempt to engage in world-making practices that speak to a more democratic futurity.

References

Ahmed, Sarah. "Imaginary Prohibitions: Some Remarks on the Founding Gestures of the 'New Materialism.'" *European Journal of Women's Studies* 15.1 (2008): 23–39. Print.

Browne, Rachel. "Parents of Disabled Want More Flexibility to Hysterectomy Ban," *Sydney Morning Herald* 28 March 2013. Web. 15 September 2015.

Butler, Judith. *Precarious Life: The Powers of Mourning and Violence.* London: Verso, 2004. Print.

Chan, Alexander. "Sympathetic Landscapes: An Aesthetics for the Leslie Street Spit." Waterloo, ON, M. Arch Thesis, University of Waterloo, 2013. Print.

Colebrook, Claire. *Sex After Life: Essays on Extinction, Volume Two*. London: Open Humanities Press, 2004. Web. 15 September 2015.

Foster, Jennifer and Gail Fraser. "Predators, Prey and the Dynamics of Change at the Leslie Street Spit." *Urban Explorations: Environmental Histories of the Toronto Region*. Ed. L. Anders Sandberg, Stephen Bocking, Colin Coates and Ken Cruikshank. Hamilton, ON: L. R. Wilson Institute for Canadian History, McMaster University, 2013. 211–224. Print.

Friends of the Spit. *Checklist of Plants on the Leslie Street Spit*. Toronto: Friends of the Spit, June 1990. PDF file.

Garside, Nicholas. *Democratic Ideals and the Politicization of Nature: The Roving Life of a Feral Citizen*. New York: Palgrave MacMillan, 2013. Print.

Grosz, Elizabeth. *Becoming Undone: Darwinian Reflections on Life, Politics, and Art*. Durham, NC: Duke University Press, 2011. Print.

Haraway, Donna. "Manifesto for Cyborgs: Science, Technology, and Socialist Feminism in the 1980s." *Socialist Review* 80 (1985): 65–108. Print.

Latimer, Heather. *Reproductive Acts: Sexual Politics in North American Fiction and Film*. Montreal: McGill-Queen's University Press, 2013. Print.

Luciano, Dana and Mel Y. Chen. "Introduction: Has the Queer Ever Been Human?" *GLQ: A Journal of Lesbian and Gay Studies* 21.2–3 (2015): 183–207. Print.

Milne, Anne. "Fully Motile and Awaiting Further Instructions: Thinking the Feral into Bioregionalism." *The Bioregional Imagination: Literature, Ecology and Place*. Ed. Tom Lynch, Cheryll Glotfelty, and Karla Armbuster. Athens: University of Georgia Press, 2012. 329–344. Print.

Puar, Jasbir. "'I Would Rather Be a Cyborg Than a Goddess': The Becoming-Intersectional of Assemblage Theory." *PhiloSOPHIA* 2.1 (2012): 49–66. Print.

Sandilands, Catriona. "Dog Stranglers in the Park? National and Vegetal Politics in Ontario's Rouge Valley." *Journal of Canadian Studies* 47.3 (2013): 93–122. Print.

——. "Feminism and Biopolitics: A Cyborg Account." *Routledge International Handbook on Gender and Environment*. Ed. Sherilyn MacGregor. London: Routledge, forthcoming. Print.

Schopf, Heidy and Jennifer Foster. "Buried Localities: Archaeological Exploration of a Toronto Dump and Wilderness Refuge." *Local Environment* 19.10 (2014): 1086–1109. Print.

Shonk, K. A., S. D. Kevan and D. V Weseloh. "The Effect of Oil Spraying on Eggs of Double-Crested Cormorants." *The Environmentalist* 24 (2004): 119–124. Print.

Swyngedouw, Erik. "The City as a Hybrid: On Nature, Society, and Cyborg Urbanization." *Capitalism, Nature, Socialism* 7.2 (1996): 65–80. Print.

Taylor, Bernard and Gail S. Fraser. "Effects of Egg Oiling on Ground Nesting Double-crested Cormorants at a Colony in Lake Ontario: An Examination of Nest-attendance Behaviour." *Wildlife Research* 39.4 (2012): 329–335. Print.

Taylor, Bernard, Dave Andrews, and Gail S. Fraser. "Double-crested Cormorants and Urban Wilderness: Conflicts and Management." *Urban Ecosystems* 14 (2011): 377–394. Print.

Toronto and Region Conservation Authority. *Tommy Thompson Park: Toronto's Urban Wilderness*, 2015. Web. 15 September 2015.

Watt-Meyer, Ben. "A New Archaeology for the Leslie Street Spit." *Cargo Collective*. Urbanworm Design, 2015. Web. 15 September 2015.

Wires, Linda R. *The Double-crested Cormorant: Plight of a Feathered Pariah*. New Haven: Yale University Press, 2014. Print.

43

BIOCITIES

Urban ecology and the cultural imagination

Jon Christensen and Ursula K. Heise

Cities are the habitat par excellence for *Homo sapiens*. Call us *Homo urbanus*: more than half of us live in cities now. Virtually all of the human population growth expected on Earth this century will effectively end up in cities. Urban populations will double, and the urban built environment will double to accommodate the growth. *Homo urbanus* is thriving in the Anthropocene.

Rapid urbanization is considered by many to be one of the twenty-first century's grand environmental challenges, along with climate change, biodiversity loss, and sustainability in a world of inequality that is itself unsustainable (Lee et al.). Paradoxically, urbanization is also seen as one of the most important solutions to these other grand challenges. At the same time, ecology is coming back to the city, recognizing a nature that was largely ignored for the past century, and in the process fashioning an emergent new field of urban ecology.

The so-called "Chicago School" of urban sociology attempted to apply emerging concepts in ecology to the city in the early twentieth century (Light). It was discredited, abandoned, and largely forgotten as ecologists moved on from the simplistic ideas of succession and climax communities the sociologists imported, and sociologists themselves turned away from ecology. Memories are short, however, and the dream of scientific explanations of the city remains powerfully enticing.

Today, a new "Baltimore School" attempts to apply contemporary theories of ecological "patch dynamics" to the city, as part of a long-term ecological research project funded by the National Science Foundation (Grove et al.). Its counterpart in Phoenix focuses largely on "ecosystem services." A new "science of cities" is emerging from the Santa Fe Institute, driven by basic theories in biology and physics (Bettencourt and West), and from the Center for Advanced Spatial Analysis at University College London, where the science is based on network theory (Batty). And the Ecological Society of America has embraced "urban ecology" (Cressey).

If nothing else, we are witnessing a tremendous surge in interest in both the nature of cities (what is their "essential quality and character?") and the nature in cities ("the material world itself, taken as including or not including human beings"), to follow two meanings of "nature" teased apart by Raymond Williams in *Keywords: A Vocabulary of Culture and Society*. Both meanings of "nature" are, of course, inextricably intertwined, and full of history, as Williams wrote in an essay entitled "Ideas of Nature." And yet, there is, curiously, a disconnect between the scientific discourse around urban ecology and the nature of cities, and the discourse around these same important ideas in the humanities and the culturally inflected social sciences, particularly in the emergent fields we call "environmental humanities" and "urban humanities."

Where do they meet? Perhaps in what we might call urban environmental humanities. Ecological studies and theories of the city have their utility, particularly in the data that they produce, but we will never have an adequate understanding of nature and the city without the humanities.

In this chapter, we trace a cultural history of the imagination of nature and cities, scan prevailing theories of urban nature, survey ideas about nature and the city today, and look ahead at "biocities" to come—through science fiction, speculative nonfiction, and scientific predictions and modeling—in an effort to map fissures and potential common ground for thinking about this human habitat, increasingly revealed as a multispecies habitat.

Nature in the city: from garden cities to biocities

The notion of the "biocity," a term we use to embrace the city as at once a human and more than human creation, may well be as old as the city itself, and the village and town before it. There are intimations of such thinking in ancient times; the first clear articulation that we are aware of comes in Plato's discussions of the city as a body in which different citizens carry out their different functions. By the sixth century in Palestine, we find records of the regulation of nature in cities, the articulation of rules for how property owners can make use of the bounties of nature within the city, what plants can be planted where, and the regulation of design and construction to preserve valuable views of the ocean (Hakim). The roots of the "sanitary city," the regulation of water and waste, go back even further, though we think of the sanitary city coming to full fruition in the West in the nineteenth and twentieth centuries in modern public works and plumbing.[1]

Like much environmental thinking, concern about nature in the city has many roots and branches—William Penn's original conception for Philadelphia was of a garden city in colonial America—but the pace and pitch really pick up at the same time as concern about nature outside of the city in the nineteenth century, in reaction to rapid industrialization and modernization. As the first country in the world to experience the demographic shift of the majority of its population to cities, England sits at the beginning of this complicated transition, which sees ideas about the country and the city become so intertwined that it often seems as if it was hard to think or talk about one without reference to the other (Williams, *The Country and the City*). Now the whole world is going through this transition.

Perhaps it should come as no surprise, then, that some of the most prominent ideas of the times, such as Ebenezer Howard's "Garden Cities" (Hall 87–141) and the grand parks and parkway proposal for Los Angeles by the landscape architect Frederick Law Olmsted and his sons (Hise and Deverell), still seem to hold such imaginative power, although many of their plans were never carried out, and others had unintended consequences. Howard's late nineteenth-century proposal for a "Garden City" had more to do with creating a just and egalitarian society of good, righteous, and healthy citizens than with anything we might call ecology today, and it resulted in leafy suburbs that played a role in making cities more segregated and unequal. Still, the dream lives on in proposals for new cities such Dongtan in China and Masdar in Abu Dhabi. And just so, Olmstedian visions, and, indeed, sometimes old Olmsted proposals, such as one for a necklace of parks and parkways for Los Angeles in the early twentieth century, still animate public discourse about nature in the city today.

The battle between Robert Moses and Jane Jacobs in New York in the early 1960s reveals a dissonant strand in this thinking. The early Moses, the "good Moses," engineer of great parks and parkways, is sometimes neglected in the aftermath of the late Moses, builder of highways, bridges, and tunnels, destroyer of neighborhoods. The Jacobs who is most remembered is the advocate of human-scale, walkable neighborhoods, with diverse street-level storefronts. Less often remembered is the Jacobs who could be ruthlessly critical of parks—not to mention empty lots or other marginal areas of the city, where nature might thrive—as dead spaces that too often add nothing to the city, and, indeed, sometimes subtract from it (Jacobs 116–145).

In the late 1960s and early 1970s, an imaginary that we might call mapping the biocity arose in the work of Ian McHarg and an army of geographic information systems specialists who now arguably dominate nearly all real planning and implementation of major projects that affect nature in many cities worldwide, at least where formal planning has force, if not yet in areas where more informal processes prevail. Here procedural democracy and bureaucracy, spatial partitioning, and a God's eye view of the city come together to imagine, if not a perfect harmony, then everything in its right place: a park here, green infrastructure there, industry over there, housing, shops, schools in other places. And the map becomes something to argue about.

A few more compelling developments seem worth touching on in this admittedly limited survey of the cultural history of the biocity. The first is biomimicry and biophilic design, the related notions that the built environment, at the building scale if not the whole city, should mimic nature in form or function (Kellert et al.). Here we see nature not as separate, in its own place, within the city, but as mixed up in hybrid natural and built forms, such as living walls, breathing buildings, and skyscraper farms.

The second notion is what we might call "climate urbanism," already a very real phenomenon in which adaptation to climate change is beginning to drive decisions in cities (Arbona). Here we see nature barging back into the city in a big way all over the world.

Finally, cities have recently come to be envisioned as one solution to climate and other environmental problems. Cities are already "green" by virtue of being dense, having a smaller per capita carbon footprint, and occupying less space on Earth. By crowding together in cities, we are saving nature out there. This perspective begs the question of which nature we are saving out there and what happened to nature in the city. In an effort to get our arms around that, we turn to a survey of theories of biocities.

Nature and social justice: theories of biocities

History does not always wear theory on its sleeve, and is, indeed, sometimes quite theory-averse. But in contemporary urban environmental histories, we can see important theories about nature and the city at work, sometimes explicitly, sometimes implicitly. William Cronon's *Nature's Metropolis: Chicago and the Great West* is an important landmark in the field, and one that makes its theory quite central to its historical narrative and argument. In Cronon's view, Chicago is "second nature," fashioned from "first nature"—wheat, timber, and cattle—produced, commodified, and exchanged within the city's expanding economic sphere over time. The city and its region cannot be understood without understanding these relations. The country and the city are co-produced in this history, as first nature is transformed through supply chains that bring commodities into the city and send rail cars filled with plows and other finished goods into the country.

Matthew Gandy goes further in proposing a "metropolitan nature" in his environmental history *Concrete and Clay: Reworking Nature in New York City*. Metropolitan nature includes not just the second nature of a modern water supply system that brings clean water to New York City from upstate, but also the ways in which ideas of nature are used by Olmsted and others to fashion Central Park and new public spaces in the city, and how political, economic, and cultural developments shaped Robert Moses's parkways and transformed the metropolitan region. Gandy also considers how understanding radical politics in a Puerto Rican barrio requires extending our understanding of an ecological frontier far beyond the immediate and familiar economic frontier of food, timber, and water. Finally, the social dimension of this ecological hinterland extends to a contemporary battle against a waste-processing facility in a working-class neighborhood and its connections to the broader global environmental justice movement.

These two histories have clear connections to another strand of theories that collect under the umbrella of "political ecology." Centered in geography with roots in Marxist political economy, urban political ecology features "metabolism" as a central concept for thinking about nature and the city. But this is not the corporeal metaphor of metabolism in the city conceived as an organism—the simple analysis of inputs such as water and energy, their consumption, and outputs of waste that result. Instead, this is metabolism as Marx and Engels used the term, the transformation of nature through work, the alienation of labor from the products of that work, and the accumulation of capital and power that results (Heynen et al.).

Ironically, there is very little "ecology" in political ecology, though it is, like Gandy, centrally concerned with the social construction of "nature" in the city and beyond. But political ecology emphasizes that there is nothing "natural" about this process at all. Indeed, political ecology wants us to resist any impulse to "naturalize" the city, that is, to make it seem the result of nature and natural processes. Instead we are urged to see the city as constructed by social, economic, and political forces (Gandy, "Urban Nature").

Studies in political ecology that resist the "naturalization" of urban processes, while enormously productive intellectually, have so far failed to have any discernible impact on policy, planning, and building in cities. On the other hand, there are a wide variety of pragmatically oriented works in public policy, planning, and urban studies that, at least for their own purposes, assume that there are certain natural elements of the city—parks, trees, rivers, streams, lakes, clean water, clean air, beaches—that can be measured and studied in a straightforward way to understand their current distributions and make recommendations of policies for better and more fairly distributing these environmental goods (e.g., Loukaitou-Sideris; Wolch et al.). Likewise with environmental risks and harms. And it is important to note that some of these studies have had real impact in cities.

The field of risk theory—which has been so productive on larger scales—is strangely underrepresented in theories of nature in the city. However, sometimes these different strands of theory come together in remarkable works such as Jared Orsi's *Hazardous Metropolis: Flooding and Urban Ecology in Los Angeles*, which dissects the political ecology of building a city in a landscape prone to flash floods and debris flows. The city was made possible by the channeling of the Los Angeles River, which enabled a metropolis to grow in this unstable landscape. In the latest phase of this history, environmental justice debates have emerged alongside efforts to restore the river and bring nature back to the city, prompting Orsi to speculate on how politics itself might become more ecological and adaptive. Mike Davis's *Ecology of Fear* offers a more dystopian vision of the differential distribution of risk throughout the city.

455

Environmental justice is concerned with the differential distribution of environmental goods and harms, according to class, race, gender, and other social and economic dimensions. In scholarship and activism that focuses on environmental justice, the material impacts of the uneven distribution of natural resources and ecological risks take center stage, for example, in fights over the impact of polluting industries. Culturally different constructions of nature can also turn into central objects of political struggle. Different understandings of land, natural resources, and humans' relations to other species, for instance, have shaped confrontations between local and indigenous communities, on the one hand, and national land and biodiversity management regimes, on the other, in many parts of the world. But this kind of culturally complex understanding of the cultural varieties of nature in cities is only sporadically represented in the literature on urban environments.

Jenny Price's seminal essay "Thirteen Ways of Seeing Nature in LA" uses different strands of the theories already discussed here to encourage us to see culturally inflected varieties of nature where we have been led to think nature is absent. Here nature is in the river that does not look like a river. It is even in the mango skin cream she finds at the Beverly Center Mall, the supply chains that go into the product, and getting it on the shelf. Without explicitly embracing the theory, Price sidles up to the "assemblages" of Bruno Latour and "Actor-Network-Theory," configurations of the human and nonhuman—with its own forms of agency—that do work in the world. There has yet to be a fully realized account of nature in the city based on Actor-Network-Theory, but we suspect we may see one soon.

The closest we come are accounts of the "multispecies" city or "zoöpolis," in the coinage of Jennifer Wolch. Here animals in cities are seen to have their own agency, and perhaps even claims to environmental justice as well, as they live their own lives in their own habitat, which is coincidentally shared by humans who might think of the city as a human creation (Van Dooren and Rose).

Oddly enough, the science of ecology has had a rough time in the city, even as "urban ecology" has become a popular emergent term in the field and well beyond. Despite decades of efforts in the long-term ecological research stations in Baltimore and Phoenix, the primary lesson is that the laws of ecology—mostly fashioned in research in study sites that excluded human variables as much as possible—simply do not apply to cities (McIntyre). Very basic empirical measurements and simple predictions, about the effects of nitrogen in waterways, for instance, are productive. But theories about the population dynamics of species in urban settings, for example, will have to be built from the ground up, if they can be built at all.

Likewise, the "new science of cities," which has very successfully imported theories of scaling from biological systems to explain the scaling of energy, consumption, and even creativity in cities, cannot explain the distribution of nature in cities (Christensen, "Contingent Ecological Urbanisms"). It cannot account for natural areas or species, let alone the ecosystem services that nature provides for people directly, such as air to breathe and water to drink, as well as indirectly, through such services as the feeling of well-being provided by the tree outside our window, or the small joy of seeing a possum trundling through our yard on its nightly rounds.

Today, in a city such as Los Angeles, we see pieces of this history, these theories, and these ideas all around us. The Natural History Museum of Los Angeles has developed a "nature lab" as well as a "nature garden" in the heart of the city and is supporting citizen science and research on biodiversity throughout the city. The city now has a plan to catalog biodiversity, too. The fate of the Los Angeles River is at the center of debates over the future of nature in the city. The Los Angeles Neighborhood Land Trust is building parks and community gardens in neighborhoods that have very little access to anything resembling what we might call "first nature." An environmental coalition is pushing for a multi-billion-dollar, multi-decade transportation plan to include green infrastructure and access to parks, as well as clean, green transportation alternatives.

Like many cities worldwide, Los Angeles is also in the throes of attempting to reduce its carbon footprint dramatically, as well as adapt to the inevitable rise in temperature and sea level already baked into the global atmospheric system. Researchers are studying the city's "metabolism"—in the sense of its inputs, consumption, and outputs, if not in the deeper sense in which political ecologists use the term—to figure out how the city can operate more efficiently and more equitably. Resilience is a watchword here, as it is around the world, though in Los Angeles resilience to earthquakes is at the top of the list of concerns about natural hazards. And we see instances of biomimicry and biophilic design popping up here and there throughout the city.

These contemporary discourses offer openings for cultural imagination. However, these openings come through the very "naturalization" of nature in the city that the political ecologists warn us to be wary of. It is to the future, and science fiction and speculative non-fiction, that we look for more complicated versions of how these histories and theories and cultural imaginings might play out.

Biocities and the imagination of the future

Scientific predictions of the future of nature in cities have proliferated in the current era of climate change modeling. Oddly enough, these model-based futures are unlikely to come true because they are forced to assume that the future will be very much like the past, except for a changing climate. Current scientific models are unable to cope with one of the most basic features of human interactions with the environment: people change, too. Climate models can crudely estimate how temperatures will change if greenhouse gas emissions continue under "business as usual" scenarios, if emissions are cut by varying degrees, or if they are completely curtailed. The computer modeling is complicated and requires many hours of data crunching by supercomputers, particularly when the models are downscaled to a scale that people can relate to: the city and the neighborhood.

But the models cannot predict how people will respond to rising temperatures. So instead, scientists often assume that people simply will not adapt. More people will die in heat waves when there are more heat waves. More houses will go up in flames when there are more wildfires. Wines will become sweeter and heavier with warmer growing seasons, assuming that wine production does not move with the changing climate, which in turn assumes that vintners are less adaptable than small mammals and birds, which have already begun to shift their ranges in response to global warming (Christensen, "Environmental Prospects"). Similarly, predictions of mass migrations of people from cities such as Los Angeles seem to willfully ignore that the city was made possible by people adapting to a warm climate, and that plenty of cities have existed for a very long time in even hotter climates (e.g. Kahn).

Sunnier versions of the future sometimes make the same mistake. *Mannahatta: A Natural History of New York City* (Sanderson) brilliantly shows how the habitat of the island of Manhattan was utterly transformed in the four hundred years after Henry Hudson first sailed up the river that would be named for him. But when it tries to look forward four hundred years into the future, it suffers a failure of imagination. People are riding around Manhattan on bicycles with kids in child seats. Windmills sit on the tops of skyscrapers, the city has a lot more vegetation, and some outlying areas have been converted to farming. But it looks like a green dream stubbornly stuck in the early twenty-first century rather than a vision of the twenty-fifth century. By focusing on more near-term futures, *Rising Currents: Projects for New York's Waterfront* (Bergdoll), a 2012 exhibition at the Museum of Modern Art, provides a more compelling and useful vision of the future of the biocity. While few if any of the projects will ever be built, the speculative proposals and drawings of a different, more porous and embracing urban relationship with water provide generative examples of hybrid natural and built forms—wetlands, oyster beds, kelp forests, aquaculture, bird habitat—that will very likely contribute to creating a more flexible, adaptive waterfront in the centuries to come.

Alan Weisman's *The World Without Us* is even more daring, on the one hand, imagining the cityscape, well, without us, as habitat for other species in the future. Lots of species, he speculates, with a firm grounding in science, would do just fine adapting to our habitat, particularly as it crumbles not just from the deferred maintenance that plagues many cities, but from absolutely no maintenance. On the other hand, the move is too disanthropic, a term ecocritic Greg Garrard coined to describe the enduring trope of "worlds without us." While it might be useful to think about what happens when people retreat from parts of cities, such as Detroit, it does not help much in thinking about how we might reimagine sharing our urban habitats with the more than human world around us.

The line between nonfiction and fiction is blurred in all of these efforts to speculate about the future of the biocity—from scientific models to architectural drawings and nonfiction counterfactuals. The range of possibilities is curiously similar in fictional accounts of the future of the city and its relationship with nature. Much like *Mannahatta*, *Ecotopia* represents the future of the city in Northern California as a slightly more stoned, crunchy-granola version of the Berkeley Ernest Callenbach knew when he was writing it in the early 1970s. Ursula K. Le Guin's *Always Coming Home*, on the other hand, published a decade later, is set in a San Francisco Bay Area utterly transformed by rising sea levels and social change. But for Le Guin, the past is not just prologue. The future is not one without us, but instead a return to the kinds of tribal societies that existed in California hundreds of years ago, even though a few, often troublesome modern technologies, such as guns, still exist.

Nathaniel Rich's *Odds Against Tomorrow*, published in 2013, is so near-term in its account of the devastation of New York City by a superstorm and flooding that it could be a nonfiction account of Hurricane Sandy, if it had been worse. While the descriptions of the storm, the evacuation of the city, the deterioration of human relationships and civility, and the resulting violence are all harrowing, it is the development of the main character's relationship to the future that is most deeply troubling. At the beginning of the novel, he is a preternaturally confident futurist. At the end, he lives only in the moment, scrabbling in the dirt of the Bronx, flattened into a clean slate by the hurricane. For the protagonist, the present in this habitat is all there is. Roland Emmerich's film *The Day After Tomorrow*, released in 2004, relies on scientific predictions, now largely abandoned, that global warming could change circulation patterns in the Atlantic Ocean, bringing a great freeze to New York City rather than warming. Here, nature is threatening once again, destabilizing and even destroying civilization, which huddles in the New York Public Library, burning books, while wolves roam outside.

In contrast to these portraits of a metropolis doomed by floods and rising sea levels, Kim Stanley Robinson's vision of New York City as an aquatic city in *2312*—which was published in 2012—is spectacularly imaginative and optimistic. The city has become a glittering, hypermodern Venice, with canals instead of streets and avenues. For the protagonist, who is used to living in space suits and domed habitats, it is one of the most natural places she has ever encountered. This is no gloom-and-doom vision of the future of nature in the city: "Indeed, it was an oft-expressed cliché that the city had been improved by the flood" (100).

Few cities are as prone to destruction in speculative fiction and films as our very own Los Angeles (Davis 275–355). And nature in the form of earthquakes especially is often the *deus ex machina* for driving plot lines from *Escape from L.A.* to *San Andreas*. But two recent films, Neill Blomkamp's *Elysium* and Spike Jonze's *Her*, both released in 2013, not only frame very different visions of the future of the city and nature, but also have become part of the discourse of thinking about that future in the city. In *Elysium*, Los Angeles has come to look like just one more urban outpost on a planet of slums. The bleak urban landscape on Earth stands in contrast to the space station orbiting earth, which looks very much like an upscale version of an Ebenezer Howard-inspired garden city, albeit ruled over by a tyrannical Jodie Foster, far from the egalitarian society Howard hoped his vision would construct. Here, nature is displaced from the city. It is the other. Back to the future, that is, to the most uncomplicated version of nature in the city imaginable.

Curiously, *Her*, while quite opposite in its portrait of Los Angeles as a dense hypermodern city of high-rise living with pleasant public transportation, also represents nature as other. A small patch of a rooftop garden and an owl on a big screen are just about all we see of nature in the city. The natural, the beach, the forests, the mountains, are all outside of the city.

And so we often recur to *Blade Runner*, released way back in 1982, to think about the future in the City of Angels. The film is set in 2019, which is soon enough. You might say there is very little of nature in *Blade Runner*'s bleak neo-noir Los Angeles. A steady rain falls on a thoroughly hardened cityscape. There are no parks, no trees, no live animals outside of the few seen in captivity. And, yet, this is a thoroughly hybrid landscape. The few animals we see are products of nature and technology. Indeed, we are left to think, so may we be in the future. And who is to tell the difference?

It is here, in the imagining of a biocity that is thoroughly hybrid, a product of nature and humanity, and a habitat for such hybrids and their odd, evolving, and adapting assemblages, that such histories, theories, narratives, and speculations might usefully be applied as we think through the future of urban environmental humanities and life in this human and more-than-human habitat.

Note

1 In this chapter we survey literature from a variety of fields and disciplines, all in English, and mainly covering Anglo-American examples. This work had its origins in "BioCities: Urban Ecology and the Cultural Imagination," a graduate seminar we developed as part of the Urban Humanities Initiative at UCLA. We thank our students for their insights and inspirations in our discussions. We are currently expanding this survey of biocities to literature in other languages and geographic areas.

References

Arbona, Javier. *Climate Urbanism*. Blog. Web. 6 Feb. 2016.

Batty, Michael. *The New Science of Cities*. Cambridge, MA: MIT Press, 2013. Print.

Bergdoll, Barry, ed. *Rising Currents: Projects for New York's Waterfront*. New York: Museum of Modern Art, 2011. Print.

Bettencourt, Luis and Geoffrey West. "A Unified Theory of Urban Living." *Nature* 467 (2010): 912–913. Print.

Blade Runner. Dir. Ridley Scott. Perf. Harrison Ford, Rutger Hauer, and Sean Young. Warner Bros., 1982. Film.

Callenbach, Ernest. *Ecotopia*. New York: Bantam, 2009. Print.

Christensen, Jon. "Contingent Ecological Urbanisms." *Now Urbanism: The Future is Here*. Ed. Jeffrey Hou, Benjamin Spencer, Thaisa Way and Ken Yocom. London: Routledge, 2015. 78–91. Print.

——. "Environmental Prospects in the Twenty-First Century." *A Companion to California History*. Ed. William Deverell and David Igler. Malden, MA: Wiley-Blackwell, 2008. 483–498. Print.

Cressey, Daniel. "Ecologists Embrace their Urban Side." *Nature* 524 (2015): 399–400. Print.

Cronon, William. *Nature's Metropolis: Chicago and the Great West*. New York: Norton, 1991. Print.

Davis, Mike. *Ecology of Fear: Los Angeles and the Imagination of Disaster*. New York: Metropolitan Books, 1998. Print.

The Day After Tomorrow. Dir. Roland Emmerich. Perf. Jake Gyllenhaal, Dennis Quaid and Ian Holm. 20th Century Fox, 2004. Film.

Elysium. Dir. Neill Blomkamp. Perf. Matt Damon, Jodie Foster and Sharlto Copley. TriStar Pictures, 2013. Film.

Gandy, Matthew. *Concrete and Clay: Reworking Nature in New York City*. Cambridge, MA: MIT Press, 2002. Print.

——. "Urban Nature and the Ecological Imaginary." *In the Nature of Cities: Urban Political Ecology and the Politics of Urban Metabolism*. Ed. Nik Heynen, Maria Kaika and Erik Swyngedouw. London: Routledge, 2006. 63–74. Print.

Garrard, Greg. "Worlds Without Us: Some Types of Disanthropy." *Substance* 41.1 (2012): 40–60. Print.

Grove, Morgan J., Mary L. Cadenasso, Steward T. A. Pickett, Gary E. Machlis and William R. Burch, Jr. *The Baltimore School of Ecology: Space, Scale, and Time for the Study of Cities*. New Haven, CT: Yale University Press, 2015. Print.

Hakim, Besim S. "Julian of Ascalon's Treatise of Construction and Design Rules from Sixth-Century Palestine." *Journal of the Society of Architectural Historians* 60.1 (2001): 4–25. Print.

Hall, Peter. *Cities of Tomorrow: An Intellectual History of Urban Planning and Design since 1880*. 4th ed. London: Wiley-Blackwell, 2014. Print.

Her. Dir. Spike Jonze. Perf. Joaquin Phoenix, Amy Adams and Scarlett Johansson. Warner Bros., 2013. Film.

Heynen, Nik, Maria Kaika and Erik Swyngedouw. *In the Nature of Cities: Urban Political Ecology and the Politics of Urban Metabolism*. London: Routledge, 2006. Print.

Hise, Greg and William Deverell. *Eden by Design: The 1930 Olmsted-Bartholomew Plan for the Los Angeles Region*. Berkeley, CA: University of California Press, 2000. Print.

Jacobs, Jane. *The Death and Life of Great American Cities*. New York: Modern Library, 1993. Print.

Kahn, Matthew W. *Climatopolis: How Our Cities Will Thrive in the Hotter Future*. New York: Basic Books, 2010.

Kellet, Stephen R., Judith H. Heerwagen and Martin L. Mador. *Biophilic Design: The Theory, Science, and Practice of Bringing Building to Life*. Hoboken, NJ: John Wiley and Sons, 2008. Print.

Le Guin, Ursula K. *Always Coming Home*. Berkeley, CA: University of California Press, 2001. Print.

Lee, Kai N., William R. Freudenburg and Richard B. Howarth. *Humans in the Landscape: An Introduction to Environmental Studies*. New York: Norton, 2013. Print.

Light, Jennifer S. *The Nature of Cities: Ecological Visions and the American Urban Professions, 1920–1960.* Baltimore, MA: Johns Hopkins University Press, 2009. Print.

Loukaitou-Sideris, Anastasia. "Green Spaces in the Auto Metropolis." *Planning Los Angeles.* Ed. David C. Sloane. Chicago, IL: American Planning Association, 2012. 191–203. Print.

McHarg, Ian L. *Design with Nature.* New York: John Wiley & Sons, 1992. Print.

McIntyre, Nancy E. "Patterns and Drivers of Biotic Population and Community Structure in Built Environments." Ecological Society of America Annual Meeting. Baltimore, MA. 12 August 2015. Conference presentation.

Orsi, Jared. *Hazardous Metropolis: Flooding and Urban Ecology in Los Angeles.* Berkeley, CA: University of California Press, 2004. Print.

Price, Jennifer. "Thirteen Ways of Seeing Nature in LA." *Land of Sunshine: An Environmental History of Metropolitan Los Angeles.* Ed. William Deverell and Greg Hise. Pittsburgh, PA: University of Pittsburgh Press, 2005. 220–244. Print.

Rich, Nathaniel. *Odds Against Tomorrow.* New York: Farrar, Straus, and Giroux, 2013. Print.

Robinson, Kim Stanley. *2312.* New York: Orbit, 2012. Print.

Sanderson, Eric W. *Mannahatta: A Natural History of New York City.* New York: Abrams, 2009. Print.

Van Dooren, Thom and Deborah Bird Rose. "Storied-Places in a Multispecies City." *Humanimalia* 3.2 (2012): 1–27. Web. 6 Febuary 2016.

Weisman, Alan. *The World Without Us.* New York: St. Martin's Press, 2007. Print.

Williams, Raymond. *The Country and the City.* Oxford: Oxford University Press, 1973. Print.

——. "Ideas of Nature." *Culture and Materialism.* London: Verso, 1980. 67. Print.

——. *Keywords: A Vocabulary of Culture and Society.* New York: Oxford University Press, 1976. Print.

Wolch, Jennifer. "Zoöpolis." *Animal Geographies: Place, Politics, and Identity in the Nature-Culture Borderlands.* Ed. Jennifer Wolch and Jody Emel. London: Verso, 1998. 119–138. Print.

Wolch, Jennifer, Travis Longcore and John Wilson. "Unpaving Paradise: The Green Visions Plan." *Planning Los Angeles.* Ed. David C. Sloane. Chicago, IL: American Planning Association, 2012. 230–239. Print.

44

ENVIRONMENTAL HUMANITIES
Notes towards a summary for policymakers

Greg Garrard

In 2010, I co-organized an AHRC-funded network entitled "The Cultural Framing of Environmental Discourse." One of the requirements of the grant was that networks should include both diverse disciplines and participants from outside universities. After a day-long discussion of the public role of the environmental humanities, an attendee from an environmental consultancy firm commented that she found our work fascinating and thought it could play a genuinely constructive role in policymaking: "Could you summarize it on two sides of A4? Because that's about the attention span of most politicians and CEOs."

There are already a number of collaborative statements of the role and significance of the environmental humanities. Given the range of disciplines included under this "umbrella term," it makes sense that manifestos would be written by groups of experts. The objective here is not to summarize the summaries, or to supplement them with one more, inevitably less complete, account. Rather, by encapsulating the environmental humanities in a succinct, concise, memorable fashion, this chapter paves the way—admittedly in more than two sides of A4—for the kind of executive summary the consultant wanted. If the IPCC can do it, so can we.

We have a working definition of the field: "The environmental humanities is a term for a range of multifaceted scholarly approaches that understand environmental challenges as inextricable from social, cultural and human factors" (Åsberg et al. 70). We suppose that three disciplines—environmental history, environmental philosophy, and ecocriticism—make up the heart of the field, with important contributions in art history, theology, and those subdisciplines of the social sciences that are effectively cognate with the humanities: cultural geography and cultural anthropology. We might construct a genealogy of EH, but, in addition to the hazards of any such approach, outsiders would have little use for it. What we need, I suggest, is a Janus-faced *characterization*: recognizable and acceptable to those working in the field, but concerned primarily to tell people who have no idea what we do and no prior commitment to the humanities, why we deserve their attention.

I argue that, for this purpose, we need a limited number of illuminating examples rather than long lists of citations. We also need an organizing principle that links them together in simple, legible terms, for which I suggest the chiasmus "ecologizing humanity/humanizing ecology." As we will see, these distinct projects—which are deliberately framed in dynamic, transitive terms—actually coalesce as we approach the most radical implications of the environmental humanities.

Ecologizing humanity

The first activity of the environmental humanities, *ecologizing humanity*, has taken three forms, which are considered here not in the chronological order of a genealogy but in order of increasingly profound and dislocating ramifications.

First, the humanities, as a specific configuration of disciplines in the Anglo-Saxon academy, were ecologized. In some cases, scholars engaged directly with the science of ecology, but more often "ecologization" meant incorporating concepts and claims from popularized ecology, as embodied, for instance, in Ernest Callenbach's *Ecology: A Pocket Guide* (1998). Ecologization took distinct forms according to discipline—traditional ethics was challenged by environmental ethics in philosophy, for instance—but in general it involved *decentering* human beings from their previously unchallenged position as the focus of exclusive interest and analysis in the humanities. While the majority of humanities research and teaching remains purely anthropocentric, practitioners cannot but be aware that a perspective exists from which such work henceforth appears, in species terms, parochial.

In ecocriticism, Ursula Heise's work is an example that resonates immediately with environmental experts outside the humanities. "Risk" is a concept around which numerous scholars from disciplines across the natural and social sciences already cluster. Moreover, the very different place of nuclear power in the energy economies of two neighboring countries, Germany and France, demonstrates that conceptions of environmental risk are strongly enculturated. Heise argues that literature provides evidence of how differing cultures of environmental risk develop, and so "narrative analysis should … play an important role in examining the way risk perceptions are generated by and manifest themselves through various forms of representation, from documentaries and journalism to fiction and poetry" (Heise 138). Heise discusses specific narrative techniques in major examples of environmental fiction, which suggest that literature *stages*, as well as reflects and influences, cultural change in environmental risk perceptions. In contexts such as teaching, where such cultural work is highlighted for critical analysis, the reflexivity of literature means it can function as a tool for thinking about how risk emerges and is negotiated within a given culture. In short, literature works as a metanarrative of environmental risk.

Second, human history is ecologized. This development is closely related to the first: the introduction of environmental perspectives to history inevitably raises questions about whether there is such a thing as "human history" at all. Jared Diamond's *Guns, Germs and Steel* is considered a work of popular science, but is better understood as a brilliantly successful example of environmental history. Alfred Crosby had already argued that European colonization was made possible by the "portmanteau biota"—dogs, smallpox, wheat, horses and so on—that the colonizers took with them. Diamond extends the analysis to include a host of other contingent factors that explain, on his account, why Europeans colonized the Americas rather than the other way round:

The modern United States is a European-molded society, occupying lands conquered from Native Americans and incorporating the descendants of millions of sub-Saharan black Africans brought to America as slaves. Modern Europe is not a society molded by sub-Saharan black Africans who brought millions of Native Americans as slaves.

(Diamond 24–25)

It is Europeans' resistance to epidemic disease, historic capacity for rapid technological development, and fortuitous array of domesticable animals and plants in Eurasia, not racial superiority, that best explains this reality. For postcolonial scholars in the humanities, Diamond no doubt neglects the social and ideological specificity of European colonialism. As long as his account is not considered exhaustive, though, there seems to be no objection to supplementing more familiar cultural accounts with biogeographical explanations of European dominance in the colonial period.

Third, humanity itself is ecologized. A range of prefixes signals this radical conclusion: *multi*species ethnography; *trans*corporeality; *inter*sectionality; *intra*-action. Much of the technical language, such as "assemblage" and "biosemiosis," extends conceptions of community beyond the human or emphasizes nonhuman capacities for agency and signification. Epistemic terms claim a qualified version of realism or materialism, as in "material feminisms," "critical realism," "agentic realism," and so on, collocations that sublate the divide between classic scientific realism and social constructionism. It is unfortunate, then, that scholars in closely related fields such as posthumanism and materialist feminism have not been able to agree on terminology because it limits the reception of their startling ideas. Timothy Morton's *The Ecological Thought* has had a substantial impact thanks to the simplicity of its central formulation and a plethora of soundbites, rather than because his ideas are wholly original. For Morton, the "ecological thought" characterizes a future culture in which even sceptics become helplessly aware of the intimate interconnectedness he calls the "mesh":

The ecological view to come isn't a picture of … a closed system. It is a vast, sprawling mesh of interconnection without a definite center or edge. It is radical intimacy, coexistence with other beings, sentient and otherwise—and how can we so clearly tell the difference? The ecological thought fans out into questions concerning cyborgs … and the irreducible uncertainty over what counts as a person. Being a person means never being sure that you're one. In an age of ecology without Nature, we would treat many more beings as people while deconstructing our ideas about what counts as people.

(Morton 8)

In Critical Animal Studies, the ecological thought implies a heterarchy of vulnerable beings, in contrast to both the anthropocentric hierarchy of Reason over nature and the animal liberationist hierarchy of sentient beings deemed worthy of moral consideration. Donna Haraway has promulgated concepts from evolutionary biology in the humanities such as "symbiogenesis" and "co-evolution," which suggest that "[t]o be one is always to become with many" (Haraway 157). In these ways, both the predominance and the distinct identity of humanity are challenged.

Richard Kerridge summarizes the appeal of new materialist and qualified realist efforts to ecologize humanity:

> The proposal is that we should move from traditional ideas of agency as an exclusively or predominantly human attribute to a concept of "agentic assemblage," in which human agency is bound up with that of "microbes, animals, plants, metals, chemicals, word-sounds, and the like," as Jane Bennett ... puts it. This proposal does not necessarily weaken the demands that can be made of that human agency, since, arguably, the recognition that agency is distributed in these assemblages and across whole ecosystems points not to determinist fatalism but to a subtler and more realistic sense of responsibility: one that does not rebound between extremes of hubris and resignation.
>
> (Kerridge 367)

While he commends the new materialism for its fundamental challenge to dualistic conceptions of culture and nature, he cautions that "there does seem to be something paradoxical about dispersing and qualifying our notion of human agency at the very moment we need to make an unprecedented demand upon that agency" (Kerridge 367).

Humanizing ecology

On the other side of the chiasmus, ecology is humanized. This does not mean, in the popular sense of "humanizing," that ecology is made friendlier or more approachable; the environmental humanities are not a scholarly PR department for the natural sciences. What it does mean varies according to the discipline (history, literature, philosophy, etc.) that posits itself as progenitor and reflexive illustrator of the "human." In each case, though, scientific ecology is subjected to a *dynamic* process of revision—indicated here by the use of active verbs—that situates its truth claims in a wider cultural context without, ideally, undermining their political efficacy by relativization.

Ecology can be humanized, in this sense, by being *historicized*. Ecosystem ecology, which dominated the science throughout the middle of the twentieth century, depicted ecological change as primarily cyclical and deterministic. The end-point for ecological change was envisaged as the "climax community," which would persist indefinitely in the absence of random disturbance once achieved. Ecology conceptualized in terms of deterministic processes tending toward equilibrium has no need of history. Post-equilibrium ecology, as popularized in Daniel Botkin's *Discordant Harmonies*, takes disturbance and change (including, crucially, evolutionary change), rather than a pre-destined balance of nature, as the norm. Equilibrium ecology treated climax ecosystems, such as mature broadleaf forests, as communities in a strong sense, but, as John Kricher argues, "Examined on a long-term time scale, the various species that today form natural associations are really accidents of history" (Kricher 90). As Simon Pooley points out:

> The displacing of the cyclical time of systems ecology has favoured a return to a historical conception of time in ecology, where irregular natural disturbances and periodicities of natural variation regain their importance. This implies it is

necessary to investigate the environmental history of an ecosystem (a place, not a generic space or system) in order to manage it sensibly. Not to discover some past pre-human (or pre-European) baseline or "natural state," but rather to reconstruct histories of major events and shifts, and to help establish [the] "historical range of variability."

(Frawley and McCalman 246).

Pooley's essay appears in a collection that establishes the vital contribution of the environmental humanities to understanding invasion ecologies.

While the gap between scientific ecology and environmental history is now closing, the humanities are also uncovering the historical roots of concepts deployed by scientists. Donald Worster's seminal work on the history of ecology has been joined by research that, in Heidi Scott's phrase, "traces the ancestry of ecological science to find lurking literary forebears" (Scott 8). Scott observes that "the paradigms of classic balanced nature and postmodern chaotic nature are two ways of portraying an immensely diverse and complex natural world. Ecological paradigms are never strictly objective: they are colored by the cultural conditions of their emergence" (2). For Scott, Romantic lyric poems foreshadow what would become the closed microcosm of ecosystem modeling, while apocalyptic fictions of the same era convey a cultural anxiety about entropy and chaos that provide the initial conditions for the later emergence of postequilibrium ecology. Botkin and Kricher provide compelling empirical evidence that ecological processes are better accounted for primarily in stochastic and evolutionary terms rather than deterministic drift towards equilibrium, but they also acknowledge, like Scott, the historical and conceptual significance of the metaphors scientists continue to think with.

Such reflections have profound consequences outside the academy. My first encounter with the environmental humanities was in the field of ethics, which began to *philosophize ecology* by looking for moral grounds for conservation. Much debate in the early years focused on the "intrinsic value" of animals, plants, and even nonliving components of habitats such as rivers. It turned out, though, that such moral questions were inseparable from the more technical and conceptual interests of philosophers of biology. As Kim Sterelny and Paul Griffiths aver:

Species are the focus of conservation efforts all over the world. But many of the types of organisms that people try to conserve do not count as species under the most scientifically well motivated definitions. New Zealand's black stilt and North America's red wolf are often cited as examples of "mere varieties" that are the subject of expensive conservation programs. Whether this matters depends on the source of concern for the environment. If conservation is seen as a human-centered activity, then we can justify our concern for a favorite color morph on aesthetic grounds. If we want to spend the conservation dollar to preserve biodiversity in some more objective sense, then we will be more concerned with the proper definition of species.

(Sterelny and Griffiths 20)

Biodiversity, too, as an object of legal frameworks and conservation efforts, is philosophically questionable: are we trying to save diversity of genes, species, or ecosystems? Are these all equally susceptible to estimation or conservation? Are they pragmatically or conceptually compatible with one another, given that preservation of specific types of habitat will inevitably prefer some species and alleles to others thanks to natural selection? In practice, as Sahotra Sarkar explains, conservationists are forced to choose "true surrogates," such as IUCN-listed endangered species, in place of a complete account of biological diversity. Such choices cannot be explained or defended in purely scientific terms:

> From a scientific perspective there is … a conventional element in such a definition of biodiversity based on true surrogates in the sense that the definition is not entirely specified by scientific facts. However, from the normative perspective these are not conventional choices. Instead, they reflect deep cultural judgments of what is worth preserving in nature and thus these choices must be carefully made. For instance, [many prominent] choices of true surrogates … mark a fairly recent shift in cultural values away from charismatic species to a more inclusive set of taxa.
>
> (Sarkar 634)

To philosophize ecology and to historicize it is to recognize that such problems were implicit all along. If history has always been environmental, albeit unrecognized thanks to anthropocentric prejudice, it is also the case that ecosystems have always been shaped by specific histories and have been philosophically questionable. We therefore have grounds for recommending, pragmatically, that multidisciplinary teams to research ecosystem restoration or conservation include scholars of the environmental humanities.

So much is straightforward to articulate and justify. By contrast, there is no word in English for the specific activity of literary writers. The German verb "dichten" is usually translated as "to write poems," but is treated—in the work of Martin Heidegger, for example—as a stronger, more active verb, "to poet" or "to poetize." Artists have a much harder time accounting for their participation in collaborative projects in terms other than congenial communication or beautification of pre-existing scientific findings. Art as a form of productive research that *poetizes ecology* is both unfamiliar and easy to mock.

As we have seen, Ursula Heise shows how literary and other narratives can function as metanarratives of environmental risk. In addition, there is a wide range of artistic activities that engage energetically and constructively with various natural sciences. Adam Dickinson's remarkable collection *The Polymers* demonstrates comfortable familiarity with the chemical processes by which long-chain organic molecules are formed, but goes further to explore a range of analogues in the techniques by which poems are formed. Dickinson's poetry is "experimental" in the two senses that conventionally distinguish the arts and the sciences: it is unique and adventurous, but it is also rigorously iterative and procedural. Dickinson proclaims his indebtedness to the "imaginary science" of "pataphysics," a species of early twentieth-century surrealism, but demonstrates, at this historical juncture, a more thorough interpenetration of poetic and scientific methodologies than his predecessors.

One of the most striking of these ingenious poems, entitled "Cigar? Toss it in a can. It is so tragic," explores malapropism as a kind of "linguistic isomer":

> Needles to say, at the pentacle of patriarticle politics
> we cannot phantom the depths to which
> battering eyelids skewer the results to make ends meat.
> …

It is perhaps a blessing in the skies that the hewn cries
sound like flaws in the ointment as we cease the day,
udderly disappointed by the ludicrust bowl in a china shop
and its new leash on life.

(Dickinson 23)

The analogy is suggestive: chemical isomers share a chemical formula but are structurally distinct, while Dickinson's linguistic isomers are comically inappropriate homophones or near-homophones (i.e., similar phonemes but different lexemes). The punning poem can be read in relation to Dickinson's scholarly work, which articulates a "pataphysical poetics" that "engage[s] the environment as a complex set of semiotic and symbiotic relationships where diverse forms of signification ... interact" (Dickinson, "Pataphysics" 137). Malapropism recalls the crucial role of error, or mutation, in biological evolution, while the analogy of pun and isomer suggests the relationship posited, in the field of biosemiotics, between linguistic and biochemical signification.

Other poems are composed of a title and a molecular diagram of a polymer, such as "Che Guevara delighted to see his face on the breasts of so many beautiful women," which reproduces the structure of Lycra, the polyester-polyurethane copolymer used in sports clothing. The title alludes to a comment made by Guevara's daughter when she was asked how her father might feel to see his image reproduced so widely. In this context, in conjunction with a polymer that has facilitated the use of clothing to promote erotic and athletic bodily display, it seems to allude to the unexpected forms "revolution" (be it materialist or material) can take.

Environmental theology seeks to *spiritualize ecology*. The need for this development was identified long ago in Lynn White Jr.'s 1967 essay on "The Historical Roots of our Ecologic Crisis," a well-known early contribution to the environmental humanities. White identifies the Middle Ages as "the period in which our technological and scientific movements got their start, acquired their character, and achieved world dominance" (Glotfelty and Fromm 7). In search of the root cause of the global environmental crisis that has since emerged, White locates it in medieval Western Christianity, which he describes as "the most anthropocentric religion the world has seen" (9). His solution to the problem is quoted less often than his analysis, however: "More science and more technology are not going to get us out of the present ecologic crisis until we find a new religion, or rethink our old one" (12). It is an unpalatable suggestion to a militant atheist such as myself, but not invalid for that reason alone. Margaret Atwood's invention, in her *MaddAddam* trilogy, of a hybrid Darwinian–Christian religion, the "God's Gardeners," suggests she, for one, may have come to the same conclusion.

Two conceptions of this "new religion" predominate in the environmental humanities: reformed, anti-anthropocentric versions of Christianity, such as the new Franciscanism promoted by White, and "neoanimist" spiritualities informed by both indigenous belief systems and Western philosophies such as phenomenology. Ecofeminist philosopher Val Plumwood was a powerful advocate for spiritual beliefs that attribute mind to many more beings than anthropocentric religions have allowed (what philosophers call "weak panpsychism"), albeit without considering the universe itself as purposive or conscious. For Plumwood, as for White and Atwood, purely secular ethics lacks both the metaphysical depth and the visceral moral torque to motivate a shift to environmentally sustainable societies. At the same time, she emphasizes that neoanimism does not require belief in a distinct "supernatural" or "spiritual" world: "[The] oppositional formulation of spirit versus matter renders invisible the important concept of a materialist spirituality which does not invoke a separate spirit as an extra, independent individualised ingredient but rather posits a richer, fully intentional non-reductionist concept of the earthly and the material" (Plumwood 222). Plumwood's neoanimism is a persuasive, coherent spiritualization of ecology, in many respects consonant with the posthumanist ecologization of humanity described above. As Atwood implies when she writes hymns for her God's Gardeners, though, any new religion needs more than a clever philosophical argument; it needs symbols, rituals, commandments, and taboos, which are sustained by and embodied in institutions. It is not impossible to invent a religion deliberately rather than adopting one that is purportedly "revealed"—L. Ron Hubbard did just that—but it is hard to imagine that many scholars of the environmental humanities would be willing to shift from critical analysis to deliberate indoctrination.

The project of *democratizing ecology* is, by contrast, already well under way. In the most general sense, the decline in unequivocal acceptance of expert (invariably scientific) evidence noted by sociologists will tend to work against the "scientification" of environmental problems. Michel Callon, Pierre Lascoumes, and Yannick Barthe observe the proliferation of environmental and health issues that cannot be managed or constrained by the traditional alliance of scientific experts and democratically delegated decisionmakers:

> GMOs, BSE, nuclear waste, mobile phones, the treatment of household waste, asbestos, tobacco, gene therapy, genetic diagnosis—each day the list grows longer. It is no good treating each issue separately, as if it is always a case of exceptional events. The opposite is true. These debates are becoming the rule. Everywhere science and technology overflow the bounds of existing frameworks. The wave breaks. Unforeseen effects multiply. They cannot be prevented by markets, any more than by the scientific and political institutions.
>
> (Callon et al. 9)

They advocate "hybrid forums" that extend scientific and political deliberation beyond existing institutional circuits.

Informal hybrid forums already exist. Rob Nixon's *Slow Violence and the Environmentalism of the Poor* brilliantly illustrates the protracted, unequal struggles, especially but not only in the global South, required by the democratization of ecology. For example, the explosion at the Chernobyl nuclear power plant exemplifies the difficulties inherent in slowly unfolding disasters:

> Maintaining a media focus on slow violence poses acute challenges, not only because it is spectacle deficient, but also because the fallout's impact may range from the cellular to the transnational and … may stretch beyond the horizon of imaginable time. The contested science of damage further compounds the challenge, as varied scientific methodologies may be mobilized to demonstrate or discount etiologies, creating rival regimes of truth, manipulable by political and economic interests.
>
> (Nixon 47)

The affected people of the Ukraine, whose geopolitical world was shifting even as the radiological story continued to unfold, acquired varying levels of "biological citizenship" as they sought official recognition from medico-juridical bureaucracies as "Chernobyl sufferers": "The ground rules for being counted and discounted kept changing. Even the boundaries of the pollution zones were unstable, shrinking and dilating through a mixture of bureaucratic caprice, economic expediency, and slippery science" (Nixon 50). By contrast with hierarchical modernist models of the dissemination of scientific expertise to a passively grateful population of ignoramuses, Chernobyl is prime example—albeit heightened by the unique conditions of the end of the Soviet empire—of the forbidding complexity of postmodern risk scenarios. The science is not only "slippery," but is also available to lay people, some of whom constitute themselves as "victims" in the light of their selections of its findings. Hierarchies of expertise continue to exist, but often in competition with one another, and all are vulnerable to accusations of self-interested bias.

This situation prevails on a global scale in relation to climate change. Five successive reports of the Intergovernmental Panel on Climate Change—a UN-constituted politico-scientific organization designed to establish a consensus about anthropogenic climate change—had, until COP21 in Paris in December 2015, failed to provide a sufficiently compelling expert basis for an international agreement, under the auspices of the United Nations Framework Convention on Climate Change, to limit greenhouse gas emissions. Mike Hulme, who contributed research to the IPCC reports and took part in its expert panels, concludes that "this particular way of framing climate change (as a mega-problem awaiting, demanding, a mega-solution) has led us down the wrong road" (Hulme 332). While he accepts the importance of the scientific work of the IPCC, he suggests that over-reliance on this model of scientific expertise has resulted in a "political log-jam of gigantic proportions, one that is not only insoluble, but one that is perhaps beyond our comprehension" (333). By calling for a wider dialogue about the meaning and significance of climate change, Hulme is arguing for the democratization of ecology. The meanings of climate, as of nature more generally, are being opened up beyond the sciences, beyond universities, and beyond the privileged social groups that overwhelmingly populate them.

The environmental humanities are participating in this work, complementing scientific climatological research with anthropological, historical, cultural, and ethical knowledge from a wide range of sources. Much of it, such as Timothy Clark's *Ecocriticism on the Edge*, a powerful study of the cultural implications of the Anthropocene, relies on the IPCC findings and accepts its consensus implicitly or explicitly, but there is also a need for research that takes seriously sceptical and anti-environmentalist perspectives. Axel Goodbody, George Handley, and I have embarked on a study of the cultural manifestations of climate scepticism in Germany, the USA and the UK. While we are all "warmists," as the skeptics would say, we believe it is vital to overcome oppositional stereotypes by trying to understand the world from the various points of view of those unpersuaded by climate science. We are finding that they are a more diverse, internally fractured, and unpredictable bunch than environmentalists tend to imagine. While we share the bias of the liberal humanities toward emphasizing the importance of contributions by marginalized (indigenous, racialized, non-Western, queer, for example) people to hybrid intellectual and political forums, we consider it vital that the environmental humanities acknowledge as subjects of research, and admit as participants in a democratized ecology, even those whose social identities include overt hostility to environmentalist objectives.

Our project therefore involves, in the words of Astrida Niemanis, Cecilia Åsberg, and Johan Hedrén's important overview of the environmental humanities, "acknowledging the differences and diffractions in worldviews, histories, subjectivities, relations and practices that various communities (both human and non-human) engage in, with respect to their environment" and "cultivating an environmental humanities that is well-placed to research and analyze these differences." The plurality of "practices of environing" or "cultures of nature" that anthropology, history, and cultural study reveal is a standing challenge to attempts, scientific or moral, to recruit "nature" in the interests of any particular humans. Our most valuable contribution could be to facilitate constructive democratic dialog within that plurality.

Chiasmic conclusion

When explored to their most radical extent, ecologizing humanity and humanizing ecology can start to look a lot like each other. For example, Val Plumwood's "neoanimist spirituality" is close kin to biosemiotics and the new materialism. Thus it may appear that the chiasmus collapses when pushed to its logical extreme. Yet the contrastive parallel between the two sides is worth sustaining for the productive tension and continual mutual correction each provides for the other. At the very moment that ecological knowledge is democratized, for instance, our conventional notion of "democracy" as restricted to sane, adult, human agents comes to seem intellectually and morally unsustainable. It is worth keeping these dynamics provisionally distinct even as we recognize that their contrast disguises a great deal of continuity.

The chiasmic characterization I propose is deliberately simple and (as university managers call it) "outward-facing," unlike Niemanis, Åsberg and Hedrén's "version of this field that self-reflexively acknowledges and even nurtures its own contradictions, variances, and necessary open-endedness." For one thing, there seems to be no danger of a congeries of humanities academics failing to nurture contradictions amongst ourselves anyhow; for another, the attention we can reasonably expect of outsiders is unlikely to stretch to encompass our many fascinating disagreements. There is, of course, no need for this particular characterization to prevail, nor is there an obvious forum, hybrid or otherwise, in which an alternative might be discussed. Even so, a coherent, legible account of the things we agree about and the difference they might make to environmental sustainability is, in my view, the most vital work for the environmental humanities right now.

References

Åsberg, C., Astrid Neimanis, and J. Hedrén. "Four Problems, Four Directions for Environmental Humanities: Toward Critical Posthumanities for the Anthropocene." *Ethics and the Environment* 20.2 (2015): 67–97. Project Muse. Web. 1 September 2015.

Callon, Michel, P. Lascoumes, and Y. Barthe *Acting in an Uncertain World: An Essay on Technical Democracy.* Trans. G. Burchell. Cambridge, MA: MIT Press, 2009. Print.

Diamond, Jared M. *Guns, Germs, and Steel: A Short History of Everybody for the last 13,000 Years.* 1997. London: Vintage, 2005. Print.

Dickinson, Adam. "Pataphysics and Postmodern Ecocriticism: A Prospectus." *The Oxford Handbook of Ecocriticism.* Ed. G. Garrard. New York: Oxford University Press, 2013. Print.

———. *The Polymers.* Toronto, ON: House of Anansi Press, 2013. Print.

Frawley, Jodi and Iain McCalman. *Rethinking Invasion Ecologies from the Environmental Humanities.* London: Routledge, 2014. Print.

Glotfelty, Cheryl and Harold Fromm. *The Ecocriticism Reader: Landmarks in Literary Ecology.* Athens: University of Georgia Press, 1996. Print.

Haraway, Donna J. *When Species Meet*. Minneapolis, MN: University of Minnesota Press, 2008. Print.

Heise, Ursula K. *Sense of Place and Sense of Planet: The Environmental Imagination of the Global*. New York: Oxford University Press, 2008. Print.

Hulme, Mike. *Why We Disagree about Climate Change: Understanding Controversy, Inaction and Opportunity*. Cambridge: Cambridge University Press, 2009. Print.

Kerridge, Richard. "Ecocritical Approaches to Literary Form and Genre: Urgency, Depth, Provisionality, Temporality." *The Oxford Handbook of Ecocriticism*. Ed. Greg Garrard. New York: Oxford University Press, 2014. 361–376. Print.

Kricher, John C. *The Balance of Nature: Ecology's Enduring Myth*. Princeton, NJ: Princeton University Press, 2009. Print.

Morton, Timothy. *The Ecological Thought*. Cambridge, MA: Harvard University Press, 2010. Print.

Nixon, Rob. *Slow Violence and the Environmentalism of the Poor*. Cambridge, MA: Harvard University Press, 2011. Print.

Plumwood, Val. *Environmental Culture: The Ecological Crisis of Reason*. London: Routledge, 2002. Print.

Sarkar, S. "Norms and the Conservation of Biodiversity." *Resonance* 13.7 (2008): 627–637. Springer Link. Web. 1 September 2015.

Scott, H. C. M. *Chaos and Cosmos: Literary Roots of Modern Ecology in the British Nineteenth Century*. University Park, PA: Pennsylvania State University Press, 2014. Print.

Sterelny, Kim and P. E. Griffiths. *Sex and Death: An Introduction to Philosophy of Biology*. Chicago, IL: University of Chicago Press, 2012. Print.

45

THE HUMANITIES AFTER THE ANTHROPOCENE

Stephanie LeMenager

At the core of the humanities lies the question of human nature, a question of "who are we?" that has been reanimated by debates about the proposed geological epoch of the Anthropocene. "Who is the 'anthro' in Anthropocene?" the cultural theorist Stacy Alaimo asks, archly. When not altogether under erasure, "human nature" presents a troubling indebtedness to dualistic thought. Writing of the conceptual tension between humanity and animality for anthropologists, Tim Ingold notes that "we are, according to [human nature], constitutionally divided creatures, one part immersed in the physical condition of animality, the other in the moral condition of humanity" (21). The politics and artistic production associated with environmentalism as a social movement in the global North grows out of the dualistic legacy that Ingold describes. Environmentalists and environmental scholars in the humanities seek in different degrees what Greg Garrard in this volume calls an "ecologizing of the human," by which we emphasize our deep interconnection with other life. Yet environmentalists and, more covertly, environmental scholars cling to fantasies of moral self-determinism.

The Anthropocene idea insults environmentalists and humanists alike by reinscribing the human above "nature" as an isolate subject and prime mover—but one whose legacy is the unforeseen result of its species-being rather than a product of moral or rational decisionmaking. As Dipesh Chakrabarty suggests in his much-discussed glosses on historiography in the Anthropocene, in the shadow of global climate change humanity's "brute force" has begun to challenge what once was conceived as human subjectivity ("Brute"). Climate change seems to demand that we rethink human agency as blown out to the scale of *Homo sapiens*. To my mind, this scalar shift from the subject to the species could make history a largely irrelevant pursuit, as perhaps any moral or even ecological education might be in a world where the human is inscribed in stone. The Anthropocene idea threatens education—a source of hope reclaimed along with the term "Anthropocene" by some public intellectuals who imagine events such as desertification and extinction as provocations to smarter technology and design (see Ackerman and Marris et al. for examples of the hopeful Anthropocene). More subtly than either the Anthropocene doomsayers or optimists, Chakrabarty registers a poignant—and rarely remarked—concern about history, and by implication the humanities, as techniques for making meaning in the era of climate change ("Climate"). His attempt to imagine historiography that integrates the scales of the economic and the geological falters, suggesting just how difficult it is to be a scholar in one of the humanities' core disciplines, at the edge of Holocene climate. Yet the popularization of the Anthropocene concept complements a veritable humanities renaissance, in the emergence of the environmental humanities as an interdisciplinary field and activist pedagogy.

This seeming paradox resolves itself if we consider the environmental humanities primarily as a form of practice, pedagogical and civic. The environmental humanities has been tried out on the ground, in college and university classrooms, and in public spaces like museums and even beaches, prior to its articulation as an academic field. There is—and has been—an "everyday" environmental humanities, worked out in earnest and at times awkward collaboration among artists and scholars, reaching across the so-called two cultures, from the humanities into the social and natural sciences. An "ecologized humanities" comes into being in interdisciplinary practice such as, for example, Anna Tsing's anthropological forays into forest ecologies with indigenous and Western scientists, or Jenny Price's dual career as an environmental historian and activist performance artist with the Los Angeles Urban Rangers, or Sverker Sörlin's long-term commitment to a "biologically and scientifically informed" Arctic humanities, as described in this volume. Whether we call such practice the environmental humanities or the posthumanities is less important than the fact that new forms of humanities scholarship are happening in response to environmental matters of concern that cannot be addressed by the skill set of a single discipline.

When we sit down in interdisciplinary working groups to address, say, environmentalist representations of the Niger Delta, as I recently did, field designations such as that of "ecocritic" make sense only insofar as they indicate the kind of knowledge that we bring to the table. As my colleague Stephanie Foote emphasized in conversation about our co-authored essay, "The Sustainable Humanities," we must learn to speak *from* our disciplines, rather than *to* them (see also "Editor's Column"). In the United States, the capaciousness of the humanities as fields still tied to the integrative model of a liberal arts education rather than to the professional trajectories of STEM curricula means that the humanities are a good place from which to speak, or to host interdisciplinary conversation. For example, a curatorial consultancy that I held at the Blaffer Museum in Houston, Texas, brought together historians, curators, literary and cultural studies scholars, and the artist Zina Saro Wiwa to create an exhibition meant to function as a virtual world through which museum-goers recognize not just the environmental wounds but the ecological abundance, *in potentia*, of Ogoniland. Of course the artist was the primary creator, but everyone in the room contributed a distinct archive, and therefore a dimension of time, to the project, in addition to narrative intelligence about how to imagine its diverse storylines. Some of the most fundamental skills that we humanists impart to our students, such as the training in argumentation offered in composition and rhetoric programs, are being reimagined by both students and found publics through "citizen humanities" projects like that of co-writing bilingual press releases with farm workers—this latter example from an undergraduate class in food studies.[1]

In my own experience, the environmental humanities has been in many respects about breaking the fourth wall, getting outside of the clerical and insular model of the professoriate summed up in the damning phrase "ivory tower." In this sense, the work of the environmental humanities follows other forms of activist scholarship, from feminist studies to race and ethnic studies. Catriona Sandilands reminds us in this volume that as we attempt to move beyond "human-centered thinking/identity/power configurations," we also must remember and embody those configurations. Problems of social and environmental justice led the environmental humanities toward curricular models that incorporate service learning or "citizen humanities" direct action, as noted above. In the realms of research and production, what some of us do has begun to approach social practice or public art—creating installations, performances, digital archives, and apps, such as Jenny Price's *Malibu Beaches* iPhone app that unlocks access to some of the most secluded beaches in the world, making an interactive argument for the maintenance of public lands. Imre Szeman's Petrocultures Research Group in Alberta moves between the museum contexts of the public environmental humanities and interface with government and corporate actors in an attempt to intervene in Canadian energy policy.

The practice of the environmental humanities responds not to the Anthropocene idea but rather to the Anthropocene *avant la lettre*, or at least *avant* the philosophical arguments that tease out its implications for humanism, species being, and ecological thought. Like the environmental humanities, the Anthropocene also manifests itself in the everyday. In the wealthiest parts of the world, the Anthropocene occupies the everyday as infrastructurally induced disease—the toxic fumes of another pipeline spillage, the bad faith of a Houston drill bit manufacturer who paints frack tools pink in support of the Susan G. Komen Fund for breast cancer, the heaps of filthy snow piled up in May in Boston's Seaport district after the great winter storms of 2015. In the poorer parts of the world, north and south, the Anthropocene can be infrastructurally sustained violence, fast and slow: the privatization of water for the purpose of uranium mining, the rise of a rare bile duct cancer among Dene, Cree, and Métis peoples along the Athabaska River, the horsewhipping and murder of young men and women suspected of trespassing at flow stations and well-head sites in Ogoniland.

What I call the everyday Anthropocene expresses our latest geological epoch as one of profound social failure. This social failure manifests itself chemically, in atmospheric carbon levels, and in lived experience through mis-attuned or deadly infrastructures, from un-retrofitted pipeline to mono-crop farmland whose toxic soils run to rivers and streams. I use the word *infrastructure* here as an homage to urban theorist and historian Mike Davis, in whose work on "ordinary disaster" infrastructure becomes the interface between macroscalar systems—the global financial system, for instance—and ordinary people living in Los Angeles (221). The director of the Center for Land Use Interpretation, Matt Coolidge, coined the phrase "the poetics of infrastructure" to name the practice of feeling, mapping, and critically inhabiting what he calls "the connective tissue" of modernity, "meaning the strands, the pipelines, the aqueducts, the electric lines, all the way back through the reservoirs, the power plants, the coal mines."

These systems—hidden in plain sight—constitute precisely those objects that make a world, to paraphrase Henry David Thoreau. Yet like works of fiction carefully knit together to create a sense of totality and yet always holding within themselves paths not taken, infrastructures can be revised. This work of revision has compelled environmental humanists for whom the aesthetic is social practice, a way to gain clarity about and even intervene in systems that seem inevitable but are not.

Genre trouble

Both critical infrastructure studies and the more explicitly literary topic of genre are developing research foci in the environmental humanities. And both are invested in counterfactuals, in thinking through what is not (yet) real. What I mean by the "real" here is habitual practice responding to the perceived parameters of material and social environment. The Center for Land Use Interpretation's infrastructure tours of, say, hydraulic systems along California's Highway 5 show us how modernity is put together, and in the process the tours dis-aggregate "the modern" so that it becomes possible to imagine other ways of being. Popular literary and film genres such as the Western establish habitual feeling-states about national belonging, gender performance, racial loyalty, and social transgression. These genres constitute an emotional and social infrastructure. As cultural theorist Lauren Berlant writes, "genres provide an affective expectation of the experience of watching something unfold, whether that thing is in life or art" (6–7). The study of genre exposes how affective expectations are put together, in the process foregrounding opportunities for innovation within existing genres. Ideally, such innovation might shift the structures of feeling that undergird hegemonic understandings of nationhood and the good life. As we live into the everyday Anthropocene, literary scholars within the environmental humanities and a broader environmentalist public have seized upon genre as a means of innovating new socio-ecological relations.

As a literary scholar and environmentalist, I have found myself on the front lines in the past year and a half of a scramble for a new genre that supposedly speaks to the conditions of the everyday Anthropocene. Dan Bloom, a freelance journalist based in Taiwan, began writing to me in 2013, urging me to declare "cli fi," or climate fiction, as a new literary genre. For many, "cli fi" is essentially an offshoot of environmentally attuned "sci fi" with a focus on global climate change. Bloom began using the term "cli fi" in 2008, and since then he has written to hundreds, perhaps thousands, of academics, authors, and artists, attempting to raise awareness about climate fiction (see Bloom for an overview of his activism). When I taught a class titled "Cultures of Climate Change" at the University of Oregon in spring 2014, Bloom alerted the *New York Times*, and their education reporter came calling. One of the reporter's key interests was whether or not we could establish cli fi as a genre—and similarly, when I spoke to *ClimateWire* and *Time Magazine* and then wrote a short article on cli fi for the scholarly news outlet *The Conversation*, concerns about what makes cli fi a genre, how to identify its conventions and to create it as a persistent object, were paramount. Was the film *Godzilla* cli fi? What about satiric novels like Ian McEwan's *Solar*? The fate of Earth seemed to hang in the balance. Never, in my lifetime, has culture been so invested with the burden of ecological continuance.

The project of creating a taxonomy for cli fi gets both climate and culture into the newspaper and in doing that, it is a *project that acts*, summoning a national and even international climate change public. Genres intend to call publics into being. This is what I understand to be the larger goal of activists such as Bloom. My own position toward cli fi as a genre remains anarchic. Whenever I have been asked if a book or film is cli fi, I have said "yes," hoping to keep the parameters of the genre open, which I also think the discussion of what it means to be a climate change public ought to be, to keep that discussion lively. Hannes Bergthaller writes in this volume of the failure of countercultural movements of the 1960s and 1970s to move away from libertarian notions of "freedom" to establish "socio-ecological relations" that encourage "collective decisions that constrain individual choice." A similar preference for individual, private experience haunts popular enthusiasm for cli fi, as if the new genre might offer therapeutic catharsis to silent readers, who will privately cope with—and perhaps put aside—the climate debacle. My own interest in cli fi as a genre dovetails with Berlant's investigation in *Cruel Optimism* of what I call *genre trouble*. Genre trouble comes about when the affective expectations we hold for how things unfold, in art and life, do not make sense anymore. The sociological desire for cli fi stems from genre trouble, and for me cli fi remains most interesting as a sociological phenomenon.

When I use the term "genre trouble" I intend two distinct emphases, touching upon social contexts and their relationship with artistic form. First, I mean the representational impasse posed by climate change and voiced by Ursula Heise in her important bid for an "eco-cosmopolitan environmentalism" that might "effectively engage with steadily increasing patterns of global connectivity, including those created by broadening risk scenarios" (210). Of climate change in the cultural imaginary, Heise writes, "Like other processes of global systemic transformation, ecological or not, climate change poses a challenge for narrative and lyrical forms that have conventionally focused above all on individuals, families, or nations, since it requires the articulation of connections between events of vastly different scales" (205). The multiple scales of space and time involved in global climate change have, since the time of Heise's writing, inspired digital climate change media that in some respects bear out her predictions about how global systems might best be known. Marina Zurkow's and Una Chaudhuri's *Dear Climate*, with its meditative audio experiences and striking agit-prop posters, and Ken Eklund's transmedia cli fi storytelling project *FutureCoast*, offer two responses to the problem of creating multi-sited narrative that also incorporates vastly distinct temporal frames. Were I to imagine cli fi in terms of formal innovation, my focus would be on how new media expand the possibilities of interactive and multi-sited authorship.

The second emphasis I bring to *genre trouble* is, again, primarily a sociological one: artistic genres are fraying, recombining, or otherwise moving outside of our expectations of what they ought to be because life itself is moving outside of our expectations for what it ought to be. It is worth considering how life itself begins to encourage new representational regimes.

I offer an example that struck me hard at the time of its publication, which was right on the heels of Hurricane Sandy, in 2012. Nonfiction literature is everywhere, and like fiction, it builds complex worlds through the delineation of concrete relationships. Consider a special "Weather" section that the *New York Times* ran on November 18, 2012, which might be called nonfiction or *found* literature. The section treats, in detail, forty-three casualties in New York City associated with Hurricane Sandy. Descriptions of the deaths are awkwardly distant, focused more on the placement of bodies in space than their integrity or injury. For example, in the subsection titled "Queens: Tony Laino, 30, was crushed while inside his home on Oct. 29th, when a tree fell into his second floor on 166th Street in Flushing." The flat tone marks a rethinking of the weather report as obituary, and the obituary as weather report. Something of the objectification of the body as inanimate object comes through here, suggesting modernist "comedy" as described by Henri Bergson. Except here humans are not cogs in the industrial machine—we are caught up in postindustrial weather. None of the accomplishments or human relationships of the "43" who died in New York City proper as a result of Sandy are noted by the *Times*, defying the expectation of obituary as commemoration of a life. Yet the fact that the story is laid out as a list of names, ages, and places of residence makes it feel altogether distinct from a typical account of weather. The dead emerge *faintly* as individuals, within a strongly drawn storm schema. Weather, it seems, is more real (or realized) than people.

In something like the comic art tradition of Hergé, author of *Tintin*, the environmental conditions of the storm victims' postures in death come into high focus while they, and the sting of mortality, blur. In his most well-known Anthropocene essay, Chakrabarty raises the question of whether or not it is possible to feel like a species—a theoretical provocation. I would argue that the *New York Times* weather section described above offers one example of how such a feeling might enter representational regimes. Depersonalization accompanies a kind of systemic thinking that is both uncomfortable and interesting. Sites of theory are everywhere, in the newspaper and in the conversation of nonexperts, such as undergraduate students (see Musiol on the sites of theory). One of the more startling admissions from my undergraduates in the Introduction to the Environmental Humanities course I taught recently—in answer to a pop quiz asking them if they "feel like a species"—was that, for some, they feel like a species only insofar as they imagine themselves at the edge of extinction. Even more startling, for many students this feeling brings relief. They did not evince what I recognize as pessimism, but rather a complex acceptance akin to a posthumanist embrace of being thrown off-center, into the thick of other forces, other lives.

In a sense, this is precisely the "ecologized humanity" that many of us in the environmental humanities have been hoping for. But being ecological might not always feel good. Bodily integrity assumes an individual, rather than a systemic or ecological project. When the newspaper intermixes its obituary and weather genres, we have a symptom of a radical recalibration of what it means to be human. At such a moment environmental humanities practice needs to emphasize what has always been an implicit if not explicit goal of higher education—the development of more adaptive ways of being social, and, in our moment, of understanding that social relationships are also ecological ones.

Making the social

A former colleague of mine once said, in passing, that what we in the humanities do is to create affect. But given how fleeting affect can be, and how individualistic, I would prefer to consider the ways in which we create social artifacts and socialities. Of particular importance to me in the making of the social are inter-generational memory, speculative fabulation, and mapping. All of these practices have been fundamental to the development of the environmental humanities, and they have been realized through community-embedded pedagogies and collaborative transmedia projects such as those described earlier in this chapter.

Inter-generational memory. In literary studies, Lawrence Buell initiated a discussion of "the possibility that literature and other expressive media might act as carriers of environmental memory" (32). Buell's scholarly exploration of how expressive media work to combat "inter-generational environmental amnesia," in the phrase of psychologist Peter Kahn, Jr., remains one of the most significant pragmatic arguments for the teaching of literature as a means of ecological survival. The Inscribing Environmental Memory in the Icelandic Sagas (IEM) project that literature professor Steven Hartman and others have shepherded responds, in part, to Buell's theorization of literature's value as memory practice. IEM scholars examine the sagas for evidence of climate shift, and shifts in socio-ecological values, over millennia.

Native North American writers such as Linda Hogan and Thomas King remind us with strong and often ironic emphasis that not *all* cultural memory need bring us to the present tense that I have called the everyday Anthropocene. If we consider King's *The Back of the Turtle* (2014) as a cli fi novel—which I do—we can recognize in it an absurdist, indigenous humor, drawn from traditional knowledge and the survival of colonial violence, that suggests pragmatic alternatives to the West's apocalyptic self-indulgence. The artist Marko Peljhan has created an intergenerational environmental memory project that to me serves as an example of how active and interactive the making of archives can be. Peljhan created a high-tech "memory box," called after an Inuit phrase the *SiNunI*, that allows Inuit hunters to log geolocated recordings of wildlife observations, greenhouse gas levels, metereological data, and place-based story. A group of Inuit elders commissioned the memory box, and they decide what knowledge from it goes where, what is proprietary traditional knowledge, and what might be shared with non-Native scientists.

Speculative fabulation. I use speculative fabulation to mean imagining and narrating collaboratively into the possibility space of the future. As an example we might consider Eklund's *FutureCoast*, which calls upon people throughout the United States and Europe to upload "voicemails from the [climate-changed] future" onto an interactive, transmedial website—what Stephen Siperstein aptly calls "D.I.Y. cli fi."[2] The scholar of science and technology studies Donna Haraway coined the term "speculative fabulation" in theoretical arguments whose playfulness approaches the tenor of speculative fiction. Haraway's most succinct gloss on speculative fabulation comes about in her account of "taking care of country" as an Australian practice of long-term, careful inhabitation, associated most prominently with Australia's Aboriginal communities. As Haraway writes, "to care is wet, emotional, messy, and demanding of the best thinking one has ever done. That is one reason why we need speculative fabulation." The ethic of care she associates with long-term dwelling in a place—indigeneity—dovetails into a discussion of caring for the "unexpected country" brought into being by technoscience, for example transgenic life.

With different emphases, the anonymous collective Uncertain Commons asks, in their manifesto *Speculate This!*, that readers consider what they call "affirmative speculation" as a mode of caring about what cannot be anticipated as a future in our present moment. "What we affirm is something that has the potential to undo us: this is not, in other words, a self-congratulatory affirmation of what we are" (*Speculate*). What speculative fabulation entails is thinking through how to care for the products and by-products of unmanageable risk, which might include undesirable forms of life.

Finally, mapping. The DIY mapping of occluded infrastructures that have begun to manifest in social media through, for example, Emily Ferguson's Ontario-based Line9Communities blog and the Twitter feed #oiltrainwatch, suggest that both narrative intelligence and GIS technologies can be tools for cultivating eco-social relationships outside of our current extractive cultures. Maps—typically associated with the strategic imagination of empire and war—can be tactics, serving grassroots movements for environmental and social justice. As an undergraduate geography student at McMaster University, Ferguson mapped a pipeline whose route the energy corporation Enbridge refused to disclose at a public meeting, and then she built an online social community with the map at its center, bringing people from Ontario to Maine into conversation with each other and sparking what she calls "communities along the line." Such pipeline communities form a centerpiece of writer Naomi Klein's hopeful understanding of resistance to fossil fuel corporations in *This Changes Everything* (2014).

I conclude with something that Emily Ferguson said to me at the end of an interview I recently conducted with her. I asked Ferguson what was the most important advice she had to give to people interested in energy politics—activists and scholars. Her answer was clear: "The number one thing is research, research, research. ... Educate yourself but then do not hold the knowledge for yourself. Share it." These words are especially poignant coming from a person only recently emerging as an activist and also only recently graduated from university. The value of scholarly research and teaching sometimes gets shoved to the margins of our discourse in the environmental humanities as we scramble to find ways of acting within an institutional culture that may not value what we do and of speaking against broken systems that foster environmental degradation, climate change, and violations of environmental justice. Living in the end times can seem like a bad time to be doing painstaking, thoughtful work. And the skills that we need in the humanities *are* shifting. However, I think it a mistake to consider the environmental humanities, which owes debts to field imaginaries like ecocriticism, feminist science studies, environmental history, and environmental philosophy, as indicative of radical, rather than cumulative, change.

Research, a commitment to thinking *long*, in community with others, remains a crucial interface between our vulnerable bodies and the asocial or even antisocial command of energy that is power. Rethinking "human tendency"—a phrase my undergraduates asked me to substitute for "human nature"—still requires some slow reading, slow writing, and slow talk. The vibrant point-scoring that lights up the Twitterverse will not save us. As the cultural studies scholar Dick Hebdige notes in the context of his work in desert studies—work that feels especially crucial in the drought years we are enduring in the North American West—critique has to be about moving, together, through crisis. What were once pastoral offshoots of major humanities fields—ecocriticism, environmental history—are now in the thick of a broad, international public discussion of biological and cultural survival. The very name "the environmental humanities" indicates that the humanities have met the world and been changed by it.

Notes

1 The term "citizen humanities" arose in conversation with Joni Adamson and others at the Western Observatory steering committee meeting for the Mellon-sponsored "Humanities for the Environment" workshop series. The food studies class referred to here was taught by Professor Sarah Wald.
2 Stephen Siperstein quoted here in the context of his brilliant, media-savvy undergraduate course on climate change.

References

Ackerman, Diane. *The Human Age: The World Shaped By Us*. New York: Norton, 2014. Print.
Alaimo, Stacy. "Your Shell on Acid: Material Immersion, Anthropocene Dissolves." *Anthropocene Feminism*. Ed. Richard Grusin Minneapolis, MN: University of Minnesota Press, forthcoming.
Berlant, Lauren. *Cruel Optimism*. Durham, NC: Duke University Press, 2011. Print.
Bloom, Dan. Interview with Hannah Gal. "From Noah's Ark to Superstorm Sandy—the Rise and Rise of Cli Fi." *The Huffington Post (UK)* 23 October 2013. Web. 1 July 2015.
Buell, Lawrence. "Uses and Abuses of Environmental Memory." *Contesting Environmental Imaginaries: Nature and Counternature in a Time of Global Change*. Ed. Steven Hartman. Leiden, the Netherlands: Brill, 2017. 31–56. Print.
Chakrabarty, Dipesh. "Brute Force." *Eurozine* 7 October 2010. Web. 29 June 2015.
——. "The Climate of History: Four Theses." *Critical Inquiry* 35 (2009): 197–222. Print.
Coolidge, Matt. Interview with Stephanie LeMenager. "A Poetics of Infrastructure." *Resilience: A Journal of the Environmental Humanities* 1.1 (2014). JSTOR. Web. 14 July 2015.
Davis, Mike. "Los Angeles after the Storm: The Dialectic of Ordinary Disaster." *Antipode* 27 (1995): 221–241. Print.
Ferguson, Emily. Interview with Stephanie LeMenager. "Communities Along the Line." *Resilience: A Journal of the Environmental Humanities* 2.2 (2015). JSTOR. Web. 18 June 2015.

Haraway, Donna. "Speculative Fabulations for Technoculture's Generations: Taking Care of Unexpected Country." *Australian Humanities Review* 50 (2011). Web. 14 July 2015.

Hebdige, Dick. Interview with Stephanie LeMenager. "High and Dry: On Deserts and Crisis." *Resilience: A Journal of the Environmental Humanities* 1.1 (2014). JSTOR. Web. 14 July 2015.

Heise, Ursula K. *Sense of Place and Sense of Planet: The Environmental Imagination of the Global.* New York: Oxford University Press, 2008. Print.

Ingold, Tim. "Humanity and Animality." *Companion Encyclopedia to Anthropology.* Ed. Tim Ingold. New York: Routledge, 1994. 14–31. Print.

LeMenager, Stephanie, and Stephanie Foote. "Editor's Column." *Resilience: A Journal of the Environmental Humanities* 1.1 (Winter 2013). JSTOR. Web. 1 October 2015.

——. "The Sustainable Humanities." *PMLA* 127.3 (2012): 572–578. Print.

Marris, Emma, Peter Kareiva, Joseph Mascaro, and Erle C. Ellis. "Hope in the Age of Man." *The New York Times,* 7 December 2011. Print.

Musiol, Hanna. "Sites of Human Rights Theory." *Routledge Companion to Human Rights and Literature.* Ed. Sophia McClennen and Alexandra Schultheis-Moore. London and New York: Routledge, 2015. 389–397. Print.

Uncertain Commons. *Speculate This!* Durham, NC, and London: Duke University Press, 2013. e-Book.

INDEX